MATERIALS IN CONSTRUCTION
PRINCIPLES, PRACTICE AND PERFORMANCE

MATERIALS IN CONSTRUCTION PRINCIPLES, PRACTICE AND PERFORMANCE

SECOND EDITION

G. D. TAYLOR
PhD, CertEd, MCIOB

An imprint of **Pearson Education**

Harlow, England · London · New York · Reading, Massachusetts · San Francisco
Toronto · Don Mills, Ontario · Sydney · Tokyo · Singapore · Hong Kong · Seoul
Taipei · Cape Town · Madrid · Mexico City · Amsterdam · Munich · Paris · Milan

The CHARTERED
INSTITUTE OF
BUILDING

Pearson Education Limited
Edinburgh Gate
Harlow
Essex CM20 2JE, England
and Associated Companies throughout the world

Co-published with The Chartered Institute of Building through
Englemere Limited
The White House
Englemere, Kings Ride, Ascot
Berkshire SL5 8BJ, England

First published as *Construction Materials* 1991
This edition published 2002

British Library Cataloguing-in-Publication Data
A catalogue record for this book is available from the British Library

ISBN 0 582 36934 7

Set by 35 in 10/12pt EhrhardtMT
Printed in Malaysia

CONTENTS

CHAPTER 1

INTRODUCTION

It is now more than 25 years since the author's first treatise on construction materials (*Materials of Construction* 1974), was produced as part of a response to initiatives by the UK Construction Education Inspectorate which called for the approach to the subject to be more closely related to scientific principles. There were in existence many excellent texts covering materials from a traditional viewpoint, but with rapidly changing approaches to construction and the development of new materials, there was a need for a firmer foundation to materials understanding. The original text, published in 1974, therefore focused upon the study, in some depth, of the structure of matter, presented in a way that should be accessible to students without the need for study of A-level physics or chemistry. There was at the time still a considerable gulf between the approaches of:

- The scientist, who tended to study materials from a microscopic viewpoint.
- The engineer, who predicted behaviour on the basis of quantitative techniques in conjunction with assumptions regarding behaviour.
- The practitioner, whose approach was largely based upon proven principles and techniques using materials with a track record.

While the earlier text attempted to bridge, at least partially, the gap between these approaches, writers were still faced with the fundamental problem that both educationalists and researchers tended to be categorised as above, with limited integrated activity taking place. The operational climate has now significantly changed, with research being much more focused (partly by commercial imperatives to produce tangible results quickly) and with more interplay between these groups both in research and education. A substantial proportion of work in the construction industry now relates to existing building stock and there are increasing instances where conservation or upgrading of such stock is a viable alternative to demolition and replacement on both commercial and environmental grounds. A most important consideration being increasingly applied is the question of environmental implications of given materials in view of calls for 'sustainable construction'. It is likely that there will be still greater emphasis on this in the next decade.

In addressing the above this text incorporates the following main themes:

- Materials science is included where there is an identifiable pay-off in terms of understanding aspects of building behaviour.
- Implications for sustainability/environmental concerns are included.
- Increasing attention is given to considerations of maintenance and conservation of existing structures.

The subject of construction materials is enormously diverse and many of the topics under consideration are the subject of ongoing research. It would be impossible to examine, in depth, the range of topics included within a single text. Inevitably therefore this text must be considered as a snapshot of construction materials from the author's perspective. Some areas are covered in reasonable depth, reflecting the author's experience, while others, which might be expected to be included, may be covered only briefly. While the intention has been to give a broad coverage of issues considered to be of main importance in the industry, the reader will be aware of the need to refer to other texts at times. This text might be read in conjunction with the companion text *Materials in Construction, An Introduction*, by the author, which focuses upon understanding at a basic level of a broader range of construction materials and includes experiments and self-assessment exercises to assist with assimilation of the subject matter. Additionally it would be wise to read more widely in order to view the subject from quite different perspectives.

1.1   MATERIALS TECHNOLOGY IN THE NEW MILLENNIUM

The trend towards mechanisation seen in the last 20 years seems set to continue and will probably gather pace as experience of newer building techniques is increased and imperatives for larger, more diverse buildings to be produced at higher speeds become stronger. Increased computing power, together with modelling techniques, have allowed the solution of problems that would have previously required prohibitive time allocation so that materials can now be more safely and efficiently used in an exciting range of applications. More buildings comprise an assembly of manufactured components with less dependence on *in situ* trades and the high labour input they require. *Component* buildings place heavy dependence on the ability of materials to work together and may cause problems of compatibility of differing materials. Their construction may also involve much greater complexity though many manufacturing and some site processes are now controlled by management systems which demonstrate that high standards of quality can be achieved.

There have been shifts in the types of materials used. Some designers have become more favourably inclined to the use of steel in buildings as a result of improved efficiency within the steel industry, relatively rapid turn round times and the current trend towards system building. For a time, the prospects for concrete were not enhanced by adverse publicity caused by problems associated with steel reinforcement corrosion and uncertainties in the extent of the problem of the *alkali–silica reaction*. More recently, understanding of the behaviour of materials such as admixtures in concrete and cement replacement materials, together with advances in materials and manufacturing techniques, should improve the prospects for this versatile material. The use of well established materials such as brickwork seems set to continue albeit increasingly in a cladding context, because it endows buildings with a sense of permanence.

Traditionally there has an understandable reluctance on the part of the designer/specifier to use new products or materials until they have a track record. A major reason for this attitude is increasing general awareness of legal rights and the willingness of the client to seek redress following inappropriate use of materials. This could render any member of the construction team liable to the charge of negligence if it fails to implement fully current knowledge. At the same time, with the dismantling of trade barriers in Europe, a competitive industry must make informed use of new products or technologies with the smallest delay possible.

All members of the construction team have a 'duty of care' to the building user and judicial interpretations of this concept have developed significantly in the last 50 years. The starting point concerning the discharge of such duty is a thorough and up to date awareness of current knowledge of the materials and processes involved in construction. One of the key questions likely to be asked following disputes associated with building failures is whether application of current knowledge at the time of construction could have obviated the failure. Understanding of the term 'failure' itself should be considered. It might be expressed by the concept in Figure 1.1.

The term *building pathology* is sometimes employed in the context of the performance of buildings. In medicine two terms are used to describe behaviour of the human body:

■ Physiology is the study of normal behaviour of the human body.
■ Pathology is the study of disease or abnormal behaviour.

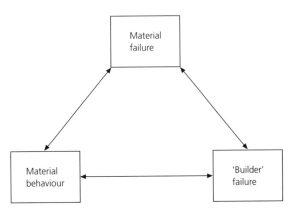

Figure 1.1   Interrelationships between the concepts of failure and behaviour

The latter term is becoming increasingly used in construction, but the point should be made that in the study of materials, pathology cannot be distinguished from physiology – they are both descriptors of normal materials behaviour. In other words, since materials obey the laws of chemistry and physics, materials failures are the result of the way they were manufactured, selected, or incorporated into the building. If they fail, they must have been produced, or used, inappropriately for the intended purpose. In that respect materials failures represent shortcomings somewhere in the design or construction process. It is, however, true that building pathology, or the study of building failures, is a valuable way of understanding materials physiology and it is in this way that important advances in materials technology have taken place. For example it was as a result of the collapses in 1973 of two roof structures formed from high alumina cement (HAC) that understanding of the behaviour of the material was enhanced.

1.2   MATERIALS SPECIFICATION

Some form of specification is often produced in order to assist the contractor in pricing and carrying out the work. The specification may be written into the contract, in which case it then forms part of the legal basis of the contract and as such is of key importance in the resolution of difficulties and the settlement of claims. There are many different approaches to specification and an increasing tendency to omit them altogether, their function being fulfilled either by detailed drawings or a bill of quantities together with requirements that the contractor should 'build well'. Alternatively, and in view of the large number of aspects to be covered in a building contract, standard specifications may be drawn up with reference to codes of practice with specific inclusions for individual projects. A further alternative is to produce a full specification under one cover for every aspect of the work which, though very time consuming, is more likely to produce a coherent whole. Whatever type of specification is produced, it should be appreciated that:

- A satisfactory specification requires a knowledge and understanding of materials and processes of building construction.
- Standard specifications should be checked for appropriateness, completeness and detail, particularly in difficult areas such as quality of work and tolerance.
- Specifications must be designed to be read and interpreted on site.

Shortcomings in specifications may be overcome by the experience and goodwill of the contractor but real difficulties arise when defects occur following operations which are inadequately described in a specification. The only way to avoid such disputes is for the specifier to be fully aware of every stage of the operation and to safeguard against all eventualities.

Where detailed materials descriptions are given together with processes or techniques, the specification could be described as a *materials* or *prescriptive* specification. The contractor is then bound to follow the operations laid down in the specification but would therefore not be liable for the performance of the final product, unless stated additionally. Such specifications are often lengthy since every aspect of detail has to be covered, though they may be preferred by contractors on account of the limited performance liability. If defects then occur, it would be the responsibility of the client to discover defects in the materials or processes and these may be quite difficult and costly to identify. To give an illustration in the case of paints, the type of paint, the number of coats, and perhaps the paint thickness might be specified. Provided they are adhered to, the contractor would not be directly responsible for the performance of the paint.

A simpler means of specification is a *performance* specification in which the materials/processes description is minimal and specific aspects of performance become of key importance, for example, the strength or durability of concrete. It should be noted in performance specifications that:

- Test regimes and procedures must be specified carefully. Very often a different method of test will produce a different result.
- Where requirements are highly demanding, contractors may not accept them or may increase the cost to cover risks.
- Provided reasonable care is taken, a contractor could not be held liable for modes of deterioration which are not known of or understood, or environmental conditions which could not be foreseen at the time of the contract.
- Performance specifications, or any other liability arising out of the contract, are only enforceable so long as the party concerned is still in business.

1.3 NATIONAL AND INTERNATIONAL STANDARDS

Standards of all types are designed to encourage and promote the use of materials or products of high quality and to provide a means of distinguishing such items from other available items of an inferior quality. Standards reflect improvements and changes and so must be updated from time to time. They are produced by committees which will normally represent the points of view of all parties with an interest in the product, for example, manufacturers, users, researchers, professional bodies and government institutions. Since there may often be a conflict of interests, standards will generally be a compromise in order that specified levels of performance can be achieved at reasonable cost by the manufacturer. Many standards, therefore, represent a basic level of performance rather than the best obtainable. They are, nevertheless, of immense value to specifiers, enabling great simplification of specifications where appropriate. Only where a higher level of performance is required or an area not covered by standards is encountered will there be a need for more detailed specifications. There follows a brief resume of some national standards authorities including Eurostandards which will, in time, replace most national standards. A word of caution is necessary as regards products or services claiming to operate to given standards. Descriptions such as 'conforms to . . .' do not necessarily imply approval by the standards authority. The best

Figure 1.2   BSI's kitemark, which indicates BSI approval that the product is manufactured in compliance with a published specification, and within a quality management system

indication of that is the relevant mark, for example, the *kitemark* in the case of British Standards. Standards authorities usually issue buyers' guides, giving lists of approved companies and products.

BRITISH STANDARDS INSTITUTION (BSI)

The need for standardisation in the UK followed rapidly from the industrial revolution in the mid-nineteenth century. The first items to be standardised were screw threads as suggested by Whitworth in 1841 (though they were not, unfortunately, adopted worldwide) and many other aspects of engineering were standardised in the years that followed. One of the earliest British Standards, BS 4, rationalising structural steel section sizes was published in 1903 by the Engineering Standards Committee – the forerunner of the British Standards Institution. Since then, a vast range of British Standards/BS Codes of Practice has been produced, of value to manufacturers and purchasers alike in standardising sizes and in defining and maintaining levels of quality. It may be a requirement of compliance with a standard that products should be marked with the appropriate standard number. In some cases (especially with regard to product safety in the domestic market), the use of the kitemark (Figure 1.2) under licence may be authorised. This also implies that the product is covered by a quality management system. There are now many thousands of British Standards in operation, many of them used worldwide, though they are now gradually being replaced by European Standards (see below), many of which are in any case adopted as 'BS EN' standards.

ISO – INTERNATIONAL ORGANISATION FOR STANDARDISATION

This was set up in 1946 and was the first truly international standardisation body, with BSI playing a major role in its development. BSI holds the secretariat of a significant proportion of the international technical committees which produce ISO standards. ISO has played a major part in the worldwide standardisation of testing procedures and product specifications which have facilitated many forms of international trade, not least in the area of construction materials and products. Many ISO standards are formally adopted as British Standards, when they will have the prefix BS ISO.

BRITISH BOARD OF AGRÉMENT (BBA)

This was set up in 1966 as a national authority for the testing and assessment of new building materials, products, systems and techniques for which:

- There is no British Standard.
- There is some innovative feature not previously covered by a British Standard.
- Performance is substantially better than the corresponding British Standard.

Products are assessed by methods similar to those used in the production of a British Standard, the testing normally being carried out by technical agents such as research departments or independent test laboratories. Assessment is specific to a certain manufacturer and product, with the result that certification can be completed in a relatively short period of time, facilitating early incorporation into construction practice.

Agrément certificates are initially valid for three years after which they are reviewed. Validity may then be extended or certificates may be renewed periodically. If, in the intervening period, products of the type in question become the subject of a British Standard then, by agreement between BBA and BSI, the Agrément certificate will not be renewed. Figure 1.3 shows the logo which is displayed in relation to products with Agrément certificates. Agrément certificates are accepted by many regulatory bodies, such as local authorites, as evidence of suitability for purpose.

Figure 1.3   British Board of Agrément certificate for innovative products (courtesy British Board of Agrément)

The British Board of Agrément (BBA) is a member of the European Union of Agrément (EUAtc) comprising similar certifying bodies in other European countries. In this way, it is possible for an Agrément certificate for a product in any one member country to 'be confirmed' by other countries in the Union. This recognition is voluntary and additional testing may be required in particular countries, for example, due to climatic differences which may exist in them.

EUROPEAN CODES AND STANDARDS

The European Common Market was originally founded in 1957, although the number of member states has greatly increased since that time, now including virtually all countries in Western Europe. A legal framework in the form of the Treaty of Rome was drawn up and Article 30 describes the aim 'to provide free movement of goods and services'.

The Single European Act 1987 was written into the EEC treaty and defines the internal market as 'an area without internal frontiers in which free movement of goods, persons, services and capital is ensured'. This includes removal of technical barriers to trade through technical harmonisation, a common standards policy and the mutual recognition of testing and certification arrangements covering many products including construction products with a deadline for implementation of 31st December 1992.

The Council of Ministers has passed *directives* or laws applying to member states which expand the articles of the Treaty. The Construction Products Directive (CPD) seeks to apply the above aims to the construction area by the introduction of European Standards (ENs). Products must satisfy certain 'essential functional requirements' relating to:

- stability
- safety in fire
- safety in use
- health
- noise
- heat retention

Full standards are preceded by draft standards (ENVs) rather like draft British Standards. They are produced by Technical Committees (TCs) of the European Standardisation Committee (CEN). Secretariats to the TCs are provided by member states (represented by their standardisation bodies, for example, the BSI in the case of the UK). This may offer an advantage to that state in influencing the content of the Standard. At the present time, Germany has the secretariat of most committees, followed by France and then the UK. Where European Standards are recognised by a member country they are incorporated into their existing system – for example:

BS EN   for the UK
DIN EN   for Germany
NF EN   for France, etc.

Figure 1.4 CE mark as evidence of conformity of a product to an EC directive via harmonised standards

When standards become harmonised across Europe they are given the prefix h EN. There are at this time no standards for construction products which are harmonised in this way, though the first may soon be published. Once harmonised standards have been produced, the compliance of construction products with the essential requirements of these standards as defined by the CPD can be assessed. Products in full compliance will be given the CE mark (Figure 1.4). (This mark can already be seen quite commonly on electrical goods.) Such compliance will require 'attestation of conformity' (A/C) which may be 'self declared' by the producer for non-critical products, while for those where there are health or safety implications, 'third party' attestation will be required.

An alternative route to the use of the CE mark is by European Technical Approvals. These are issued by national bodies in accordance with guidelines prepared by a joint European body. In the case of the UK this role is undertaken by the British Board of Agrément (BBA) (in parallel with its role in the issue of Agrément certificates). These approvals are for specific innovative products and are produced in conjunction with the European Organisation for Technical Approvals (EOTA). They are to be seen as pre-Eurostandard approvals. The CE mark may be awarded when a product gains a European Technical Approval together with a certificate of attestation of conformity (A/C) as above.

The use of products is covered by Eurocodes (for example, EC 2 – *Concrete Structures*, which will eventually replace BS 8110), together with interpretative documents, IDs, which harmonise product standards with individual member state building regulations, taking into account such things as climatic variations. Eurocodes are rather like the Approved Documents in the UK which interpret the National Building Regulations.

Since products or services which comply with harmonised Eurostandards or codes must be accepted throughout the EU, 1992 marked the beginning of a process which will lead to increased competition within the market – and it will be increasingly incumbent on producers, designers and contractors to adapt their wares to European standards as necessary. Failure to develop along these lines may lead to an erosion of 'home' markets as well as lost opportunities to exploit the wider potential of EU countries.

There are frequent references in this text to remaining British Standards, and to some obsolete standards which nevertheless still form a part of the 'currency' of construction specification – such as use in Building Regulations. Since changes continue to be made it is important that the reader consults the current version of the British Standards Yearbook if access to the current standard on any topic is required. The BS Yearbook is valuable in indicating the current status of all British Standards.

1.4   QUALITY

There have been many definitions of the term *quality* though a simple definition which will suffice for the purposes of this study is 'fitness for purpose', that is, the ability of a product or system to satisfy all the requirements it was designed to meet. *Quality control* is, broadly, the system of practical arrangements which helps to maintain quality, though it is now understood that the consistent production of a quality product is linked to the overall structure and management of any company. *Quality assurance* is a management process designed to inspire confidence in the product or process. It requires the adoption of a *quality system*. The rules for quality systems were originally laid down in BS 5750 – *Specification for Quality Management Systems* in 1979, the chief features being:

- The existence of a quality manager not involved with production or marketing.
- A comprehensive system of procedures for every process involved (quality manual).
- Detailed records of all inspections.
- Effective training arrangements for personnel.
- A formal procedure for identifying and rectifying substandard goods or operations.

In 1994 BS 5750 Parts 1, 2 and 3 were adopted by ISO and the harmonised version became the BS EN ISO 9000 series, referred to as *Quality Management Systems* with corresponding international validity and recognition.

The operation of quality systems can only satisfactorily be checked by means of a third party, that is, a party having no link with either the producer or the customer. The third party is a body rather like a BS committee, convened to assess a particular scheme. (Certifying bodies should themselves be assessed against the requirements of BS EN 45012 supported by the internationally recognised ISO/IEC Guide 62. The United Kingdom Accreditation Service (UKAS), formerly the National Accreditation Council for Certifying Bodies (NACCB), has been set up for this purpose.) The certification body first designs a quality assessment schedule against which the particular scheme can be assessed. Once the quality manual is approved, visits to works or sites are arranged to check that procedures are actually in line with it. Companies complying in every way are then *registered* and become eligible to use logos signifying their assessed capability.

Figure 1.5   Logo indicating third party registration to a quality management scheme (courtesy Lloyd's Register Quality Assurance)

Quality management systems have been in use for many years, mainly in a manufacturing context but are now increasingly being introduced in design and construction processes which are the areas traditionally giving rise to a large proportion (about 90 per cent) of building faults. Most products of the construction process – buildings – are unique, involving many different teams. Standards for construction quality are also often difficult to define; every site operates under different conditions and personnel changes are common. Progress has, nevertheless, been made and some designers and contractors now also use BS EN ISO 9000 systems. A typical logo used, including the certification body, is shown in Figure 1.5.

In the early 1990s the concept of *total quality management* (TQM) was introduced and this can be regarded as the ultimate in quality systems, since TQM is seen as pervading every aspect of a business. Key words in this context are:

■ common vision
■ organisational culture
■ teamwork
■ continuous improvement
■ customer satisfaction
■ perfection

The business goal of TQM may well be competitive edge over rival companies but studies suggest that TQM also results in more efficient organisations with better quality outputs and lower wastage.

INTEGRATED QUALITY, ENVIRONMENTAL MANAGEMENT AND SAFETY SYSTEMS

Every company has to address the questions of quality, environmental management and safety, and key aspects of these involve similar management issues. The current trend is therefore to incorporate these into a single *integrated management system* (IMS). The individual components are currently assessed separately:

quality	BS ISO 9000
environmental management	BS ISO 14000
safety	BS 8800

The British Standards Institution is currently developing integrated assessment systems and the major accrediting bodies are moving towards certification of such systems.

CHAPTER 2

MATERIALS' PERFORMANCE AND ITS MEASUREMENT

This section attempts to identify, define and discuss measurement of those properties of materials which might have to be considered when incorporating them into buildings, together with comments on interpretation where appropriate. An understanding of the subject matter of this section is essential before the properties of individual materials are examined and this understanding will be assumed in the following chapters. It will be evident that the performance of building materials often has many facets, failure to consider all of them sometimes leading to premature distress in spite of careful attention to most areas of performance. *Limit state* design codes address this situation. Each mode of failure is identified and the design process carried out so as to produce an acceptably low risk of failure in each mode. In addition to the *ultimate* limit state representing overstressing of the structure, there are various *serviceability* limit states caused by deflection, cracking, environmental effects and so on.

2.1 MECHANICAL PROPERTIES

These are associated with the effects of load.

STRENGTH

Strength may be defined as the ability of a material or component to carry load without structural failure or excessive plastic deformation. Note the key parts of the definition:

Load   This is normally expressed as stress (= load/area) in view of the fact that it is normally concentration of load that causes failure.

Structural failure   Many materials such as bricks, glass, concrete and some metals are *brittle*, generally implying that, when the limit of elasticity is reached, they are close to disintegration or failure. Acceptable working stresses for such materials are obtained by loading them to destruction and then applying a suitable

13

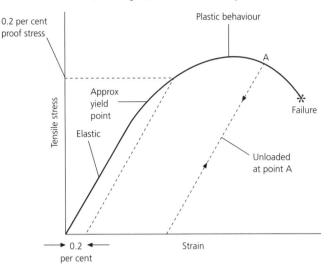

Figure 2.1   Tensile stress/strain graph for a material that yields. Although permanently distorted the material may still behave elastically if unloaded and reloaded at point A

safety factor. This factor of safety should take into account that failure may be sudden or dramatic since plastic flow in them is not normally possible.

Excessive plastic deformation   Some materials *yield* when overloaded, that is, they deform excessively (without breaking) and do not return to their original size or shape. Clearly, this is unacceptable in buildings and maximum stress must, therefore, be based on yield point stress, rather than ultimate failure stress. Many of the softer metals and some plastics exhibit such behaviour (Figure 2.1). Yield points can often be seen from stress/strain graphs, though in metals such as copper and some steels, yield is so gradual and progressive that it cannot be observed from a graph and a given amount of plastic movement must be used to provide an index of strength instead. In the case of metals, it is usually 0.2 per cent plastic movement and the corresponding stress is called the 0.2 per cent proof stress (Figure 2.1). In safety terms, some plastic flow prior to failure is advantageous. For example yield in a mild steel would not cause total failure of a reinforced concrete beam but only excessive deflection. Indeed, on yielding, the strength of many metals increases by *work hardening*, so they become better able to support the applied loads. Materials which flow plastically before failure are described as *ductile*. Ductility is normally measured by the amount of elongation that occurs prior to failure. Figure 2.2 compares the tensile behaviour of some common metals.

Testing for strength

Materials are normally tested in a manner which roughly simulates their operation in the building, though *in situ* stress patterns are often complex. Chief modes of testing are compression, tension, flexure and shear. Brief comments follow on each.

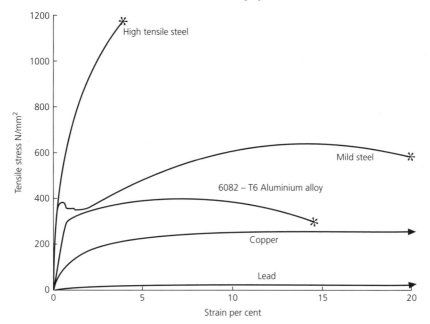

Figure 2.2 Stress/strain curves for mild steel, high tensile steel, structural aluminium alloy, copper and lead. The last two reach very high strains before failure

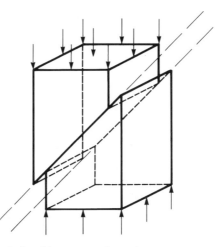

Figure 2.3 Shear stress induced by compressive stress

Compression

Compressive forces in materials act in the same manner as atomic bonding, forcing atoms together and this action in itself cannot cause failure. However, compression induces shear stresses (Figure 2.3) and tensile strain leading to tensile stress by the

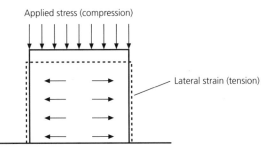

Figure 2.4   Poisson's ratio effect in compression; the tensile stress is at right angles to the applied compressive stress. Poisson's ratio is equal to the horizontal movement as a proportion of the vertical movement

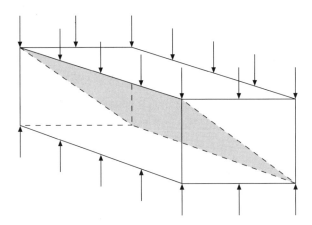

Figure 2.5   Plane of maximum shear in a brick during compression. The shear stress is less than that in the prism of Figure 2.3 because the plane is not at 45° to the applied stress

Poisson's ratio effect (Figure 2.4). Depending on the material type, specimen shape and loading arrangement, compression may cause shear or tensile failure or some combination of the two. The compression test is quite easy to carry out and is often preferred for any situation in which compressive forces predominate, as in most *ceramic type* materials. However, in view of the complications indicated, there can be no absolute compressive strength and it is therefore imperative that standard procedures are followed exactly. Illustrating this is the effect of shape (Figure 2.5) The maximum shear obtained by compressive forces is at 45° to the applied force and, since in the brick the plane of maximum shear is much less than 45°, the sample will indicate a higher strength than a prism shaped sample of the same material which contains 45° planes (Figure 2.3) The brick shape is also restrained from the Poisson's ratio effect by friction with the very stiff steel platen surfaces

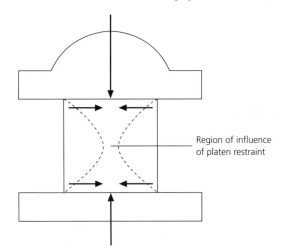

Figure 2.6 Stresses caused by platen restraint on a cube during testing

which effectively reinforce the material, even though load is being applied through them. The region of restraint in the prism is smaller, giving a second reason for it to fail at a smaller stress. The area of platen restraint in a cube is illustrated in Figure 2.6. Insertion of rubber pads above and below test specimens reverses the Poisson's ratio effect as the rubber squeezes out sideways causing a tensile failure in the form of a vertical crack at very low stress. Hence, the details of the testing machine should also be standardised, bearing surfaces nearly always being of machined steel. Further comments on this important topic are made under 'Concrete testing'.

Tension

Tension testing is used mainly for metallic or fibrous components designed to carry tensile stress, though it is occasionally used for materials such as concrete where some degree of tensile performance is required, or simply to give a strength index. As with compression testing it must not be assumed that the mode of failure corresponds directly to the stress applied. It is well known that metals deform by flow along crystal planes and this is a shear rather than tension effect. Hence ductile metals yield by the effect of induced shear stresses produced by tension. For this reason the performance of ductile metals in compression and tension is similar (assuming buckling does not take place in compression) since the induced shear stresses are the same. In brittle materials tensile failure is usually caused by stress concentrations at cracks or imperfections resulting from Griffith's considerations (see page 133).

The main problem in tensile testing is the provision of an effective gripping system on the specimen. Jaws must either be threaded (as in some metals testing

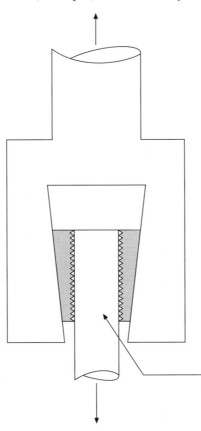

Figure 2.7   Method of gripping a smooth tensile specimen. Poisson's ratio stresses at the point indicated sometimes cause failure within the grips, especially in ceramic samples

Figure 2.8   Waisted tensile specimen designed to ensure failure in the centre portion. The concave surfaces must be gently curved to avoid stress concentrations at these positions

applications) or have some arrangement such as tapered inserts which grip the specimen more tightly as load is increased (Figure 2.7). This in itself causes further problems as Poisson's ratio tensile stresses within the jaws due to lateral compression add to those already in the specimen. Hence, some sample types tend to fail within the jaws and must be waisted to avoid the problem (Figure 2.8). The waist may be produced by casting in a special mould (non-metals) or by machining (metals). Fortunately, many steel components do not suffer from this difficulty so that reinforcing bars, for example, can generally be tested as received.

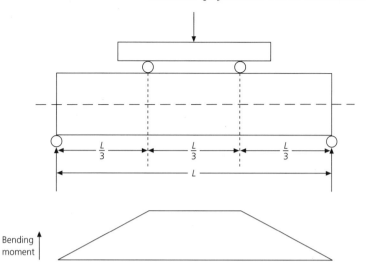

Figure 2.9 A common method of loading in flexural testing. It results in a constant bending moment in the centre section of the beam

Flexural

This can be quite easily carried out on small or medium sized samples by an attachment to compression machines. Loads are applied through rollers which can be pivoted to align accurately with bearing surfaces. Contact at each position is, however, through a narrow line and to avoid stresses at these positions surfaces should either be very smooth, or a load spreading arrangement in the form of bearing pads or plates may be necessary. In the case of short spans of deep section, checks should be made to ensure that the bearing stresses at the points of loading and support do not influence the test result. In some tests, for example, concrete testing, two loading rollers are employed at 1/3 and 2/3 of the span and this gives a constant bending moment in the centre third of the span so that failure should be at the weakest point in this section (Figure 2.9). In simply supported flexure, there is compression at the top of the beam and tension at the bottom (soffit). Hence, there is a plane of zero stress (neutral axis) and in a beam of rectangular section, this is assumed to be at mid-depth. Note that there is shear acting horizontally at the neutral axis position – if a beam were cut horizontally at this position, it would be significantly weakened, even though there is no compressive or tensile stress at this position. It may be convenient to introduce the concept of principal stresses at this point in order to demonstrate how shear in a beam can lead to failure. The principal stress concept is that:

> Any combination of stresses can be represented by a simple system of pure compression stress and pure tension stress on planes at right angles to each other.

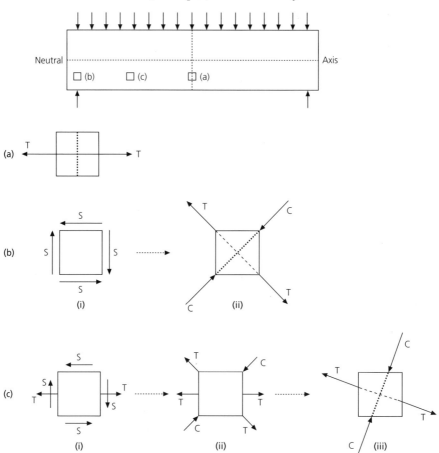

Figure 2.10   Principal stresses resulting from application of a uniformly distributed load to a simply supported beam

Figure 2.10 shows a beam in simple bending subject to a uniformly distributed load. At the centre of the beam (a) the shear stress is zero but the soffit of the beam is subject to maximum tension due to bending. The principal stress is therefore a tensile stress parallel to the beam axis (acting on a vertical plane, in bold).

Adjacent to the support (b) the bending stress is almost zero but there is a shear stress (vertical arrows (i)). In any section subject to shear there is a tendency for the beam to rotate so there must be a 'complementary' shear stress as shown (horizontal arrows) to balance this. The shear stress and complementary shear stresses can be combined to give the principal stress arrangement shown in (ii). Note that there is a tensile stress at 45° to the beam axis and it is this stress that tends to cause a 'shear' failure on the plane in bold near to the beam supports.

In case (c) there are both tensile and shear stresses, though these are of reduced magnitude compared with (a) and (b), respectively. The shear stresses can be combined as before (ii) and then the resultant of them can be combined with the tensile stress (iii).

Failure is most likely in the tensile plane (bold dotted lines) of the principal stress in each case. Note that in (a) this plane is vertical, in (b) it is at 45° to the vertical and in (c) it is at an intermediate angle. Hence the failure plane rotates on moving from the centre to the ends of the beam. Observed shear failures tend to be at angles of around 45° to the horizontal.

In materials such as concrete the flexural tensile stress is normally taken by steel reinforcement near the soffit of the beam. Such reinforcement would not, however, be able to cater for the effects of shear near the beam ends so that special shear provision would need to be made, for example in the form of links. Shear stresses tend to be larger relative to tensile stresses in shorter beams.

Stresses are usually calculated on the assumption that classical bending theory applies; that is, the strain in the material increases linearly with distance from the neutral axis. The flexural strength or *modulus of rupture* is taken to be the tensile (or compressive) stress at the lower (or upper) surface, respectively. However, since in most materials bending theory is not obeyed, the stress obtained is only an index rather than the true stress in the material. For example, in concrete the neutral axis tends to rise during the test, increasing the area of the lower tensile zone which is where failure initiates by crack propagation. This gives a flexural result which suggests that the material is stronger in tension than it really is. In fibrous materials the compression zone is more critical since fibres tend to buckle here. It is found that the neutral axis tends to drop in the test, increasing the area of the critical compression zone.

It will be evident that care is necessary when interpreting such results for design purposes.

STIFFNESS

The stiffness of a material or component may be defined as its ability to resist deformation (as distinct from failure) under stress. Stiffness is measured by the term *modulus of elasticity*, E, the ratio of the stress applied to a component to the strain resulting from that stress. Modulus of elasticity is measured by obtaining a stress/strain graph for the material concerned, the E value being obtained from the gradient of that curve in the elastic region. Although Hooke's law (stress is proportional to strain) is roughly obeyed by virtually all materials at stresses which are low compared with failure stress, stress/strain graphs are not usually exactly linear, especially in non-metals, so some care is necessary in measuring E values. The highest value obtained is the initial tangent modulus (Figure 2.11) representing the E value at zero load. For estimating strains under service loads, it is more appropriate to measure an average E value up to the appropriate stress and this could be obtained from the secant modulus at the desired stress (Figure 2.11). Stiffness checks are an important part of structural design in buildings, and deflection of large span beams in particular may be more critical in design than

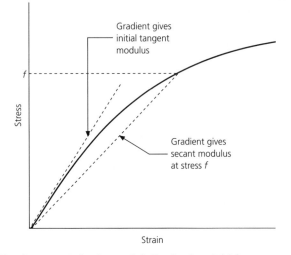

Figure 2.11   Non-linear stress/strain graph indicating how initial tangent modulus, and secant modulus at stress *f*, would be calculated

Table 2.1   Moduli of elasticity of some common building materials

Type	Modulus of elasticity kN/mm²
Rubber	0.001–0.02
Thermoplastics	0.1–4
Timber	3–30
Concrete	15–40
Glass	70
Aluminium	70
Steel	210
Carbon fibres	200–450

their load bearing capability. Table 2.1 shows *E* values of some common building materials.

Building elements may be subject to a conflict of requirements in stiffness terms. In order to retain shape and to minimise disturbance to finishes (especially wet applied coatings such as plasters or renderings) high stiffness is desirable. Brick, concrete or steel buildings would fall into this category. However, environmental changes generate strains in the material which, if not free to move, give rise to stress:

stress = E × strain

For a given strain, stiff (high *E*) materials become subject to higher stress. The problem is particularly serious when environmental changes lead to tensile stress in ceramic type materials which have high stiffness, yet low tensile strength and are, therefore, quite vulnerable.

When designing buildings generally, the question of stiffness should be addressed. Permanence is traditionally associated with rigidity, but the latter often brings with it the problem of high building weight, possibly high building cost and the need to design carefully to avoid movement problems. Timber frame construction is a promising form of lower stiffness building, displaying some advantages over more rigid methods, both in construction and performance terms in low rise building. This trend may continue as more lightweight materials or composites are developed. Fabric structures might be regarded as the ultimate in low stiffness buildings.

Where stiff materials subject to movement must be joined and sealed, there will be distinct advantages in using low stiffness materials for the purpose of forming the seal. Sealants are one example, modern elastomeric sealants having much larger movement capability than older, stiffer materials, such as cements or hardened putty.

A further illustration of the effects of stiffness is in post-tensioning wires for prestressed concrete. Glass fibres have strengths higher than that of steel but with only one third of the stiffness of the latter (Table 2.1). In consequence, they must be stretched three times as much to reach their working stress and subsequent shrinkage in the concrete would, therefore, cause a smaller loss of prestress than in the stiffer steel. (There are, of course, other aspects of performance that must also be considered in this case.)

TOUGHNESS (IMPACT RESISTANCE)

The toughness of a material reflects its ability to absorb energy suddenly by impact. An impact is a rapidly applied stress, though it is not the stress itself which causes the damage so much as the energy associated with its application, which must be dispersed or absorbed by the material. Toughness can be quite easily measured by pendulum type machines, such as the Charpy tester (Figure 2.12). A standard sample, containing a notch to initiate failure, is carefully placed in the machine and subjected to impact by the weighted pendulum. The energy absorbed is equal to the loss of energy of the pendulum as indicated by the difference in height between the pre- and post-impact height positions.

The test result depends upon the specimen size and geometry – especially the root of the notch which concentrates the stress. Temperature can also have a marked effect especially in the more brittle metals, which absorb much less energy at lower temperatures. The test must therefore be taken as an index of performance rather than an indication of performance of real structures. A good indication of impact performance can be obtained by reference to stress/strain graphs for the material concerned (Figure 2.13). The area under such a graph has the units of stress times strain and these are the same as energy per unit volume:

$$\text{stress} \times \text{strain} = \frac{\text{force}}{\text{area}} = \frac{MLT^{-2}}{L^2} = ML^{-1}T^{-2}$$

$$\frac{\text{energy}}{\text{volume}} = \frac{\text{force} \times \text{distance}}{\text{volume}} = \frac{MLT^{-2} \cdot L}{L^3} = ML^{-1}T^{-2}$$

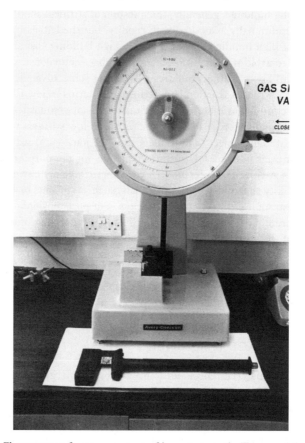

Figure 2.12   Charpy tester for measurement of impact strength. Two pendulums are provided, of 5 joules and 15 joules. Absorbed energy is recorded by a 'lazy pointer'

Hence, the area under the curve is related to the energy absorbed in stressing the material to failure. Ceramic type materials exhibit curves such as A (Figure 2.13) and are, therefore, usually poor in impact (brittle). Softer metals and wood give rise to curves similar to B which suggests that they should have very much better performance in impact conditions; they are *resilient* or tough. Note that high toughness does not necessarily imply high strength; indeed, as metals are strengthened or hardened by alloying or heat treatment, they tend to become much more brittle because plastic flow occurs less readily. The ability to flow plastically appears to be the main prerequisite for tough materials. Fibrous composites, whether natural, such as wood and leather, or synthetic, such as fibre reinforced concretes, gain their toughness from the work needed to destroy the bond between fibre and matrix. Similarly, in materials of a given type, softer varieties, for example, building blocks, are usually tougher because, under impact, they yield locally, producing a small area of damage, rather than transmitting energy through the unit by crack propagation and possibly causing overall failure. The impact behaviour of

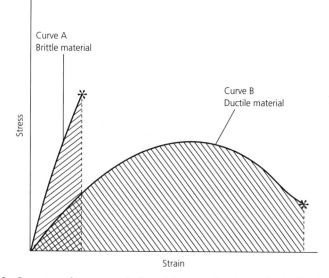

Figure 2.13 Impact performance as indicated by stress/strain graphs for brittle and ductile materials. The area under the stress/strain graph for the brittle material is small, indicative of inability to absorb much energy

plastics depends on their molecular make-up, the toughest ones containing long molecular chains, lightly bonded, as in polyethylene and some other thermoplastics.

RESILIENCE

This is similar to toughness in that it relates to energy absorption. However, in this case, the energy must be absorbed elastically. Resilience may be measured in a similar way to toughness, except that the pendulum should not break the sample, but should rebound. The amount of rebound (representing the elastic energy returned to the pendulum from the material) compared to the initial pendulum energy represents the resilience. Resilient materials should have a long elastic range – rubber is a good example. High tensile steels are resilient, though they may show brittle failure if overstressed.

CREEP

Creep is defined as time dependent strain resulting from prolonged application of stress. It may be regarded as very slow yielding of a material at stresses which, in the short term, would result only in elastic deformation. Creep occurs mainly in three classes of materials:

1 Metals at temperatures fairly close to their melting points (see 'Creep in metals'). Conversely, when metals are well below their melting points, as with steel at room temperature, creep will not be a problem.

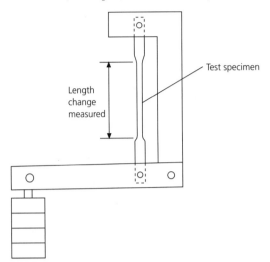

Figure 2.14　Simple creep rig

2　Moisture susceptible materials. Materials which, for example, swell when damp
　are likely to exhibit creep and this may be related to moisture flow in the
　material. Hence, more porous ceramics, such as concrete, are subject to creep.
　Timber is another material in this category; deflections, especially in joists and
　rafters, progressively increase with age.
3　Lightly bonded or fibrous materials. Creep in these may result from breaking and
　reforming of atomic bonds or, in fibrous composites, progressive slip of fibres in
　the matrix. Examples include bitumen, some thermoplastics and many fibrous
　composites. Timber would fit into this category as well as the one above.

　Provided stresses are well within yield stresses (say 50 per cent), creep should not
lead to failure, though in some aspects of structural design it should be allowed for,
as in timber or prestressed concrete beams. One of the best known illustrations is in
lead roofing. When deformation becomes excessive, lead must be stripped, recast
and reapplied.

　There is no short cut to creep testing; specimens must be loaded to their service
stress which should be maintained, preferably for a number of years, strain being
monitored. Forces are often applied using weights and some leverage device
(Figure 2.14).

FATIGUE

Fatigue damage results from repeated application of stress. The phenomenon
was not identified until the 1950s when fatigue failures in aircraft wings caused a
number of disasters. Fatigue fracture in many materials will occur at stresses well
below normal yield or failure stress if loads are reapplied sufficiently. The number
of cycles to failure depends on the applied stress (see Figure 8.20 in Chapter 8)

which may have to be less than half the yield stress if a great number of load cycles is anticipated. The unfortunate aspect of fatigue is that no yield is normally visible, failure often being catastrophic, as cracks propagate at highly stressed positions.

Fatigue performance can be simply measured in metals by rotation of circular section specimens supported at each end in roller bearings and from which a weight is centrally suspended by means of a further bearing. In materials such as timber or concrete, tests are normally carried out by repeated hydraulic loading on beams. For full size section this can be a very expensive process.

Fatigue failures are rare in buildings, since dead loads are usually much larger than live loads. Some building components may, however, be affected. Failures are quite common in metallic components such as hinges or brackets subject to movement, especially where corrosion weakens the metal or increases friction.

Behaviour of structures subject to vehicular traffic, such as roads and bridges, is in marked contrast to buildings and design must be centred upon fatigue effects. The *design life* of roads is measured in standard axles, equivalent to the passage of one axle of mass 8.2 tonnes. For example, if a life of 15 years is required for a road, the projected number of standard axle passes equivalent to anticipated traffic types and intensity must be estimated. This can be very difficult and it is perhaps not surprising that actual lives of roads may differ greatly from those intended. Some motorways are designed for lives of over 100 million standard axles. It should be noted that fatigue effects are highly load-sensitive. It would require passage of several thousand small family cars to produce the effect of one standard axle, while passage of a heavily overloaded commercial vehicle would be equivalent to passage of over one hundred thousand standard axles or several million small cars.

A related problem in buildings is ground floor slabs which in warehouses or retail outlets may be subject to passage of forklift trucks or similar vehicles. With solid tyred vehicles there may be the added element of vibration transmitted to the floor. Failure in such instances may well be fatigue related.

HARDNESS

This may be defined as resistance to indentation under load and is often used to measure the resistance of a surface to local damage by stress or impact. Tests for hardness normally involve measuring the diameter of the permanent depression left by application of a hardened ball to the surface under a standard load, for example, Brinell hardness used in metal testing (Figure 2.15). In this case, the hardness test is used because it correlates well with the yield strength of metals, hence, it is used to assess the effects of heat treatment. The Schmidt hammer is similarly used to estimate the compressive strength of concrete from its hardness.

Hardness tests are very relevant in the performance assessment on walling and flooring materials. The hardness of many materials can be increased by treatment of surface layers and this technique is widely used to improve performance of many surfaces subject to wear and tear. Tests are necessarily subjective; for example, larger indenting tools under higher loads will tend to measure hardness at lower

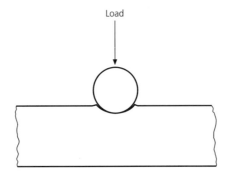

Figure 2.15 Measurement of hardness of a surface by use of a hardened steel ball. The combination of load and ball diameter should be such that the diameter of the permanent impression is 30 to 60 per cent of the ball diameter

levels in the material so that standardised methods of test are essential where properties of the material vary with depth.

The nature of the dent will to some extent characterise the material; for example brittle materials produce a clear indentation without a lip; soft metals tend to produce a clear lip as material is displaced sideways, while metals which work harden may cause some sinking of the metal around the impression.

WEAR OR ABRASION RESISTANCE

This is defined as the ability of a surface to resist wear due to frictional contact with moving objects or materials. Hard surfaces will generally be abrasion resistant but the abrasion resistance of softer surfaces can also be greatly improved by use of hard, well bonded coatings such as epoxy resin bonded calcined bauxite used to improve the anti-skid properties of roads around hazard spots. Wear resistance requirements will depend on the type of traffic encountered. For example, steel tyred vehicles place much greater demands on surfaces than rubber tyred traffic. Generally, harder tyres and bigger loads require harder and better bonded surfaces to resist wear. Related to wear and abrasion is *cavitation* caused by formation and collapse of tiny voids in fast moving liquids, such as water. Very high local surface stresses can be produced leading to breakdown, especially in materials such as concrete. Strong, hard surface layers are again the best way of maximising life in situations where rapid liquid movement cannot be avoided.

2.2 ENVIRONMENTAL EFFECTS

The term 'environmental' will be taken to mean 'non-mechanical', or not associated with applied loads, though some of the effects lead to stresses in the material which may cause eventual failure. Environmental effects often result in movement in materials so that measurement of movement may indicate the seriousness of their

Table 2.2   Tensile failure/yield strains for some common building materials

Material	Tensile failure/yield strain $\times 10^{-6}$
Concrete (12 hours of age)	10–20
Mature concrete	100–200
Mild steel	1000–2000
Aluminium	2000–4000
Thermoplastics	5000–20 000
Glass fibres	50 000
Rubber	Up to 700%

action. Hence, a few comments will be made on movement initially. Movements, as measured by strains (proportionate length changes) are usually very small so that a common and convenient unit is *microstrain*, 10^{-6}, that is 1 mm movement per km of length. Strains measured in more flexible materials may alternatively be measured as percentages. Tensile failure (or yield) strain values of some common materials are given in Table 2.2.

Movement or strain measurement is often required when monitoring structures and Table 2.3 indicates some methods of obtaining this information. Great care is necessary in taking such measurements to separate actual movement from measurement or random errors, especially when long term testing is involved. Errors may result from limitations of the equipment and/or the user. There is no substitute for experience in use of each type of equipment. However simple experiments can be carried out to determine confidence levels including the effect of the operator if required.

SHRINKAGE/MOISTURE MOVEMENT

These take place in virtually all materials which contain accessible (as distinct from sealed) pores. They include bricks, concrete, stone and timber. Susceptibility is due to the fact that water is quite strongly attracted to solid surfaces and exerts a swelling action on wetting, especially when surfaces are close together as in fine grained materials. For a given materials group, stronger materials are less affected than weaker ones since bonding restrains movement. The extent of moisture movement in common porous building materials is summarised in Table 2.4.

Many natural materials vary greatly; for example, there are natural stones fitting into each category and timber properties depend very much on species. A most important point to be considered when using porous materials is their *starting point*. Materials which are baked prior to use, for example, clay bricks, will take in moisture on exposure, even to dry atmospheres, while those which start life saturated, for example, timber and concrete, will lose water in the same atmosphere. The situation may be represented as in Figure 2.16.

Hence, clay bricks and concrete undergo opposite changes in size during/after construction and, in the case of, for example, brick panels in a reinforced concrete

Table 2.3 Some methods of measuring movement, given in order of increasing sensitivity

Type	Principle	Approximate resolution mm or strain	Cost	Applications	Comments
Glass slide	Cemented across crack. Breaks if crack moves	Indication of movement in excess of ~0.1 mm	Very low	Simple check for continued movement	Actual movement cannot be measured
Rule/crack tell-tales	Calibrated scale	1 mm	Very low	Large cracks only	Simple but not very accurate
Portable microscope	Magnified calibrated scale	0.1 mm	Low	Measurement of medium/large cracks	Good lighting necessary (some types have built-in illumination)
Calipers/demountable dial gauges	Calibrated scale + studs fixed to surface	0.01 mm ($= 200 \times 10^{-6}$ strain on 50 mm gauge)	Low	General testing of medium/small cracks	Calibration should be checked. Care needed when taking readings
Electrical resistance strain gauge	Change of resistance of wire bonded to surface when strained	10×10^{-6}	Medium	Widely used to measure strain in steel, timber and concrete (not for cracks)	Must be well bonded to smooth surface. Affected by damp and temperature changes
Linear variable differential transformer	Induced current between two coils depends on their relative position	1×10^{-6}	High	Research/development where high sensitivity needed	Very sensitive. Gauges expensive but can be re-used. Mainly for short term tests
Acoustic strain gauge	Frequency of vibration of fixed wire measured	1×10^{-6}	High	Long term tests in *in situ* concrete	Must be built in during construction. Very good long term stability

Table 2.4 Moisture movement characteristics as affected by grain structure/strength

Type	Fine grained/weak	Fine grained/strong	Coarse grained/strong
Amount of movement dry–wet	over 1000×10^{-6}	$100–1000 \times 10^{-6}$	less than 100×10^{-6}
Examples	Clays, weak mortars, wood across grain	Bricks, concrete, softer stones, wood along grain	Strong bricks, gypsum plaster, harder stones

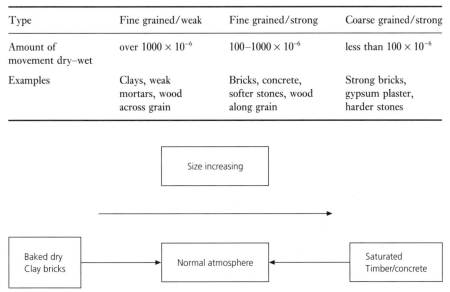

Figure 2.16 Size changes following production of clay bricks, timber and concrete

frame, there would be a tendency for the frame to compress the panel, leading to the possibility of a buckling failure. Very often, the first drying or wetting movement of newly produced materials is partly irreversible and so must be considered separately from other movements. The first drying contraction of concrete, for example, is larger than subsequent movements and is given the name *shrinkage* while subsequent movements tend to be described as *moisture movements*.

There are several possible problems resulting from moisture movement:

1 Relative movement between components affects their fit. Timber is perhaps one of the chief examples of this, swelling of flooring, doors and other close fitting components often leading to severe impairment of their function.
2 During the process of adjustment to a changing environment, there will often be differences in moisture levels between inner and outer layers of the material. This may lead to damage within the material. Ideally, all environmental changes, especially in new buildings, should be controlled carefully, especially where large timber sections are present.
3 Where movements are restrained, for example, by non-moisture susceptible materials, or the ground, stress and subsequent failure may occur. One of the main reasons for provision of joints in buildings is in order to accommodate environmental changes of this sort.
4 Special care is needed in the use of weak, fine grained materials, for example, foundations in clay are seriously affected by moisture content changes. Deep foundations overcome this problem because weather linked moisture changes decrease greatly with depth.

POROSITY/PERMEABILITY

Porosity can be defined as

$$\frac{\text{volume of pores in a given sample}}{\text{bulk volume of sample}} \times 100$$

or

$$\frac{\text{bulk volume of sample} - \text{solid volume of sample}}{\text{bulk volume of sample}} \times 100$$

Bulk density (D_B) and solid density (D_S) can be related as follows:

Bulk density $D_B = \dfrac{M}{V_B}$ M = mass

$\qquad\qquad\qquad\qquad\qquad V_B$ = bulk volume

$\qquad\qquad\qquad\qquad\qquad V_S$ = solid volume

Solid density $D_S = \dfrac{M}{V_S}$

Porosity $P = \dfrac{[V_B - V_S] \times 100}{V_B} = \left[1 - \dfrac{V_S}{V_B}\right] \times 100$

Hence,

$$\frac{V_S}{V_B} = 1 - \frac{P}{100}$$

Substituting:

$$D_B = \frac{M}{V_B} = \frac{D_S V_S}{V_B} = D_S\left[1 - \frac{P}{100}\right]$$

Pores may take a number of forms: they may be cracks, irregular voids as in granular materials, or spherical voids formed by gas generation while the material is in a plastic state.

A more significant term than porosity in the context of environmental behaviour is *permeability* which relates to the passage of gases or liquids through porous materials. Permeability results when voids are interconnected and its magnitude depends both on the physical state (liquid or gas) of the fluid concerned and on its molecular properties. Large molecules, for example, have more limited access to smaller voids.

The permeability of porous materials to liquids may be defined as:

$$\text{permeability} = \frac{\text{volume of liquid absorbed/unit area/second}}{\text{hydraulic gradient in material}}$$

The term *hydraulic gradient* means the rate of reduction of pressure with depth in the material; it is this pressure gradient that drives the fluid in. If pressure is measured as a head (m) of water, hydraulic gradient has no units and units of permeability are $m^3/m^2s = m/s$.

The permeabilities of porous materials have some bearing on durability, though other factors are involved. For example, granite which is highly durable has a water permeability of 1.6×10^{-6} m/s, whereas that of cement paste of water/cement ratio 0.5 is much lower – about 0.002×10^{-6} m/s. Nevertheless, increases in permeability for materials such as concrete do give an indication of likely damage by frost or chemicals in solution.

Permeability to gases may also be important, for example, in the design of vapour barriers, the permeability of *breathing paints* to water vapour, and the degree of impermeability of anti-carbonation coatings for concrete to carbon dioxide. It is more common in such cases for the flow rate to be measured in mass terms but otherwise the definition of permeability is similar.

Rearranging the permeability equation and using mass instead of volume:

Permeability × pressure gradient = mass of vapour per unit area per second

or

$$\text{Permeability} \times \frac{\text{pressure}}{\text{thickness}} = \text{mass of vapour per unit area per second}$$

The term

$$\frac{\text{thickness}}{\text{permeability}}$$

is known as vapour resistance – it applies to films of defined thickness. Hence,

$$\text{mass of vapour per unit area per second} = \frac{\text{pressure}}{\text{vapour resistance}}$$

Typical values of resistance to water vapour are given in Table 2.5. BS 2972 (*Methods of Test for Inorganic Thermal Insulation Materials*) indicates that water vapour resistance should not be less than 15 GNs/kg for vapour barriers; the figures below should be compared with this value. Alternatively, vapour resistances may be expressed in terms of their equivalents in metres of air. One metre of air

Table 2.5   Typical values of resistance to water vapour

	GNs/kg
Microporous paints	1–4
Gloss paints	5–8
0.15 mm polythene film	350
Aluminium foil	4000

has a vapour resistance of approximately 6 GNs/kg, hence a vapour barrier would be equivalent to 2.5 metres of air.

As a further alternative, a standard vapour pressure may be applied across films and their permeability to water vapour, or other gas, may be measured in g/m^2 per 24 hours. This is often used for the comparative assessment of surface coatings.

FROST DAMAGE

It is not always clearly understood that temperatures below 0 °C are of no consequence as regards materials unless they contain water, hence, a prerequisite of frost damage is that there should be significant water levels in the material concerned. Materials must therefore not only be porous, but these pores must be linked to produce permeability, allowing water access. It also follows that high moisture levels caused by high porosity/permeability and severe exposure increase the risk of damage when the temperature of the material falls below 0 °C. There are, nevertheless, major variations in the behaviour of porous, permeable materials which require some investigation. Several theories of the mechanism of frost have been proposed, of which one of the more attractive ones centres upon the effect of capillary type forces in pores.

It is found that water in porous materials is prevented from freezing at 0 °C by surface forces which prevent the alignment of molecules necessary for crystallisation (freezing). The effect is more powerful in finer pores, and water in these is found to remain in the liquid state well below 0 °C. A further effect occurs in materials which contain a variety of pore sizes (as do most materials); Figure 2.17. On cooling below 0 °C, water in larger voids freezes first because surface forces are relatively small in these. As cooling continues, water in progressively finer pores becomes

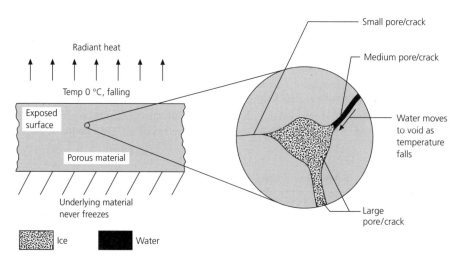

Figure 2.17   Representation of pores/cracks in a porous material subjected to freezing while saturated

Figure 2.18   Ice extrusions at a crack in a saturated brick parapet wall

unstable, possibly due to vapour pressure effects, and tends to diffuse to larger ice filled pores where it adds to the volume of ice already present. This 'feeding' process occurs continuously as the temperature falls, causing build up of ice in larger pores and water removal (desiccation) of smaller pores. (Very small pores are considered to be harmless since water in them is so tightly bound that it never moves or freezes even at very low temperatures.) It will be evident in this situation that expansion of water on freezing is not necessary to cause damage; indeed, frost damage has been found in materials saturated with liquids which do not expand on freezing. However, there is little doubt that the 9 per cent expansion of water adds to the damage produced by the above mechanism.

Some pores in many materials take the form of cracks so that build-up of ice in these cracks places severe stress on these already weakened areas. Figure 2.18 shows ice extrusions occurring at a crack in saturated brickwork on freezing. Water is fed to these positions from beneath, causing the ice columns to be pushed outwards from the surface. Note also:

1 Frost susceptible materials expand greatly as damage increases; indeed, *frost heave* is one way of monitoring the effect. Conversely, the volume of some frost resistance materials has been observed to reduce on freezing, presumably because water in smaller pores is removed in the process, causing shrinkage provided there is room in larger pores to accommodate the ice formed.
2 Damage is cumulative; each frost cycle will open cracks further or multiply cracks as ice crystals form in them.

3 It is the actual freezing process that causes damage. Steady low temperatures do not add to the problem. Hence, freezing and thawing are much more harmful than the steady low temperature encountered in more severe climates. A further related point is that rain followed by freezing is also much more serious than precipitation in the form of snow as occurs in colder climates – snow has no wetting properties. The frost resistance of materials is, therefore, measured by freeze/thaw testing, usually on a cycle of about 1 day.

4 Damage is initially a surface phenomenon since underlying layers are usually warmer. However, lower temperatures will increase the depth of the damaged region.

5 Frost damage is minimised by keeping saturation to a minimum. Hence, surfaces which shed water are far less susceptible to damage than those which permit water to accumulate and low permeability grades of materials, particularly concretes, are much less prone to frost attack.

6 Materials worst affected contain a mixture of coarse and fairly fine pores. In coarse pored materials, the feeding mechanism is absent and durability is usually good, for example, engineering bricks, some stock bricks and many sandstones and some autoclaved building blocks. Examples of finer pored materials which are not frost resistant are some common bricks and some limestones.

Experience shows that the construction period is a high risk period from the point of view of frost attack, since materials are often in a damp state and buildings do not have the benefit of a roof. Particular care is essential at this stage.

Assessing frost resistance of porous materials

It is of extreme importance that the frost resistance of any new product be accurately estimated before being incorporated into buildings, since replacement of frost damaged material can be prohibitively expensive once the construction is complete. There are several approaches to the question of prediction of performance in frost, for example, exposure tests, accelerated testing and indirect tests, such as pore size measurements.

Exposure testing   Small panels, piers or other realistic arrays of the material in question are exposed to weather in such a way as to match the maximum anticipated exposure in the proposed building. An earlier edition of BS 3921 (1974) for clay bricks indicates that such exposure should be for at least three years, the chief disadvantage of this type of test, since the producer cannot confidently guarantee performance of a new material for at least this time.

Accelerated tests   As an example of such tests, the British Ceramic Research Association has developed a test for clay bricks, built into panels, 100 cycles of freezing and thawing in a saturated state being applied. Bricks which are undamaged at the end of this test certainly merit the description 'frost resistant'. There are problems, however, in relating results of such tests to performance in normal climatic conditions, especially if a product is slightly affected by this very severe exposure test. Some additional assessment may be needed in relation to the actual exposure anticipated in such cases.

Analysis of pore size distribution   It has been explained that pore characteristics play a decisive role in determining frost resistance and, therefore, it might be expected that performance could be specified in this way. There are several ways in which an estimate of pore dimensions can be produced, of which perhaps the most widely used are gas adsorption and scanning electron microscopy. The latter has aroused particular interest in recent years; clean fracture surfaces are rendered electrically conductive by coating with a very thin layer of gold or other conductive material and then scanned in a vacuum by a focused electron beam rather like the lines of a television screen. *Secondary emission* at each point on the surface of the material modulates an electron beam and builds up a picture of the surface by simultaneous scanning of a monitor screen. Examples of scanning electron micrographs of a fracture surface in a Fletton brick (lightly vitrified) and an Otterham stock brick (well vitrified) are shown in Figure 2.19. The latter corresponds to a small number of coarse pores, though the total porosity of these two products is similar. The Otterham brick is, as would be expected, much more frost resistant than the Fletton. The problem with specifying frost performance in this way lies in the difficulty of quantifying the information given in Figure 2.19. There are also considerable variations in the appearance of fracture surfaces within and between samples so that some rapid, automated system applied to large fracture areas would be needed to obtain an index for a particular product.

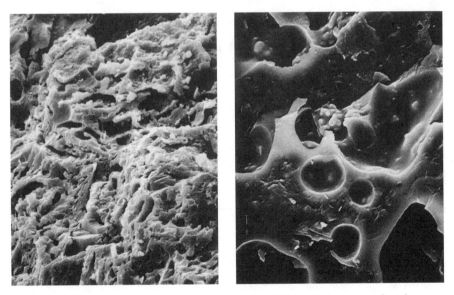

Figure 2.19   Scanning electron microscope photographs of tensile fracture surfaces in (a) Fletton brick and (b) Otterham stock brick

Saturation coefficient   This describes the extent to which pores are filled by immersion in water. The saturation coefficient is equal to:

$$\frac{\text{volume of pores filled in saturation soaking test}}{\text{total pore volume}} \times 100$$

Values may vary between 0.4 and 1.0, the latter indicating that all pores are filled on saturation. It may be argued that materials with a saturation coefficient of 0.90 or more would be at risk in frost because there is insufficient room for the 10 per cent expansion of water. From the reasoning given above, this argument clearly has little validity and damage may occur in materials having much lower saturation coefficients. The term is, nevertheless, still used as one index of durability for some more porous sedimentary rocks.

Percentage failure rates for porous products

A particular problem of masonry units is that small failure rates can result in unacceptable appearance, even though structural and weather proofing functions are unimpaired, for example, a one per cent failure rate of bricks in brickwork in a private dwelling is likely to affect the resale value of the property (Figure 2.20). Although legal liability of the producer may be limited to replacement of defective items, the question of damaged confidence in a product must also be addressed and,

Figure 2.20   Very small failure rates in bricks can lead to an unacceptable appearance. In this case the bricks most seriously affected had similar colouring. Perhaps they were subjected to a lower firing temperature

in consequence, producers are often cautious about implementing new test methods or recommending newer products for more exposed situations until an adequate track record has been established.

CRYSTALLISATION/EFFLORESCENCE

These effects are based on the same principles as frost except that the solidification process results from salts dissolved in the water and, therefore, relates to drying patterns rather than temperature change. Drying results from external humidity changes and, therefore, always commences at surfaces, dissolved salts being left behind as pure water evaporates. In many cases, a surface film of salt is left and this is described as efflorescence. Efflorescence is common on exposed or windward walls of buildings, becoming visible at times of drying – usually being worst in spring. It is unsightly, but not usually harmful. Provided salts are not replaced, for example, by atmospheric pollutants or contact with the ground, they will eventually disperse. When evaporation takes place from beneath the surface, damage similar to frost damage may occur, in the form of spalling. This is much more serious and is one of the common forms of stone deterioration. In this case, part of the problem involves chemical action between atmospheric pollutants and the stone. Sands for mortar often contain salts and form a common source of contamination. Where salts accumulate by uneven drying, they can be very difficult to remove and may perpetuate dampness due to their hygroscopic (water retentive) properties. Clearly it makes sense to keep all building materials as clean as possible.

ACID ATTACK

The effects of acids in various aspects of materials performance are considered so important that a brief resumé of terminology and properties is given here.

There are now several definitions of acids, some including references to protons or 'acceptance of lone electron pairs', though for present purposes, it will be sufficient to describe acids as substances containing hydrogen ions (H^+). Because the hydrogen atom itself has only one proton and one electron, the hydrogen ion is simply one proton. This has very high chemical reactivity on account of its extremely small mass. Most acids occur in water (aqueous solution) and the reactivity of the H^+ ion causes it to form the hydroxonium ion when water is present:

$$H^+ + H_2O \rightarrow H_3O^+$$

Nevertheless, for chemical purposes, acids are represented by H^+. Acids react with:

- metals
- bases
- basic compounds

Their action on bases is now further explained.

Bases (including alkalis) contain OH^- ions which are neutralised by acids. For example,

$$HCl + NaOH \rightarrow NaCl + H_2O$$

Most acid–base reactions take place in water which itself has acid/basic properties, since water is always partly ionised:

$$H_2O \leftrightarrow H^+ + OH^-$$

The equilibrium in this reaction is characterised by an important constant, the equilibrium constant for water dissociation K_w:

$$K_w = \frac{[H^+][OH^-]}{[H_2O]}$$

The square brackets refer to the ionic/molecular concentration measured in moles per litre. (Note one mole of a substance is an amount equal to its molecular weight in grams. Hence water H_2O has molecular weight $2 + 16 = 18$. So 1 mole of water $= 18$ g.)

One mole is a convenient quantity because it always contains the same number of molecules: 6×10^{23}, Avagadro's number.

In pure water, it is found that $[H^+] = [OH^-] = 10^{-7}$ moles/litre. This indicates that the vast majority of molecules are undissociated. Hence, the number of moles in 1 litre (which contains 1000 g) is:

$$[H_2O] = \frac{1000(g)}{18(g)}$$

$$= 55.6 \text{ moles}$$

Notice how much larger this is than $[H^+]$ or $[OH^-]$. In fact, $[H_2O]$ can be taken as constant even in acidic solutions so that, since K_w is constant:

$[H^+] \times [OH^-]$ is constant and equal to $10^{-7} \times 10^{-7} = 10^{-14}$ moles/litre

Ion concentration in acids may be measured by the pH value:

$$pH = \log \left(\frac{1}{[H^+]} \right)$$

In the case of pure water, $pH = \log 1/10^{-7} = \log 10^7 = 7$.

In an acid $[H^+]$ may be as high as 1 mole per litre, hence

$$pH = \log 1/1 = 0$$

Note that since

$$[H^+][OH^-] = 10^{-14}$$

then, in this case, $p[OH] = 14$

In bases (including alkalis) $[OH^-]$ may similarly be high, for example, 1 mole/litre. Hence,

$$[H^+] = \frac{10^{-14}}{[OH^-]} = 10^{-14}$$

and the pH is 14. Theoretically, (but not commonly) $[H^+]$ can be over 1 mole/litre, in which case the pH is negative. For the same reason, in concentrated bases, the pH may be over 14. In strong acids such as hydrochloric acid, nitric acid or sulphuric acid, ionisation is almost complete, hence the acid is *fully active* and will react quickly with any alkali present. Weak acids are partly ionised but will ionise further if more water is provided through dilution or during neutralisation.

The reactive power of an acid or base (that is, their ability to neutralise each other) may be measured by their *normality* or their *molarity*. A 1N (normal) solution of acid contains 1 mole of H^+ ions per litre and would neutralise an equal volume of 1N alkali. The following would also neutralise each other:

25 ml 0.1N acid and 250 ml 0.01N base
10 ml 5N acid and 50 ml 1N base, etc.

Normalities are widely used in industrial chemistry. A 1M acid (molar solution) contains 1 mole of the acid per litre of water though the amount of hydrogen produced depends on the acid type, for example:

$$H_2SO_4 \leftrightarrow 2H^+ + SO_4^{2-}$$

$$HCl \leftrightarrow H^+ + Cl^-$$

It follows that 1 mole of sulphuric acid releases twice as many hydrogen ions as 1 mole of hydrochloric acid. Hence, in the neutralisation of sulphuric acid and sodium hydroxide:

$$H_2SO_4 + 2NaOH \leftrightarrow Na_2SO_4 + 2H_2O$$

25 ml of 1M sulphuric acid would neutralise 50 ml 1M sodium hydroxide. When dealing with molarities, the chemical make-up of the acids and bases reacting must, therefore, be known.

It must also be added that the reactivity of an acid and base towards each other depends on the interaction of their total chemical characters and not just on their H^+ and OH^- components. For example, sulphuric acid, H_2SO_4, and nitric acid, HNO_3, are described as *oxidising acids* and are more aggressive as a result. Hence, they will attack a wider range of materials. Strong acids, which are fully ionised, tend to produce salts which are still acidic in character, the same applying to bases. The action of acids and bases is seen to be complex, and specific problems may require specialist treatment. A summary is given here of the main characteristics of acids in relation to building materials:

1 They produce H^+ ions which tend to neutralise bases or basic compounds and react with metals.
2 The H^+ concentration is measured by pH values which range from 7 for pure water to 0 for a strong acid.
3 The reactive power of acids with bases is measured by molarity or normality.

4 Strong acids are fully ionised and exhibit acid action towards a wider range of materials than weak acids.

5 Some acids have additional oxidising or other chemical effects.

The rate of attack in practice depends on:

1 The concentration, pH and rate of supply of the acid. Small quantities of a weak acid may be quickly neutralised.

2 The physical form of the material being attacked. Acids will penetrate permeable materials, operating below as well as at the surface.

3 The temperature. The ion concentration of solutions increases greatly with temperature, accelerating attack.

Further examples of acid action will be given under specific materials headings.

2.3   THERMAL PROPERTIES

An understanding of thermal properties of materials will be helpful when examining the thermal response and insulation requirements of buildings together with their performance in fire. A summary of important properties of some common building materials in order of decreasing density is given in Table 2.6.

SPECIFIC HEAT CAPACITY

Specific heat capacity (C_p) is the number of joules required to raise the temperature of 1 kg of the material by 1 °C. The units J/kg °C arise directly from the definition. It is noticeable in Table 2.6 that values for common building materials generally decrease as density rises (since there are fewer molecules in a given mass of higher

Table 2.6 Properties of common building materials which affect fire performance, given in order of decreasing density

Material	Density	Specific heat	Thermal conductivity	Thermal diffusivity mm^2/s	Thermal inertia $W^2 s/m^4 °C^2$	Coefficient of thermal expansion
	kg/m^3	J/kg °C	W/m °C	$\times 10^{-6}$	$\times 10^6$	$/°C \times 10^{-6}$
Copper	8900	390	300	86	1040	17
Steel	7800	480	84	22	310	11
Aluminium	2700	880	200	37	480	24
Dense concrete	2400	840	1.4	0.69	2.8	11
Brickwork	1700	800	0.9	0.66	1.2	8
UPVC	1400	1300	0.3	0.16	0.55	70
Lightweight plaster	600	1000	0.16	0.27	0.096	5
Wood across grain	500	1200	0.14	0.23	0.084	3
Expanded polystyrene	25	1400	0.033	0.94	0.001	70

density materials). In view of the large variations in density of many building materials, it is misleading to take C_p values on their own in assessing capacity to absorb heat.

THERMAL CAPACITY

This refers to components of buildings and is defined as the number of joules required to raise the temperature of the component by 1 °C.

Hence,

thermal capacity = specific heat × mass

Components of a certain size, having high density will, therefore, have a higher thermal capacity than those of lower density since their mass will be higher, although as indicated above specific heat capacities of denser materials are somewhat lower. Traditional buildings, employing bricks and dense concrete, will have high thermal capacity and, therefore, respond slowly to change, whether heating or cooling. They have a tendency to stay cool through hot summer days and to lose heat slowly on cold nights.

THERMAL CONDUCTIVITY

This represents the ability of a material to transmit heat by conduction. It is defined by the equation:

$$Q = \frac{kA(T_1 - T_2)}{d}$$

where k is the thermal conductivity, Q is the heat in joules/s passing through a slab of area A and thickness d with steady surface temperatures T_1 and T_2 in the steady state. The units are W/m°C and typical values are given in Table 2.6. There is evidently an enormous range of values in common building materials, values generally decreasing with density. In view of their much superior insulation properties (which increase with $1/k$), it is not surprising that there is a trend towards low density materials in building.

THERMAL DIFFUSIVITY

This is defined as

$$\frac{k}{\rho C_p}$$

where k is the thermal conductivity, C_p the specific heat capacity and ρ the density of the material. Since C_p does not vary greatly in non-metallic materials and k is approximately proportional to density, values do not vary greatly for this group.

Metals have very high k values, giving them high diffusivity. For most non-metals, diffusivities are in the range $0.2 - 1.0 \times 10^{-6}$ mm^2/s, while steel has a value of 22×10^{-6} mm^2/s (Table 2.6). Diffusivity describes the ability of the material to transmit heat in a fire, rather than become hot. In practice, the physical form must also be considered. For example, steel columns and beams, having high surface areas and limited thickness, will heat up rapidly rather than diffuse heat.

THERMAL INERTIA

This is defined as $k\rho C_p$. If a surface material with high thermal inertia is exposed to heat the rate of temperature rise of that surface is reduced, k representing its ability to conduct the heat away while C_p determines its rate of response to the applied heat. Materials such as brick and dense concrete are excellent in this respect, while insulating materials such as expanded polystyrene are very poor because they have low k and ρ values. Materials such as plasterboard and wood are intermediate.

THERMAL MOVEMENT

This may be defined as proportionate length change per °C temperature change. Hence, units are simply /°C though the submultiple 10^{-6} (microstrain) is convenient when expressing values. Table 2.6 shows typical values for common building materials. There is evidently a great range of values, though as a rough approximation, stronger materials in each of the metallic and organic groups have lower thermal movement. If movement is restrained, stresses are raised (as in shrinkage), the stress being proportional to the coefficient of expansion and the E value for the material. High movement materials generally have lower E values so thermal stresses do not necessarily rise in them, though joints should nevertheless be provided to avoid problems such as abrasion or buckling. Further problems arising from thermal movement are differential stresses where materials contain temperature gradients resulting from severe temperature changes as in fire. Materials with low thermal coefficients are generally more durable in fire.

2.4   PROBLEMS DUE TO MOVEMENT IN MATERIALS

Problems related to movements in materials have increased dramatically in the last century. This may be related to marked changes in the types of materials in common use and the types of building they are employed in. These changes include:

- Use of higher movement materials such as polymeric materials (plastics) and aluminium.
- Use of thinner lighter sections which respond more quickly to environmental changes.
- Use of larger buildings which undergo larger relative movements, for example at joints of low absorption cladding materials which shed, rather than absorb, water.

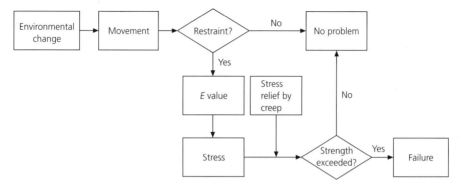

Figure 2.21 Schematic representation of processes leading to movement failure

The result of this is that there have been numerous failures due to movement. Occasionally structural failures may occur though they are relatively uncommon since most of the scenarios identified above relate to lighter materials with a non-structural function which are subject to more severe environmental changes. The most common mode of failure is usually water admission since it leads to rapid deterioration of materials, but other forms include:

- Unsightly cracking in finishes or gaps at junctions.
- Air penetration producing a draughty enclosure.
- Visual impairment due to distortion or loss of alignment.

Many of these have been in high profile buildings – indeed these might be considered to be the ones most at risk of such problems since in attempting to produce different and unique structures, the way they will behave may be very difficult to predict.

It is important when attempting to design for movement, that the stages in developing stress and failure are understood. These may be summarised as in Figure 2.21.

Step 1 Cause of movement

Material movements may originate from many sources and some of these such as settlement or subsidence are more specialist subjects which form studies in their own right. This study will be concerned with movements originating from environmental changes of temperature and humidity.

Thermal movement values were given in the previous section and outline information for moisture movements are given in Table 2.7 (for more detailed information see the respective sections).

The magnitude of such movements can usually be predicted with some confidence though detailed consequences may be more difficult to estimate. Nevertheless if anticipated movements can be predicted to within an order of magnitude, this may be sufficient for appropriate action to be taken to avoid

Table 2.7　Summary of moisture movements to be found in common construction materials

Clay brick expansion on exposure	500×10^{-6}
Clay brickwork expansion	300×10^{-6}
Concrete shrinkage on drying	200×10^{-6}
Mortar shrinkage on drying	3000×10^{-6}
Wood along grain on drying	1000×10^{-6}
Wood across grain on drying	$50\,000 \times 10^{-6}$

problems. Hence the figures given in examples below can be taken as no more than a guide but should be adequate for design purposes.

Step 2 Environmental change – critical scenario

The critical scenario must be identified. Cooling or drying produce contraction which tends to cause tensile stress. This may be important when low tensile strengths such as brickwork or concrete are concerned. Heating or wetting cause expansion and would be most likely to cause problems in slender structures which may then be at risk of buckling. In some materials such as timber, heating may cause drying so that the effects are to some extent compensating. As regards temperature a notional change must be identified and this will depend on the effective temperature at which a product was installed (assuming that creep has not altered the situation). The extreme temperature chosen will depend upon the material exposed to the change, its physical form and any effects of insulation which might prevent the dissipation of heat. With massive materials such as concrete, temperature changes under given conditions will be much smaller than those in, say, a well insulated lightweight cladding unit, especially if directly exposed to heat of cold. As regards moisture it will normally be sufficient to take movements between the wet and dry state (though initial movements for manufactured materials such as clay brick or concrete will be much bigger than subsequent movements).

Step 3 Strain calculation

The strain to be used for the determination of possible stress problems must be obtained by considering the combined effects of input movements. It will normally be measured in microstrain; see Section 2.2.

Step 4 Restraint situation

If materials are free to move, then stresses will not arise as result of movement. There must either be some external restraint, such as friction with the ground or connection to a component which has different movement characteristics, or there may be problems due to temperature or moisture content differentials within the material. The restraint may be partial – for example, friction between a structure and the ground it rests on – or absolute, where no relative movement is possible.

Ground restraint can be estimated from coefficients of friction. If in doubt it is best to assume that restraint is absolute.

Step 5 Stress calculation

Stress is calculated from the equation

$$\text{stress} = E \times \text{strain}$$

The moduli of elasticity of most materials in tension and compression are similar. Note that high stiffness materials generate stress more easily than those of low stiffness. Elastomers, the lowest stiffness materials used in building, are normally employed specifically because they are slow to raise stress when subject to movement.

Step 6 Stress relief

Some materials such as timber and thermoplastics exhibit substantial creep which can relieve stresses caused by movement. Though this avoids failure, creep may cause other problems. Visible distortion may arouse concern and neighbouring materials may be affected.

Microcracking, such as occurs in softer materials such as lime mortars, will relieve stress due to movement, though again this may result in other problems such as visible distortion.

Step 7 Failure due to stress

Two modes of failure were identified above: tensile failure, usually associated with reductions of temperature or drying of brittle materials; or buckling failure, usually associated with increases of temperature or wetting. Where calculated stresses are in excess of 50 per cent of notional short term stress levels, failure may in any case occur over a period of time due to the added effect of fatigue. Buckling failure is much more difficult to predict – in slender components this may occur at stresses as low as 10 per cent of failure stresses in test samples of cubic shape. In some structures such as masonry walls, repeated temperature or wetting and drying cycles may produce a cumulative effect. For example tension stresses lead to cracking and dirt or debris fills these cracks. During expansion cycles the filled cracks are subject to compression so that the masonry slowly 'grows'.

ILLUSTRATIONS OF COMMON MOVEMENT PROBLEMS

It will be appreciated that actual situations can be very complex, buildings appearing to behave satisfactorily in one situation while similar buildings in other situations run into problems. Many older buildings seem to survive without joints while newer ones exhibit cracking. The following are designed to illustrate the general principles.

Brickwork movement

This can be a most complex topic. As above, initial movements of around 300×10^{-6} can be expected – expansion for clay brickwork and contraction for other types. On the basis of a modulus of elasticity of, say, 20 kN/mm^2 this would give rise to a stress of $300 \times 10^{-6} \times 20\,000 = 6$ N/mm^2 if completely restrained. In the case of the clay brickwork this would be a compression stress which should not cause too many problems, though there would be a risk of buckling in restrained panels (e.g. concrete framed building). In the other types of brickwork, shrinkage is more likely and such a tensile stress would exceed strength in tension and cause cracking. There is little doubt that cracking in brickwork is commonplace. The important requirement is to avoid cracking which is structurally or visually unacceptable. Some of the factors involved in determining the severity of cracking are listed below.

Restraint If there were no restraint, brickwork would be free to move and cracking would be unlikely. However this is rare. Free standing walls are restrained at their base by friction. The magnitude of this depends on the details. Any wall on a dampproof course (DPC) has some freedom to slide, leading to low friction and lower stresses. Where there is no DPC or there is high bond, material friction will be much greater. Stresses will be largest at the centre of each panel, the greatest stress being at the foot of the wall. Rigid bonding to other materials – for example of bricks of a different type – should be avoided as it leads to very high levels of restraint, e.g. clay brick/sand-lime brick/concrete/stone.

Materials properties Stronger mortars will transmit stress rather than absorb movement by microcracking. This puts stress more onto the bricks themselves. The biggest risk to bricks themselves cracking is low strength bricks bedded in a strong mortar which bonds well to the bricks (typical stock bricks). Such structures need adequate joints (say every 6 m rule of thumb). The most critical situations are those producing tensile stress, so non-clay bricks are most at risk. Failure of sand-lime bricks in large lengths without joints is quite common for that reason. Conversely, clay bricks in a soft mortar have been laid to large (effectively infinite) lengths without visible cracking.

Pointing This can cause problems especially if a strong *pointing* mortar is used in conjunction with a weak lime *bedding* mortar. The stresses generated are proportional to the E value, hence the pointing might pick up a stress five times or more the average value. Since this occurs at the edge of the brick it can shear off the brick face at the joint. This also happens when pointing is carried over a soft DPC material such as bitumen. The surface picks up high stress and may spall the edges of the bricks at DPC level.

Exposure It will be clear that high exposure will increase movement. The biggest effect is probably rain exposure. Brickwork swells when wet and some of this movement may be irreversible. Then, on drying, tensile stresses are set up, especially if drying coincides with cold weather.

Returns   Large straight panels of brickwork are most prone to problems. Small returns greatly reduce movement problems since they allow slight rotation when movements occur.

Concrete slabs

These obey similar rules to those above. However they may need to transmit loads so that dowel bars will then be needed at joints; especially ground floor/industrial slabs. The dowel bars must be unbonded on one side and, if exposed to weather, joints must be designed to prevent water/dirt admission. Joint spacing must be calculated using the rules given above. Reinforcing steel may permit cost reductions by reducing the number of joints. Where large movement is anticipated, occasional expansion joints may be recommended, especially if the work is done in cold weather. Workmanship must be good. For example if dowel bars are badly aligned, they act as reinforcement and may prevent movement, leading to failure in adjoining concrete, especially if it is un-reinforced.

General claddings

These can involve complex combinations of many materials including brick, glass, concrete, aluminium, steel, etc. The worst potential scenario must be identified and appropriate movement allowance made. This is likely to be either very low or very high temperature according to situation. Note that surfaces subject to the sun's radiation can reach temperatures of 70 °C or more especially if of low thickness or density. Full movement must be allowed for and note that jointing materials may have to be five times as thick as the actual movement anticipated. The critical situation may occur where differentials exist, for example where dark colour glazing bars absorb heat, therefore heating the perimeter of glazing. See also Chapter 9 on sealants.

2.5   FIRE

While significant improvements are constantly being made in fire safety, reports of fatalities are still commonplace and major fire disasters are by no means unknown. Many fires are associated with the use of hazardous materials and, though ongoing legislation should result in continuing improvements, it often takes several years before older, high risk materials are replaced. Statistics show that fires often occur in buildings where financial constraints have hindered incorporation of the safest techniques or materials. In deep plan or tall buildings, the much higher potential risk resulting from reduced access of emergency services has long been recognised though, again, fires still occur, sometimes due to failure of one or two parts of the structure to prevent fire spread. Increasing attention is now being given, for example, to fire stops in cavities, ducts and roof spaces. Current legalisation is still based largely upon the observed behaviour of materials and components in simplified standard tests (which are much easier to replicate), though increasing information from larger scale fire tests is now becoming available.

In designing buildings to resist fire, the need to reduce the risk of combustion must first be addressed and then, should fire take hold, the need to restrict its spread and permit the occupants to escape safely must be considered. The topic is dealt with here in outline only, in order to establish criteria upon which the fire performance of building materials can be judged.

COMBUSTION

There are three prerequisites for a fire – fuel, oxygen and heat.

Fuel

Almost all organic (carbon) based materials behave as fuels while few other material types burn. Carbon and hydrogen are the main active constituents so that materials rich in these will be a greater hazard and especially those rich in hydrogen, such as oil products and gas, since weight for weight, hydrogen generates more heat than carbon.

Oxygen

Oxygen is not a fuel, but is the essential means by which fuels burn. It is present in the form of air, diluted with nitrogen which is inert. Pure oxygen, sometimes stored in cylinders, is highly dangerous on account of fuels' affinity for it. In some materials such as plastic foams, oxygen may be incorporated within the material, assisting combustion.

Heat

Heat has two very important influences on the combustion process:

- It causes chemical decomposition of most organic materials which then release volatile vapours. The effect is known as *pyrolysis*.
- It assists the reaction between both the solid and vapour fractions and oxygen, for example:

$$C + O_2 \rightarrow CO_2 + heat \qquad \text{solid fuel}$$
$$CH_4 + 2O_2 \rightarrow CO_2 + 2H_2O + heat \quad \text{hydrocarbon fuel}$$

These are both combustion processes though it is the reaction of vapours with oxygen together with accompanying light emission that is described as a 'flame'. Flames are not necessary for fire but their presence usually increases the severity of a fire because:

- Gases have much greater mobility than solids so that flames help spread the fire.
- The temperature in a flame is very high – usually around 1200 °C.

The process of initiation of fire, particularly flaming combustion, is described as ignition. The application of sufficient heat will initiate the combustion process which then generates more heat and ultimately, when the temperature is high enough, ignition or flaming, will occur. In many fires, the application of a spark or pilot flame, which is effectively a small zone of high temperature, will itself produce combustion without extra applied heat. This can happen only when the fuel exists in a very low thermal capacity form, as in fibres, thin sheets or foams. The ease and speed with which combustion occurs in such materials places them in a very high risk category. Combustion can be categorised into at least five types, as shown in Figure 2.22. Note:

- In type 2 combustion, flames are not necessary and fire can travel through partitions such as glass doors or walls in this way, provided they conduct or radiate heat sufficiently.
- Type 3, smouldering combustion, may occur in stored bulk materials, especially if dampness restricts the rate of temperature rise. It often makes final extinguishing of fires a time consuming process.
- Type 4, spontaneous combustion, is rare but has caused serious fires and indicates the need to take care when storing organic materials, particularly those prone to biological deterioration.

DEVELOPMENT OF FIRES

The damage resulting from a fire depends on the situation which gives rise to it and on the way in which it develops and spreads. In the early stages of most fires, spreading is largely the result of flaming and localised heat generation, hence, it requires materials which are not only combustible but in a flammable form. It is not possible, for example, to set fire to sizeable wood sections with a match, though paper, a related material, is highly flammable. Assuming suitable material is present, the hot gases released by combustion will increase the temperature of the enclosure, particularly if it is only partially ventilated. A most important event in any fire is *flashover*, Figure 2.23. This occurs when the air temperature in the enclosure reaches about 600 °C. At that point, pyrolysis (vaporisation) of all combustible materials takes place so that they all become involved in the fire. Flaming reaches dramatic proportions, limited only by the total fuel available and/or by the supply of air. In many such fires, flaming will occur mainly outside open or broken windows owing to lack of air internally and this helps spread the fire upwards to adjacent floors of the building or to adjacent buildings. A priority in design for fire resistance is to prevent or delay flashover since there is no chance of survival in an enclosure once this has occurred. The thermal inertia of the surfaces of an enclosure is an important factor. Highly conductive, heat absorbent materials such as brick, help delay the temperature rise as well as being non-combustible.

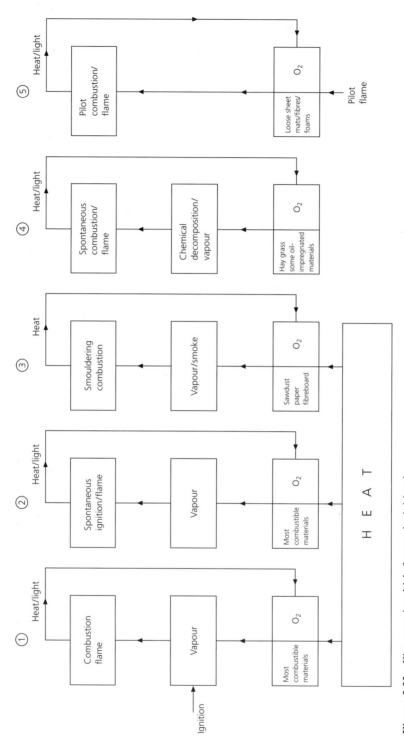

Figure 2.22 Five ways in which fire can be initiated

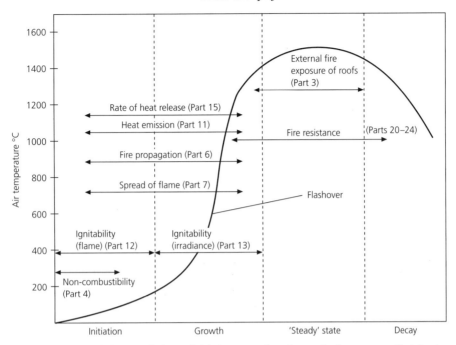

Figure 2.23 BS 476 tests relating to initiation, growth and spread of an uncontrolled fire in the compartment of origin

Flashover may be prevented in poorly ventilated enclosures due to lack of air for combustion, or delayed in very large ones where there are large volumes of air in relation to available fuel – they have a cooling effect. The fire then develops as if it were unenclosed. It is estimated at present that the fire services arrive before flashover in about 90 per cent of fires.

TOXICITY

An important aspect of safety is release of smoke and toxic gases by materials, since most fatalities result from asphyxiation rather than burning. There are complex issues, smoke production depending, for example, on oxygen availability as well as fuel type, being most severe when oxygen levels are inadequate for total combustion. Some materials release highly toxic gases, for example, polyurethanes which release HCN gas. Nevertheless, in most fires, the main problem is carbon monoxide which is present in relatively large quantities, especially, again, when air for combustion is limited. Bonded wood products such as fibreboards and chipboards are prone to smoke production as well as the less flammable plastics such as PVC. There is, at present, little legislation on smoke production and in view of the rapidity with which smoke can be produced, emphasis is probably best placed upon early detection for which smoke detectors can be employed.

DESIGN FOR FIRE PROTECTION

The ideal for habitable buildings might be considered to be avoidance of all combustible materials but this normally would be totally impracticable because organic materials are inseparably linked to human comfort – furniture, furnishings, clothing – and to human activity – books, paper and implements. In addition, there are many combustible materials which, in the form they are used in buildings, do not consitute a fire hazard.

The enclosure itself will normally involve combustible materials, for example, wooden floors, doors, window frames and partitions. The risk represented by the combustible contents of an enclosure is defined as the *fire load*, equal to the total mass of combustible contents in the enclosure, expressed as wood equivalent per unit floor area.

Tests show that high fire loads tend to produce longer duration fires so that, if a building is to stay structurally intact throughout the fire, a higher fire resistance will be needed. Hence fire loading is one of the factors used in arriving at 'purpose groups' as defined in *Building Regulations*. It will be appreciated that this is a simplification and that, in practice, the fire severity will depend on the type and disposition of combustible material and on other factors such as ventilation characteristics. Good ventilation may reduce the duration of fires by allowing heat to escape.

FIRE TESTS AND LEGISLATION

Fire tests attempt to classify materials and components in relation to fire performance and form the basis of *Building Regulations*. They cover two chief areas:

- Development and spread of the fire. These tests include combustibility, ignitibility, fire propagation, spread of flame and heat emission of combustible materials.
- Effects of fire on the structure, adjacent structures and means of escape. The first priority in any fire is the safety of the occupants, fire fighters and people in the vicinity of the building. These tests are concerned with the structural performance of buildings, their ability to contain the fire and problems associated with smoke.

Many larger buildings are divided into *compartments*, which are designed to prevent fire spread throughout the building. Escape is secured by use of *protected shafts*. Boundaries between each of these are subject to stringent fire regulations.

Many of the aspects of performance are covered by BS 476, Table 2.8 indicating parts of the Standard applicable to initiation and spread of an uncontrolled fire. A brief resumé of the contents of parts relating to fire performance in current *Building Regulations*, 1991, is given in Table 2.8 together with classifications and typical materials' behaviour. Some specific illustrations follow.

Table 2.8 BS 476 tests relating to fire development and spread referred to in current *Building Regulations*

BS 476 Part number and title	Test procedure	Results	Examples
Part 4, 1970 (1984) Test for non-combustibility	40 × 40 × 50 mm samples heated in an electric furnace. Flaming/temp. rise observed	If no continuous flaming for 10 s and temp. rise less than 50 °C – non-combustible	Non-organic materials usually incombustible – exception, some magnesium/aluminium alloys
Part 6, 1989 Fire propagation index	228 × 228 × 50 mm samples subject to increasing heat/flame. Temperature measured at 3 min, 10 min, 20 min.	Indices dependent upon temperature reached i_1, i_2, i_3 at each time $I = i_1 + i_2 + i_3$	I and i_1 usually quoted e.g. 60/40 – insulation board (unpainted), 10/5 – plasterboard
Part 11, 1982 (1988) Heat emission	45 mm diameter 50 mm thick cylinder. Method similar to Part 4.	Temperature rise in furnace and in specimen measured; flaming; mass loss recorded.	
Part 7, 1971, 1987, 1997 Surface spread of flame	900 × 230 mm sheet not more than 50 mm thick exposed to pilot flame/radiant panel	Flame spread measured at $1\frac{1}{2}$ and 10 minutes / 10 min: Class 1 not more than 165 mm / Class 2 165–455 mm / Class 3 455–710 mm / Class 4 more than 710 mm	Class 1 asbestos cement, fibre insulation board + 3 coats emulsion / Class 3 untreated softwood
Parts 20–23 Fire resistance	Full size components – for example, beams, floors, walls subjected to standard temperature time curve	Stability value stated	FD 30S (30 minute stability)
Part 3, 1958 External fire exposure roof test	Roof represented by panel. Preliminary ignition test, then flame/radiation	1958 Letter designation / 1st letter penetration time / A – D. A – 1 hour; D – prelim test. / 2nd letter flame spread A – nil; D – sustained. / Prefix F flat; S sloping	AA – best – e.g. Slates, clay or concrete tiles on rafters. / CC – bitumen felt on boarding
Part 3, (1975) as above	Similar	Prelim test: persistent or extended flaming X; otherwise P. / Penetration time. Flame spread not indicated	P60 – best – as AA above

Combustibility

Non-combustible materials are defined as:

- Products which are classified as non-combustible under BS 476 Part 4 (1970).
- Materials which when tested to BS 476 Part 11 do not flame or cause any temperature rise on either the centre specimen or furnace thermocouples.
- Totally inorganic materials such as concrete, fired clay, metals and plaster and masonry containing not more than 1 per cent by weight or volume of organic material.
- Concrete bricks or blocks meeting BS 6073 Part 1.

These may be specified in situations of highest risk.

Building Regulations allow materials of 'limited combustibility' for certain walling purposes. The designation includes composite materials with non-combustible cores not less than 8 mm thick with combustible linings not more than 0.5 mm thick. Alternatively, materials can be so designated by their performance in BS 476 Part 11 (1982).

These may be employed in some types of compartment walls and fire protecting ceilings.

Spread of flame in surface linings

Surface finishes are subject to performance criteria in addition to the above materials criteria, in order to limit flame spread. Standard tests have been subject to continuous development and improvement and there is often overlap between them. For example, the Fire Propagation Test (BS 476 Part 6) is intended to overcome shortcomings in the Surface Spread of Flame Test (BS 476 Part 7). Each measures flame/fire spread of surfaces in the very early stages of a fire but the former test provides more detailed information on the behaviour of low flame spread materials such as treated timber and is more suited to testing thermoplastic surface linings.

Building Regulations (1991) designate a *class 0* level of performance, materials complying if the material or surface of a composite are either:

- composed of materials of limited combustibility throughout (as described above), or
- composed of class 1 material to BS 476 Part 7 (see Table 2.8) and having values of I not greater than 12 and i_1 not greater than 6 in BS 476 Part 6. Hence, class 0 materials achieve a better rating than class 1 on their own.

Thermoplastics

Thermoplastics are now commonly used in suspended ceilings, diffusers and rooflights. *Building Regulations* identify three types together with limitations of application:

- **Type TP(a) rigid** include solid rigid PVC sheet, solid polycarbonate sheet at least 3 mm thick, multi-skinned sheet having class 1 rating under BS 476 Part 7, or other sheets performing to a stated level under BS 2782, Method 508A (Rate of Burning).
- **Type TP(a) flexible** refers to flexible (fabric) products with a given level of performance under flammability requirements of BS 5867.
- **Type TP(b)** includes thinner solid or multi-skinned polycarbonate sheets or other products conforming to BS 2782 at a lower performance level.

Fire resistance

Fire resistance measures the ability of elements of a building to fulfil their function for a specific length of time in a fire. For standard purposes, a furnace producing a defined and reproducible temperature–time curve is used. The function of an element may be:

- Structural (that is, carrying loads) – referred to as load bearing capacity.
- Insulation – to stop the fire spreading by passage of heat through the element.
- Integrity – to stop passage of flames or gases.

Fire resistances are now measured in BS 476 Parts 20–23. These are designed to improve the repeatability and reproducibility of results compared with the previous BS 476 Part 8. The term 'load bearing capacity' used above replaces the former term 'stability'.

Relevant aspects of performance will need to be assessed in a fire test. The applicability of each to specific components is easy to judge, for example, *stability* applies to any load bearing component, while *insulation* and *integrity* apply to floors and walls.

The structural (load bearing) fire resistance requirements of *structural elements* of buildings depend very much upon the building purpose and height. Table 2.9 shows typical values required by current *Building Regulations*. The lowest requirements are for open car parks where no additional fire protection is necessary. The major factor in determining required fire resistance is building height since upper floors of higher buildings are inaccessible to emergency services.

Fire doors were originally subject to stability, integrity and insulation requirements under BS 476 Part 8. Terms such as 30/20 used to be used in relation to fire doors implying 30 minutes stability and 20 minutes integrity and such doors would be described as *half-hour fire check* doors. A 30/30 performance level door would be described as a *half-hour fire resistance* door. The current BS 476 Part 22 test measures integrity only with a simple performance rating, e.g. 'FD 90' being given. Where used in compartmental walls the area of fire door permitted is limited unless evidence of insulation performance levels is available. In some cases – for example, protected stairways – there may be smoke leakage requirements applied to doors also. Performance in relation to smoke is measured by BS 5588 or BS 476, Section 31.1.

Table 2.9   Summary of structural fire resistance requirements from Approved Document B. England & Wales Recommendations 1992

	Height of top storey (metres)			
	<5	<20	<30	>30
Approx. number of storeys	2	5/6	8/9	9+
Residential (non-domestic)	30	60	90	120
Offices	30	60*	90*	120 plus sprinklers (floors 90 mins)
Shops, commercial assembly	60*	60	90*	
Industrial and storage	60*	90*	120*	
Car parks – closed	30	60	90	
Car parks – open sided	15	15	15	60

* Reduced by 30 mins when sprinkled

External fire exposure

Building Regulations still refer to the 1958 edition of BS 476 Part 3 – the external fire exposure roof test. Designations are shown in Table 2.8 together with those of the 1975 edition designations for comparison.

ACTIVE FIRE SYSTEMS

The installation of sprinkler (*active*) systems can play an important role in suppressing spread of fire and smoke and may make a large contribution to preserving the structural integrity of the buildings also. For this reason fire resistance requirements of buildings are often relaxed by 30 minutes when active systems are installed (Table 2.9). Sprinkler systems make a major contribution to fire safety. In one major fire in America, decorators' linseed oil impregnated rags, piled into a heap on the 24th floor, ignited spontaneously due to oxidation of the oil (type 4 combustion in Figure 2.22). The fire could not be controlled from the ground and spread uncontrolled to the 30th floor. A sprinkler system, installed on this floor, was activated and extinguished the fire even though it was well established by this stage.

There may also be reductions in insurance premiums when such systems are used. The use of many types of passive fire protection adversely affects recycling opportunities since additional materials are required and some of these can be difficult to separate at demolition stage. Active systems represent a further advantage in this respect.

CHAPTER 3

SUSTAINABILITY ISSUES

3.1 BACKGROUND

There can be few aspects of construction which in recent years have attracted greater attention than sustainability and this trend is likely to gather momentum as general awareness of matters relating to the environment increases. The question is increasingly being asked as to whether current rates of exploitation of the planet's resources can be sustained without serious implications for the future.

Sustainability itself can be quite easily defined by reference to 'sustainable' in a dictionary. The emphasis of the word lies in the future. However the term 'sustainability' may have different connotations in different spheres and indeed much time has been spent in attempting to arrive at suitable definitions in specific contexts. The problem is that the subject can be approached from technological, social, economic or political standpoints, with different interests and agendas in each case. It may nevertheless be worthwhile to suggest a definition of the term 'sustainability' which could be considered to embrace a perspective relating to the technology of construction:

> Sustainability is the extent to which current needs can be fulfilled without compromising the needs of future generations.

There are a number of aspects of the statement which might be examined prior to more detailed consideration of the construction context.

'Needs' is a subjective term which involves judgements mainly relating to standards of living. If one considers how life styles compare now with perhaps as little as 150 years ago prior to the industrial revolution, the changes will be recognised to be enormous. At that time, even in the relatively developed countries, human needs were largely fulfilled by supply from the local environment. This was particularly the case concerning any bulk or heavy materials since mechanised transport was unknown. For example every region was served by its own brickworks or other building materials supply, home grown timber and local food and fuel supplies.

In the current developed world, every household depends upon supplies from around the globe, low energy prices permitting bulk materials/products such as heavy metals, hardwoods, bulk foodstuffs and cars to be transported cheaply over enormous distances. Estimates suggest that when normal living and energy requirements are added to those relating to supply and maintenance of a 'developed world' life style, they amount to as much as 2 kW continuously per individual throughout life, whereas corresponding consumption by individuals in the 'developing world' would be a minute fraction of this figure. Examples of items which as recently as 30 years ago would have been luxuries but are now widely considered to be necessities include:

■ Central heating for most dwellings, leading to winter comfort levels that would previously have been achieved by additional clothing.
■ Increasing use of air conditioning in homes and cars.
■ Perceived need to adapt to current fashion in terms of possessions.
■ Increased mechanisation in the home and workplace.
■ Need to travel caused by:

> replacement of the village shop by superstores
> increased separation of workplace from home
> perceived need to travel abroad for holidays

■ High quality healthcare.

In providing for current needs there are many areas in which quality of life might be affected:

■ Consumption of resources.
■ Damage to the landscape (scarring by extraction; landfill).
■ Adverse effects on the natural environment such as wildlife.
■ Reduction in the quality of the visual environment, e.g., transmission lines.
■ Pollutants in the atmosphere and ground.
■ Global warming.

As far as 'future' is concerned the time scale over which needs might be considered is also difficult to identify. Perceptions or concerns relating to time are usually in a given subjective context – for example governments tend to focus mainly upon terms which are of the same order of magnitude as their period in office, that is, 5–10 years. Individuals may be mainly concerned with time scales relating to their immediate descendants which would be the order of a generation – around 20–30 years. When viewed in the context of a decade, let alone a century or a millennium, it is difficult to contemplate what might be regarded as 'future needs'. Current perception probably focuses more upon a rate of advancement of expectation in standard of living terms rather than a need at a particular point in time. It is even more difficult to imagine how such advancement could be achieved without placing additional environmental burdens upon the planet. In spite of all this some statements concerning sustainability aim at the *zero tolerance* approach:

> No activity should take place at a rate greater than that at which natural processes can restore existing balances.

Many natural decay/reinstatement processes are extremely complex and are so slow that the only way of achieving this aim would be to cease those operations which involve extraction from the ground – notably fossil fuels and mined materials. A further related concern will be disposal of waste – that this should only take place in a manner in which natural processes can disperse the effects at the same rate that they are occurring.

While in the longer term the imperatives for such an approach will increase, the view of most at this time is that some progress needs to be made in the *direction* of sustainability rather than in attempting to achieve ideals. Perhaps the best that a text such as this can achieve is to raise levels of awareness of the more technological facets of sustainability issues and, given that there is often some choice in the selection of materials for specific applications, provide a basis by which comparisons might be made between competing materials.

GREEN HIERARCHY

The green hierarchy can be applied to virtually every aspect of human need. It is as follows:

    reduce
    re-use
    recycle
    recover
    dispose

Reduce

In welfare terms there is little motivation to reduce. As suggested above, the perception of many is that continuous progress should be achieved. In the developed world most people and governments are driven by the expectation that 'things will get better'. This is fundamentally in opposition to the notion that reductions in provision of any form should be made. In the developing world, where provision for the majority is woefully inadequate and most are caught in a poverty trap, few would argue that the concept of reduction is unrealistic. Use of the term therefore tends to focus upon achievement of lower wastage rates in production while maintaining all-round improvements in provision. Even here economies are by no means always well received. For example it would seem a relatively simple step to reduce the amount of packaging used for common commodities, but there is no discernible trend towards this, mainly because attractive packaging helps sell products.

Re-use

Re-use indicates employing buildings or components for new applications with minimal alteration. There are considered to be great opportunities in the

Figure 3.1   Re-used building stone, slates and stone cobbles in a new housing development

construction industry for re-use. Much of the older building stock has considerable attraction from a re-use point of view and yet demolition is a common occurrence perhaps chiefly because living habits have changed and it can be difficult to adapt older buildings to very different modes of use. Examples of situations where re-use has been applied successfully are older structures such as factories where adaptations to offices or flats have been highly successful. There is an urgent current need to consider the provision of flexibility for possible re-use when designing buildings, from the points of view of the buildings as a whole and also the components which make it up. Some progress is being made in the latter – for example, re-use of bricks and simple structural timber sections such as joists – but there is still a long way to go.

In the house shown in Figure 3.1 the walls were built from stone recovered from demolished Victorian industrial buildings, the roof comprises slates from the same source and the external paving comprises stone cobbles which were formerly widely used for paving purposes.

Recycle

This implies conversion of the previous material into some other form, usually by production of a lower grade product, since some degree of deterioration usually occurs during service so that re-use is not a possibility. Again, progress is being made, for example, in recycling concrete as aggregates for concrete, and increasing proportions of some metals. Some recycling of plastics occurs but in view of the variability of the recycled components of products, only goods with basic performance requirements such as packaging and garden furniture are currently produced. Government policy can have a major impact here. Landfill taxes plus taxation on 'virgin' materials such as mined materials would form a major commercial incentive for developing and implementing recycling processes.

Recover

This term is normally used to denote energy recovery from materials with limited recycling options. Incineration plants are increasingly being used to produce heat for various applications. In some cases there may be advantages of using recyclate compared with fossil fuels – for example where emissions might be a problem. However, energy recovery itself may have associated environmental problems. For example, the operation of large, more efficient plants often requires transportation of waste over large distances. The energy cost of such transport must be considered and, since plants are normally situated away from densely populated areas, there may be some environmental concerns in quieter rural areas.

Dispose

This should only be considered when all of the above options have been excluded. Many local authorities are now taking steps to reduce the landfill burden of all kinds of waste. The efforts are likely to strengthened, as taxes, in particular those for *active waste*, are increased. All rubbish tips and bonfires should be seen as lost opportunities. There is a need for greater public awareness of the situation and for the use of all materials to be seen in cyclic terms rather than the use and discard approach which is largely still prevalent.

3.2 ENERGY TYPES, IMPLICATIONS AND POSSIBILITIES

Energy matters are often to be found at the heart of considerations relating to sustainability from the point of view of

- resources
- emissions
- global warming

It will therefore be appropriate to give some consideration to principles and relative merits of alternative sources.

Primary energy is the initial energy released from a given source. For example in the case of oil, gas, coal or nuclear power the primary energy is the heat released by combustion and some losses occur at this stage – for example some heat escapes. It is not possible to convert any of these fuels directly into electricity or mechanical power. In conversion of primary energy into *secondary energy* other losses must occur and these further decrease the efficiency of the sources concerned. To produce electricity, for example, heat is normally used to drive a turbine with typically 30 per cent loss of energy in the process. Further losses will occur during transmission of the secondary energy to the point of need where it can be described as *delivered energy*. It is found in general that large power plants work more efficiently so they tend to be built as near as possible to fuel sources, the secondary energy, usually in the form of electricity, being transmitted at high voltage (with therefore smaller power losses) to where it is needed. In some cases energy passes in transit through the form in which it is finally needed in order to assist in transmission – for example turbines (mechanical energy) are used to produce electricity which, as delivered energy, is often used to drive mechanical equipment.

CARBON BASED FUELS

There are vast (though finite) reserves of oil, gas and coal in the earth's crust. They are thought to be derived from a time when carbon dioxide levels in the atmosphere were very much higher that at present. A possible explanation is that primitive life forms such as bacteria, formed by photosynthesis, progressively reduced carbon dioxide levels to those found today. By the decay of these bacteria, carbon became encapsulated in solid, liquid or gas form and produced the reserves which are now being exploited. The same process still happens with all plant life, though at much reduced rates. Photosynthesis effectively increases the *embodied* energy content of atmospheric carbon dioxide by incorporating it into living matter. When fossil fuels or organic matter are burnt, the embodied energy in them is released in the form of heat and they revert to the carbon dioxide from which they were originally formed, whether recently in the form of animal or plant life, or millions of years ago in the form of bacteria.

The active ingredients of fossil fuels are carbon and hydrogen in some form. The basic reactions by which heat is released are:

$$C + O_2 \rightarrow CO_2 + energy$$

$$2H_2 + O_2 \rightarrow 2H_2O + energy$$

Hydrogen produces only water on combustion and no CO_2 and therefore the higher the hydrogen content of a fuel the 'greener' if may be said to be. The calorific value of fuels can also be judged from their contents of carbon and hydrogen.

Coal contains about 85 per cent carbon by weight with the remainder being approximately equal parts of hydrogen and oxygen. In liquids and gaseous fuels the carbon may be combined with hydrogen to form hydrocarbons and in gases there may be hydrogen gas included.

Coal also contains some sulphur which has in the past been a major cause of acid rain, though power stations now largely remove this using pressurised fluid bed combustion.

Natural gas, which is now the main source of gas supplies, comprises mainly methane. Natural gas is therefore the greenest fuel, having the highest hydrogen content, partly as gaseous hydrogen and partly as methane (CH_4).

Oils comprise mainly linear hydrocarbon chains with formulae of the type:

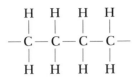

They have an atomic hydrogen-to-carbon ratio of 2 to 1 compared with 4 to 1 for methane. Note that a 2 to 1 atomic ratio is equal to 6 parts to 1 part, carbon to hydrogen, in mass terms since the carbon atom is 12 times heavier than the hydrogen atom.

A further possible energy source is wood which has the approximate composition in atomic terms of:

C	H	O
8	1	8

Oxygen is of course non-combustible (it is the means by which carbon and hydrogen burn) and in weight terms comprises almost half the total, so wood has a relatively low calorific value on a mass basis. The CO_2 emissions from the fuel types above are:

Relative to coal	(100%)
Wood	64%
Oil	62%
Natural gas	42%

Note however that if wood is derived from a renewable source its effective CO_2 emission is zero because the fuel was derived in the *short term* from atmospheric CO_2.

NUCLEAR FUEL

Nuclear fuels derive energy from nuclear reactions involving relatively small masses of very specialist fuels. The principles are similar to those upon which radioactivity is based (see page 105). The main positive benefit of nuclear fuels is that they are the only fuels in the earth's crust in which there are no carbon dioxide emissions (from the fuels themselves). However the spent fuel is radioactive and therefore potentially harmful, with sustainability implications over very long periods since the radioactivity can be very slow to subside. The seriousness of the storage/disposal problem was probably underestimated in the 1970s when there were ambitious plans for development of nuclear power stations. The situation changed dramatically in 1986 following the Chernobyl disaster and no nuclear power stations have been built in the UK since that time. A number of countries, including France, do however obtain substantial energy supplies from nuclear power.

RENEWABLE ENERGY SOURCES

Renewable energy is energy that is free from environmental implications because it arises from events which would occur anyway. There is some debate over what constitutes renewable sources, but the term might be widened to include energy derived by incineration of waste materials such as tyres and domestic waste. This is because it may be argued that this energy would in any case be released by degradation in landfill sites if not recycled in some way. It might also be considered that even fossil fuels are derived from the sun's energy, which is the case, but they are not renewable because of the very great time scale which would be required to restore them to fuels after combustion, by natural processes. There is a long list of sources and so only an overview can be given here with chief reference to those which might be involved in the manufacture of construction materials. Most renewable energy sources are derived from the sun's energy, with a few exceptions such as tidal power.

Solar energy

There is sufficient energy in the sun's rays to satisfy all foreseeable energy requirements. The main problem is harnessing this energy. A further problem is that in most parts of the globe there are seasonal fluctuations, solar energy being most abundant when the demand is least. Solar energy may be collected by:

- Solar collectors. These are quite commonly used to heat water for domestic use though they cannot generate temperatures sufficient for conversion to secondary forms of energy. In this sense they are not at present viable in the materials manufacturing context.
- Photovoltaic cells. These convert solar energy to low voltage direct current electricity. Again they can be used to drive small motors or lighting installations and can be converted to alternating current, but are not, at the moment, sufficiently developed for industrial uses.

Biomass

This term can be used to describe fuels based upon all forms of vegetable matter such as bacteria, plants and trees but also many forms of waste materials since these also usually have quite high organic contents. The calorific value of such fuels (and their carbon dioxide output) will be determined by their carbon and hydrogen contents as described above. The major advantage of both plant and waste derived fuels is that they are almost universally available so that transport costs and their associated environmental problems are reduced. Perhaps one of the most important biomass fuels is wood, which in developing countries may comprise the largest single source of energy. Such sources are however less viable in developed countries because of much higher land prices and the relatively slow rate at which timber grows, though rural areas of countries such as France depend heavily upon coppicing of hardwood plantations of oak and chestnut for fuels (Figure 3.2).

Figure 3.2 Plantations of chestnut saplings which are coppiced for fuel

The general problems of most vegetable based fuels are

1 Calorific values are reduced due to their oxygen content.
2 They are usually cellular so that appliances must be designed to accommodate their generally bulky nature.
3 Most forms contain water which must be dried off before use to avoid calorific values being further reduced.

They are therefore rarely used at the present when fossil fuels are a viable alternative.

A source of energy of rapidly increasing significance is waste material of all types. With rising landfill costs the economic viability of this resource is now much greater and the trend looks set to continue. In terms of industrial use for materials production two important illustrations are the use of waste tyres and used engine oil. Some cement manufacturing processes can now incorporate ground used tyres or oil derived *secondary liquid fuel* in place of coal for firing. These may also offer environmental advantages in emissions compared with incinerators since the very high temperatures achieved – up to 2000 °C – result in effective decomposition of potentially harmful constituents. Such fuels may also offer emissions advantages, particularly in terms of nitrogen and sulphur, compared with coal. Fuels based upon waste materials could be considered to be relatively 'green' in the sense that the materials involved would in any case decompose over time so that there is no additional CO_2 loading associated with them.

AVAILABILITY OF FUELS AND IMPACT OF FUEL SOURCE ON ENVIRONMENTAL PROFILE

It will be appreciated from the above that the environmental burden associated with the manufacture of a specific material will depend quite significantly on the fuel sources used and that most fuels pose enormous problems in sustainability terms. Choice of fuels has traditionally been largely a matter of fashion based upon commercial criteria. Current estimates suggest that coal reserves are the most substantial resource with perhaps 200 years supply remaining at current rates of use. Natural gas supplies will probably begin to run out in about half this time. Oil reserves will last a similar length of time though there may be problems of access to some in politically sensitive areas.

In developed countries awareness of environmental issues has led to improvements in the way that fossil fuels are used but the position in relation to imported materials is much less clear. Where products, or components of products, are known to be imported some research may be considered essential to determine information on the rating of the material in environmental terms.

3.3   CARBON RESERVOIRS

Carbon forms such a small proportion of the earth's crust in percentage terms that it does not register on a pie chart (Figure 5.1 in Chapter 5). It is nevertheless a vital component since carbon supports all life. It is also increasingly understood that the amount of carbon present in the earth's atmosphere has profound effects upon the climate.

Carbon could be described as occupying 'reservoirs' in a state of dynamic equilibrium as shown in Figure 3.3. These reservoirs can be grouped broadly into three levels of stability.

High stability reservoirs

Rocks   75 million billion tonnes mainly in the form of limestone
              containing about 500 billion tonnes of fossil fuel.
Deep ocean   36 000 billion tonnes of debris from once living organisms.

These are not sensitive to changes on the time scales of interest because quantities involved are so large and rates of exchange are very low.

Moderate stability reservoirs

Ocean surface layers   These contain about 600 billion tonnes of carbon
                        (dissolved carbon dioxide) and marine organisms.
Top-soil   There is about 1500 billion tonnes of carbon in top-soil.

These are in a state of dynamic equilibrium, most exchanges taking place with the low stability reservoirs.

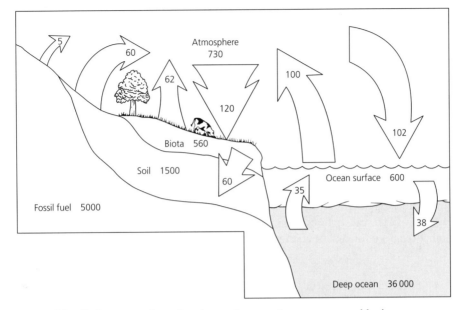

Figure 3.3 Carbon reservoirs and exchanges between them on an annual basis. All numbers are in billions of tonnes (Reproduced with permission from *New Scientist* magazine © RBI 1991, www.newscientist.com)

Low stability reservoirs

The atmosphere	This contains about 730 billion tonnes of carbon in the form of carbon dioxide.
Biota (all life forms)	These contain about 560 billion tonnes of solid carbon making up living matter.

The arrows in Figure 3.3 represent the size of the exchanges of carbon between reservoirs. It will be noticed that the largest changes occur between the medium and low stability reservoirs.

There is a broad balance overall between exchanges.

Surface layers of the earth respond to atmospheric changes in carbon dioxide and can to some extent act as a sink when atmospheric levels rise. A net input to the oceans of 2 *billion tonnes* is indicated.

There is also an approximate balance of exchanges between the atmosphere, biota and top-soil. Living matter absorbs 120 billion tonnes per year from the atmosphere. Decay of this matter releases 60 billion tonnes into the top-soil and a further 62 billion tonnes directly back into the atmosphere by decomposition. The top-soil also releases 60 billion tonnes into the atmosphere by breakdown and so there is an approximate balance (2 *billion tonnes* net release to the atmosphere) in terms of growing matter.

So far the net release to the atmosphere of living matter balances the net input to the oceans from the atmosphere.

The small arrow on the left represents the effect of human interference with the balance. This is the result of fossil fuel consumption and it releases 5 billion tonnes per year into the atmosphere. Because the atmosphere is a low capacity reservoir this results in a net addition to the notional 730 billion tonnes already present, of 5 billion tonnes per year. This might seem quite small but

1 It is at current rates of usage of fuels, which could well increase as demands increase.
2 Even on a 100 year span, this would result in an increase of carbon dioxide of 500 billion tonnes, or 67 per cent, which is a very significant increase.

The figures given above are notional but serve to indicate the great significance of fossil fuels in the delicate balance that exists in carbon balance within the atmosphere.

A further significant effect could be an increase in atmospheric carbon dioxide caused by large scale deforestation (especially by incineration) and failure to replant timber removed for construction purposes.

3.4   EMBODIED ENERGY VALUES

Embodied energy will be taken to mean the total energy involved in producing a material 'at the factory gate' including extraction and manufacturing costs. Table 3.1 shows typical values for the major materials used in construction. The values are given in order of decreasing energy input per tonne.

Table 3.1   Approximate embodied energy of some common construction materials

Materials	Energy input GJ/tonne
Aluminium	190
Polyethylene	100
Polyvinylchloride	80
Copper	50
Steel	30
Glass	20
Portland cement (wet process)	6
Softwood (imported)	5
Clay bricks/tiles (not Fletton)	4
Portland blast furnace cement	3
Aerated concrete	2
Concrete	1.3
Sand–lime bricks	1.2
Oak (green)	1.2
Softwood (home grown)	0.8
Cut stone/slate	0.8
Fletton bricks	0.6
Crushed granite aggregate	0.2

Table 3.2 Embodied energy in thermal insulation materials

Material	Embodied energy GJ/m^3
Polystyrene	4.1
Glass	2.7
Mineral fibre	0.8
Recycled paper	0.5
Sheep's wool	0.1

The figures must be interpreted with care. For example a major additional consideration is the density of the material. Aerated concrete is indicated as having higher embodied energy than concrete, but when used as, say, the inner skin of a cavity wall the mass used would be only about one third that of the dense concrete equivalent and the aerated material would also greatly improve the thermal behaviour of the wall. Other points emerging from the table include:

- Aluminium is the most energy intensive material in Table 3.1. The figure might however be adjusted according to the proportion of material which can be recycled, since there is much less energy involved in recycling. Again, aluminium requires 2–3 times the energy input of PVC. This must however be considered in the light of their relative densities (aluminium is 2.5 times as dense as PVC) and their relative E values (aluminium is 15 times as stiff).
- Inclusion of additives in cements greatly reduces their embodied energy.
- Figures for timber depend greatly on their origin since transport costs are high.
- The embodied energy in Fletton bricks is lower because the clay contains some organic matter.
- Materials such as building stone have low embodied energy. Where cut stones could be re-used they would look particularly attractive.

As regards insulation materials it will be more appropriate to measure embodied energy in terms of volume. Values for common materials are shown in Table 3.2. Valid comparisons could only be made on the assumption of equal insulation values, though the materials shown are roughly equivalent in these terms.

It will be noticed that there are large differences between those materials that are manufactured and those that are from natural sources. In the case of insulation it should be borne in mind that there may be substantial differences in cost. Durability ratings are also likely to differ considerably, especially in damper climates.

It should also be borne in mind when all the above figures are considered that there will be an additional environmental loading related to the carbon dioxide emissions corresponding to the energy values indicated on the assumption that fuels used in manufacture are predominantly fossil based.

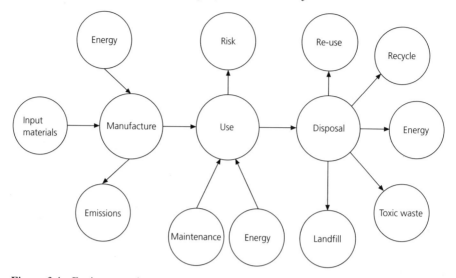

Figure 3.4   Environmental transactions associated with construction materials

3.5 ENVIRONMENTAL ASSESSMENT OF CONSTRUCTION MATERIALS

While there are many different approaches to the detailed evaluation of construction materials, in environmental performance terms there is little disagreement as to aspects of performance which should be included in the equation. Comprehensive assessments of materials performance must be in the context of complete life time (life cycle assessment, LCA). Aspects of performance can be subdivided into

- production/incorporation
- performance in use
- disposal/re-use

A simplified presentation of the situation shown in Figure 3.4. Each of the smaller bubbles in the figure can then be broken down into a similar diagram. An example of a small bubble feeding into each of the larger bubbles will be taken to illustrate the principle. When considering concrete, one of the input materials is cement which itself might be scrutinised schematically as in Figure 3.5.

Yet again the energy input of the cement can be broken down as in Figure 3.6. The complexity of the process of an overall assessment of various construction materials over their life times will be evident.

ASSESSMENT PROCEDURES

While there is widespread accord on the matter of the need to address environmental concerns, the techniques which are being developed to assess

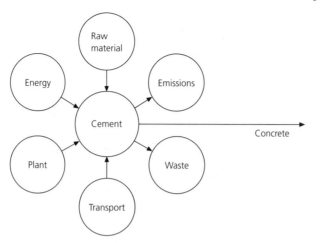

Figure 3.5 Environmental transactions associated with cement manufacture

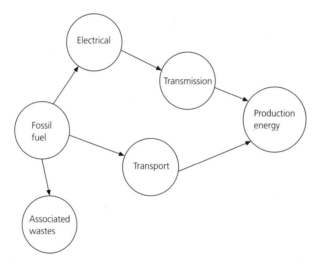

Figure 3.6 Energy transactions associated with cement production

construction materials vary enormously. Possible ways of appraising the performance of individual materials might include:

- A simple go/no go statement as to whether a material could be regarded as 'green'.
- Comparative analysis of materials which might be considered for a given function.
- A description of given performance criteria, the decision being left to the user.
- Production of an index of performance.

The method of tackling environmental performance will depend to an extent on the interest of the recipient of the information. Recipients may be:

Figure 3.7   Daisy logo awarded to products that satisfy stringent European Commission criteria for environmental performance

the designer
the client
the regulator
the public at large

For example:

■ The designer may be mainly interested in the comparative performance of materials which might compete to fulfil the same role.
■ The client may wish to convey a particular image in terms of the building.
■ The public at large might want to see some form of labelling on a product rather than having to weigh up sometimes complex evidence for or against a product.

The European Commission has instigated a scheme, at present voluntary, by which products can be described as *environmentally friendly*. Products are assessed on a number of criteria such as emissions to water and the atmosphere, associated toxins, and fitness for purpose. Successful products are badged with a *daisy* logo (Figure 3.7).

To permit comparison in performance terms *in situ* it is necessary to identify a *functional unit* for the product or material under consideration. This is probably most easily done for components which perform the function of *membranes* in buildings. Such items include flooring, walling, roofing, windows, etc. In this case the functional unit can be considered to be one square metre of the *membrane* concerned. Elements which perform a predominantly structural function may be rather more difficult to identify in functional unit terms.

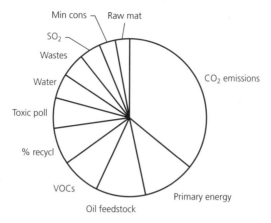

Figure 3.8 Environmental assessment weightings used in the BRE *Green Guide*

BUILDING RESEARCH ESTABLISHMENT *GREEN GUIDE* TO SPECIFICATION

This forms a good example of a practical environmental assessment method. It is intended primarily for designers but will also be of use to clients and others who may wish to be aware of the considerations involved and to examine possible options for approaching new construction. The information is a compilation of information derived from specialists with a wide range of viewpoints including local and central government, materials producers, developers and investors, construction professionals, environmental activists and lobbyists, and academics and researchers.

The guide comprises a comparative study of membrane type materials. Seven main areas are identified for consideration:

- toxicity
- primary energy
- emissions
- resources
- reserves
- wastes
- recycling

Within each category a classification A, B, C is given according to relative performance levels of alternative materials over a 60 year life cycle. It should be emphasised that because classifications are relative they only have significance within the group.

For each category of systems a final A, B, C rating is assigned according to the overall performance of the materials using a weighting system derived from other studies in which numerical ratings were assigned within each material/group. Figure 3.8 shows the weightings used to produce the profile. Some of the topics identified above have been subdivided for assessment purposes. For example, emissions include:

Carbon dioxide – effects already covered.

Volatile organic compounds (VOCs)   These are mainly associated with solvent based paints and some adhesives used in board materials (though they also arise from furnishings such as carpets). They can act as irritants but are in any case undesirable in the atmosphere.

Sulphur dioxide – causes acid rain.

Care should be taken with interpretation of data because not all aspects of performance were given numerical values in the study. It will be noted that oil as a fuel is quite heavily penalised because it features in primary energy, emissions and oil feedstock figures. Profile sheets exist for the following:

High mass	External walls
	Roofs
	Upper floors
Medium mass	Internal walls and partitions
	Windows
Low mass	Wall insulation
	Roof insulation
	Floor surfacing
	Floor finishes and coverings
	Doors
	Paint systems
	Suspended ceilings and ceiling finishes

In each case a price range is given, together with anticipated life, so that a broad overall view can be taken.

As an illustration of the *Guide* Table 3.3 shows the results for one of the relatively simple groups – single and double glazing. It will be noted that in most categories and the summary category, the full range of designations A–C is given on account of the relative nature of the study. The materials softwood, hardwood and galvanised steel are all assigned A summary ratings for energy/emissions reasons. Aluminium attracts a C summary rating for the same reasons. uPVC is given the intermediate B summary rating. The material attracts rather poor ratings in terms of energy emissions and toxicity but scores better than other materials in terms of waste and recycling.

A careful review of the information in this relatively simple assessment will give some appreciation of the complexity of environmental profiling generally. In specific cases other criteria may need to be applied in addition to those highlighted. For example, in the case of windows, aluminium has an advantage, not referred to, of quite high stiffness and thermal stability, which may give it the edge over, say, uPVC for larger openings subject to high environmental temperatures.

The methodology of the BRE assessment is published so that those wishing to consider the subject in more detail may do so. However not all the data

Table 3.3 Environmental ratings of various single and double glazing systems (from BRE *Green Guide*. Reproduced by permission of BRE. The table is only a limited excerpt from the Green Guide and is for illustration purposes. The Green Guide is updated occasionally – for full details contact BRE on 01923 664462)

	4 mm SINGLE GLAZING															6 mm DOUBLE GLAZING														
	Softwood opening			Hardwood opening			Al opening			Galvanised steel opening			uPVC opening			Softwood opening			Hardwood opening			Al opening			Galvanised steel opening			uPVC opening		
	A	B	C	A	B	C	A	B	C	A	B	C	A	B	C	A	B	C	A	B	C	A	B	C	A	B	C	A	B	C
Toxicity in manufacturing	A	B	C	A	B	C	A	B	C	A	B	C	A	B	C	A	B	C	A	B	C	A	B	C	A	B	C	A	B	C
in combustion	A	B	C	A	B	C	A	B	C	A	B	C	A	B	C	A	B	C	A	B	C	A	B	C	A	B	C	A	B	C
Primary energy	A	B	C	A	B	C	A	B	C	A	B	C	A	B	C	A	B	C	A	B	C	A	B	C	A	B	C	A	B	C
Emissions CO_2	A	B	C	A	B	C	A	B	C	A	B	C	A	B	C	A	B	C	A	B	C	A	B	C	A	B	C	A	B	C
VOCs	A	B	C	A	B	C	A	B	C	A	B	C	A	B	C	A	B	C	A	B	C	A	B	C	A	B	C	A	B	C
NOx	A	B	C	A	B	C	A	B	C	A	B	C	A	B	C	A	B	C	A	B	C	A	B	C	A	B	C	A	B	C
SO_2	A	B	C	A	B	C	A	B	C	A	B	C	A	B	C	A	B	C	A	B	C	A	B	C	A	B	C	A	B	C
Resources minerals	A	B	C	A	B	C	A	B	C	A	B	C	A	B	C	A	B	C	A	B	C	A	B	C	A	B	C	A	B	C
water	A	B	C	A	B	C	A	B	C	A	B	C	A	B	C	A	B	C	A	B	C	A	B	C	A	B	C	A	B	C
oil feedstock	A	B	C	A	B	C	A	B	C	A	B	C	A	B	C	A	B	C	A	B	C	A	B	C	A	B	C	A	B	C
Reserves	A	B	C	A	B	C	A	B	C	A	B	C	A	B	C	A	B	C	A	B	C	A	B	C	A	B	C	A	B	C
Wastes generated	A	B	C	A	B	C	A	B	C	A	B	C	A	B	C	A	B	C	A	B	C	A	B	C	A	B	C	A	B	C
Recycling % contained	A	B	C	A	B	C	A	B	C	A	B	C	A	B	C	A	B	C	A	B	C	A	B	C	A	B	C	A	B	C
% capable of being	A	B	C	A	B	C	A	B	C	A	B	C	A	B	C	A	B	C	A	B	C	A	B	C	A	B	C	A	B	C
% currently recycled in UK	A	B	C	A	B	C	A	B	C	A	B	C	A	B	C	A	B	C	A	B	C	A	B	C	A	B	C	A	B	C
energy required to	A	B	C	A	B	C	A	B	C	A	B	C	A	B	C	A	B	C	A	B	C	A	B	C	A	B	C	A	B	C
Summary rating cost range (£/m²)	130–170			180–250			190–253			134–173			165–215			170–200			230–290			230–310			175–210			175–225		
replacement interval (yrs)	20			20			20			20			20			20			20			20			20			20		

accumulated in the study is published and the validity of some of it has been challenged by parties such as manufacturers with a vested interest. Computer based profiling techniques are now available, though implementation of such profiles will always be influenced by other performance or economic based factors and client preferences.

In guides of this type assumptions must be made concerning, for example, the fuels used in production and some generalisations must be made. The situation may vary considerably according to the place of manufacture of a product and advances in emissions controls may also affect the outcomes of assessments of this type over quite short time scales. Regular reappraisal will therefore be essential and there may be instances where information available on specific products will lead to different verdicts.

3.6   ILLUSTRATION OF LIFE CYCLE ENERGY ASSESSMENT (LCEA) IN DWELLINGS

As explained above, LCEA can be regarded as a subset of sustainability. It focuses upon the energy content of buildings from cradle to grave. It can be regarded as comprising a number of components:

■ Embodied energy in materials used to construct a facility.
■ The energy needed to maintain the facility over its working life.
■ The running cost of the building.
■ The cost (or energy recovery) associated with disposal at the end of its life.

In each case the energy identified should be primary energy since this represents the environmental loading associated with the material. Note:

■ Figures given must be regarded as very approximate since individual processes may vary considerably in their efficiency, and transport costs will also depend on the proximity of materials/energy source and point of use.
■ Interpretation of all such forms of assessment is difficult.

An indication of the main areas to which attention should be paid will be given by reference to published data for a specific case.

It is quite important to place embodied energies into the context of the building life time and, in view of the fact that domestic construction forms a major component of the construction market, this will form the basis of the case study.

Embodied energy

Figure 3.9 shows the proportionate LCEA values for the components of a typical two storey detached timber framed domestic building of moderate size (130 m²) with brick cladding, single glazed timber windows and moderate insulation levels (Fay *et al.*, *Life Cycle Energy Analysis of Buildings: A Case Study*; see the references for this chapter at end of the book for further details).

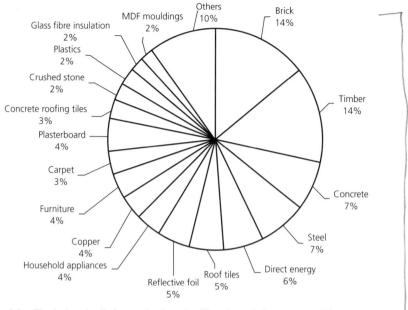

Figure 3.9 Typical embodied energies in a dwelling (rounded to nearest %)

The information is shown in the from of a pie chart because it is the relative values in a real life situation that are of interest. The embodied energy information is indicated in order of decreasing amounts clockwise from the 12 o'clock position. The figure includes all 'hardware' items, including those that are involved in setting up a home. The climate is a fairly warm temperate, perhaps a little warmer than the southern UK with 10 frosts per year and significant air conditioning requirements for two months per year. The embodied energy figures used in the study accord well, in general, with figures given in Table 3.1 though the timber embodied energy quoted (per m³) corresponds to a figure of around 20 GJ/tonne which is well in excess of that given in Table 3.1 even for imported timber. The following comments arise:

■ The largest single figure (before rounding) is brick, indicative of the high energy input for this type of cladding material. It would be difficult to justify such a material for cladding on environmental grounds unless the bricks were laid in a lime mortar so that they could be re-used.
■ The second highest figure is for the timber frame and roof. The embodied energy figure behind this is considered excessively high as indicated above. If the figure shown in Table 3.1 for imported softwood were used the proportion attributable to timber would be much less; and less still if the home grown figure of Table 3.1 were used. Nevertheless the figure for timber should be read in the context of the fact that timber forms the structural frame of the dwelling.

- Concrete and steel are next in order of embodied energy. The concrete figure is based upon the use of 30 m³ of concrete for foundations and floors, which is quite a high figure.
- Note that the direct energy (for example in transport, plant and machines) in building a home is quite significant.
- The figure also shows that the provision of internal fittings for a house can consume substantial energy.
- The embodied energy of the glass fibre insulation (albeit installed in rather basic amounts) is quite small in relation to the other energies.
- The 'other' category refers mainly to items such as small appliances and china necessary for running a home.

The total embodied energy cost involved in setting up the home was found to be 14.1 GJ/m². A higher insulation level option was also considered in the study and this increased the figure to 15.2 GJ/m².

Maintenance costs

In the illustration considered an annual maintenance energy cost was included and this varied through the 100 year period. The average annual value was:

Maintenance energy requirement   0.21 GJ/m²

Running costs

The figures assumed in this study are, on an annual basis:

Thermal energy (basic insulation)   0.30 GJ/m²
Thermal energy (extra insulation)   0.21 GJ/m²
Non-thermal energy   0.75 GJ/m²

Total life time energy cost

The total energy cost of operating the home over various periods from 0 to 100 years can be obtained by taking the embodied energy cost with increments for maintenance and adding annual operating costs. The effect is shown in Figure 3.10. Curves C and D show the embodied energy values, with increments for maintenance, the difference between them being due to extra energy involved in the additional insulation option. The total energy costs on the basis of the above are also shown, curves A and B crossing at about 12 years. This would be the 'pay back' time (in energy terms only) of adding extra thermal insulation. It is perhaps surprising that the operating costs of the building, even with the additional insulation, form a major part of the total energy cost. This might appear to diminish the value of the considerations given above in this section in which embodied energy considerations play a major part. The following points might, however, be noted:

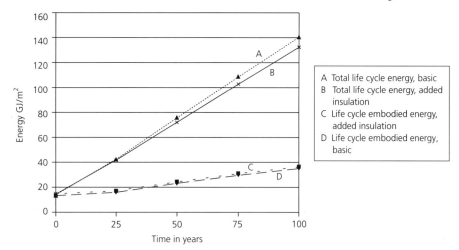

Figure 3.10   Life time energy requirements of a dwelling

- Carbon dioxide emissions and other green/sustainability issues are also important and their effects must be overlaid onto the pure energy considerations given here.
- If thermal insulation levels were increased, the differences between the materials options would then become more pronounced.
- There is a need to change perceptions concerning energy use in the home. In winter additional layers of clothing would reduce the need for high heating levels while in summer well designed buildings with high thermal mass could greatly reduce the need for air conditioning. At the present time energy is still relatively cheap and does not encourage conservation.
- The 100 year energy figure of 140 GJ/m^2 might be considered in the context of individual energy requirements in the developed world. Supposing that we each need about 25 m^2 to operate 'domestically', the total energy needed for 100 years for a person would, on this basis, be

$$140 \times 25 = 3500 \text{ GJ}$$

per person. This can be reduced to a continuous power input by dividing by the number of seconds in 100 years:

$$\frac{3500 \times 10^9}{100 \times 365 \times 24 \times 60 \times 60} \text{ watts}$$

This gives 1.1 kW as the steady average energy requirement associated with the home, throughout life. This excludes the cost of food, healthcare, holidays, transport and provisions in the workplace and might be compared with the figure of 2 kW given in Section 3.1. The cost of living, in energy terms, in a developed civilisation will be appreciated.

3.7   SOME OBSERVATIONS ON SUSTAINABILITY AND CONSTRUCTION MATERIALS

It was concluded in the previous section on embodied energy that the energy content itself of construction materials may not be the dominant factor when examining buildings in a life time framework. Nevertheless where options exist, without other major implications (of which cost may at present be very important), significant reductions in embodied energy may be achieved.

It is suggested that the focus regarding sustainability should be upon the following:

- Use of materials and building *designs* which minimise overall lifetime energy costs. On this basis there is enormous scope for very large improvements in thermal insulation values for buildings in order to reduce heating loads in winter conditions. Factors having a major bearing are building shape (square buildings having a lower surface area for given volume) and orientation in relation to sun and sizing of windows.
- The use of heavyweight materials which absorb heat in the day, allowing it to be dissipated at night, should be seen as a step forward in this respect.
- Preference should be given to those materials which

  1 are relatively abundant with minimum impact upon the landscape
  2 involve minimum emission levels during manufacture
  3 are amenable to recycling
  4 pose minimum risk when finally disposed of

There are currently three principal ways of producing a building frame – steel, reinforced concrete and timber. Brief comments about these follow.

Steel

Steel as a material has the highest embodied energy of the three major structural materials, though this energy forms in general a small part of total energy on a LCEA basis. Steel forms an essential part of almost all structures. Even in timber frame construction, it essential for connections and fixings. The following points should be considered in environmental terms:

- The material performs extremely well in catastrophic events such as earthquakes or explosions, so that it offers human safety advantages.
- Steel has recycling advantages because it rarely deteriorates seriously with time and, being magnetic, is relatively easy to extract from other loose waste.
- Re-use of the material depends upon the form in which it is used. Bolted steel joists are easily dismantled and therefore eminently re-usable. Welded hollow sections are somewhat more difficult to re-use.
- Fire protection systems may pose problems, either because there may be some environmental problem with protection systems themselves, or, if concrete is used, reclamation of sections becomes more difficult.

Concrete

For structural purposes, concrete must be reinforced or prestressed. Hence steel is required, though quantities as little as 1 per cent steel (by volume) produce required structural performance. In this respect reinforced concrete structures could be regarded as very satisfactory, with the lower energy concrete carrying compressive stress, while the higher energy steel is needed only in tensile areas. Note the following:

- Concrete provides thermal mass to buildings, assisting in control of temperature in warm conditions.
- Fire protection can usually be built in without need for additional materials.
- Demolition of reinforced or prestressed concrete can be difficult and may involve safety issues and/or disturbance. Separation of steel from concrete can be difficult and quite energy intensive. Crushed concrete for recycling may be contaminated with other material.

Timber

From a number of criteria, timber might be considered to be the winner in environmental terms. However this initial view must be tempered by the following:

- Preference should be given to home grown timber since substantial costs are incurred by transportation over large distances.
- Use of green (unseasoned) timber avoids energy costs associated with drying.
- Use of naturally durable species avoids the need for preservatives, many of which have quite serious environmental questions associated with them.
- Timber framed buildings work on a 'cellular' principle so that clear spans and building heights are limited (see also Chapter 10).
- There is still a preference for brick cladding of timber framed buildings. While this offers advantages such as thermal mass and rigidity to structures, it adds an environmental burden to the building type. Tile clad building offers better re-use prospects.

3.8 SOME OTHER OPTIONS FOR 'GREEN' BUILDING

The above considerations lead to the suggestion that emission, pollution and depletion issues may be as important as pure energy questions as regards environmental friendliness, at least in the short term.

Some suggestions as to how to achieve some progress towards sustainability are as follows:

- Use natural, low energy insulants, such as cellulose fibres where feasible. Some forms actually absorb gases such as formaldehyde which might be emitted from internal board materials and furnishings which incorporate synthetic adhesives. Care would need to be taken that such materials are durable in the situation identified.

- Use surface coatings derived from natural sources with low levels of VOCs.
- Use floor coverings of natural origin. These include linoleum. Some rubber based coverings may be VOC free.
- Use lime for masonry, external coatings and finishes where appropriate. Use of lime mortars may increase the longevity of some types of masonry as well as improving prospects of re-use. Lime washes enable masonry to breathe, reducing the risk of damp problems.
- Avoid more complex combinations of materials which would be difficult to separate and recycle at the point of building demolition. For example, while the use of lightweight blocks for the inner leaf of walls has produced great benefits in thermal insulation terms, the recovery of such materials for re-use is difficult, especially if plastered finishes are applied. The use of timber frame construction with cellulose fibre internal boarding would greatly increase recycling opportunities.
- Rammed earth construction as the infill in timber framed buildings offers exciting prospects. Unstabilised material would be preferred though low strength versions, produced using small quantities of binder, should also break down fairly easily at the point of demolition. See Chapter 5.
- There is perhaps a need for a culture change towards greater flexibility in building designs to cater for possible future changes of use or even dismantling and reassembly. The use of relatively simple precast units fixed together in a

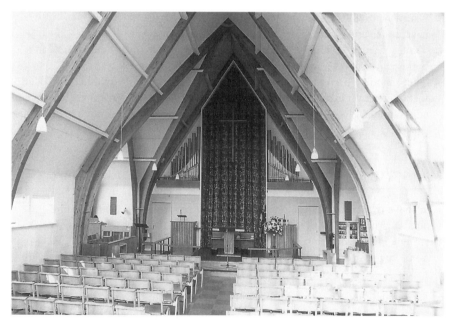

Figure 3.11   Glued laminated timber church, assembled by bolting together of simple portal frames. The church was dismantled to be replaced by a larger structure. Because no buyer could be found it was cut up for waste

simple way by mechanical fixing would be a step in this direction. The technique could be applied to structures with timber, steel or concrete frames. It is recognised, of course, that this could pose a severe design constraint, and it is in this respect that a change of attitudes to building design would be called for. To illustrate the point, Figure 3.11 shows a glued laminated timber building, built as a church, comprising just two types of structural elements in the form of portal frames. The more slender frames form the nave and transepts, while those of broader section are arranged diagonally on plan to provide the chancel (all of the same height). All connections were bolted. The church, although in excellent condition, was replaced after 20 years, since a larger building was required. It was easily dismantled without damage. However, because no buyer could be found, it was cut up as waste.

CHAPTER 4

STATISTICAL APPROACH TO MATERIALS EVALUATION

The importance of a statistical approach to design and the measurement of parameters involved in design has been recognised for many years and some design techniques employing statistical methods are well established. An example is design for wind loading in which the wind load allowance is increased for long design life of buildings, reflecting the increased probability of very high wind speeds over a longer time span. Other design parameters may be more difficult to describe in statistical terms though research is continually yielding the information required for design purposes. Central to an integrated and rational design philosophy is the need to identify and control all sources of variability. Such techniques also form a key part of quality management systems (BS EN ISO 9000), providing a clear quantitative basis upon which to judge the need for intervention when changes occur. Not surprisingly, there is an increasing tendency for new standards and codes to use statistical terminology and methods. A brief resumé of the basic principles of statistics in materials performance, therefore, follows.

4.1 VARIABILITY AND ITS MEASUREMENT

It was explained in Chapter 2 that in the assessment of a material's suitability for a specific task in construction, it might have to be appraised in many ways. Each mode of assessment will involve a testing procedure and, even if that is flawless (this point will be discussed later), there will almost inevitably be variability in the results caused by inherent materials variations. For any one property, these can be represented as histograms and as increasing testing is carried out, their shape becomes better defined and takes the form of a curve (Figure 4.1). The exact form depends on many factors; for example, in the case of strength, it may be truncated at either end due either to a ceiling strength for the material or the absence of 'negative' strengths. For practical purposes, a standard bell shaped curve, called the normal or Gaussian distribution is assumed. The curve is symmetrical and defined by the value of the measured property at its peak (or mean value) μ (pronounced 'mu') and the width at a specified position. The standard deviation, σ, pronounced 'sigma' is equal to the difference between the mean value and the value below or

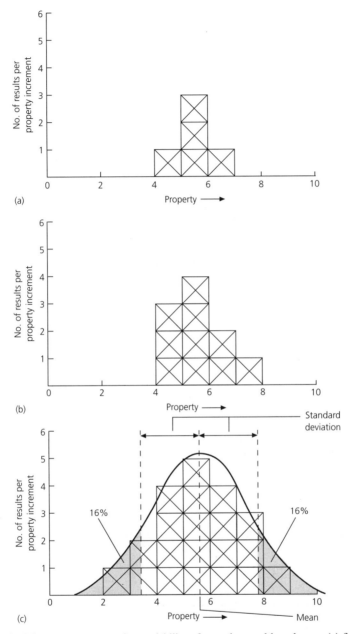

Figure 4.1   Histograms representing variability of a product and based upon (a) five results, (b) 10 results, and (c) 20 results. As the number of results increases the histograms approximate to bell shaped curves

above which 16 per cent of results fall (Figure 4.1.) Both μ and σ could only be measured exactly if every item in question were tested and this is usually impracticable. In practice, if a sample size n is tested for variations in property x, the standard deviation estimate S (rather than σ) is given by

$$S = \sqrt{\frac{(x_i - \bar{x})^2}{n - 1}} \qquad \text{where } \bar{x} \text{ is the measured mean of individual results } x_i$$

If, for example, the results of three tests were 8, 10 and 12, then S would be

$$\sqrt{\frac{(8 - 10)^2 + (10 - 10)^2 + (12 - 10)^2}{3 - 1}} = \sqrt{\frac{8}{2}} = 2$$

Note that the standard deviation is about equal to the differences between consecutive results, 2 in this case (though it is not usually exactly equal). Most scientific calculators can be used to calculate S values quite quickly. Having obtained the mean and standard deviation, the proportion of results above or below any specified level can be predicted by use of a simple mathematical formula and z values given in Table 4.1:

$$C = \bar{x} \pm zS$$

C is often described as the *characteristic value* and it represents a 'safe' limit for the property, though the level of safety (that is, the number of acceptable failures) must be specified and there is always some risk that a value outside it may occur. Take, for example, the previous values $\bar{x} = 10$ and $S = 2$. The characteristic value of x below which 5 per cent of results would be expected to lie would be (Table 4.1)

$$\bar{x} - zS = 10 - 1.64 \times 2 = 6.72$$

A similar number would lie above $10 + 3.28 = 13.28$. Descriptions of this sort are far more valuable than simply quoting a mean value since the latter gives no indication of variability and associated risk.

Standard error

If each result is obtained as an average of n readings, it can be shown that the standard deviation of the average result is then

Table 4.1 Table of z values for various values of percentage failures

Percentage of failures permitted	z factor
16	1.00
10	1.28
5	1.64
2.5	1.96
2	2.05
1	2.33

$$S_e = \frac{S}{\sqrt{n}}$$

where S is the standard deviation of individual readings. This is described as the standard error S_e. Measurements are often carried out in pairs so that the standard error would be

$$S_e = \frac{S}{\sqrt{2}}$$

$$= 1.4 \quad \text{approximately}$$

To obtain the proportion of the means of pairs above or below a certain value, the previous formula would be used, substituting standard error for standard deviation.

Variance

This is simply the square of the standard deviation. It is sometimes used to permit addition of errors or variations from two or more sources. Standard deviations cannot be combined by simple addition.

Coefficient of variation

This is defined as

$$V = \frac{S}{\bar{x}} \times 100$$

In other words, this is the standard deviation expressed as a percentage of the mean. Errors or variations are often expressed in these terms because it may be argued that when the quantity being measured is larger, there is larger scope for variation.

4.2 PRODUCT QUALITY

Figure 4.2 illustrates a simple non-mathematical method for obtaining statistical data as well as the advantages of products of low variability. Curves A and B represent the results of 50 concrete cube compression tests at two works. As in Figure 4.1, each square represents one test result. Curve A signifies a product of lower mean strength than curve B but it is also of reduced width. The effect is that concrete from source B has a greater probability of very low strengths – below about 30 N/mm^2, as illustrated on the graph. This may have implications for safety in the structure in which concrete B is used because the pattern of failures in practice appears to be associated with very low strengths rather than exceptionally

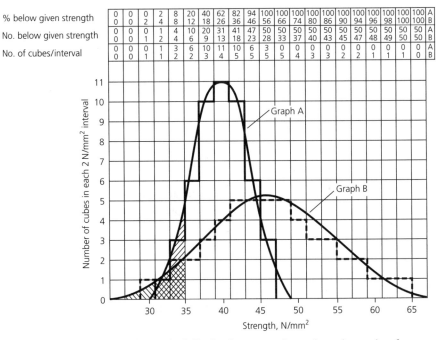

% below given strength	0	0	0	2	8	20	40	62	82	94	100	100	100	100	100	100	100	100	100	100	100	A
	0	0	2	4	8	12	18	26	36	46	56	66	74	80	86	90	94	96	98	100	100	B
No. below given strength	0	0	0	1	4	10	20	31	41	47	50	50	50	50	50	50	50	50	50	50	50	A
	0	0	1	2	4	6	9	13	18	23	28	33	37	40	43	45	47	48	49	50	50	B
No. of cubes/interval	0	0	0	1	3	6	10	11	10	6	3	0	0	0	0	0	0	0	0	0	0	A
	0	0	1	1	2	2	3	4	5	5	5	5	4	3	3	2	2	1	1	1	0	B

Figure 4.2   Histograms and smoothed distribution curves drawn from the results of two sets of 50 concrete cube tests. Graph A represents good quality control and Graph B poor quality control. The total area under each graph is the same. Also, the area under each curve to the left of the 35 N/mm² line is the same, implying equal numbers of results beneath this strength. Curve B is, however, indicative of a greater probability of very low cube results than curve A

high stresses. The information in these graphs has been summarised at the top of Figure 4.2. The number of cubes at or below each strength is obtained by adding the number of cubes in each strength interval from the left. For example, the number of cubes at or below the peak strength of 40 N/mm² for A is

$$1 + 3 + 6 + 10 + 11 = 31$$

The top row gives these figures as percentages, that is, ×2 in this case. Specially weighted graph paper is available to plot the percentage of cubes below a given strength so that a straight line graph is obtained. Results are also shown in Figure 4.3. Note that the characteristic strength of concrete B with less than 10 per cent failures is lower than A, the difference becoming larger as the acceptable failure rate decreases. On consideration of the fact that works B must use more binder (cement in this case) than works A to achieve the higher mean strength, it will be evident that the situation would warrant an investigation into quality procedures which might more than repay the cost if a significant reduction in variability were achieved.

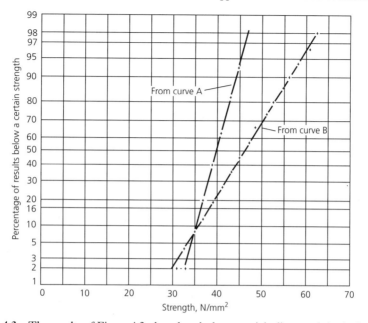

Figure 4.3 The results of Figure 4.2 plotted such that a straight line graph is obtained. In each case, the standard deviation can be obtained by subtracting strengths corresponding to the 50 per cent and 16 per cent lines. For example, Graph A corresponds to a standard deviation of $40.0 - 36.5 = 3.5$ N/mm^2. Note the absence of 0 and 100 per cent values on the vertical axis

ASSESSMENT OF RISK – EXPECTATION

When designing a product or component for a specific situation, a margin of safety against failure must be built in and this should be chosen with regard to:

- the risk of failure in that situation
- the consequences of failure

The second factor may be measured in different ways – for instance in safety or economic terms. Where, for example, the failure of a single component could lead to structural collapse, or where replacement of the component would be difficult, it would be advisable to 'overdesign' it to reduce the risk of failure. The product of the above two terms is called the *expectation*:

expectation = risk of an event occurring × consequences
(in economic or other terms)

Expectation is thus an index of liability or risk associated with the component and it should be kept at a uniformly acceptable level for different parts of a composite structure.

To implement this on a quantitative basis requires an estimate of the risk of failure as well as an analysis of possible consequences. Each of these can be difficult to arrive at, for example, when making an engineering judgement on a part of a structure found to be of poor quality. A further illustration is found in the use of preservatives for structural timber. Both the likelihood and consequences of decay are considered in arriving at an appropriate specification for preservatives. There are many other areas within building design and construction where expectation principles can be profitably applied.

ACCEPTANCE TESTING

Where large quantities of any item are supplied on a regular basis there must be some agreement between supplier and purchaser as to acceptable levels of quality. A testing plan is, therefore, devised by which the producer can be confident that the purchaser will not reject goods of a reasonable quality while the purchaser can be confident that goods of an unreasonable quality need not be accepted. Acceptance testing, as it is known, must be very clearly defined and agreed between the two parties as it will usually form part of a contract between them. It is, however, rarely based on testing all items in a batch or 'lot' because:

- Where many similar items are produced testing would be unnecessarily expensive.
- In some cases, tests must be to destruction, as in strength testing, so that items tested cannot be used.

Almost all testing is, therefore, based on a sampling plan in which each batch (lot) is represented by a sample of appropriate size and the whole batch is rejected if the sample fails to reach the required standard. This brings a further problem. Since the defect level of samples will vary at random around the true defect level of the lot, testing of the sample may lead to deceptive information. There is a possibility that in a lot which is of acceptable quality, tests on a sample might lead to rejection of the whole batch. This is described as the *producer's risk*. There is a corresponding risk that the lot might be accepted when defective (*consumer's risk*). In a simple sampling plan, the following must be agreed:

1 sample size
2 level of defectives at which a lot is rejected
3 producer's risk (often taken at 5% or 0.05)
4 consumer's risk (often taken at 10% or 0.10)

Figure 4.4 shows the ideal testing plan in which there is 100 per cent probability of a lot being accepted up to an agreed failure rate (6 per cent in this case) and then zero chance of acceptance. The graph is known as the *operating characteristic* (OC). In practical sampling schemes, there is no chance of such a good OC curve being produced. To illustrate, take the following case:

Sample size = 50
Agreed acceptable percent defective = 5%

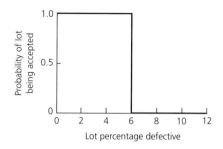

Figure 4.4   An ideal operating characteristic (OC) curve. The probability of acceptance of a lot drops from 100 per cent to 0 per cent when the percentage defective rises above the agreed level of 6 per cent

In this case, since $5\% \times 50 = 2.5$, it might be decided to accept the lot with up to two failures in the batch but to reject the lot if there are more than two failures. The OC curve will be obtained for various lot percentage defectives (LPDs) (a simple method without mathematical formulae is given here though other methods can be used where knowledge of statistics permits).

1 Lot percentage defective = 0. There will be no defectives in the sample so $p(\text{accept}) = 1.0$ or 100%.
2 Lot percentage defective = 2% or 0.02.

Find first the probability of no defectives – written as $p(0$ defectives) in the sample of 50.

This is $(1 - 0.02)^{50} = (0.98)^{50} = 0.364$. The probability that just the first one in the sample is defective is:

$p(\text{first sample defective}) \times p(\text{remaining 49 samples are good})$
$= 0.02 \times (0.98)^{49}$ (since $p(\text{accept}) = 0.98$)

This equals 0.00743.

However, this could happen in any one of 50 ways since any one in the sample might be defective so that

$p(\text{one defective}) = 50 \times 0.00743$
$= 0.372$
$p(\text{two defectives})$ – just the first two in the sample $= 0.02 \times 0.02 \times (0.98)^{48}$

But the two could be distributed throughout the sample in any combination. The number of combinations of 2 in a sample of 50 is $\dfrac{50 \times 49}{2}$. Hence, $p(2$ defective) is

$$0.02 \times 0.02 \times (0.98)^{48} \times \frac{50 \times 49}{2} = 0.186$$

The probability of 0, 1 or 2 defectives in the sample is

$0.364 + 0.372 + 0.186 = 0.922$

The probability of accepting the sample with an LPD of 2 per cent is, therefore, 0.922.

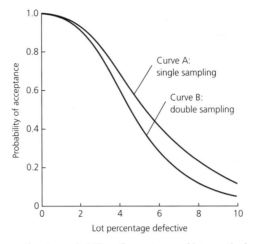

Figure 4.5   OC curves for the probability of acceptance of lots on the basis that up to two failures in a sample of 50 are acceptable (curve A), and a retest is carried out if two failures in the sample of 50 occur (curve B)

Note that, in this case, there is a chance of $1 - 0.922 = 0.078$ or 7.8 per cent that the lot would be rejected (producer's risk), even though the LPD is below the acceptable level of 5 per cent. The probabilities of 0, 1 or 2 defectives for LPDs up to 10 per cent are given in Table 4.2 in the same way.

This treatment assumes that the probability of a defective item is not altered by taking the sample, which in turn requires that the number of defectives in each lot is small in relation to the lot size.

Notice that there is an 11.2 per cent chance of accepting a lot with twice the stipulated failure rate (consumer's risk). The OC curve is shown in Figure 4.5, and it can be seen that it is a poor approximation to the ideal curve of Figure 4.4. In particular, the producer may be unhappy with the prospect of a rejection rate of 0.323 or 32.3 per cent for an LPD of 4 per cent, while the consumer might be unhappy with a pass rate of 0.416 or 41.6 per cent of lots with LPD 6 per cent.

The scheme can be improved by use of a double sampling technique. For instance, if in this case there are found to be two or three defectives in the sample, either:

1 a fresh sample could be taken from the batch and the same criteria applied as before, or
2 every item in the sample could be tested for acceptance (provided samples were not destroyed in the process!)

These procedures improve the OC curve in the critical region. Taking, for example, a retest, with a new sample when two defectives per batch are found, accepting the batch if not more than two defectives are found in the second sample (Table 4.3.), the new OC curve is shown in Figure 4.5. This form of double sampling will suit the consumer because the risk of acceptance at high LPDs is now much lower. However, the risk of rejection at low LPD has increased slightly and the producer, to offset this,

Table 4.2 Calculation of the probability of acceptance of lots with varying proportions defective on the basis that up to two defectives in a sample of 50 are acceptable

LPD	p (0 failures)	p (1 failure)	p (2 failures)	p (accept) $= p(0) + p(1) + p(2)$
0	1.00	0	0	1.000
2	$(0.98)^{50} = 0.364$	$0.02 \times (0.98)^{49} \times 50 = 0.372$	$(0.02)^2 \times (0.98)^{48} \times \dfrac{50 \times 49}{2} = 0.186$	0.922
4	$(0.96)^{50} = 0.13$	$0.04 \times (0.96)^{49} \times 50 = 0.271$	$(0.04)^2 \times (0.96)^{48} \times \dfrac{50 \times 49}{2} = 0.276$	0.677
6	$(0.94)^{50} = 0.045$	$0.06 \times (0.94)^{49} \times 50 = 0.145$	$(0.06)^2 \times (0.94)^{48} \times \dfrac{50 \times 49}{2} = 0.226$	0.416
8	$(0.92)^{50} = 0.015$	$0.08 \times (0.92)^{49} \times 50 = 0.067$	$(0.08)^2 \times (0.92)^{48} \times \dfrac{50 \times 49}{2} = 0.143$	0.225
10	$(0.90)^{50} = 0.005$	$0.10 \times (0.90)^{49} \times 50 = 0.029$	$(0.10)^2 \times (0.90)^{48} \times \dfrac{50 \times 49}{2} = 0.078$	0.112

Table 4.3 Double sampling scheme in which a new sample of 50 is tested if two failures occur in the first sample

LPD	$p(0)$	$p(1)$	$p(2)$ $= p(\text{retest})$	$p(\text{retest})$ $\times p(\text{accept on retest})$ $= p(2)\,[p(0) + p(1) + p(2)]$	$p(\text{accept})$ (double sampling)	$p(\text{accept})$ (single sampling)
0	1.000	0	0	0	1.000	1.000
2	0.364	0.372	0.186	0.186×0.922	0.907	0.922
4	0.130	0.271	0.276	0.276×0.677	0.588	0.677
6	0.045	0.145	0.226	0.226×0.416	0.284	0.416
8	0.015	0.067	0.143	0.143×0.225	0.114	0.225
10	0.005	0.029	0.078	0.078×0.112	0.043	0.112

might request, say, a second sample test on batches from which an initial sample test reveals three defectives, in the knowledge that some could pass the second test.

In all acceptance testing, the cost and complexity of the test plan must be considered against the benefits to the consumer and producer. For instance, use of large batches reduces the testing cost per batch but leads to bigger rejection levels should the sample be unsatisfactory. A common practice is that the producer works to a higher standard than that required in the test plan in order to reduce the risk of failures with their consequent implications, such as the cost of replacements and, possibly, damaged confidence in the product. Testing should be carried out by an independent body, though with provision for inspection by producer and user alike.

4.3 QUALITY CONTROL

This is the process of checking and monitoring of production which leads to a product of a consistently satisfactory standard. It is essentially an internal process (that is, operated by the manufacturer) though where a product is quality assured, documentary evidence of all quality control procedures is required. Much time, effort and expense is often invested in quality control, especially where components are of critical importance, for example, where failure could cause a catastrophe. Conversely, quality control will also be concerned with saving of cost where appropriate; there would be little point, for example, in employing a very strong hook on a weak chain – all components should be of similar performance. The hallmark of a good quality control system is that every aspect of the process should be inspected with special attention given to those stages that have a greater risk of producing faults or where the consequences of faults are more serious.

An essential part of any quality control process is product assessment. This will normally be quite independent of acceptance testing, although the tests carried out may be very similar. The manufacturer tests the product prior to despatch in order to ensure that there is a good chance of acceptance. Whatever test is carried out, it is important to ensure that errors associated with the testing process itself are acceptable; this is described under 'Testing precision'. It is also important that

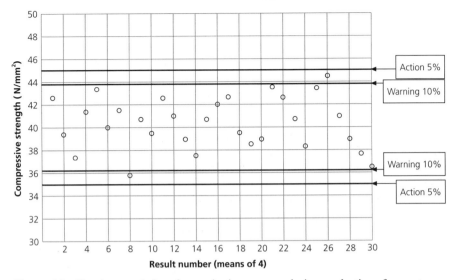

Figure 4.6 Simple control chart for monitoring progress during production of concrete. In this case progress is satisfactory since there are no results outside 'action' limits and results are equally distributed about the mean

when a change of quality occurs, it should be detected as soon as possible so that the cause can be investigated. Use of histograms, referred to above, is not a good way of detecting such changes because results are treated in large groups which tend to disguise short term trends.

One simple system is by control charts as in Figure 4.6. Strength testing of concrete is used here, though the process could be applied to any property, for example, size, cover to steel in reinforced concrete products, toughness, hardness and so on. The process relates to concrete of characteristic strength 35 N/mm² with 5 per cent failures permitted. If the expected standard deviation is 6 N/mm², then the target mean strength should be

$$35 + 1.64 \times 6 = 39.8 \text{ N/mm}^2$$

A value of 40 N/mm² will be assumed. In this scheme, means of 4 will be plotted so that standard errors of $= 6/\sqrt{4} = 3$ N/mm² are used. Hence, no more than one result in 20 (5 per cent) should be above

$$40 + 1.64 \times 3 = 44.9 \text{ N/mm}^2$$
$$(z)$$

or below

$$40 - 1.64 \times 3 = 35.1 \text{ N/mm}^2$$

These would be regarded as *action* lines (especially if results fell below the lower line). *Warning* lines might be positioned $3 \times 1.28 = 3.8$ N/mm² above and below the mean of 40 N/mm². One in 10 results would be allowed above and below these

lines. Many manufacturers play safe by erring upwards but the important feature
of the chart is that trends can be identified as results are added on a regular basis,
though the exact point at which action is necessary may not be clear.

CUSUM TECHNIQUES

These are a more sophisticated approach to quality control and are amenable to
computerisation. In their simplest form, cusum methods can be applied in many
ways, for example, if a car journey of 120 miles is planned with 3 hours allowed,
an average distance of 10 miles should be travelled in each 15 minute interval.
The mileometer can be checked at these intervals to ascertain progress against the
10 mile increments expected; if, for example, a higher average speed is maintained,
there will be an increasing difference between actual and planned mileage.

A further, more complex example is shown in Table 4.4. for concrete cube
testing, a procedure often monitored by cusum techniques. Strengths are compared

Table 4.4   Application of cusum technique to daily concrete cube test results. The target mean
strength (TMS) is 40 N/mm^2. The target mean range (TMR) is 5.5 N/mm^2

Cube number	Date cast	28 day strength	28 day strength – TMS	Cusum of mean	Range	Range – TMR	Cusum of range
		N/mm^2	N/mm^2	N/mm^2	N/mm^2	N/mm^2	N/mm^2
1	May 2	48.0	8.0	8.0	–	–	–
2	2	45.5	5.5	13.5	2.5	–3.0	–3.0
3	3	40.5	0.5	14.0	5.0	–0.5	–3.5
4	4	45.0	5.0	19.0	4.5	–1.0	–4.5
5	5	41.5	1.5	20.5	3.5	–2.0	–6.5
6	5	43.5	3.5	24.0	2.0	–3.5	–10.0
7	6	37.0	–3.0	21.0	6.5	1.0	–9.0
8	9	38.0	–2.0	19.0	1.0	–4.5	–13.5
9	9	41.0	1.0	20.0	3.0	–2.5	–16.0
10	10	29.0	–11.0	9.0	12.0	6.5	–9.5
11	11	36.0	–4.0	5.0	7.0	1.5	–8.0
12	12	32.5	–7.5	–2.5	3.5	–2.0	–10.0
13	12	34.5	–5.5	–8.0	2.0	–3.5	–13.5
14	16	36.0	–4.0	–12.0	1.5	–4.0	–17.5
15	17	44.5	4.5	–7.5	8.5	3.0	–14.5
16	17	50.5	10.5	3.0	6.0	0.5	–14.0
17	18	42.5	2.5	5.5	8.0	2.5	–11.5
18	18	48.5	8.5	14.0	6.0	0.5	–11.0
19	19	40.0	0	14.0	7.5	2.0	–9.0
20	20	51.0	11.0	25.0	11.0	5.5	–3.5
21	20	45.5	5.5	30.5	5.5	0	–3.5
22	23	36.0	–4.0	26.5	9.5	4.0	0.5
23	23	46.0	6.0	32.5	10.0	4.5	5.0
24	24	34.5	–5.5	27.0	11.5	6.0	11.0
25	25	45.5	5.5	32.5	11.0	5.5	16.5
26	25	40.0	0	32.5	5.5	0	16.5

with the target mean strength (TMS) and, just as important, the difference between successive results (range) is compared with an average range (target mean range – TMR) for the expected standard deviation. In this case, a characteristic strength of 32 N/mm^2 is required (5 per cent failures permitted) and with an anticipated standard deviation of 5 N/mm^2, the TMS is

$$32 + 5 \times 1.64 = 40 \text{ N/mm}^2 \qquad \text{(to the nearest 0.5 N/mm}^2\text{)}$$

It can be shown that the mean range for results having a given S is $1.128S$. Hence, the TMR is $5 \times 1.128 = 5.640$ N/mm^2 (5.5 N/mm^2 to the nearest 0.5 N/mm^2).

In the table, strengths are compared with 40 N/mm^2 and ranges with 5.5 N/mm^2, the difference being accumulated and then drawn as in Figure 4.7. Note that a downward turn in the cusum of the mean represents falling strength, while a downward turn in the cusum of the range represents improving quality control. By use of transparent masks, usually of truncated V shape, the progress of work can be accurately checked. The essential features of a mask are the *decision line* and the *decision interval* (Figure 4.8). The decision lines have a chosen gradient corresponding to a significant departure from planned performance. The decision interval effectively delays the detection of a change; if the decision interval were zero, many small random changes would be read as significant. The design of the masks will depend on individual conditions and the property being assessed. In general, masks for mean and range will be different. In the example in question, the decision intervals have been made equal to $5S$ (= 25 N/mm^2) while the decision lines have slopes of $S/6$ and $S/10$ for mean and range respectively, that is, changes of $S/6$ and $S/10$ respectively for each successive result. The datum point of the mask is placed over the latest result as it is indicated on the chart and a change is signified when the sloping arms intersect the cusum curve. In Figure 4.7 the first indication that a change has occurred is at result 13, it being evident that the mean strength has decreased significantly although range values are satisfactory. At this point, the mean strength was adjusted (by increasing the cement content in the case of concrete) and the first result reflecting this change is result 15. (In the case of concrete, these tests would be accelerated tests but would still take one or two days to yield results.) One advantage of cusum is that the date of the original change can often be ascertained. In the case of this example, it was around result 6. Investigations could then be conducted if necessary into the source of the change. The cusum of the mean increases steadily subsequent to the decision to increase the strength. Perhaps the change was a little too large but, more importantly, the cusum of range now begins to rise; this corresponds to the rather variable results obtained after result 16 as shown on the simple control chart at the top of Figure 4.7. By result 25, a change is signified in the cusum of the range, the mask intersecting the cusum at around result 14. This signifies that the standard deviation S has increased so that both mean and range masks must now be changed to ones corresponding to a larger S value, perhaps 6 N/mm^2 (the masks are not changed when the cusum of *mean* indicates the need for action). Production, therefore, continues on the basis of higher S values, though the cause of the change would again be investigated.

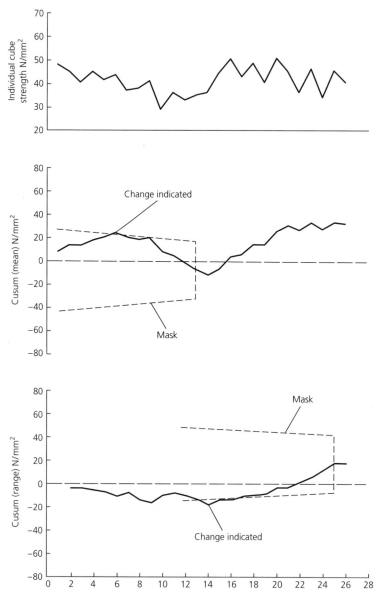

Figure 4.7   Cusum technique applied to cube test results

Cusum techniques can be tailored to specific requirements, masks being designed to detect changes in the product performance before there are problems of acceptance while not being oversensitive to small changes. The whole process can be computerised, including calculation of the exact change required. Further guidance on cusum techniques is given in BS 5703 Part 3.

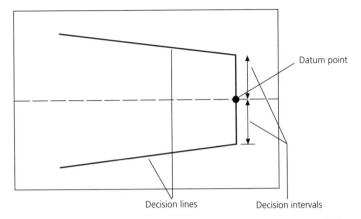

Figure 4.8   Truncated V mask for monitoring performance using cusum techniques. The decision line and control interval are chosen to suit the property being monitored

4.4 SAMPLING AND TESTING PRECISION

In any test scheme, whether for quality control or acceptance purposes, it is important to know what uncertainties are associated with the actual sampling and testing procedure, since no amount of effort in quality control terms will compensate for variations resulting from inaccuracies in testing.

Samples must be selected carefully from the whole of the batch so that they are representative, particularly in the case of loose materials such as aggregate for concrete, where poor sampling can negate the value of an otherwise sound testing scheme. Such sampling errors can be measured; for example, a method is given in BS 1881 Part 101* for measuring the sampling error in concrete testing. Two samples are taken, using identical techniques, from each of 20 batches of concrete (say, ready mixed truck loads) of the same specification and pairs of cubes made from each sample (a total of 80 cubes). Testing and sampling variances are clearly included in the expression

$$V_{ts} = \frac{\sum(M_s - M_d)^2}{40} \tag{4.1}$$

where M_s is the mean strength of the pair of cubes in the first sample in each case and M_d is the mean strength of the pair of cubes in the duplicate sample. The variations associated with the machine and operator must be separated from this and these can be obtained from the differences within each pair (D_s and D_d), hence,

$$V_t = \frac{\sum D_s^2 + \sum D_d^2}{80} \tag{4.2}$$

---

* No longer current, but used to illustrate technique.

Application of the laws of variance to equations (4.1) and (4.2) leads to an expression for the sampling variance:

$$V_s = V_{ts} - 0.5V_t$$

The error expected from sampling or testing is defined as the square root of the variance in each case but is normally expressed as a percentage of the mean:

$$\text{Testing error (\%)} = \frac{100}{M}\sqrt{V_t}$$

$$\text{Sampling error} = \frac{100}{M}\sqrt{V_s} = \frac{100}{M}\sqrt{V_{ts} - 0.5V_t}$$

BS 1881 Part 101 gives ceiling values of 3 per cent for testing and sampling errors, above which procedures should be checked. The values taken in other situations clearly depend on the material or property being tested and the precision of the testing machine/operator.

REPEATABILITY AND REPRODUCIBILITY

'Repeatability' relates to 'testing error' which is an indication of differences that can be expected between two single test results on identical test material with the same operator and laboratory with only a short interval between the tests. The situation is shown in Figure 4.9. It can be shown that if test results obey a normal distribution with standard deviation S, then differences between these results will

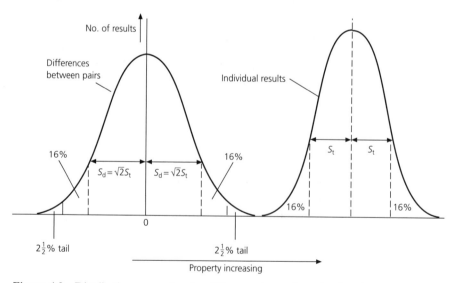

Figure 4.9   Distribution curves showing differences in results, related to the distribution of the results themselves

also obey a normal distribution. The mean of this second distribution will be zero and the standard deviation will be $S_d = \sqrt{2}S_t$. Repeatability, r, is defined as the difference between two single test results which would be exceeded on only 5 per cent of occasions. Half these differences would be positive and the other half negative (Figure 4.9) and the two 'tails' form the 5 per cent, hence each tail contains 2.5 per cent of results for which $z = 1.96$ (Table 4.1) Hence:

$$r = 1.96\sqrt{2}S_t = 2.8S_t \qquad \text{approximately}$$

An example of the application of this rule is given in BS 1881 Part 116 on testing concrete for compressive strength. The repeatability for pairs of 150 mm cubes made from the same sample, cured in similar conditions and tested at 28 days, expressed as a percentage of mean strength is required to be not more than 10 per cent at the 95 per cent probability level. Suppose the mean strength is 40 N/mm². This would mean, in simple terms, that on only one occasion in 20 should differences in pairs exceed 4 N/mm². The corresponding standard deviation S_t is given by $2.8S_t = 4$ or $S_t = 1.4$ N/mm², quite a low value. Accuracy checks of this type are clearly important when the implications of unsatisfactory test results are considered.

It should be noted that in tests which cannot be repeated on the same sample, as in destructive testing of any material, repeatability will inevitably contain a component that is due to variation of the material itself, even if it is from the same sample. Hence, target repeatability values must, in such cases, be based upon the material being tested as well as on the test equipment and operator skills.

An equally important term is *reproducibility*. This is of similar definition to repeatability except that it relates to different operators on different machines at different times. Reproducibility values will always be larger than repeatability values though the difference will depend on the relative skill of different operators, machine variability and any effects of time on the property being measured. When results obtained at different times from different laboratories are compared, the validity of differences may be difficult to establish, unless both repeatability and reproducibility data are available. Such data can be obtained from precision experiments which must be designed around the property being measured.

CHAPTER 5

THE STRUCTURE OF MATERIALS

In much the same way as the structural properties and performance of a completed building depend on the individual units which it comprises and on the way these are assembled, so the behaviour of a single component, for example, a brick or steel joist, depends on the 'building units' which it contains. These units or atoms are entirely responsible for every property of the material, whether it be physical, chemical or mechanical. Atoms are made up of still smaller units, the three main ones being protons, neutrons and electrons. These particles are important in their role as building units in atoms and, in some cases, are used to test materials. Table 5.1 shows their properties.

Many other subatomic particles have been discovered during recent research but they are of little consequence concerning the behaviour of bulk materials, since they are produced only by relatively rare subatomic interactions and are generally extremely shortlived; they soon revert to some combination of the more common particles. Atoms can be conveniently described in sections, concerned first with the nucleus and then with the electron shells.

5.1 STRUCTURE AND PROPERTIES OF THE NUCLEUS

The nucleus contains the protons and neutrons and, therefore, is by far the heaviest part of the atom. While it would be wrong to regard nuclear particles as completely motionless, they nevertheless occupy an extremely small space compared to that occupied by the atom as a whole. The protons and neutrons are held together by

Table 5.1 Properties of the fundamental particles which comprise atoms

Particle	Charge (coulomb)	Mass (kg)	Relative mass
Proton	$+1.602 \times 10^{-19}$	1.672×10^{-27}	1.000
Neutron	0	1.675×10^{-27}	1.002
Electron	-1.602×10^{-19}	9.109×10^{-31}	$\dfrac{1}{1836}$

very short range forces which disappear at separations greater than about 10^{-15} m. These forces are peculiar to nuclear particles – they are neither electrostatic nor gravitational in origin. The stability of nuclei stems from the fact that, when protons and neutrons come together, there is a small release of mass (M), this mass representing the *binding energy* (E) of the nucleus. According to Einstein's equation:

$$E = Mc^2$$

where c is the velocity of light. A particularly stable unit is the proton–neutron pair and, in lighter atoms, the number of protons is often equal to the number of neutrons. The binding energy per nuclear particle tends to increase with nuclear size, reaching a maximum at element number 26, iron (Fe). Thereafter, binding energy decreases, due to mutual repulsion of protons until, after lead (Pb), element number 82, nuclei are not fully stable. Hence, there are 82 basic types of stable element in the earth's crust, containing between 1 and 82 protons. Although, in some lighter atoms, the number of neutrons is equal to the number of protons, in heavier atoms, extra neutrons become essential to minimise proton–proton repulsion. In many elements (that is, atoms with a specified number of protons), a variable number of neutrons is possible; for example, carbon, which has six protons, may have six, seven or eight neutrons. The latter are known as isotopes of carbon and the relative quantities of each in naturally occurring carbon give an indication of their relative stabilities. In general, the neutron excess is only sufficient to stabilise the nucleus; if the excess becomes any greater than this, neutrons would tend to decompose, since free neutrons are not, in themselves, highly stable particles – they break up into protons and electrons.

Since the iron nucleus has the highest binding energy per unit mass it is the most stable nucleus. Smaller nuclei have a tendency to grow (fusion) and larger nuclei tend to break up (fission). It may therefore be argued that all matter should eventually change to iron but, although theoretically this is the case, the stability of all 82 nuclei is, in practice, sufficiently great for rates of change to be infinitesimally small. Nuclear power is, nevertheless, derived from the increase of nuclear stability that occurs on the fusion of light elements or the fission of heavy elements.

It is possible to change one element into another by nuclear physics and this might, for instance, be considered as a method of producing more useful or valuable elements from other elements but this could never be carried out on a commercial scale.

The use of materials must therefore be heavily geared to the existing balance of elements in the earth's crust. This balance is affected by relative nuclear stability but by other factors as well. The number of elements which are abundant in the earth's crust is quite small. These elements include oxygen, silicon, aluminium and iron. Those making up 98 per cent of the total are indicated in Figure 5.1.

RADIOACTIVITY

It has been established that the 82 stable elements form the basis for virtually the entire range of materials used on the planet. However, a number of heavier

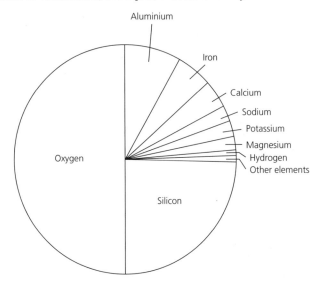

Figure 5.1   Distribution of elements in the earth's crust by mass

elements do exist naturally and, in addition, it is possible to produce 'artificial' elements, that is, very large elements, or isotopes with an abnormal proton/neutron ratio. Such elements are described as radioactive, since they undergo changes in structure which eventually result in the formation of one or more of the 82 stable elements. It should be emphasised that radioactive (nuclear) changes in materials are relatively rare. Nevertheless, radioactivity does have a number of important applications in the construction industry and so a brief description of this type of behaviour is given.

The main cause of radioactivity is excessive nuclear size, resulting in disintegration by proton–proton repulsion. Since the proton–neutron pair is itself a highly stable group, heavy elements normally decay by rejection of a particle comprising such pairs, in fact, two protons and two neutrons. This group constitutes a helium nucleus and is known as an alpha (α) particle. These particles have a positive charge due to the two protons which, combined with their substantial mass, causes them to be rapidly absorbed by normal materials. In consequence, they do not find application in construction. A further cause of radioactivity is proton–neutron imbalance, caused by an excessive number of neutrons – for example, when a heavy nucleus decays by emission of alpha particles to a new element which requires a smaller neutron excess for stability. Neutrons in this situation are converted into the more stable form of a proton and an electron (known as a beta (β) particle):

$$n \quad \rightarrow \quad e^- \quad + \quad p^+$$

neutron   beta   proton

particle

The beta particle is emitted, often at speeds up to 99 per cent of the speed of light, while the proton remains in the nucleus, improving the proton/neutron balance. It is quite common, therefore, for beta particles to be emitted by nuclei undergoing radioactive decay after they have emitted two or three alpha particles.

A further emission of energy in the form of short wavelength electromagnetic waves called gamma (γ) rays often accompanies alpha or beta particle emissions. The emission is due to the nucleus ending up in an excited or high-energy state as the result of the decay process. Gamma rays are similar to X-rays and, although they damage human tissue, they are useful for testing materials, since they can penetrate materials as dense as concrete. A common source is the isotope of cobalt containing 33 neutrons:

$$^{60}_{27}\text{Co} \rightarrow {}^{60}_{28}\text{Ni} + \text{beta particle} + \text{gamma ray}$$

In each case, the superscript refers to the nuclear mass and the subscript to the number of protons.

The most recent form of radioactivity to be discovered is neutron emission. Neutrons may be produced in a number of ways, for example, by bombardment of light nuclei with alpha particles:

$$\begin{array}{ccccc} ^{9}_{4}\text{Be} & + & ^{4}_{2}\text{He} & \rightarrow & ^{12}_{6}\text{C} & + & ^{1}_{0}\text{n} \\ \text{beryllium} & & \text{alpha} & & \text{carbon} & & \text{neutron} \\ & & \text{particle} & & & & \end{array}$$

Neutrons are particularly useful since, being uncharged, they can penetrate to the nuclei of atoms, unlike alpha or beta particles. In consequence, they are slowed down or *moderated* by nuclear collision. Hydrogen has a much greater moderating effect on neutrons than any other element, since it has similar nuclear mass to the neutron (the neutron merely rebounds off heavier nuclei).

Half-life

The emission of radioactive particles occurs on a statistical basis and, as one particular element decays into another, the emission of the former material will decrease, since less of it remains. The relationship is exponential:

$$\frac{dN}{dt} = -\lambda N$$

where N is the number of particles at time t and λ is a constant. The minus sign signifies that this is a decay process.

Separating variables and integrating:

$$\log_e N = -\lambda t + C \qquad (C = \text{constant})$$

If, when $t = 0$, $N = N_0$, this gives

$$N = N_0 e^{-\lambda t}$$

Table 5.2   One route by which uranium can decay into lead

New name	Old name	Atomic number	Mass	Particles emitted	Half-life
Uranium 238	Uranium 1	92	238	α	4.5×10^9 y
Thorium 234	Uranium X_1	90	234	β, γ	24.1 d
Protactinium 234	Uranium X_2	91	234	β, γ	1.17 m
Uranium 234	Uranium II	92	234	α	2×10^5 y
Thorium 230	Ionium	90	230	α, γ	8×10^4 y
Radium 226	Radium	88	226	α, γ	1622 y
Radon 222	Radon	86	222	α	3.82 d
Polomium 218	Radium A	84	218	α	3.05 m
Lead 214	Radium B	82	214	β, γ	27 m
Bismuth 214	Radium C	83	214	β, γ	20 m
Polonium 214	Radium C^1	84	214	α	1.6×10^{-4} s
Lead 210	Radium D	82	210	β, γ	22 y
Bismuth 210	Radium E	83	210	β	5 d
Polomium 210	Polomium	84	210	α, γ	138 d
Lead 206	Lead	82	206	Stable	–

y = years d = days m = minutes s = seconds

The time $t_{1/2}$ at which half the material $1/2N_0$ remains is called its *half-life*:

$$t_{1/2} = \frac{\log_e 2}{\lambda}$$

The term 'half-life' gives an idea of the useful life of a particular isotope. The emission of cobalt 60, for example, would decrease by 50 per cent over a period of about 5 years, its approximate half-life. Sources having half-lives in this region are often used – they correspond to a useful rate of emission combined with a reasonably long active life. Isotopes with very long half-lives correspondingly change at a very low rate and have small emission. Table 5.2. shows one means of decay of uranium into lead. Note that half-lives vary greatly.

APPLICATIONS OF RADIOACTIVITY IN THE CONSTRUCTION INDUSTRY

The two main areas of use involve the absorption of gamma rays by materials, according to their density, and the detection of hydrogen using fast neutrons. These techniques form a rapid and convenient way of *in situ* assessment of materials used for road bases.

Gamma radiography

Concrete up to a half a metre thick can be tested in this way. A small-diameter gamma source is placed in front of the structure to be photographed and an X-ray

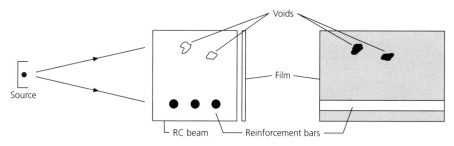

Figure 5.2 Gamma radiography of reinforced concrete. Higher density areas produce lower transmission and, therefore, lighter areas on the film

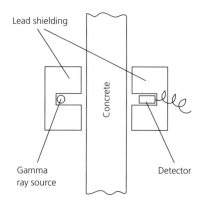

Figure 5.3 Transmission technique for measurement of the density of concrete using gamma rays

film enclosed in a flat cassette behind it (Figure 5.2). Lead-intensifying screens are normally used in front of and behind the film. These emit electrons when irradiated and they effectively make the film more sensitive. The gamma rays are not focused and, for maximum sharpness, the source should be as far as possible from the specimen, subject to obtaining satisfactory exposure times. The technique is particularly useful for examining grouting in prestressed concrete and for locating the position of steel.

Density measurement

This is a direct application of the effect of density of materials on their gamma ray absorption. The density may be obtained by a transmission technique or by measuring back-scattered radiation. In the transmission technique, a source of gamma rays is placed on one side of the structure and a detector on the other (Figure 5.3). The transmitted intensity decreases as density increases (Figure 5.4). In the back-scatter method, the source is placed on the surface with the detector adjacent to it but shielded from direct radiation by a lead screen (Figure 5.5). The detector then measures back-scattered radiation, the detector response depending

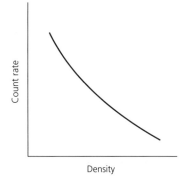

Figure 5.4   The relation between transmitted intensity of gamma rays and density

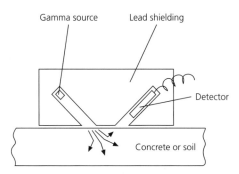

Figure 5.5   Back-scatter method for density measurement using gamma rays

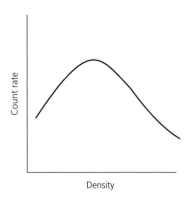

Figure 5.6   Relationship between back-scattered gamma radiation and density

on the instrument geometry. The variation of response with density is shown in Figure 5.6. Some instruments are designed so that the density of concrete occurs on the positive gradient portion and others so that it occurs on the negative gradient portion. The back-scatter method is particularly suitable for density measurement of structures with only one accessible surface, such as concrete slabs. Material between 50 and 100 mm from the surface contributes towards back-scattered radiation and corrections may be necessary if other materials lie within this region. There is also a tendency for different types of material to produce different density response characteristics. Greatest accuracy is obtained by calibration of each material used.

The detection of gamma rays used for density measurement is usually carried out using a geiger counter, which is based on the ionising effect of the rays. The radiation enters a partially evacuated glass tube containing electrodes at a high DC voltage. It ionises the gas and causes a pulse of current which may be detected by a loudspeaker, giving the familiar crackling noise, or, when quantitative measurements are required, measured using a counter.

Moisture content measurement using nuclear methods

The moderation of fast neutrons by hydrogen atoms provides a ready means for moisture content measurements in inorganic materials since, in these, moisture would be the only source of hydrogen. A generator is placed on the material and a detector which responds only to moderated or *thermal* neutrons is positioned near it. The response of the detector will then be dependent on the moisture content of the material. Rapid, *in situ* measurements can be made. For greatest accuracy, calibration graphs may be produced, using results obtained by conventional methods. *In situ* measurements can be of great advantage, for example, in soil compaction where optimum moisture content depends on the compaction method, so that a small hand-compacted sample may have a different optimum moisture content from the correct value corresponding to compaction by machine. A further application is the measurement of binder content in asphalt, the binder consisting of hydrocarbons and, therefore, behaving in a manner similar to water.

Thermal neutrons may be detected using boron trifluoride tubes in which alpha particles are produced on irradiation:

$$\begin{array}{ccccccc} {}^{10}_{5}\text{B} & + & {}^{1}_{0}\text{n} & \rightarrow & {}^{7}_{3}\text{Li} & + & {}^{4}_{2}\text{He} \\ \text{boron} & & \text{neutron} & & \text{lithium} & & \text{alpha particle} \end{array}$$

The alpha particles are detected as previously.

Instruments are available that contain both gamma ray and fast neutron sources so that density and moisture content measurements can be conveniently measured.

NATURAL RADIOACTIVITY

Small amounts of radioactivity are present both in the atmosphere and in the ground and these are generally harmless. An exception has been found, however, when dwellings are built over granite bedrock: granite contains small amounts of uranium 238 which eventually decays into lead as indicated in Table 5.2. The problem is caused by radon, a radioactive gas which seeps out of the rock, through the soil and into dwellings, sometimes assisted by pressure differentials, especially if extract fans lower the air pressure in a building. Radon, though radioactive, is chemically inert (see Table 5.4 below) and is not specially harmful to lung tissues in the quantities breathed.

Two of its 'daughters', polonium 218 and polonium 214, are however not inert; they form clusters or become attached to dust or smoke particles which can lodge in the lungs where they decay by alpha particle emission. This radiation is much more damaging than gamma rays on account of the large particle size and energy content.

There is a risk of contamination in all dwellings built over granite rock; radon gas can even seep through cracks in and around oversite concrete and concrete ground floor slabs. Radon gas levels can be reduced by incorporation of membranes, such as polythene, under dwellings in risk areas, effective ventilation of underfloor spaces and avoidance of negative pressures within dwellings. Parts of south-west England appear to be most affected by radon gas emissions but other areas containing granite rocks may also be at risk.

HAZARDS OF RADIOACTIVE RADIATION

Although the advantages of radioactive techniques are evident, radioactive materials are potentially extremely dangerous, causing burning and destruction of living tissues when excessive doses are received. They should be stored in lead containers (or in paraffin wax or polythene in the case of fast neutron sources); clearly labelled and handled with care. Personnel should keep out of direct line with the sources. When used regularly, a badge containing sensitive film should be worn by the operator and developed periodically. This will give the accumulated radiation received over that period and should be compared with permissible levels. Badges are, of course, no use in preventing severe exposure due to careless use of radioactive materials. All such sources are dangerous but fast neutrons are particularly harmful.

5.2 ELECTRON SHELLS – THE PERIODIC TABLE

In practical terms, the behaviour of electrons which orbit the nucleus is far more important than the behaviour of the nucleus itself, provided, of course, the nucleus is stable. The outer electrons, in particular, form the interfaces between atoms and determine almost every property of the material which the atoms build up to form. These properties include all mechanical properties such as strength and stiffness together with thermal properties, chemical stability and so on.

The electrons orbit the nucleus, and, rather like the subnuclear particles, obey special laws which do not apply to materials in bulk. For example, the movement of electrons is usually associated with the production of energy in the form of electromagnetic waves and yet this could not apply to electrons in orbit around the nucleus. Such an emission would cause the electrons to spiral downwards into the nucleus as they lose energy. We find that the energy of an electron orbiting the

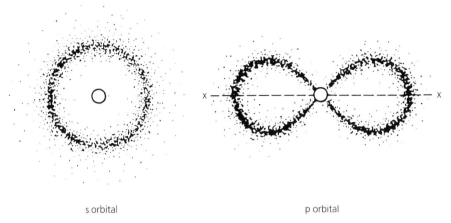

s orbital   p orbital

Figure 5.7   Charge clouds corresponding to s and p orbitals of an electron

atom is quantised, that is, it may take certain discrete values only, rather like the potential energy corresponding to the various steps on a staircase. The quantisation of the electron's energy gives rise to a number of orbits (called *orbitals*) around the nucleus, the orbitals of greater energy being further away from it. The electron shells are designated by the letters K, L, M, N, etc. in order of increasing energy. The numbers 1, 2, 3, 4, etc. are also often assigned to electrons in these orbitals. It has been found, in addition, that there are two other aspects of an electron's motion which can be identified:

1 Although circular orbitals (designated s orbitals) are possible, other shapes also occur, especially in higher energy orbitals and these more complicated orbital shapes form families. For example, figure of eight (p) orbitals are possible and there are three orbitals for any one energy content, corresponding to the three dimensions of space (Figure 5.7). In M and larger shells, d orbitals are possible, the shapes of these being complex with five different possibilities. In the N and larger shells, a still more complicated family of seven orbitals (f orbitals) is possible, and so on. (If it would appear, perhaps, unlikely that electrons could travel in these complex orbitals, it may be worthwhile to comment that the orbitals have, in fact, been calculated by considering the electron to be a wave rather than a particle. The wave–particle duality concept of matter is now well established and is essential to current explanations of atomic structure.)
2 The final aspect of the electron is *spin* – the electron can spin in the clockwise or anticlockwise direction in any one orbital.

Table 5.3 summarises the electon orbitals and shows the total number of different orbitals in the first four shells. The numbers of suborbitals are all in multiples of two on account of the two spin possibilities of the electron.

Before the placing of electrons in shells can be considered, it is necessary to note that the suborbitals described complicate the energy situation – the more complex

Table 5.3   Types of electron orbital

Shell	K	L	M	N
(number)	1	2	3	4
Shape permitted	s	s p	s p d	s p d f
Number of suborbitals	2	2 6	2 6 10	2 6 10 14
Total electrons per shell	2	8	18	32

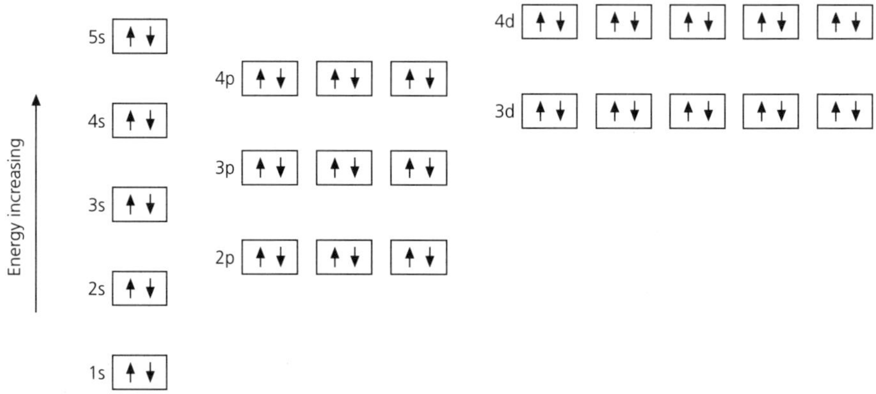

Figure 5.8   Orbital energies of electrons. Note that for any one shell, the energy increases in the order s, p, d, . . . The arrows represent the opposite spins of each pair of electrons

orbital shapes correspond to higher energy. These energies are represented in Figure 5.8 and it is evident that orbitals 3d and 4s; 4d and 5s, respectively, have similar energy. Orbitals 4f (not shown) would clearly have greater energy than 5s, hence, it would be erroneous to consider that the outermost shells automatically correspond to greater energies.

The build-up of the periodic table can now be considered and it is subject to several important conditions:

■ The number of electrons in a stable atom must be equal to the number of protons in the nucleus. The corresponding negative and positive charges, therefore, balance.
■ Only one electron is allowed in any one suborbital. This is known as Pauli's exclusion principle.
■ The lowest, most stable, orbitals fill first.

The initial formation of matter in the universe is likely to have involved the formation of stable nuclei, which would then gather electrons around them in orbitals satisfying the conditions stated. Eighty-two stable atom types were formed in this way (Table 5.4), the simplest involving one proton and one 1s electron – hydrogen. The first ten elements have the following configurations:

Table 5.4 Periodic table after Mendeleev. Elements below/to the left of the bold line are metallic when pure

IA	IIA	IIIB	IVB	VB	VIB	VIIB	VIIIB	VIIIB	VIIIB	IB	IIB	IIIA	IVA	VA	VIA	VIIA	0
1 **H** 1s																	2 **He** $(1s)^2$
3 **Li** 2s	4 **Be** $(2s)^2$											5 **B** 2p	6 **C** $(2p)^2$	7 **N** $(2p)^3$	8 **O** $(2p)^4$	9 **F** $(2p)^5$	10 **Ne** $(2p)^6$
11 **Na** 3s	12 **Mg** $(3s)^2$											13 **Al** 3p	14 **Si** $(3p)^2$	15 **P** $(3p)^3$	16 **S** $(3p)^4$	17 **Cl** $(3p)^5$	18 **Ar** $(3p)^6$
19 **K** 4s	20 **Ca** $(4s)^2$	21 **Sc** $(4s)^2 3d$	22 **Ti** $(4s)^2(3d)^2$	23 **V** $(4s)^2(3d)^3$	24 **Cr** $4s(3d)^5$	25 **Mn** $(4s)^2(3d)^5$	26 **Fe** $(4s)^2(3d)^6$	27 **Co** $(4s)^2(3d)^7$	28 **Ni** $(4s)^2(3d)^8$	29 **Cu** $(4s)(3d)^{10}$	30 **Zn** $(4s)^2(3d)^{10}$	31 **Ga** 4p	32 **Ge** $(4p)^2$	33 **As** $(4p)^3$	34 **Se** $(4p)^4$	35 **Br** $(4p)^5$	36 **Kr** $(4p)^6$
37 **Rb** 5s	38 **Sr** $(5s)^2$	39 **Y** $(5s)^2 4d$	40 **Zr** $(5s)^2(4d)^2$	41 **Nb** $5s(4d)^4$	42 **Mo** $5s(4d)^5$	43 **Tc** $5s(4d)^6$	44 **Ru** $5s(4d)^7$	45 **Rh** $5s(4d)^8$	46 **Pd** $(4d)^{10}$	47 **Ag** $5s(4d)^{10}$	48 **Cd** $5s^2(4d)^{10}$	49 **In** 5p	50 **Sn** $(5p)^2$	51 **Sb** $(5p)^3$	52 **Te** $(5p)^4$	53 **I** $(5p)^5$	54 **Xe** $(5p)^6$
55 **Cs** 6s	56 **Ba** $(6s)^2$	57 71 ●	72 **Hf** $(6s)^2(5d)^2$	73 **Ta** $(6s)^2(5d)^3$	74 **W** $(6s)^2(5d)^4$	75 **Re** $(6s)^2(5d)^5$	76 **Os** $(6s)^2(5d)^6$	77 **Ir** $(6s)^2(5d)^7$	78 **Pt** $6s(5d)^9$	79 **Au** $6s(5d)^{10}$	80 **Hg** $(6s)^2(5d)^{10}$	81 **Ti** 6p	82 **Pb** $(6p)^2$	83 **Bi** $(6p)^3$	84 **Po** $(6p)^4$	85 **At** $(6p)^5$	86 **Rn** $(6p)^6$
87	88 $(7s)^2$	89 103 ●															

Row labels (left margin): 1s, 2s, 3s, 4s, 5s, 6s, 7s
d-block row labels: 3d, 4d, 5d, 6d
p-block row labels: 2p, 3p, 4p, 5p, 6p

57 71 — Lanthanides (4f filling)
89 103 — Actinides (5f filling)

The atomic number is indicated above each element. Subshells above or to the left of each element are complete. Lead (atomic number 82) is the largest stable element

1 Hydrogen (H)   1s
2 Helium (He) $(1s)^2$ (i.e. two 1s electrons)   K shells complete – inert
3 Lithium (Li) $(1s)^2 2s$
4 Beryllium (Be) $(1s)^2 (2s)^2$
5 Boron (B) $(1s)^2 (2s)^2 2p$
6 Carbon (C) $(1s)^2 (2s)^2 (2p)^2$
7 Nitrogen (N) $(1s)^2 (2s)^2 (2p)^3$
8 Oxygen (O) $(1s)^2 (2s)^2 (2p)^4$
9 Fluorine (F) $(1s)^2 (2s)^2 (2p)^5$
10 Neon (Ne) $(1s)^2 (2s)^2 (2p)^6$ L shell complete – inert

The build-up of electrons in elements of atomic number 11 (sodium) to 18 (argon) continues as would be expected, argon having the configuration $(1s)^2(2s)^2(2p)^6(3s)^2(3p)^6$. However, as indicated in Figure 5.8 the 3d subshell contains orbitals of slightly greater energy than the 4s subshell, the latter, therefore, filling first, giving elements 19 and 20, potassium and calcium. The ten 3d orbitals fill subsequently, giving a familiar range of elements known as the transition elements. In element 36 (krypton), the 1s, 2s, 2p, 3s, 3p, 3d, 4s and 4p subshells are full, the 4d subshell filling after the 5s. Further complications occur as the f sections of the fourth and fifth shells fill, producing the lanthanides and actinides, respectively. The process could theoretically continue indefinitely, except that the nuclei of heavy elements are, as already explained, unstable.

It may be tempting at this stage to try to predict the properties of materials from the periodic table (Table 5.4). However, although some general rules regarding behaviour of the elements can be determined in this way, most practical properties of materials depend as much on the interaction of elements with one another as on their own basic properties, though the former follow from the latter. The most important basic property is the stability of each element formed. It is found that the most stable elements are those in which electron shells are complete. These elements are shown in Table 5.5.

It will be noticed that the last four correspond to completed s and p sections of shells rather than full shells but this is due to the delayed filling of d subshells explained above. Elements 4 (beryllium) and 12 (magnesium) are not inert, although they have complete s subsections, because there is still room for more electrons in those shells. The rule, therefore, is that, when the outermost shell

Table 5.5   Elements with complete electron shells

Atomic number	Element
2	Helium
10	Neon
18	Argon
36	Krypton
54	Xenon
86	Radon

can accept no more electrons, the stability of the corresponding element reaches a maximum, the next heavier element being of much reduced stability. The periodic form of Table 5.4 reflects the variation in atomic stability. The most stable (inert) elements are referred to as Group O. Groups IA to VIIA, in turn, indicate increasing stability. For instance, in a Group VIIA element, the electrons are held more powerfully to the nucleus than in the Group IA orbit of the same shell, since there are more protons in the nucleus of the former. The stability of elements in general decreases as they become larger, due to electron–electron repulsion – the same effect that produces relatively high energy states in the more complex d and f orbitals. The energy needed to ionise (or remove an electron from) argon is, for example, less than that needed to ionise helium, the most inert of the *inert* gases. A further effect in heavy elements is that their density increases, since the electron shells are attracted more strongly towards the larger nuclei. Heavy elements therefore exist normally as denser materials than light elements.

Those elements in which a d or f section is filling behave quite differently from the others, since these elements do not achieve a stable state by loss of such electrons or by completely filling that particular subshell. They are, therefore, given the B designation, the most important examples being elements 21–30. These elements are quite similar to one another in a number of respects, since they contain either one or two electrons in their outermost shell.

Excitation and ionisation

It is possible that, if energy is supplied to an atom, for example, by irradiation, electrons may be raised to higher orbitals. This occurs in a discharge tube when the excitation is by electron collisions. When an electron returns to the ground state from a certain excited state, there will be an emission of the appropriate wavelength of light, the wavelength being shorter for a bigger jump. In a sodium discharge tube, for example, there is a substantial amount of the yellow colour of wavelength 589 nm (1 nm = 10^{-9} m). This occurs when an electron jumps from the 3p level to the 3s level (that is, the ground state), hence, the yellow colour of sodium lights. In some cases excitation may take place to the extent that an electron completely leaves the atom, causing it to be ionised. Ionisation also plays an important part in the operation of discharge tubes.

5.3 BONDING OF ATOMS

Bonding is that process by which atoms become linked together, forming molecules.

As regards the atom, bonding occurs because it increases the stability (or reduces the energy) of atoms taking part. Many types of bond are based on the stable octet comprising two s and six p electrons, which is characteristic of the inert elements. The octet is obtained by the sharing or transfer of electrons. In other bond types the mechanism by which stability is increased is less straightforward.

From the point of view of the materials technologist, bonding is vitally important, since it is the only means by which cohesion, producing liquid and solid materials, can arise. In fact, it is true to say that, in any one lump of solid, there must be some link between virtually every atom involved. In some cases, this link may be in the form of mechanical interlock between molecules although, without atomic bonding to form large molecules of complex shape, such interlock would clearly be impossible. The properties of all solid materials are derived from the properties of these bonds, although other effects often modify their strength.

When bonding unites atoms only in small groups, then the molecule would probably exist in the form of a gas at normal temperatures, for example, oxygen (O_2) and nitrogen (N_2). Each molecule in these examples comprises two atoms. As molecular size increases, the material is more likely to exist in liquid and, finally, in the solid form, especially if cooled, since weak bonding tends to occur between virtually all atoms and molecules leading to the liquid and then the solid state if thermal energy is low enough. Large molecules or groups of molecules often exist in the form of patterns known as crystals, the type and properties of these depending on the nature of the bonds which are involved. To summarise:

- In gases, the molecules are largely unaffected by bonds between them, obeying the laws of motion which constitute the kinetic theory.
- In liquids there are bonding forces sufficient to produce cohesion but not rigidity.
- In solids, atoms become rigidly orientated in relation to one another. In many solids, there is some degree of crystallinity. The strength of bonding in a solid can often be ascertained from its melting point and in some cases by its mechanical strength. The strongest solids must be heated to very high temperatures before thermal energy can overcome the bonding effect.

Valency

This is a most important property of atoms in determining the nature of bonds formed and the elements with which they will combine. It will be recalled that, with the exception of the innermost (K) shell, the most stable atoms are those in which a stable octet occurs – a group comprising two s electrons and six p electrons. Those atoms which have a small number of electrons in their outer shell will have a tendency to lose them, leaving them with a positive charge. Hence, they are termed *electropositive*. Those in which the outer shell has a small electron deficiency will tend to gain them, so that they have a negative charge. They are termed *electronegative*. The number of electrons by which an atom exceeds or falls short of an electron octet is described as its *valency*. Clearly, no elements can have a simple valency greater than four, corresponding to half complete s/p sections, carbon and silicon being the most important examples. Group IA, IIA and IIIA elements have valencies of one, two and three respectively, as have elements in Groups VIIA, VIA and VA, respectively. The valencies of elements in which d sections are filling are much less well defined since, on losing the electrons in this section or gaining electrons to fill the section, inert elements are not obtained. Many of these elements may exhibit two valencies, for example, iron, Fe, number 26, may have valencies of two or three.

Oxidation and oxidation number

Those who have followed a modern chemistry course may be more familiar with the above terms than with the term 'valency'. Originally, the term *oxidation* referred to the addition of oxygen to substances to form new compounds, for example, changing of iron to iron oxide. The term has now been widened to include any reaction in which an element loses electrons. Conversely *reduction* is a reaction by which a substance gains electrons. Hence, many chemical reactions become oxidation/reduction (redox) reactions, for example, sodium and chlorine:

$$2Na + Cl_2 \rightarrow 2NaCl$$

The sodium has been *oxidised* by giving electrons to the chlorine which has, in turn, been reduced.

The term *oxidation number* is linked to this concept. It is simply the charge on the element or compound concerned in its particular state of chemical combination. In many cases, oxidation number and valency are numerically equal, although electropositive elements have a positive oxidation number and electronegative elements a negative oxidation number. For example, oxygen has valency 2 and oxidation number −2 (usually). The oxidation number of uncombined elements is 0, hydrogen 1 (usually) and chlorine −1 (usually). The oxidation number of iron may be 0 (metallic), 2 or 3. Oxidation numbers can be helpful in identifying the role of atoms in compounds (assuming they are ionic) and are now generally preferred to *valency*. For example, in calcium sulphate $CaSO_4$, we have:

	Ca	S	O_4
oxidation number	2		$4 \times (-2) = -8$

The sulphur, therefore, has an oxidation number of 6 to produce neutrality in the whole compound. Hence, the valency of sulphur, −2, due to two electrons in its outer shell, is not relevant in this case.

PRIMARY BONDS

These bonds satisfy the basic bonding needs of all atoms and some form of primary bond is present in every material.

Ionic bond

This occurs when, for example, an atom containing a single outer electron encounters another atom with a single electron missing in its outer shell. The electron in the former atom transfers to orbit the nucleus of the other atom. Hence, they have empty and complete shells, respectively. But the previously uncharged

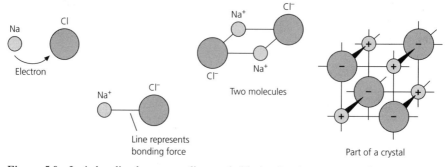

Figure 5.9   Ionic bonding between sodium and chlorine forming common salt

atoms will now be charged and, therefore, attract one another, coming together until this attraction is balanced by the repulsion of their respective electron clouds. If there are further atoms of each type available, the process will continue and a crystal will form. Common salt is an example (Figure 5.9). Sodium has a single outer electron and chlorine has seven outer electrons. Similar reactions will take place between divalent elements, for example, calcium and oxygen (CaO); between divalent and monovalent elements in appropriate proportions, for example, calcium chloride ($CaCl_2$), and so on.

It is important to appreciate that the bonding between the atoms results purely from their electric charges, hence, ionic bonding is non–directional and the crystal structure of the resulting compound is such as to minimise the distance between opposite charges and maximise the distance between like charges. The configuration of the ions depends on their size. There is a tendency for positive ions to be smaller than negative ions. For example, K^+ is smaller than Cl^-, although they both have the same number of electrons, because K^+ has two more protons in its nucleus and these attract the electrons more strongly. It is found that, if ions are of equal size, one positive ion can be surrounded by up to eight negative ions without the latter touching, giving an arrangement called the body-centred cubic crystal (see Figure 8.1 in chapter 8). More commonly, however, the negative ions are bigger and fewer can pack around the positive ion without touching, hence differing crystal structures are found. In sodium chloride, for instance, in Figure 5.9, each ion has six near neighbours of opposite sign. In some ionic crystals, ion packing and valency requirements can result in solid materials having densities which are low compared to normal solid densities.

The presence of ions in water often interferes with the force between the ions of crystals so that many ionic compounds dissolve in water, producing a dispersion of ions.

Covalent bonding

This results when certain atoms come together so that one or more electrons orbit both nuclei and each atom has a stable octet for a part of the time. In methane, for

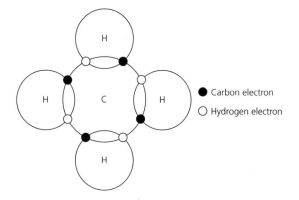

Figure 5.10 Covalent bonding between carbon and hydrogen forming methane, CH₄

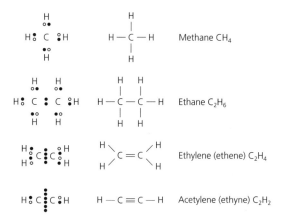

Figure 5.11 Various ways of representing the structure of some organic compounds (i.e., formed from carbon). Note that, in each compound, hydrogen atoms attain two electrons and carbon atoms attain eight electrons

example, four hydrogen atoms, each having one electron in its shell, approach a carbon atom having four electrons in its outer shell and each hydrogen electron encompasses both its own nucleus and the carbon nucleus. In return, the four electrons from the carbon atom each encompass a hydrogen atom. The arrangement is shown schematically in Figure 5.10. The carbon–hydrogen bond is, in fact, very versatile, forming the basis of many organic materials. Some important examples are shown in Figure 5.11.

Note that it is possible for two or three electrons to orbit both atoms, as in ethylene (ethene) C_2H_4 and acetylene (ethyne) C_2H_2, forming double and triple bonds, respectively. In both natural and synthetic organic materials, the carbon may form long-chain molecules containing many thousands of atoms. The covalent bond may therefore give rise to fibrous materials. In contrast, many materials comprise very small molecules bonded in this way, for example, oxygen (O_2), nitrogen (N_2) and carbon dioxide (CO_2).

Another common feature of the covalent bond is that, unlike the pure ionic bond, bond directions are well defined, because orbits of bonding electrons must fit in with those of non-bonding electrons. This leads to great rigidity of which, perhaps, the most notable case is diamond. The bonding in diamond is pure covalent, on account of the fact that only one element is involved and the s and p orbitals in carbon are *hybridised*, or averaged, to form four symmetrical bonds in the shape of a tetrahedron (Figure 5.12.) The water molecule also contains bonds having well

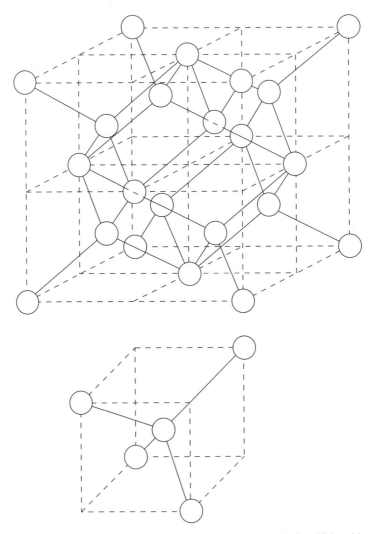

Figure 5.12　Build-up of diamond crystal from carbon atom tetrahedra. All bond lengths and bond angles are the same. Some of the many planes in diamond are evident in the crystal

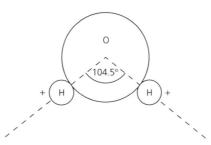

Figure 5.13 Well defined bond direction in the water molecule

defined directions (Figure 5.13.). Some salts cystallise, taking water molecules into their bond structure. The consequent expansion if this happens within a porous material may lead to extensive damage.

It would be misleading to suggest that all bonds are either ionic or covalent. Many bonds are a combination of the two. For example, in water, each hydrogen atom shares its electron with the oxygen atom and the oxygen shares one electron each way, in reverse. However, since the oxygen contains more protons than the hydrogen, the electrons tend to spend most time near to the more strongly charged nucleus, producing what is described as a semipolar bond.

In silica, one of the earth's commonest materials, the bonding is partially covalent, each silicon atom sharing an electron with each of four oxygen atoms. A tetrahedral shape is formed (see Figure 6.5 in chapter 6) with silicon at the centre and an oxygen atom at each apex. Each oxygen atom attains stability by sharing another electron with a further silicon atom. The repeated tetrahedra form a crystal and, since the oxygen atom at each apex of the tetrahedron is bonded to two silicon atoms, the chemical formula is SiO_2. The structure of other common materials may be more complex, for example, anhydrous calcium sulphate (the basis of gypsum plaster) consists of a sulphate ion, SO_4^{2-}, in which the sulphur (six outer electrons) shares two of its electrons with each of three oxygen atoms (each with six outer electrons), the two extra electrons which are necesary being borrowed ionically from the calcium which forms the Ca^{2+} ion, as in Figure 5.14. Note that the sharing between the sulphur and the oxygen is in one direction only since, with the two ionically borrowed electrons, the sulphur has eight electrons and has no need to borrow from the oxygen atoms. Each oxygen is, on the other hand, two electrons short and must, therefore, share with the sulphur. Such covalency is known as coordinate covalency. The one-way sharing often results in charged dipoles which,

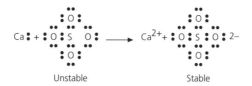

Figure 5.14 The formation of calcium sulphate from calcium, sulphur and oxygen

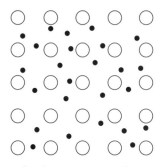

Figure 5.15   Simple two dimensional representation of a metal showing valence electrons forming a cloud around symmetrically positioned metal ions

as in ionic compounds, tend to cause crystals to grow. The arrangement for calcium sulphate given above does not, of course, consitute set gypsum plaster. On addition of water, the crystals reform to include water within their structure. Hence the plaster 'sets'. The formula for calcium carbonate ($CaCO_3$) is produced in a similar way to that of calcium sulphate. Various crystalline forms occur naturally and calcium carbonate crystals are also produced by the action of carbon dioxide on calcium hydroxide (slaked lime) in the presence of water. This is the action by which non-hydraulic building limes harden.

Metallic bond

When atoms of a single type and containing a few valence electrons approach one another, the electron orbits may change to a nondescript nature – they move around the nuclei rather like a gas and cause an overall attraction between the positively charged remainders of the atoms and the electrons themselves (Figure 5.15). Such bonding is clearly non-directional and the electrons therefore pack in tight patterns, producing crystals. An electric potential will cause an overall drift of the electron gas through the metal, constituting an electric current. Hence metals conduct electricity. The thermal conductivity of metals is also high, since electrons carry energy from one point to another by collision.

The great majority of elements form metals when they are present in the pure state; all those to the left of or below the bold lines of Table 5.4 being metallic when pure. This is not so much an indication of the stability of the metallic bond, but a result of the fact that ionic and covalent bonds are not easily formed by these elements in the pure state. Indeed, many of these quickly combine with oxygen in ordinary atmospheres, producing more stable covalent or ionic bonds. The occurrence of most of these elements as oxides or other compounds is an indication of their instability in metallic form. Tin (Sn), element number 50, can, in fact, change to a covalent form simply by cooling, hence, it reverts slowly to a grey powder below 18 °C. There is probably some tendency to form covalent bonds in most metals and, according to the extent of this, three different metallic crystal types may be formed. In all cases, however, the tight packing of atoms results in materials which are of much higher density in metallic form than their non-metallic counterparts. The metallic bond is not, in general, as strong as ionic and covalent bonds (actual bond strengths vary between about 10 per cent and 50 per cent of ionic/covalent values, according to the material). However, the density of packing of ions in metal crystals, combined with certain other properties considered later, result in metals being a most important bulk structural material.

SECONDARY BONDS

In the bonds considered so far, the atoms have combined to produce a more stable arrangement by interaction of their electrons. In some cases, the materials produced are crystalline and, therefore, solid. In others, a stable arrangement is produced by formation of small groups of atoms, for example, oxygen O_2 or ethylene C_2H_4. It is commonly known, however, that the latter materials may be liquefied or even solidified if the temperature is lowered sufficiently. This can be explained by the *van der Waals* bonds. In any one molecule, the electrons in orbit around it produce small instantaneous eccentricities of charge which induce similar charges in neighbouring molecules causing them to be attracted to it. This type of bond is operative between any adjacent molecules although it would, on average, have a strength of only about one per cent of ionic/covalent bonds and it is easily overcome by thermal energy. Nevertheless, it may produce a solid with a certain amount of strength, especially in the more *polar* molecules. The softer plastics are examples of molecular chains made solid in this way (Figure 5.16). Van der Waals bonds are also important in concrete and may be used to explain the phenomenon of creep. Colloidal properties of materials such as clay are similarly explained. Van der Waals bonds are responsible for the solid nature of graphite – they cause the attraction between covalent bonded carbon sheets. If the materials timber, bitumen, paints and adhesives are added to the list, the importance of the bond will be appreciated. It may be argued that the bonds, which will solidify almost all

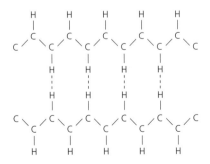

Figure 5.16 Polyethylene (polythene) chains linked together by van der Waals bonds (broken lines)

non-crystalline, microcrystalline or fibrous materials, manifest themselves in very different ways in the materials listed. The differences are, however, not as great as may at first be supposed. Essentially, there are two groups: those which absorb water, including clay, concrete and, perhaps, timber; and the remainder, which are impervious to water. When water is absorbed into porous materials, it tends to cause swelling by being attracted to solid surfaces, reducing the effectiveness of the van der Waals bonds. For this reason, there may be an initial increase in the strength of clay, concrete and timber if water is removed by heating. This is, however, followed by a decrease of strength at higher temperatures. Materials in the remaining group all lose strength immediately on heating.

A similar bond is the hydrogen bond in water. Each small positively charged hydrogen atom (Figure 5.13) can approach quite close to bonded oxygen atoms in other water molecules causing a weak bond which, at low temperatures, is sufficient to crystallise water into ice.

LIQUID STATE

In the gaseous state atoms or molecules are in a condition of disorder, being more or less free to move as projectiles in space. Their velocity is dependent on temperature and the pressure exerted is the consequence of collisions with the walls of the containing vessel. The space occupied by the molecules of a gas is much larger than the size of the molecules themselves.

In a solid, there are rigid links extending throughout any one piece of material, resulting from chemical bonds or interlocking of molecules. Solids are, therefore, ordered structures, amorphous solids exhibiting short range order and crystalline solids long range order.

Liquids are intermediate in nature. Cohesion results from the tendency for molecules to bond with one another. This causes molecules to come into contact with their neighbours but without forming rigid bonds. Hence, liquids are fluid and do not have strength, though their volume is comparable with that of the corresponding solid that would be formed if thermal energy were decreased by cooling. The cohesive effect of liquids results in surface tension, which tends to cause liquids to assume the minimum possible surface area – spherical drops, if suspended in space (Figure 5.17). In addition, molecular affinity for most neighbouring solid materials results in capillarity, which may cause powerful attraction of liquids towards solids (Figure 5.18).

There is, strictly, no clear division between liquids and solids since, in glass, there are no well defined bond directions or crystals, although, if glass is to be regarded as a liquid, it is certainly a very high viscosity liquid and can be treated, for practical purposes, as a solid. Bitumen and cellulose putties are also intermediate – they have short term strength but flow under sustained stress. These are examples of more complex groups of materials referred to as suspensions or dispersions, in which different types of material are intimately mixed.

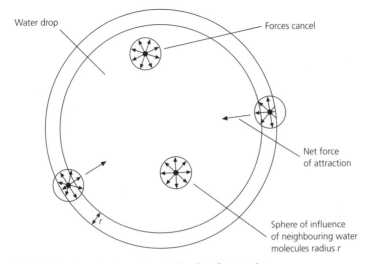

Figure 5.17   Molecular cohesion – the origin of surface tension

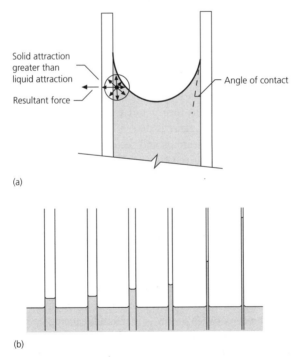

Figure 5.18   Capillarity: (a) Attraction of water to the solid surface; the greater the attraction, the smaller the angle of contact. (b) Effect of tube size on capillary rise

The term *suspension* generally refers to a solid dispersed in a liquid or gaseous medium. The particles are, however, relatively large (greater than 1 μm) and would separate with time. Alternatively, separation from the dispersion medium can be accelerated by centrifuging. When the dispersed particles become smaller – the order of 1 μm or less – they are known as *colloidal dispersions*. Examples are fog, smoke and clay, the dispersed phase and dispersion medium being, respectively, liquid/gas, solid/gas and solid/liquid. The properties of colloids depend very greatly on the particle size and shape, and on interactions between the dispersed phase and the dispersion medium. In true colloids, separation due to gravity does not occur. In fact, observation under the microscope shows that colloidal particles are in a state of continuous random motion, known as *Brownian motion*. Important examples of colloids in construction are clays and paints, and these will be considered in more detail under those headings.

5.4 ELASTIC DEFORMATION

A solid material may be defined as one which is able to support some stress without plastic deformation; that is, a degree of elasticity exists. This does not, of course, preclude the possibility of time dependent plastic movement known as *creep*, which occurs in a number of solid materials. *Elasticity* describes the ability of the material to return to its former shape on removal of the stress. It may be worthwhile to clarify, first, the molecular basis for elastic deformation and then to consider how, on overstressing materials, they deform further, causing eventual failure. It has been shown that, in most solids, there are bonds between virtually all atoms involved and that, in many cases, the atoms form regular arrays, as in crystals or fibres. The position of any atom in this array is determined by two influences: first, the bonding force holding it to other atoms and second, a force of repulsion which prevents atoms from approaching too closely. The situation between any two atoms is represented from an energy point of view in Figure 5.19(a). Note that although there is an energy release or decrease of energy due to the bonding force as atoms approach, there is an even greater increase of energy, due to atomic repulsion, if atoms approach too closely. Adding these energy curves gives rise to what is known as a *potential well*. The equilibrium position of the atoms is at the lowest point of this well (Figure 5.19(a)).

The elasticity of materials under stress is best visualised from the force/separation curve of Figure 5.19(b), which is obtained by differentiating (finding the gradient of) the net energy curve of Figure 5.19(a). On applying tension to the material, the distance between atoms is increased, causing an increase in the bonding force corresponding to an increase of the distance between atoms. Compression produces the reverse effect. There is, however, an important difference between tensile and compressive stresses. If a tensile force sufficient to separate the ions completely is applied (Figure 5.19(b)), the bond and hence the material will fail (tensile failure of this type is known as *cleavage*). There is no such direct effect in compression, since greater compression simply causes greater

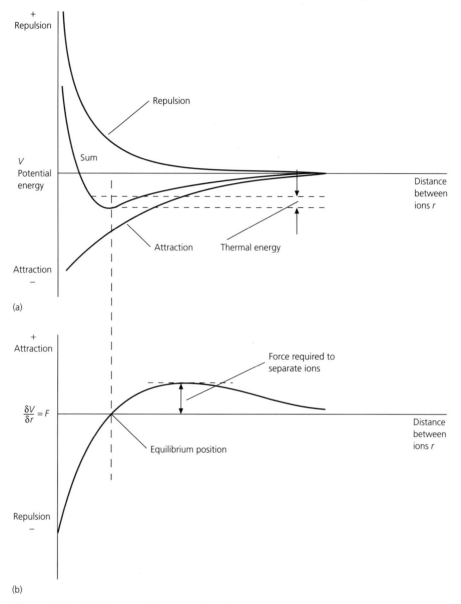

Figure 5.19 (a) The potential energy of two ions as a function of their separation. The negative energy curve represents the metallic bonding tendency. The positive curve represents short range repulsion. (b) The force–distance curve obtained by differentiating curve (a)

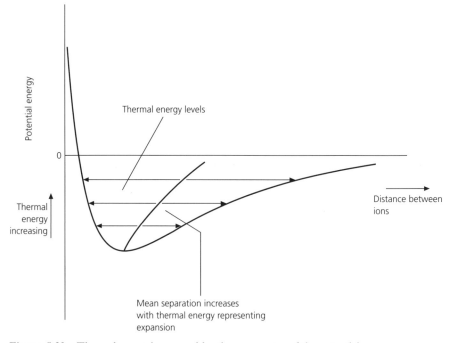

Figure 5.20   Thermal expansion caused by the asymmetry of the potential energy curve

repulsion of atoms. Hence pure compression cannot cause failure. The stiffness or resistance to deformation of materials is determined by the slope of the force/separation curve at the equilibrium position (Figure 5.19(b)). Although the relationship is non-linear at this point, practically observed strains correspond to extremely small displacements on the scale shown, so that the non-linearity is not apparent and observed stress/strain relationships are usually sensibly linear. The elastic moduli of most materials in tension and compression are also equal, since there is no appreciable change of slope on the tensile and compressive sides of the equilibrium position, provided the displacement is small.

The expansion of materials on heating is due to the fact that the sides of the well in Figure 5.19(a) are not symmetrical. If thermal energy is added to the minimum equilibrium value, the atom vibrates, rather like a spring, between positions corresponding to the intercepts of its new energy value on the total potential energy curve (Figure 5.20.) The new equilibrium position is mid-way between these points and atom separations increase, corresponding to expansion, as the material is heated. If heating continues, the positive thermal energy would eventually exceed the negative bond energy and melting would occur, as bonds become ineffective.

The modulus of elasticity of a material increases with the proportion of primary bonds in it; the approximate effect is shown in Figure 5.21.

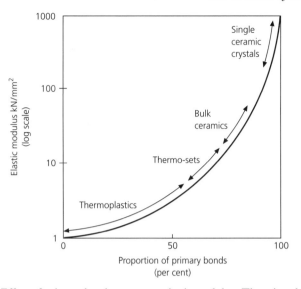

Figure 5.21 Effect of primary bond content on elastic modulus. There is a dramatic rate of reduction of E as the primary bond content decreases initially from 100 per cent. (Broad effects only are indicated here)

5.5 PLASTIC DEFORMATION

The application of load to structural components results in stresses which can be categorised into three groups – compression, tension and shear. In atomic terms, their effects are quite different, since they place quite different loadings on atomic bonds. The effects may be visualised by consideration of a simple crystal under stress. To begin with, a metallic crystal is considered (Figure 5.22).

Application of a compressive stress directly to the crystal merely reinforces the bonds and could not, on its own, cause failure. It should be remembered, however, that a uniaxial compressive stress can be resolved into shear stresses of maximum value at 45° to the direction of the former and that it also induces tensile stress by Poisson's ratio effects.

Application of a tensile stress acts directly against atomic bonds; though, to cause tensile failure, a large number of bonds would have to rupture at the same instant. Tensile stresses induce shear stress in the same manner as compressive stresses.

The effect of shear stresses in metals is different from tensile stresses, since plastic deformation can take place without total bond failure, ions merely swapping partners, as in Figure 5.23. Hence, the effect often occurs at lower stresses than would cause tensile failure, resulting in a distorted, though still intact, crystal. The ability for crystal planes to 'slip' gives metals the unique property of ductility. This important characteristic (together with the effects of imperfections in metal crystals which reduce their yield strength) is considered in more detail in Chapter 8.

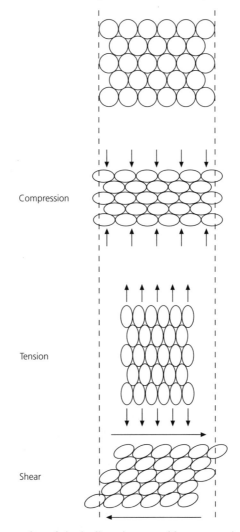

Figure 5.22   Representation of elastic distortion caused by compression, tension and shear in a metal

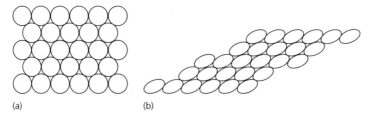

(a)   (b)

Figure 5.23   Metal crystals: (a) zero stress; (b) under shear which has caused both elastic and plastic deformation. Each atom has slipped by one spacing

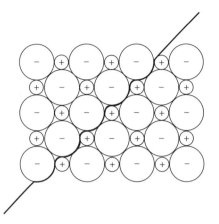

Figure 5.24 Ionic crystal showing one possible slip plane. Note the increased irregularity of this plane compared to a similar one in a metal crystal (Figure 5.22)

Consideration is now given to ceramic crystals, which are represented by the arrangement of Figure 5.24. The essential difference from metallic crystals is that adjacent ions are now oppositely charged and, generally, of different size. In pure tension or compression, the situation is not fundamentally different from that of metals but we find that, in shear, the ability of crystal planes to slide over one another is much reduced on account of the greater irregularity in the surfaces of possible slip planes and the more rigid bonding involved. Slip planes do exist in some ionic crystals and a possible plane is indicated in Figure 5.24 but ductility is never exhibited in practice in ceramics which contain ionic or covalent bonds, since they often contain amorphous (non-crystalline) areas and other imperfections which tend to cause fracture rather than plastic flow. These imperfections take the form of microscopic cracks or voids which amplify the stress at localised points and which, beyond a certain stress, propagate without plastic flow to cause rapid (*brittle*) failure.

5.6 IMPERFECTIONS AND THEIR EFFECT ON STRENGTH – GRIFFITH'S THEORY

A great many materials contain imperfections of some sort on a microscopic scale – these may originate from the formation process, as in many ceramics, or they may result from extensive working in metals. The effect of these imperfections, which may take the form of cracks or voids, is to concentrate an applied tensile stress at certain points in the material. The essential difference between ductile materials, such as pure metals and brittle materials, is that in the former, plastic flow is possible, at least in the early stages of deformation. Stress relief is therefore possible in the areas which, as a result of imperfections, are under the highest stress. The stresses in the various rivets of a riveted connection are equalised by this process, the most heavily stressed rivets yielding and, therefore, allowing others to carry a share of the load. In brittle materials, plastic flow is not possible and the concentrated stresses tend to increase the size of the imperfection or flaw, leading to failure of the material as a whole. In consequence, observed failure stresses in bulk materials are usually many times lower than values which would be suggested by bond strengths. The plastic flow considerations given earlier would tend to predict, for example, that shear or tensile failure should not occur until very substantial strains are undergone, perhaps of 10–20 per cent. In practice, tensile failure strains in bulk ceramics are more typically about 0.0001 (or 100×10^{-6}).

Griffith proposed that the likelihood of propagation of a crack in a given material depends on the nature of the crack and on certain properties of the material. Since solids as well as liquids have surface energy (or tension), the propagation of a crack requires energy to be applied to the solid in order to create new surfaces. However, as the crack propagates, stress relief occurs in a triangular region above and below the new part of the crack, such that stored elastic energy is released in the material. The energy released for a given additional length of crack is found to increase as the crack gets larger, since larger cracks produce a much greater area of stress relief, in much the same way as the area of a triangle of a given shape increases as the square on the length of its base. Griffith proposed that cracks would not propagate unless the length of the crack exceeds a critical length, so that the elastic energy released by its propagation is greater than the energy required to form new surfaces. If this length is exceeded, the crack could be referred to as *unstable* since, once propagation is initiated by stress or impact, the energy required to continue propagation will be provided by the elastic energy released. The critical crack length depends on the type of material: in ceramic materials, it is much smaller than in pure metals because, in the latter, additional energy is needed to cause the metal to flow plastically at the crack tip. Hence in metals cracks usually reach observable sizes well before brittle fracture occurs, while in materials such as glass, even microscopic surface imperfections may exceed the critical length. The critical crack length decreases as the applied stress increases, since greater applied stress results in greater stored elastic energy. Hence, materials such as normal bulk glass can only be used safely provided the applied stress is small enough for the cracks which are already present to be below the critical value for that stress. The basic Griffith criterion for crack propagation from energy considerations is that:

$$f = \sqrt{\frac{2E\gamma}{\pi L}}$$

f being the stress required to propagate a crack of length L in a material of modulus of elasticity E, having fracture surface energy γ per unit area (γ is larger where extra energy is absorbed by plastic flow).

An alternative approach is to consider the extent to which a crack concentrates the stress. It has been found that the curvature at the tip (root) of each crack has an important influence on the resultant stress (Figure 5.25.) The stress f_R at the root is related to the applied stress f_A by the equation:

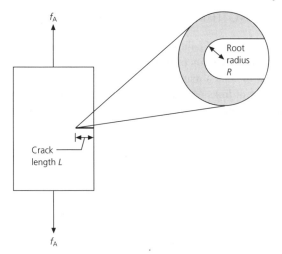

Figure 5.25 Griffith's model for estimation of stress concentration due to a small crack in a brittle material

$$f_R = 2f_A\sqrt{\frac{L}{R}}$$

where L is the crack length and R its root radius. If f_R is assumed to be the bond fracture stress and R is taken to be constant (R tends to have atomic dimensions in brittle materials – the order of 10^{-10} m), then:

$$f_A = \sqrt{\frac{k}{L}}$$

where k is a constant. Hence, this equation has a similar form to that obtained by energy considerations.

Crack widths vary in practice, obeying statistical laws, and the *weak link* theory must be applied to brittle materials, the largest crack in a particular sample of material being the cause of failure. It should be noted that, as soon as cracks begin to propagate under stress, the stress concentration effect becomes more pronounced. Failure can, therefore, be very rapid, the crack propagating at a velocity approaching that of sound in the material. The fracture strengths of brittle materials would also be expected to vary quite considerably, according to the size of the largest flaw in any one sample – see below.

One consolation of the effect of flaws in concentrating stress is that some brittle materials are easily cut to size by the introduction of continuous flaws in desired positions, such that a small stress propagates a crack in the flaw direction. This is the principle of glass cutting, though there is some skill in producing a continuous flaw with the cutter and, for reliable cutting, the glass surface itself should be free from additional flaws resulting from, for example, dirt or weathering.

STATISTICAL STRENGTH MODELS – THE WEIBULL MODEL

Weibull applied the theory of failure caused by cracks or flaws to bulk materials which would contain many flaws of different sizes.

A simplified view of such a material might be that it is likened to a chain comprising a number of links or elements, each with the same probability of survival P_{se}. If any one link fails, the whole chain (the bulk sample) will fail. Weibull proposed that P_{se} for any one element would be related to the stress applied f_A by the equation:

$$P_{se} = \exp\left[-\left(\frac{f_A}{f_0}\right)^m\right] \tag{5.1}$$

where f_0 and m are constants, that is:

$$P_{se} = e^{\left[-\left(\frac{f_A}{f_0}\right)^m\right]}$$

where $e = 2.718 \ldots$ is the base of natural logarithms.

f_0 represents the stress above which survival is likely (approximately the mean failure stress) while m represents the variability of the material. Figure 5.26 shows the effect of variation of m on P_{se}. For small m values, that is, high variability, there are quite large probabilities of failure at low stress and survival at higher stress. Many ceramic materials behave in this way (m is typically about 5). When m is

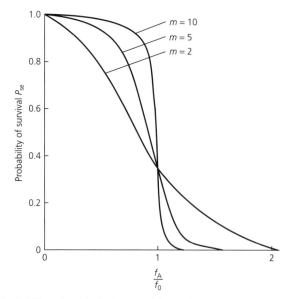

Figure 5.26   Probability of survival of a material as a function of applied stress f_A and variability factor m

high, these probabilities are each reduced – failure at or near f_0 is much more likely, for example, metals where failure stresses are much less variable (m is about 100). Note the similarity of this curve to the operating characteristic curve of Figure 4.5 which is based on similar principles.

If a sample of bulk material contains n elements, then each must survive if the material as a whole is to survive. Hence,

$$P_{sb} = (P_{se})^n \qquad (5.2)$$
(bulk material)

The number of elements in the material can be taken as proportional to its volume V, hence,

$$n = \frac{V}{V_e}$$

where V_e is the volume of one element.

Hence, from equation (5.2):

$$P_{sb} = (P_{se})^{V/V_e}$$

$$= \exp\left[\frac{V}{V_e} \log_e (P_{se})\right] \qquad (5.3)$$

(since $a^x = \exp(x \log_e a)$). Substituting equation (5.1):

$$P_{sb} = \exp\left[\frac{V}{V_e} \log \exp -\left(\frac{f_A}{f_0}\right)^m\right]$$

$$= \exp\left[-\frac{V}{V_e}\left(\frac{f_A}{f_0}\right)^m\right] \qquad (5.4)$$

The important point about this equation is that the volume of the bulk sample is seen to be significant. The negative sign means that increases of volume or applied stress each decrease the probability of survival of the bulk sample.

Although models of this sort cannot fully describe the behaviour of real materials, they do give an insight into patterns of behaviour. The volume effect on strength certainly exists – the failure stresses of concrete, timber and metals all tend to decrease as volume increases, though there may be effects other than the action of flaws involved, especially in metals. The statistical analysis of timber strength test results has traditionally been based on the Weibull model.

CHAPTER 6

SILICEOUS MATERIALS AND CERAMICS

Silica (SiO_2), an oxide of silicon, is the most common compound in the earth's crust and it is not surprising, therefore, that it forms the basis of many building materials. Most clays contain a high percentage of silica. Fired clay products are traditionally termed ceramics, although modern use of this term includes many other siliceous materials with similar properties, such as glass and mica, as well as some non-siliceous materials such as metallic oxides used in electrical insulators and glazes. Even most building limes contain some silica. The chapter includes a review of limes, together with applications.

6.1 STRUCTURE OF SILICATES – MINERALS

These are based on the SiO_4 group which, in itself, is not stable because, although the silicon (tetravalent) is satisfied by means of one electron from each oxygen atom, making a stable octet in the silicon atom, each oxygen atom has a deficiency of one electron. The situation may be represented as in Figure 6.1. The structure is, like silica, tetrahedral, that is, each oxygen atom is at the vertex of a pyramid, the silicon being at the centre (Figure 6.5(a) below). Hence, four electrons are required and there is a tendency for the silicate units to form ionic bonds with metals by borrowing electrons from them. A large number of metals may combine in many different ways, producing the wide range of minerals which exists in the earth's crust. The silicate unit itself may form seemingly unrelated types of molecule. For example, two silicate tetrahedra may be linked by having a base oxygen atom in common. Figure 6.2 shows the production of a chain in this way. In the repeat distance shown, there are two silicon atoms and six oxygen atoms (four of which require an extra electron). Hence, the formula will be $(Si_2O_6)^{4-}$ or, in its simplest form, $(SiO_3)^{2-}$. The two minus signs represent two ionically borrowed electrons, provided by metals such as magnesium or aluminium which become bonded into the chain. These chains may also be linked by metals to make the ceramic solid. Double chains are possible, as in Figure 6.3. There are four silicon atoms and eleven oxygen atoms in the repeat distance, six of which have one electron missing. Hence, the formula is $(Si_4O_{11})^{6-}$. These chains may be packed together in regular

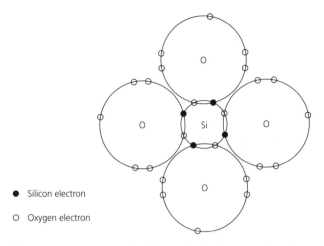

Figure 6.1 Electron arrangement in the silicate unit. Each oxygen atom has only seven electrons

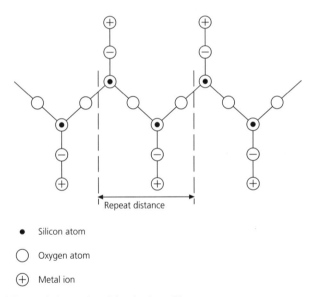

Figure 6.2 Silicate chain produced by sharing of base oxygen atoms

arrays by means of metal ions, when they produce crystalline forms which occur naturally as minerals. Alternatively, if the fibrous nature is retained, materials such as asbestos result. It is also possible for the silicate unit to form a sheet-like structure as in Figure 6.4. The various forms for silicates are represented by models in Figure 6.5.

Another material which is common in minerals is hydrated aluminium oxide (gibbsite), $Al(OH)_3$. In this compound, one Al^{3+} ion is ionically bonded to three $(OH)^-$ ions. A sheet structure is again formed, consisting of upper rows of

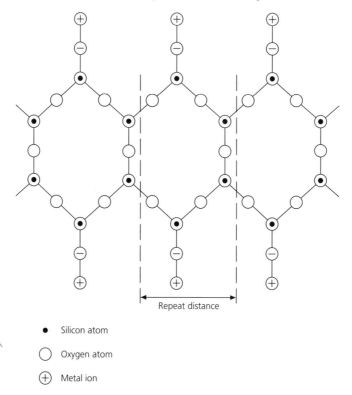

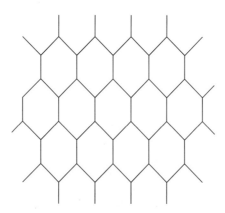

Figure 6.3   Formation of a double chain by sharing of base oxygen atoms. Hexagonal shaped patterns are formed. Note that apex oxygen atoms (those above silicon atoms) must be bonded to a metal ion

Figure 6.4   Sheet structure resulting from repetition of the hexagonal shaped units of Figure 6.3

(a)

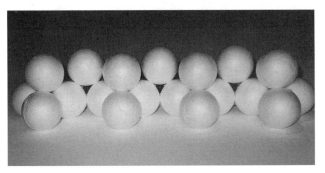

(b)

(c)

Figure 6.5 Various combinations of silicon and oxygen to form silicates. Dark spheres represent silicon; light spheres represent oxygen. 'Atoms' are not to scale. (a) Basic silicate unit. All four oxygen atoms require one electron. (b) Silicate chain. Apex and side oxygen atoms require one electron. (c) Hexagonal silicate unit. This could form a double chain by repeating the hexagons in one direction, or a sheet structure by repeating the hexagons in both horizontal directions

(a)

(b)

Figure 6.6   (a) Gibbsite sheet with upper layer of hydroxyl ions (dark) omitted to show aluminium ions (light) occupying two thirds of available sites. (b) A complete gibbsite sheet showing how upper hydroxyl ions fit between lower ones

$(OH)^-$ ions, then a layer of Al^{3+} ions (filling two-thirds of available gaps) and then a lower sheet of $(OH)^-$ ions, similar to the upper sheet but displaced, so that lower $(OH)^-$ ions fit between upper $(OH)^-$ ions (Figure 6.6). Many minerals are formed when sheets based on silicon and aluminium come together. Each apex oxygen atom of the former replaces one $(OH)^-$ ion of the latter, producing a stable formation. Kaolinite, formula $Al_2(OH)_4Si_2O_5$, is a simple example (Figure 6.7). Other minerals may form, for example, montmorillonite when the lower sheet of the gibbsite layer is attached to a silicate sheet in the same way as the upper layer, producing $Al_2(OH)_2Si_2O_5$. These sheets tend to pack together by means of van der Waals bonds and may form many layers in some cases. Normally, the sheets are very small, due to strains imposed by the arrangement of the base oxygen atoms. These strains can be relieved by means of further metal atoms and the sheets then become bigger, forming the mica group of materials.

Properties of the silicates

In all these materials, covalent and ionic bonds play an important part. Hence, there are no free electrons and the materials are good insulators of heat and electricity.

(a)

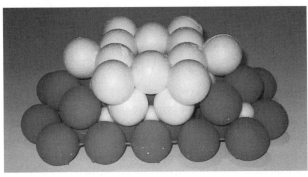

(b)

Figure 6.7   (a) Gibbsite sheet with an upper hexagon of hydroxyl ions removed.
(b) The apex oxygens of a silicate sheet (Figure 6.5(c)) fitting into the hexagonal space in (a),
producing kaolinite

Mechanically, properties depend on the particular structural forms. The crystalline
forms, for instance, in minerals, are strong and rigid but have planes of weakness
(cleavage planes), where bond densities are lower. The plate forms, if bonded by
van der Waals forces, tend to slip over one another easily, as in clay. They may also
tend to absorb moisture. The fibrous forms have tensile strength in the fibrous
direction but little strength in the other directions. The basic structures of all
silicates are very stable, so that they are unaffected by the atmosphere and many
acids. This, together with high melting points, makes them a natural choice as
building materials.

6.2   CLAY AND ITS PROPERTIES

Clay is normally described as a soil with particle size less than 5 μm, (cf. 150 μm,
the smallest BS sieve for fine aggregate for concrete). Hence, clays need not, by
definition, consist of minerals, although most clays contain a substantial proportion
of mineral materials. The formation of kaolinite from silica and hydrated aluminium

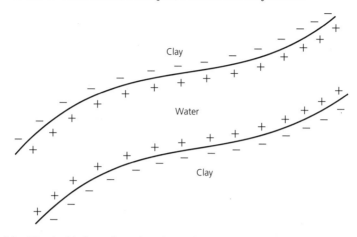

Figure 6.8   The double layer formed at the surface of clay particles as a result of their negative surface charge

oxide has already been described, this being, in fact, one of the commonest minerals in clay. There are a number of others and many of these differ from kaolinite only in the arrangement of the silicon and aluminium based platelets. Many properties of clays derive from the fact that they have a strong affinity for water, owing to a negative charge which exists on the surface of clay particles. Possible causes of this are:

- An aluminium ion (trivalent) occupying the place of a silica ion (tetravalent). The silicate sheet would then have one negative charge.
- A divalent ion, for example, magnesium, occupying the place of the aluminium ion in the alumina sheet.
- Broken bonds, as at the edge of a sheet.
- Adsorption of negative hydroxyl $(OH)^-$ ions to the clay surface.

The negative charge causes positive ions in the water to be attracted towards the clay surface, forming a double layer (Figure 6.8) and water is, therefore, bonded to the surface. The state which the clay assumes depends on the relative magnitudes of the repulsion of adjacent surfaces, due to the double charged layer and the attraction of surfaces by van der Waals forces. In most clays, the charged layer is insufficiently strong to prevent coagulation due to van der Waals forces and the clay *gels*, taking up solid properties. An exception is, however, montmorillonite, which is hydrophilic, having a strong affinity for water. It absorbs large quantities, which prevent van der Waals forces from causing coagulation of the particles. They can, therefore, remain in the liquid (*sol*) form for long periods of time, if sufficient water is available. Montmorillonite is the main mineral in bentonite, an altered volcanic ash, which, within a certain range of moisture contents, behaves as a *sol* on agitation, reverting to a soft *gel* on being allowed to stand. In gel form, the material has a high resistance to water flow and has, therefore, been used for the stabilisation of soils during excavation at levels below the water table. The bentonite may be injected by grouting, or excavation may take place through a slurry of bentonite which penetrates soil surfaces and then gels, forming an impermeable film.

It is well known that there is an optimum moisture content for compaction of clay. This is evident from the above, since, when large quantities of water are present, the material will behave almost as a liquid while, at low moisture contents, bonding increases the stiffness of the clay so that air voids cannot be removed. Furthermore, the fact that the water is adsorbed – held to the clay – makes it difficult to drain saturated clay.

SOIL STABILISATION TECHNIQUES

There are at least two methods of stabilising soil which depend on the principles given above. The first, *electro-osmosis*, is a direct consequence of the presence of the negative charges on the surface of clay particles. The gaps between particles in a saturated clay will be filled with water containing positively charged ions, as shown in Figure 6.8. If a voltage is applied to the clay by means of electrodes, the positive ions in the water will be drawn to the negative electrode (cathode), carrying moisture with them. This phenomenon gives a simple, if expensive, means of draining saturated clay or silty soils. A further effect connected with electro-osmosis is that colloidal particles move simultaneously in the opposite direction (toward the anode). This is known as *electrophoresis* and has been used for the injection of bentonite into soils. The bentonite then gels and stabilises the soil, forming a waterproof diaphragm. Alternatively, bentonite may be used in this way to waterproof walls and basements.

Another method utilises the fact that coagulation of sols can be obtained by the addition of a suitable salt in solution. This causes a contraction of the double layer of charge at the surface of particles, enabling them to approach more closely, so that van der Waals forces cause coagulation.

The injection of salts into clays is not practicable, due to their low permeability. However, coarser grained soils, such as sands, can be consolidated by injecting a concentrated solution of sodium silicate, which can then be made to form a gel by injection of a solution of a salt such as calcium chloride.

A further technique in soil stabilisation is using either quicklime or hydrated lime. For road construction the stabilisation of surface layers by lime addition is effective in both cost and environmental terms since it obviates the need to import more expensive crushed material. The lime may stabilise clay in one of the following ways:

■ Quicklime may dry clay, partly by the slaking process itself (in which water becomes chemically combined) and partly by the heat produced on slaking, to produce a material more amenable to mixing and compaction. Care should be taken not to over-dry the soil since this might prevent optimum compaction being achieved.

- Ion exchanges may greatly reduce the plasticity of the clay, reducing the sensitivity of compaction characteristics to moisture content. The soil therefore becomes less vulnerable to the effects of wet weather.
- Lime reacts with silicates in the soil in much the same way as pozzolanas operate. This produces a long term increase in strength as well helping the soil to resist frost heave and the effects of moisture.

As an example, hydrated lime, typically in a proportion of about 4 per cent, may be added by metred spreaders to surface layers and mixed in place using large rotary machines. An initial reaction occurs, causing stiffening and the material may be remixed several days later. This causes further breakdown and mixing of the lime and soil and produces a product of suitable stability for road subbases. Where higher levels of performance are required, pozzolanas such as pulverised fuel ash or ground granulated blast furnace slag can also be added.

6.3 EARTH CONSTRUCTION

Building with earth might be considered to be the greenest approach to construction, albeit for restricted applications, since both the origin and disposal site for the material may be immediately adjacent to the building. Prior to the industrial revolution earth was the most universally available bulk material for construction and its use was widespread. It was often mixed with some fibrous material such as straw to provide additional stability. In the nineteenth century there followed a rapid decline in its use in the developing world as bulk transport provided the means to move materials and fuels over large distances. The building method nevertheless continued in less developed countries or where suitable supplies of manufactured materials were unavailable. With changing perceptions and sustainability issues becoming more important, there are moves to re-introduce the material to developed countries on a much wider scale and much information has been supplied by those countries which might be considered to have been left behind by the industrial revolution, and which have continued to use earth as a construction material.

Earth buildings have the following attractions:

- The energy costs in production are extremely low.
- Earth products have high thermal mass, but better insulation properties than those of dense masonry, and provide temperature stability to the internal environment in both winter and summer conditions.
- Good fire resistance and noise insulation are achieved.
- Disposal at the end of service life is extremely simple.

The main problems relating to the material are limited structural performance and susceptibility to the effects of weathering or damp. The provision of a good roof with generous eaves projection is always advised. Where there is a possibility of wind driven rain, surfaces can be stabilised by application of calcium silicate solution or lime-based washes or coatings.

It is important that earth should be free of organic material since it is relatively unstable, traps moisture and provides a source of nutrition for living organisms that may move in. In some cases this may mean excavation of 1 metre or more of surface material.

The following are the chief forms of earth construction:

wattle and daub
cob
rammed earth construction (pise, from *pise de terre*, the French for rammed earth)
stabilised rammed earth construction
adobe (sun-dried mud bricks)

Wattle and daub

This form of construction is probably the most amenable to use with a minimum of skill and mechanised equipment and there are probably thousands of buildings still in existence in the UK produced by this technique. Typically, a simple timber shell for the building is produced and then panels of wattle are formed by weaving thin sapling twigs onto somewhat stouter sapling uprights. Weatherproofing is achieved by application of a mud or mud/straw mixture to the framework (Figure 6.9). For good results the walls should be clear from ground level. Protection can be achieved by application of lime wash or lime render to the panels. Provided adequate protection is achieved, very long service life is obtainable, though such structures clearly have limited load bearing capacity and are at risk from impact damage.

Figure 6.9 Wattle and daub construction (Source: Houben and Guillaud, *Earth Construction*)

Figure 6.10   Cob construction. Walls are built in lifts of about 1 metre. The walls in this photograph have been allowed to dry and then pared back to form vertical, flat surfaces after consolidation with a mallet

Cob

Cob differs from the above technique in that walls can be built with little or no timber frame. To be suitable, soil must be cohesive, containing appreciable contents of clay, but with some other grittier material to improve dimensional stability. It is mixed with straw to a consistency suitable for compaction by treading underfoot. It is built freehand into the approximate required shape, lift by lift, perhaps rising up to 1 m at a time (Figure 6.10). Timber frames for openings can be built in by attaching broad, wedge shaped fillets of timber to uprights, which key into the cob. Lintels (of width equal to the overall wall width) can be made of substantial timber sections and similar sections can be used to form bearing for floor joists. Ceiling heights must allow for 2–3 per cent shrinkage of the material on drying. When the earth has hardened, mainly by drying, the excess material is cut back to give the correct profile. Substantial thicknesses, typically of around 600 mm, are required for the technique, weather protection again being by a lime based coating of some form. The method is claimed to produce a building with good thermal insulation as well as high thermal inertia. Compressive strengths are slightly lower than those of lightweight insulation blocks – around 1 N/mm^2. Internal fixings can be quite easily made using plastic plugs, or for heavier fixings, 25 mm timber dowels. Constructions of this type must be built clear of the ground, should have good protection during construction and have a roof with good eaves projection. Experimentation is currently under way to investigate the possibility of re-introduction of the technique for domestic applications.

Rammed earth construction

In rammed earth construction the cohesive qualities of a suitably graded earth based material are utilised without any form of fibre or other reinforcement. It is one of the most promising techniques for current use. In many cases earth can be used 'as dug' from the ground to produce satisfactory wall structures with a minimum of framing. A fairly dry earth mix of suitable composition and moisture is well compacted to form walls in lifts of about 600 mm between firmly held timber forms. The forms are internally braced by struts, which are removed as work progresses. Tensile forces on formwork are carried by ties which are removed at the end of each lift.

The particle size range for earth can be broadly classified as in Table 6.1. All mineral materials have natural bonding capabilities as explained above but as they become finer the number of points of contact increase so that they become more cohesive. Hence gravels have virtually no cohesive ability while clays when dry can become quite hard. Unfortunately the finer materials also become much more susceptible to the effect of moisture content changes. The sand and silt components shown in Table 6.1 embrace a particle size range of about 300 times and are often subdivided into fine, medium or coarse. The gravel and sand can be considered to provide the bulk of the material. The silt acts as a void filler and the clay binds the whole together. In broad terms the proportions of the constituent materials should lie in the ranges indicated in Table 6.2.

Table 6.1 Particle size range of components of soil

Material	Particle size range (decreasing)
Gravel	6–2 mm
Sand	2–0.06 mm (60 μm)
Silt	60 μm–5 μm
Clay	Less than 5 μm

Table 6.2 Guide to particle size ranges required for rammed earth construction

Material	Percentage required
Sand and fine gravel	50–75
Silt	15–30
Clay	5–25

For analysis purposes the sand and silt can be separated from the soil by wet sieving through 2 mm and 60 μm sieves, respectively. The silt can be separated from the clay by shaking vigorously in a water filled measuring cylinder and siphoning off the water (which contains the clay) after allowing to settle for 4 minutes. This method is approximate but gives an indication as to whether the ratios of materials are in the range indicated above.

A critical factor in achieving a satisfactory result is the moisture content of the soil. It is normally impracticable to dry the soil in large quantities prior to use and this is not normally a problem, but water can be added to achieve the optimum moisture content. The technique is the same as that employed for normal soil compaction. The moisture content may be increased until the maximum density obtainable by a particular method of compaction is achieved, values of around 10 per cent being typical. At lower values than this the material has a high void content, while at higher levels the density of the material is reduced on account of water having a relatively low density. The soil mix and moisture content for any one application will be dependent upon the compaction technique and formwork to be used. For the strongest product, thick formwork boards are employed, suitably restrained with steel bracing, compaction being achieved by a combination of vibration and pressure. These techniques can lead to a product with remarkably good strength and hardness.

A further key requirement for rammed earth is that compressive strength be adequate. Figures in the region of 1.0 N/mm^2 are suggested for single storeys, higher figures being needed for two or more storey work. This may be compared with the compressive strength requirement for the higher insulation value aerated blocks of BS 6073 of 2.8 N/mm^2. Since the compressive strength depends upon the particle size grading within the clay fraction as well as the proportion of clay, various other rule of thumb tests have been used to ascertain suitability of the material prior to manufacture. One such test is the *roll test* which determines when the material has appropriate plasticity. Figure 6.11 shows the roll test. A sample of

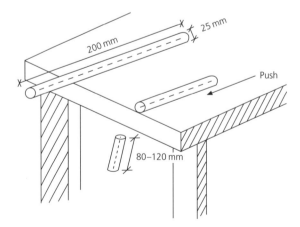

Figure 6.11   Roll test for determining the suitability of soil for rammed earth construction

Figure 6.12   Rammed earth construction used in conjunction with steel frame to produce a sports hall. This wall is 320 mm thick

appropriate moisture content is rolled into a cylinder of 25 mm diameter and length 200 mm and slowly pushed off the edge of a table. The unsupported end should break off between 80 mm and 120 mm from the end. If the break occurs before 80 mm the clay content is too low, while if the break occurs more than 120 mm from the end there is too much clay. Since it is not possible to adjust the particle size proportions within a given sample of soil the normal way of controlling clay content is to import a suitable sand and to blend this as required. Various other empirical tests such as the *drop test* (in which the fracture pattern of a dropped sample on hitting the ground is observed) can be used to obtain further evidence of suitability of the earth to be used in terms of both composition and moisture content.

Openings in such structures can be formed without lintels, though concave corners above openings are best avoided owing to the risk of cracking at such positions (Figure 6.12). If desired, the natural finish of the earth can be retained, though it is essential to afford some weather protection (for example by provision of a good roof projection) as well as protection from damp – for example by provision of a masonry plinth. For improved surface serviceability a lime based wash or render can be applied internally or externally. Surprisingly strong fixings can be made – for example a 100 mm nail provides a firm fixing provided it is driven into a well compacted material.

Use of stabilisers

These may have the following benefits:

- They may have increased compressive strength for improved load bearing capacity, including fixings to walls.
- They may have increased weathering and damp resistance.
- They may permit the use of otherwise unsuitable earth.
- They may permit the use of material without any external finish.

Conversely, use of stabilisers increases costs, and makes disposal of the material more difficult once service life is finished.

Various stabilisers can be used according to availability. In the UK cement is the most common but limes can also be used; non-hydraulic limes can be combined with pozzolanas such as brick dust to ensure a reasonably prompt set, or hydraulic limes can be used if available. Optimum moisture contents must be evaluated at the desired cement content in order to ensure maximum compaction. High cement contents are not required. A proportion of 4 per cent by weight of total materials is likely to double strength levels from around 2 N/mm^2 to 4 N/mm^2. Experimentation is currently underway to investigate the feasibility for more widespread use with the aid of mechanised techniques now available and rammed earth probably represents the most viable form of construction for use on a wider scale.

Adobe

This operates much along the same lines as rammed earth except that convenient size units (usually larger than standard clay bricks) are formed and allowed to dry in the sun. Clearly such techniques depend upon suitable drying conditions and are therefore more suited to hotter countries. However there may be value in using adobe bricks in conjunction with rammed earth to form certain details of structure such as gables (where compaction could be difficult) and openings or returns. Since the material is allowed to dry before use, the particle size requirements are less strict than for rammed earth though the same principles apply.

6.4   FIRED CLAY PRODUCTS

Many products used in construction are formed by heating naturally occurring minerals, such as clay or sand. Examples are clay bricks, blocks, tiles, pipes, terracotta, faience and sanitary ware. Materials with broadly similar properties or structure also exist naturally in the form of stones (many of which have been subject to heat at some stage in their history) and asbestos. These materials have a certain similarity in their properties and, hence, are considered as a group, but attention is first given to the effects of firing natural minerals.

The essential feature of most clays is that they contain a variety of intimately mixed minerals. The balance of the various types depends on the origin of the clay and widely differing properties result accordingly. On account of the predominantly very small particle sizes, substantial quantities of water are normally trapped in clays and this water may contain salts in solution. The firing process for clay products is heavily dependent on the diversity of mineral types, since clay products are invariably removed from their moulds before firing and are, therefore, required to retain their shape during the process. Heating causes certain minerals to fuse, while others remain solid, the overall effect being that bonding is extended throughout the whole unit without loss of shape. Total fusion could not be permitted unless the product were heated within its mould (and this would greatly increase manufacturing costs), while inadequate fusion would result in inferior strength and durability.

The first minerals to fuse are normally metallic oxides and these act as fluxes, absorbing at the same time some of the salts present. On cooling, a glassy matrix is formed, enclosing those particles which did not fuse. The process is known as *vitrification*, or, in some cases, *sintering*, according to the method used. The term 'glass' is given to any material in which extensive bonding occurs in such a way that crystals do not form. It may not be immediately obvious why, unlike most materials, the fused fractions of clay products do not crystallise on cooling. The reason is, again, based on the complexity of the fused portion, which itself contains a great variety of minerals. The bonds between them become directionally rigid on cooling but without orientation to form regular patterns. The structure can hence be regarded as a three dimensional irregular array of molecular chains. Even where the range of minerals is restricted, as in ordinary glass in which silica is the predominant mineral, crystallisation is normally prevented unless the material is *soaked* within a particular temperature range. This temperature usually corresponds to dull red heat in the region of 900 °C. Above that temperature, thermal energy is too great for crystallisation and below that temperature the liquid is too viscous to permit the organisation of atoms to form regular arrays. Hence, ordinary glass may be regarded as a super-cooled liquid, its transparency resulting directly from the absence of crystals, the boundaries of which would tend to scatter light. Most clay products are opaque because the degree of vitrification in them is very limited – it is only sufficient to bind the remaining solid material. Chemically, most ceramics are very stable, since the raw materials have been in existence over very long periods of time.

CLAY BRICKS AND OTHER PRODUCTS

Bricks may broadly be described as building units which are easily handled with one hand, though BS 3921 defines bricks as units not exceeding 337.5 mm in length, 225 mm in width and 112.5 mm in height. (Units which exceed any of these dimensions are referred to as building blocks.) By far the most widely used size at present is the standard metric brick of actual (work) size $215 \times 102.5 \times 65$ mm. Note that the length is twice the width plus one mortar joint (10 mm) while the thickness is one third of the length plus two mortar joints – the traditionally accepted brick proportions. The coordinating size (size including mortar joints) is, therefore, $225 \times 112.5 \times 75$ mm. It is these dimensional relationships which give rise to the large number of brickwork bonds which can be used. A new format of $200 \times 100 \times 75$ coordinating size has been selected (BS 6649) for those requiring a dimensionally coordinated brick. This will give a slightly stumpier appearance than the traditional format.

For the past 100 years, clay bricks have dominated the UK market as building units but more recently concrete blocks have provided strong competition, especially since thermal insulation regulations were tightened and lightweight concrete blocks became available. The development of coloured calcium silicate and concrete bricks has provided further competition though clay bricks continue to be the most widely used building unit, combining excellent durability (when they are selected and used correctly) with, in the case of facing bricks, lasting aesthetic properties.

Clay bricks are made by pressing a prepared clay sample into a mould, extracting the formed unit immediately and then heating it in order to sinter (partially vitrify) the clay.

Many different types of brick may be produced, depending on the nature of the clay used, the moulding process and the firing process. There are three basic subdivisions of type:

Common bricks   These are ordinary bricks which are not designed to provide good finish appearance or high strength. They are, therefore, in general, the cheapest bricks available.

Facing bricks   These are designed to give an attractive appearance, hence, they are free from imperfections such as cracks. Facing bricks may be derived from common bricks to which a sand facing and/or pigment has been applied prior to firing.

Engineering bricks   These are designed primarily for strength and durability. They are usually of high density and well fired.

Many clay bricks are designated by their place of manufacture, colour or surface texture. Examples are:

- Fletton – a common brick manufactured from Oxford clay, originally in Fletton, near Peterborough.
- Staffordshire blue – an engineering quality brick produced from clay which results in a characteristic blue colour.
- Dorking stock – the term 'stock', originally a piece of wood used in the moulding process, now denotes a brick characteristic of a certain region.

Indentations and perforations in bricks

Indentations (frogs) and perforations (cylindrical holes passing through the thickness of the brick) may be provided for one or more of the following reasons:

■ They assist in forming a strong bond between the brick and the remainder of the structure.

■ They reduce the effective thickness of the brick and hence reduce firing time.

■ They reduce the material cost and hence the overall cost of the brick without serious *in situ* strength loss.

For greatest strength, bricks with a single frog should be laid frog-up, since this ensures that the frog is filled with mortar. When perforations occupy less than 25 per cent of the total brick volume (as is usually the case), the brick can be regarded as 'solid' both from the point of view of strength and fire resistance properties.

Manufacture of clay bricks

There are four basic stages in brick manufacture, though many of the operations are interdependent; a particular brick will follow through these stages in a way designed specifically to suit the raw material used and the final product.

Clay preparation

After digging out (winning), the clay is prepared by crushing and/or grinding and mixing until it is of a uniform consistency. Water may be added to increase plasticity (a process known as *tempering*) and in some cases chemicals may be added for specific purposes, for example, barium carbonate which reacts with soluble salts producing an insoluble product, therefore reducing efflorescence in the final product.

Moulding

The moulding technique is designed to suit the moisture content of the clay, the following methods being described in order of increasing moisture content.

Semi-dry process   This process, which is used for the manufacture of Fletton bricks, utilises a moisture content in the region of 10 per cent. The ground and screened material has a dry granular consistency which is still evident in fractured surfaces of the fired brick. The clay is pressed into the mould in up to four stages. The faces of the brick may, after pressing, be textured or sandfaced.

Stiff plastic process   This utilises clays which are tempered to a moisture content of about 15 per cent. A stiff plastic consistency is obtained, the clay being extruded and then compacted into a mould under high pressure. Many engineering bricks are made this way, the clay for these containing a relatively large quantity of iron oxide which helps promote fusion during firing.

The wire-cut process   The clay is tempered to about 20 per cent moisture content and must be processed to form a homogeneous material. This is extruded to a size which allows for drying and firing shrinkage and units are cut to the correct thickness by tensioned wires. Perforated bricks are made in this way, the perforations being formed during the extrusion process. Wire-cut bricks are easily recognised by the perforations or the *drag marks*, caused by the dragging of the small clay particles under the wire.

Soft mud process   As the name suggests, this process utilises a clay, normally from shallow surface deposits, in a very soft condition, the moisture content being as high as 30 per cent. Breeze or town ash may be added to provide combustible material to assist firing or improve appearance. The clay is pressed into moulds which are sanded to prevent sticking. The green bricks are very soft and must be handled carefully prior to drying. Hand made bricks are produced by a similar process, except that *clots* of clay are thrown by hand into sanded moulds. This produces a characteristic surface texture which has great aesthetic appeal and is the reason for the continued production, on a limited scale, of hand made bricks.

Drying

This must be carried out prior to firing when bricks are made from clay of relatively high moisture content. Drying enables such bricks to be stacked higher in the kiln without lower bricks becoming distorted by the weight of bricks above them. Drying also enables the firing temperature to be increased more rapidly without problems such as bloating, which may result when gases or vapour are trapped within the brick. Drying is carried out in chambers, the temperature being increased and the relative humidity progressively decreased as bricks lose moisture. The process normally takes a number of days, higher moisture content bricks requiring a greater time. Wire-cut bricks and those produced by the soft mud process must be dried prior to firing.

Firing

The object of firing is to cause localised melting (sintering) of the clay, which increases strength and decreases the soluble salt content without loss of shape of the clay unit. The main constituents of the clay – silica and alumina – do not melt, since their melting points are very high; they are merely fused together by the lower melting point minerals such as metallic oxides and lime. The main stages of firing are:

100 °C	evaporation of free water
400 °C	burning of carbonaceous matter
900–1000 °C	sintering of clay

The latter stages of firing may be assisted by fuels whether present naturally in the clay or added during processing.

The control of the rate of increase of temperature and the maximum temperature is most important in order to produce bricks having satisfactory strength and quality. In particular, too-rapid firing will cause bloating and overburning of external layers, while too low a temperature seriously impairs strength and durability. Stronger bricks, such as engineering bricks, are normally fired at higher temperatures.

The following are the main processes.

Clamps Bricks are stacked in large special formations on a layer of breeze, though the bricks also contain some fuel. The breeze base is ignited and the fire spreads slowly through the stack, which contracts as the bricks shrink on firing. The process may take a week or more to complete and the fired product is variable, some underburnt and overburnt bricks being obtained. After firing, the bricks are sorted and marketed for various applications. Well fired bricks are extremely attractive, with local colour variations being caused by temperature differences, variations in oxygen access and points where fuel ignition occurred. Clamps are still used, though now on a limited scale, owing to the difficulty in controlling these kilns and the relatively high wastage involved.

Continuous kilns These are based on the Hoffman kiln and comprise a closed circuit of about 14 chambers arranged in two parallel rows with curved ends (Figure 6.13). Divisions between the chambers are made from strong paper sealed with clay and, by means of flues, the fire is directed to each chamber in turn. Drying is carried out prior to the main firing process and is achieved by warm air obtained from fired bricks during cooling. The kilns are described as continuous, since the fire is not extinguished – it is simply diverted from one kiln to the next, the cycle taking about one week. Coal was traditionally used but firing now may be by oil or gas. These kilns are very widely used for brick production.

Tunnel kilns These are the most recently introduced kilns and they can reduce firing time to a little over one day. Units are specially stacked onto large trolleys incorporating a heat resistant loading platform. The trolleys are then pushed

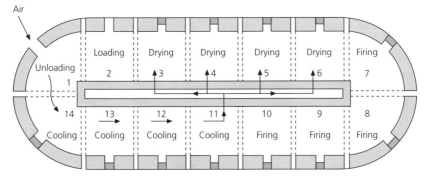

Figure 6.13 Continuous kiln showing sequence of zones and movement of air for drying purposes

end-to-end into a straight tunnel with a waist that fits the loading platform closely. The bricks pass successively through drying, firing and cooling zones, firing normally being by oil or gas. The process provides a high degree of control over temperature, so that the process is suited to the production of high strength, dimensionally accurate bricks. Perforated bricks are often fired in this way, the perforations allowing relatively rapid heating without undue risk of bloating.

PROPERTIES OF CLAY BRICKS AND THEIR MEASUREMENT

Strength

Since bricks are invariably used in compression, the standard method of test for strength involves crushing the bricks, the direction of loading being the same as that which is to be applied in practice – normally perpendicular to their largest (bed) face. Attention is drawn to the following aspects of the BS 3921 test.

1 The treatment of frogs or perforations should be designed to simulate their behaviour in practice, hence, the perforations in perforated bricks should not be filled prior to testing and frogs should only be filled if the bricks are to be used frog up. When frogs are filled, the strength of the mortar used should be in the range 28–42 N/mm^2 at the time of testing.
2 Strength varies with moisture content and, since the saturated state is the easiest state to reproduce, bricks should be saturated for 24 hours (or by boiling) before testing.
3 In order to obtain representative results, 10 bricks should be tested, these being carefully sampled either during unloading or from a stack.
4 Bricks are tested between 4 mm thick plywood sheets which reduce stress concentration resulting from irregularities in the brick surfaces.
5 The strength of the brick is equal to

$$\frac{\text{maximum load}}{\text{area of smaller bed face}}$$

(the two bed faces may not always be identical in size).
6 In earlier editions of BS 3921, bricks were designated as class 1, 2, 3, 4, 5, 7, 10 and 15, according to their average compressive strength. These classes, which may still be used, derive from formerly used Imperial units, the numbers being multiplied by 6.9 to give minimum crushing strength in N/mm^2. Hence, a class 10 brick has a minimum compressive strength of 69 N/mm^2.
7 There is normally good correlation between density and strength, though the precise relationship depends on the method of forming and firing the brick. Underfired bricks would have much reduced strength for a given density, as would bricks which are damaged by firing too rapidly.
8 The strength of brickwork is usually quite different from that of the brick, as measured according to BS 3921. It depends on the mortar used (though this would not normally need to be stronger than 10–20 per cent of brickwork

strength) and in particular on the shape of the masonry unit. A very common mode of failure in walls is by buckling, especially when they are tall and slender with little lateral restraint.

Water absorption

Water absorption may be an important property of clay bricks, since bricks having very low absorption are invariably of high durability (though the converse is not always true).

In the BS 3921 water absorption test, oven dried bricks are boiled for 5 hours in order to ensure maximum possible penetration of water (even in this test, there will normally be some voids which remain unfilled). Alternative methods for control purposes only are to soak the brick in water for 24 hours or use a vacuum method (no longer in BS 3921).

The water absorption is:

$$\frac{\text{mass of water absorbed}}{\text{mass of oven dried brick}} \times 100 \quad \text{(per cent)}$$

It should be noted that absorption does not give an indication of the porosity of bricks because:

- Absorption is a ratio of masses of materials having different densities. For example an absorption coefficient of 5 per cent would correspond to more than 10 per cent on a volume basis since the clay fraction is normally about twice as dense as water.
- Many smaller pores are inaccessible to water by normal soaking procedures.

Note that absorption measurement of clay bricks is now described in BS EN 772-7. The method is similar to the BS 3921 method.

Efflorescence

This is the name given to the build-up of white surface deposits on drying out. It results from dissolved salts in the brick and quite commonly spoils the appearance of new brickwork, especially if exposed to weather, as in parapets, or to the prevailing wind (Figure 6.14). The effect is most noticeable after periods of wet weather.

There is little that can normally be done about efflorescence. Brushing is ineffective on account of the surface texture of bricks, while wetting merely re-dissolves the salts so that they are re-absorbed into the brick. Weathering will normally remove the deposits over the first year or so. It is, of course, possible that the salts may originate from other sources such as the sand used for the mortar, or from soil contamination.

The 1974 edition of BS 3921 included a test for efflorescence in which bricks are saturated with distilled water in order to dissolve any salts present and then allowed to dry such that salts are carried to one exposed face. The test has now been

Figure 6.14   Unsightly efflorescence on a newly completed building. It weathered away over the first year

deleted because it has proved difficult to relate the outcome of the test to the actual risk of efflorescence of brickwork.

Expansion on wetting

Fired clay products of many types undergo a progressive irreversible expansion as moisture penetrates pores and is absorbed onto internal surfaces. Over a period of years, this expansion may amount to over 1000×10^{-6}, especially in more porous bricks, though the movement of the brickwork is normally only about 60 per cent of this unless, for some reason, the mortar is expansive. The expansion roughly follows a negative exponential law, the rate of expansion depending on the condition of exposure. The time taken for 50 per cent of the movement to occur is typically given as one week, though it may vary considerably, according to brick type. Perhaps, surprisingly, soaking bricks before laying is not effective in reducing expansion. There is no British Standard test for moisture expansion but general recommendations are that clay bricks should not be used fresh from the kiln; a period of at least one week should be allowed before laying, unless bricks are known to have very small expansion. The reversible moisture movement of bricks is small, being less than 200×10^{-6}. Movement joints should be provided at least every 12 m for clay brickwork and more frequently where openings, such as windows, might act as crack initiators.

Thermal expansion

The coefficient of thermal expansion of clay brickwork is approximately 7×10^{-6} per °C, considerably less than that of most other building materials, so that thermal movement is not normally a problem.

SOME WELL KNOWN CLAY BRICK TYPES

Fletton bricks

The pinkish coloured Fletton with its deep frog and 'kiss' marks caused by
stacking in the kiln is very widely used in the UK. These bricks originated at
Fletton near Peterborough and their cost is very low, partly on account of the
possibility of firing the semi-dry clay without pre-drying and because the clay
contains some carbonaceous material which significantly reduces the amount of fuel
required. Even including distribution costs, they are still very competitive over a
radius of more than 100 miles from their source. Fletton bricks have a number of
attractive properties:

- They are easily cut, even parallel to their bed face (the largest face). This makes
  them popular for small infill areas such as in window reveals, or to produce half
  courses of bricks.
- They have good suction properties for rendering. (Keyed varieties are,
  nevertheless, preferred.) Plain Flettons would normally be finished in this way.
- They are available in *sand faced* and *rustic* forms for use in face brickwork. (Note
  that the facing may eventually be worn away by severe weathering or abrasion.)

It is important to appreciate that Fletton bricks are not frost resistant and
should not, therefore, be used in highly exposed situations. They should always
be protected by a roof or coping when used externally.

Stock bricks

The original stock bricks were the yellow London stocks, manufactured in millions
for use around London and still sought after in the second hand market since
traditionally many of them were bedded in lime mortar, permitting relatively
easy recovery on demolition. The term 'stock' is now synonymous with a brick of
interesting colour and/or texture of which there are many varieties, especially in
southern Britain. The overall colour of a stock brick from a particular works is the
result of the clay mix and firing process used but much of the attraction of these
bricks is that no two are exactly the same, colour often varying greatly around the
average. For this reason, it is very important that stock bricks for the whole project
are ordered together, with bricks from each pack being mixed to produce random
colour differences. Failure to do this may detract from a first class appearance by
formation of bands of bricks of a given colour. When selecting a brick, reference
to colour charts is recommended; an enormous range is available, including reds,
yellows, browns, blues, etc. Many stock bricks are frost resistant and can, therefore,
be safely used for parapets or other exposed details.

Engineering bricks

These bricks have specific requirements relating to absorption and strength in
addition to those such as dimensional tolerances and durability requirements which

Table 6.3   Engineering brick requirements (BS 3921, 1985)

Class	Minimum compressive strength N/mm^2	Maximum water absorption per cent by mass
Engineering A	70	4.5
Engineering B	50	7.0

apply to ordinary quality bricks. Table 6.3 indicates these requirements for engineering class A and class B bricks. They are intended primarily for situations of high stress. It cannot in general be assumed that engineering quality bricks are suited to severe exposure.

HIGHER DURABILITY CLAY BRICKS

Durability ratings are now largely based upon salt content and behaviour in frost as follows.

Low salt content bricks

Contents by weight of the following soluble radicals should not exceed the percentages given:

Sulphate	0.5%
Magnesium	0.03%
Potassium	0.03%
Sodium	0.03%

The sulphate content is restricted in order to reduce the risk of sulphate attack in the cement mortar. The other radical contents are restricted to reduce crystallisation damage and efflorescence.

Frost resistance ratings

Manufacturers are required to classify bricks into the categories:

- frost resistant (F)
- moderately frost resistant (M) or
- not frost resistant (O)

This classification is normally on the basis of proven satisfactory performance of the bricks over a number of years, though other tests might be employed, especially for new products (see page 36).

Table 6.4 Durability designations (BS 3921, 1985)

Designation	Frost resistance	Soluble salt content
FL	Frost resistant (F)	Low (L)
FN	Frost resistant (F)	Normal (N)
ML	Moderately frost resistant (M)	Low (L)
MN	Moderately frost resistant (M)	Normal (N)
OL	Not frost resistant (O)	Low (N)
ON	Not frost resistant (O)	Normal (N)

Table 6.5 Exposure capabilities of clay bricks

Degree of exposure	Designation BS 3921 (1985)	Examples of use
Mild	OL, ON	Internal walls, partitions. Any walls protected from rising damp or rain. May need protection on site during winter
Normal	ML, MN	External walls in most buildings
Severe	FL, FN	Parapet walls, external walls of buildings subject to extreme exposure, paved areas, brickwork below DPC

DURABILITY OF CLAY BRICKWORK

The durability of clay brickwork is much more likely to be a problem than its strength, since in most situations, clay bricks are very much stronger than is required structurally.

Virtually all durability problems are associated with moisture penetration and it is therefore of paramount importance that bricks be suited to the degree of moisture likely to be found in any one position and that exposure to moisture in this position does not exceed the intended value. Table 6.4 shows typical designations for clay brick based upon soluble salt content and frost ratings while examples of three broad exposure categories to which the various designations are suited are given in Table 6.5.

As a general rule, the resistance of brickwork to frost will be greatest when all possible means are taken to prevent moisture penetration. In the O or M type bricks, the highest risk may be present during construction. It is important to keep bricks dry and to avoid high moisture contents in partially complete structures which do not have the protection of a roof, especially during cold weather.

A further problem, crystallisation damage, is associated with the crystallisation of salts beneath the brick surface. The extent of damage is directly related to the soluble salt content of bricks, hence, the best way of avoiding damage is to use type L bricks. Salts from other sources may also contaminate brickwork. Such sources include poor quality sand in the mortar, contaminated groundwater and wind borne spray from the sea. Nitrates or chlorides are more common than sulphates, though they all have largely the same effect. Some salts may even be released from the cement, though the effect of these should be short lived. One problem of salt crystallisation is that it tends to occur locally at boundaries between damp areas of brickwork, such as parapet walls and adjacent drier areas. Salts in solution are drawn to these drier areas, and deposits therefore build up, causing local damage. Many salts are hygroscopic (they absorb water), so that they may tend to perpetuate dampness where they occur. Hence, in order to avoid these problems:

- Use well fired bricks or, in situations where dampness is likely, higher performance bricks as described above.
- Use clean materials and avoid contact between brickwork and ground-water or soil.
- Design and detail brickwork in such a way as to minimise moisture penetration.

The mortar in brickwork may also be affected by sulphate attack. This occurs in damp situations, especially when the cement contains appreciable quantities of tricalcium aluminate, though the sulphates responsible often originate in the bricks. It results in expansion and eventual disruption of the brickwork. Avoidance of sulphate attack is achieved by the same precautions as given above to prevent crystallisation damage, though the use of sulphate-resisting cement in the mortar will reduce the severity of attack where these precautions cannot be fully met, such as in earth retaining walls.

A further situation often causing problems is when bricks are used for retaining walls. Figure 6.15 shows an example in which the bricks have performed satisfactorily but moisture has penetrated into cracks or voids in the wall from behind and has caused lime leaching. Calcium hydroxide in the mortar has been dissolved and brought to the surface where it carbonates, forming an unsightly insoluble coating which will not wash off. The effect also occurs in concrete if cracked and is one of the ways in which the presence of cracks comes to light. To avoid this happening a dampproof membrane should be used behind the brickwork to prevent moisture penetration. Dilute acids can be used to dissolve the material, though substantial deposits may be slow to dissolve and normal precautions are necessary.

OTHER FIRED CLAY PRODUCTS

Although the use of preformed plastic and concrete units has increased tremendously, many different types of fired clay products are still used and accepted as first rate materials. They have the advantage that they have been proved by use in buildings over many years. Although only clay roofing tiles and pipes are described here, the range of fired clay products is extensive, including, as well as the above, sanitary ware, glazed porcelain tiles, flooring tiles, terracotta (unglazed clay units used for decorative or architectural purposes, for example, Figure 6.16) and faience (glazed decorative units). The products to be described are similar in composition and manufacture to clay bricks, though it may be worthwhile to indicate the chief criteria on which clay tiles and pipes may be judged.

Figure 6.15 Lime leaching in a clay brick retaining wall caused by moisture penetration from behind

Figure 6.16 Novel use of terracotta (fired clay). The hollow fired units are reinforced with stainless steel

Figure 6.17   Delamination of machine made clay roofing tiles due to frost action. Deterioration on this north facing aspect was accelerated by moss growth which delays drying in fine weather

CLAY TILES FOR ROOFING

The first requirement is durability, since:

- Exposure of roofing tiles is much more severe than that of clay units in vertical surfaces.
- Tiles are of necessity much thinner than bricks.

The most severe conditions are those in which rain in winter months is quickly followed by frost. The temperature of horizontal surfaces can decrease very rapidly on clear nights, causing rapid freezing of absorbed water. Key properties in this respect are impermeability and frost resistance which are specified by BS EN 1304; the latter according to climatic area. It has been found, in the past, that machine made tiles often had a shorter life than hand made tiles, since they had a greater tendency to delaminate (Figure 6.17). Modern manufacturing methods may reduce this tendency and the use of hand made tiles is now decreasing, owing to the difficulty in obtaining labour at reasonable cost. It seems likely, however, in view of the aesthetic properties of hand made clay tiles, that these units, which have first class durability, will continue to be made available for high quality applications.

Clay pipes

The continued use of clay pipes, perhaps like tiles, reflects their proved reliability in use. Modern plastic jointing methods have reduced fixing costs and allow greater movement of pipes. Salt glazing has been replaced by ceramic glazing or by vitrifying pipes, since, during the salt glazing process, hydrochloric acid is emitted in gaseous form into the atmosphere, causing pollution. The British Standard for clay pipes (BS 65) does not require any type of glazing and this is due to the comparatively smooth abrasion resistant surface which is produced by modern manufacturing methods. Water absorption is of little consequence, except where aggressive fluids are encountered, since pipes are not normally subject to frost attack.

BS 65 gives dimensional tolerances and loads for pipes of *standard strength* and *extra strength*. A pressure test is specified, with two classes, corresponding to surface and underground pipes. The latter also require an impermeability test. Alkali- and acid-resistance are also specified.

SOME EXAMPLES OF THE TRADITIONAL USE OF CLAY PRODUCTS

On account of the remarkable durability of clay products there are many quite ancient buildings displaying notable and sometimes ingenious examples of the use of fired clay. Many of them require minimal maintenance though occasionally sympathetic repairs may be required. A few illustrations are given here to indicate possible problems of repair or maintenance.

Decorative brickwork

Figure 6.18 shows how great elegance can be achieved through simple functional components of buildings. In centuries past, chimneys were often made an architectural feature of buildings, sometimes being a dominant feature even in small scale structures. The example shown in Figure 6.18 involves the use of many 'specials' which could have been produced at low cost in the past but which would be very expensive to manufacture today. The chimneys remain in good condition in spite of a high exposure level, largely due to use of durable bricks with a good quality mortar, well pointed. The only maintenance required will be occasional inspection of the pointing and use of a suitable mortar for repointing if required. Use of excessively strong mortars must be avoided, a mix based upon lime being preferred.

Mathematical tiles

Mathematical tiles were introduce around the end of the eighteenth century following the introduction of a *brick tax* on new bricks in 1784. The shape of the tiles is such that, when nailed to battens in the same way as modern tile cladding would be carried out, the appearance of brickwork would be created. The gaps between the tiles would be pointed with a lime mortar. The tiles used for the

Figure 6.18   Use of 'specials' to create architecturally elegant chimneys from clay bricks

building shown in Figure 6.19 have been glazed to assist in resisting the sea air near to the south coast of England. The corners and reveals in the building are one indication that tiles, rather than bricks, form the walls. The tiles are subject to movement and mortar may be dislodged by vibration – for example, caused by traffic. When this occurs repointing should be carried out using a soft mortar. Mathematical tiles are still produced on a small scale and are considered a viable option for modern building claddings where a 'brick' finish is required without the weight and energy input associated with conventional brickwork.

Tuck pointing

This type of pointing was introduced in the nineteenth century to create the impression of quality brickwork when in reality low cost bricks were being employed (Figure 6.20). Flush joints were produced using a mortar coloured to

Figure 6.19 Use of mathematical tiles to create the impression of brickwork. The use of tiles is indicated by corners and reveal details

Figure 6.20 Use of tuck pointing to create quality appearance using low cost bricks

Figure 6.21   Use of local flints in modern domestic construction

match the bricks. Lines were then ruled into the joints and they were filled very carefully with a lime mortar which was then ruled to a uniform thin finish, typically only 5 mm wide. The brickwork is very pleasing in appearance but weathers away slowly in more exposed positions. It can be replaced using the same technique but this is expensive due to the very high labour costs now prevailing together with the very slow rate of progress (1 to 2 m² per day).

Flint construction

In many parts of the country local materials may be used as a decorative feature of brickwork. In Figure 6.21 beach flints have been incorporated into clay brickwork to good effect. The brickwork nevertheless forms the frame of the structure since stones of small size do not form good structural bonds. The area of flint panels should also be limited for the same reason and they need to be built against a solid backing. Flints are almost indestructible and so might be considered to be an environmentally attractive material.

6.5   OTHER BRICK/BLOCK TYPES

CALCIUM SILICATE (SAND-LIME) BRICKS

These are made by blending together finely ground sand or flint and lime in the approximate ratio 10 : 1. The semi-dry mixture is compacted into moulds and then

Table 6.6 Brick equivalents

Clay brick designation	Calcium silicate brick strength class
OL, ON	1
ML, MN	2–3
FL, FN	4 or greater

autoclaved, at about 170 °C and 10 atmospheres pressure for several hours. A surface reaction occurs between the sand and lime, producing calcium silicate hydrates which 'glue' the sand particles into a solid mass. The main properties of calcium silicate bricks are:

- A high degree of regularity, with smooth surface texture and sharp arrises.
- Very low soluble salt content, so that efflorescence is not a problem.
- Fairly high moisture movement.
- Compressive strength in the range 7–50 N/mm^2 (strength classes are employed by BS 187 as for clay bricks).
- Good overall durability in clean atmospheres, though they may deteriorate slowly in polluted sulphur-containing atmospheres.

Selection should, as with clay bricks, be according to purpose but, unlike clay bricks, strength is the best guide to durability. The equivalents shown in Table 6.6 can be applied for most purposes.

Additionally, on account of their increased moisture movement, joints in calcium silicate brickwork are recommended every 7 m or so and they should not be laid wet. Since movement characteristics are different, they should not be bonded directly to clay brickwork. The use of calcium silicate bricks has increased, due at least in part to the fact that they are now obtainable in a great variety of colours, such as pastel shades of red and blue, as well as the original off-white colour. These produce very different aesthetic effects from fired clay products with their more traditional, natural colours.

CONCRETE BRICKS

These are now manufactured at a number of geographic locations, often having natural colours obtained from pigments which, together with carefully textured surfaces, can be quite difficult to distinguish from clay bricks. Possible advantages of concrete bricks include:

- freedom from efflorescence
- no expansion on exposure (a possible disadvantage of clay bricks)
- generally high strength and durability

Aspects where the future is less certain include the effect of long term weathering on appearance and the availability of matching bricks for repairs or alterations.

CONCRETE BLOCKS

Concrete blockwork has developed into a major form of construction for the following reasons:

- Rates of production during construction are substantially greater with blocks than with brick size units, for example, a 100 mm thick block of size 440 × 215 mm is equivalent to approximately six standard bricks.
- A great variety of sizes and types is available to suit purposes ranging from structural use to lightweight partitions.
- Modern factory production methods ensure consistent and reliable performance.
- High quality surface finishes are obtainable, obviating the need for rendering or plaster coatings in many cases.

Concrete blocks are covered by BS 6073 which also covers concrete bricks. They may be solid, hollow or cellular.

Solid blocks are largely voidless but may have grooves or holes to reduce weight or facilitate handling. These must not exceed 25 per cent of the gross volume of the block.

Hollow blocks have voids passing right through. They can be *shell-bedded*, that is, the mortar is laid in two strips adjacent to each face, so that there is no continuous capillary path for moisture through the bed. Load bearing capacity is, of course, reduced when blocks are laid in this way. The strength of hollow blockwork can be increased by filling the cavities with concrete, especially if reinforcement is included. Sound insulation is also improved in this way.

Cellular blocks are a special type of hollow block in which the cavities are closed at one end. The solid edge would normally be laid upwards and, in the case of thin blocks, this makes it easier to produce an effective bed joint.

A great variety of aggregate types is used in the production of concrete blocks; these include natural aggregates, air cooled blast furnace slag, bottom ash from boilers, and milled softwood chips. The density and strength of the resultant product vary accordingly and they are also influenced by the manufacturing technique.

Denser types of block can be used for load bearing purposes and decorative or textured facings may be applied.

Lightweight types are now in particular demand, due to the need for high thermal insulation walling. Many types include pulverised fuel ash and may be aerated and/or autoclaved (see page 260). BS 6073 does not lay down strength classes but gives minimum strengths for all block types. These are:

- Thickness not less than 75 mm – average strength of 10 blocks not less than 2.8 N/mm^2. No individual block less than 80 per cent of this value.
- Thickness less than 75 mm – average transverse strength of five blocks not less than 0.65 N/mm^2.

BS 6073 also stipulates drying shrinkage: this should be not more than 0.06 per cent (except autoclaved blocks), and not more than 0.09 per cent (autoclaved aerated blocks).

It is important, on account of their high shrinkage, that autoclaved aerated blocks are not saturated prior to laying, otherwise shrinkage cracking may result. Handleability is also much impaired when wet due to their greatly increased weight in this condition.

No guidance regarding durability of blocks is given in BS 6073. The principles as given in Chapter 7 on concrete apply, though it may be added that some open-textured or autoclaved blocks are often more resistant to frost than their strength would suggest. This may be due to the type of void present and the fact that saturation is rare. Nevertheless, where severe exposure or pollution is likely, blocks of average compressive strength not less than 7 N/mm^2 should be specified. Other types can, of course, be protected by rendering.

6.6 NATURAL STONE FOR BUILDING

In the 1950s and 1960s, there was a marked decline in the use of natural building stones due partly to the great variety of new manufactured materials being produced and partly to greatly increased labour costs in this craft orientated industry. There are now indications that this trend is slowing down and the aesthetic appeal of building stones with their unique colouring and texture properties, combined with more efficient extraction techniques and greater understanding of behaviour, should lead to current levels of usage at least being maintained. Reconstructed stones continue to be widely used, especially for flooring and the introduction of polymers has improved durability of such stones for external applications, though attention here is concerned primarily with the main groups of solid stone, their properties and applications.

The essential feature of building stones is that they are simply extracted, cut to size and used so that natural properties (and variations) within the material must be accommodated during use. The energy input into such materials is very low but there are a number of constraints relating to use if they are to perform satisfactorily.

CLASSIFICATION

It is convenient to classify stones according to their composition, method of formation and age, since these are closely related and have an important bearing on behaviour and geographic location. The geological time scale, together with the main rock groups, are shown in Table 6.7. The earliest rocks are igneous (*ignis* meaning fire in Latin), formed from *magma*, the fluid at the centre of the earth which produced a solid crust as the earth cooled. Igneous rocks have also been formed more recently by volcanic action, as a result of rapidly cooled lava. Such rocks are glassy (non-crystalline) and are described as *extrusive igneous* – they formed by extrusion through the earth's crust, hence the rapid cooling; for example, pumice. Many igneous rocks are formed by cooling below a previously formed crust. They are *intrusive* or *plutonic* and, because they cool very slowly, they have

Table 6.7   Main geological age divisions with examples of stone types

Geological age	Millions of years	Stone type (examples)
Tertiary		
Cretaceous	100	Chalk
Jurassic		Oolitic limestone
Triassic	200	
Permian		Millstone grit
Carboniferous	300	Carboniferous limestone
Devonian	400	Devonian and old red sandstone
Silurian		
Ordovician		
Cambrian	500	
	600	Igneous
Precambrian		

a coarse crystalline structure, of which granites are a good example. Granite is described as an *acid igneous rock* because it is mainly composed of silica. Those with a minority of silica (for example, basalt) are described as *basic igneous.*

Weathering of some igneous rocks over millions of years caused them to break down into small fragments in the form of sand or mud. When such fragments or others such as shells of living organisms become bonded, sedimentary rocks are formed. Those formed from igneous rocks are described as sandstones, and those formed from shells (calcareous materials) are known as limestones. Sedimentary rocks vary greatly in age and form the bulk of rock found in the UK with the exception of northern Scotland.

Where rocks or other material are modified by heat and/or pressure, the product is described as metamorphic. Some metamorphic rocks are very old, formed from igneous rocks, but the most important ones in building are more recently formed, notably marble, formed from calcareous rocks, and slate, formed from clay containing flaky minerals such as mica.

Each stone type has its advantages and disadvantages and all can be employed with great aesthetic effect in building, provided their behaviour is understood and accommodated in design and construction.

Igneous rocks

By far the most commonly used igneous rocks in building are granites which contain a predominance of silica in the form of quartz, feldspar and mica. Some stones containing larger quantities of feldspar (and, therefore, not strictly granites) are imported, especially from Scandinavia. The feldspar and other minerals often results in very attractive blue, green, black or white light-reflecting inclusions, making them popular for use in shop fronts and entrance foyers.

The essential characteristics of granites and related rocks are high strength, intense hardness and low porosity, with a medium to coarse grain structure. These crystals often result in very attractive colouring on account of the range of minerals they contain. Sources in the UK are mainly restricted to the west and north; notably Cornwall (mainly grey), Peterhead (mainly pink) and Aberdeen (mainly grey). Individual quarries in each of these areas may produce stones of quite different colour and texture.

On account of their high strength and the fact that rocks do not have bedding planes, granites are difficult and expensive to produce. Blasting is essential for the initial extraction process and cutting is carried out by large diamond tipped saws. Use as a building stone is now, therefore, generally restricted to thin slabs as facing materials. They are eminently suitable for this purpose, thicknesses as low as 25 mm being feasible, and finishes ranging from a coarse texture (in thicker sections) to highly polished (Figure 6.22). On account of their low absorption, granites are largely self-cleaning. The absence of natural planes of weakness leads to unrivalled

Figure 6.22 Use of polished granite for cladding purposes. The curved sections are very expensive; they cannot be cut as thin sections

durability, even in polluted atmospheres, though granites containing orientated mica should be cut across these planes to avoid localised surface deterioration by flaking.

Sandstones

Sandstones are predominantly quartz rock fragments, often originating from weathered granite, and of size up to about 2 mm cemented together with natural iron, calcium or silica compounds. The properties of sandstones depend on the coarseness of the grains and the type of cementing material, while their colour depends on the content of minerals such as iron, which produces the sandy or brown colour of many sandstones. They are laid down in well defined strata of differing composition, the planes of weakness between them producing bedding planes in the final rock. These planes greatly facilitate extraction but limit the thickness of stone which can be obtained.

There are more sandstones currently quarried than any other type, most sources being in the north Midlands or northern England, with a few scattered widely elsewhere. Durability depends more on cement type than porosity or geological origin. Calcareous sandstones are susceptible to pollution by dissolution of the calcium carbonate cement and are not, therefore, suited to use near industrial areas. Porosities vary greatly in the range 1–25 per cent and, though strengths vary accordingly, this is not normally a problem in modern applications where stress levels are generally low. The relative coarseness of grain of sandstones relative to limestones is reflected by generally lower saturation coefficients – mostly in the range 0.5–0.7. Frost resistance of such stones is rarely a problem. It is, nevertheless, generally advisable, if using stones of different types, as in repair work, to match porosity to avoid less porous stones shedding water to give higher levels in the more porous stones.

A further consequence of coarse grain in sandstones is that they cannot generally be polished. Polluted atmospheres may cause damage by crystallisation beneath the surface as well as detracting from its aesthetic value by causing progressively darker colouring as grime penetrates the pores. Many sandstones turned almost black in the 1950s, though these have responded well to cleaning and are likely to remain relatively clean in this age of greater environmental awareness. Figure 6.23 shows variations in the rate of weathering of different sandstones used in the same building.

Limestones

These are predominantly calcium carbonate and are formed when calcareous deposits are cemented together underwater by further calcium dissolved in the water. Perhaps the best known are oolitic limestones. Oolites are egg-like grains formed by chemical deposition, under water, of calcium carbonate on a nucleus of shell or sand. The deposits may be larger shell fragments or eroded limestone particles. Impurities such as clay are often present and these will affect the colour of the stone. The age of limestone rocks varies greatly. Older ones, such as

Figure 6.23   Differing rates of weathering of two types of sandstone used in north
Scotland. Bedding planes are apparent in each type. The column which is eroding has a
much finer grain, the bedding planes intersecting the vertical face

carboniferous limestones, are generally denser, more crystalline and, therefore, more
durable than more recent stones, such as cretaceous stones. The chalk found in
southern England is, in fact, of cretaceous origin and is rather soft for most
building purposes, though more northern varieties are hard enough to be used for
building. Older, more crystalline stones are also more amenable to polishing;
indeed, many 'marbles' are, in reality, limestones. As with sandstones, limestones
are laid in beds of limited size and thickness, hence, use must relate to the material
thickness available. Most stones in the UK are found in a broad band, running
north-east from south-west to east England and there are a few sources in northern
England and Ireland.

Limestones are, on account of their make-up, much more chemically active
than sandstones. This activity begins on exposure of the stone after quarrying.
Limestones contain water in which salts and calcium carbonate are dissolved (*sap*).

On exposure, this migrates to the surface and crystallises, forming a hard shell. The stone is then said to be *seasoned*. Dressing (carving) stone prior to seasoning is easier, though there is another opposing viewpoint which is that any harmful salts should be allowed to crystallise at the surface by seasoning so that they can then be removed by dressing. It is generally agreed, irrespective of processing procedure, that a seasoned stone has enhanced frost resistance, partly on account of its lower moisture content. Frost is not, however, the main problem with limestones but rather acid attack, mainly from the atmosphere in the form of sulphurous or sulphuric acids found in polluted atmospheres. These produce surface deposits of calcium sulphate on exposed limestone surfaces which form a hard skin, eventually flaking or crumbling away. Crystallisation of salts also may add to the problem.

The porosity of limestones varies greatly, from as little as 1 to over 40 per cent, though, in common with other materials, pore size distribution is more significant than total porosity, coarse pored stones being most durable. The proportion of pores below 5 μm is commonly taken as the yardstick (see page 37) and stones are considered durable where this is less than 30 per cent, while those with a microporosity of over 90 per cent are not durable. For intermediate microporosity values, other tests are necessary – a further indirect test is the saturation coefficient (see page 38) and stones with values of less than 0.6 are again considered durable. Many limestones fall into the 'risk' area when tested as above and a further direct test, the crystallisation test, can be carried out in such cases; this test is particularly appropriate in view of the normal mode of deterioration of limestones. Crystallisation is induced by application of 15 wetting and drying cycles using cubes of size 40–50 mm in sodium sulphate with six classes, A to F, being assigned according to performance, A being the best (see BRE Digest 269). The Building Research Establishment has defined four exposure zones appropriate to these classes (Table 6.8).

Frost may have some effect in marginal cases and the table also usefully illustrates the range of exposures of various building components in differing environments. There are examples of commonly quarried stones in all six classes obtained from crystallisation tests. Different beds within the same quarry may vary considerably.

Table 6.8 BRE exposure zones for limestones according to their durability class

Zone	Examples	Lowest classification suitable	Worst exposure level
1	Paving, steps	A	Any
2	Copings, cornices open parapets etc.	B C	Pollution and/or coastal exposure Neither of above
3	Quoins, strings, plinths, tracery, mullions, sills	C D	Pollution or coastal exposure Neither of above
4	Plain walling	E/F	Neither pollution nor coastal exposure

Figure 6.24 Stone claddings. The upper panels are limestone and the ones at ground level are granite. Careful attention to detail is necessary to give a satisfactory result

Notable examples of limestones include Portland stone (Dorset), an oolitic stone with properties greatly dependent on the bed:

- Roach (uppermost bed) yields a rock with varying texture and pronounced shell formation suitable for ashlar (regular coursed construction) or cladding, Figure 6.24; class A–D depending on source.
- Whitbed (next bed) is finer grained and easier to work, Figure 6.25; class A–D.
- Basebed (deepest bed) is very fine grained (class D–E). It is called *freestone* since it can be freely worked in any direction.

Other areas famous for limestone are:

- Bath – oolitic, usually buff colour; Figure 6.25 (class C–F).
- Purbeck – jurassic; grey blue or buff, including freestones; some can be polished and are wrongly described as marble. Class A–E; for example, the columns of Westminster Abbey.
- Ketton – oolitic; cream/buff/pink, including freestones. Class A–B; for example, some Cambridge colleges.
- Hopton Wood – carboniferous; Derbyshire, cream/grey. Class A; for example, the Victoria and Albert Museum, London.

Figure 6.25   Natural stone cladding to a new building with a steel/reinforced concrete frame. String courses are in Bath stone and remaining stone is Portland (Whitbed). Large brackets support the string courses at each storey while smaller brackets restrain other units which are bedded in mortar and dowelled together. Stainless steel is used for all fixings. Thermal insulation and a dampproof membrane are provided behind the stone (Grand Buildings, Trafalgar Square, London)

Marble

This consists of limestone recrystallised by heat or pressure, though the attraction of marble results from the presence of impurities which form into veins or mottled bands during the process. An important material in many marbles is serpentine which produces a green colour, though other colours, such as pink, yellow and blue can be obtained. The beauty of marbles is best revealed by polishing and since this would be lost by weathering, especially in exposed or polluted atmospheres, most applications are for internal use, usually in prestige building. Most true marbles are imported.

Slates

This is formed by modification of clay, shale or other material, by heat and/or pressure to form a stone with very pronounced laminations. After extraction, the rock is sawn to the desired length and width and then split along the laminations, normally by hammer and bolster, though mechanised methods have also been tried and are in any case used for thicker, architectural applications. For the traditional roofing application, the thinnest possible slates are preferred; probably the best in this respect are Welsh slates of thickness 3–5 mm, of colour grey, green, blue or red. Other important sources are Cornwell (grey/green; 4–7 mm) and Cumbria which produces, amongst others, a very attractive green slate of thickness 8–11 mm. The latter are derived from volcanic ash mud. Some slates, generally of inferior quality, are imported.

Roofing applications of slates are covered by BS 680 which includes tests for water absorption, acid resistance and resistance to delamination by wetting and drying. Although these give a guide to durability, conformity is not a guarantee of performance and some slates failing the acid resistance test have been shown to perform satisfactorily. Experience of use is probably the best guide to suitability. Larger thicknesses of slate can be used for paving, cladding, window sills and other applications, their attractions being their hardness and their interesting texture if cleaved by hand.

Resin bound synthetic equivalents are now available with a simulated natural texture. These are virtually indistinguishable from the natural material, though long term weathering properties have yet to be established.

The best quality natural slates are virtually indestructible and there is therefore an active second hand market resulting from building demolition (Figure 3.1). In this respect slates can be considered to be highly rated in environmental terms.

SELECTION OF BUILDING STONES

Though a detailed treatment of this subject is beyond the scope of this book, it may be appropriate to draw attention to some of the important basic points concerning selection. Building stones are invariably used for aesthetic effect and it is important to appreciate that weathered stone may have a quite different appearance from newly quarried material. Examples of weathered stone in the environment concerned should be inspected prior to selection. Durability may be a problem in sedimentary based stones. Selection of stone type should be based on the worst exposure anticipated. Durability classifications are themselves often difficult to arrive at, though there are excellent publications from the Building Research Establishment for limestone and sandstone (see references) which should be consulted, as well as inspecting examples of the material concerned in use.

Current external applications of building stones are likely to be in two main areas.

Alterations/repairs to existing buildings

Many older buildings use stone as masonry for structural support. A common example is ashlar – coursed stone, rather like brickwork except with a much bigger unit size, typically 600 × 300 mm, and very thin mortar joints. A further attractive form is random walling, often used with *rag* stones – stones which are not easy to work. This may be combined with dressed stone at corners and openings.

Very often there would be several types of stone in the same building, either for aesthetic effect, in order that some areas could be carved, or from durability considerations. When addressing questions of repair, alteration or conservation of such stones, there are some important principles which should be followed, for example:

- Sedimentary based stones should be used wherever possible with natural bedding planes in compression, since even quite small tensile stress may cause eventual delamination. Great care is necessary in decorative tracery for windows and in treatment of projections such as cornices; in the latter case, the bedding planes should generally be vertical and run at right angles to the building face.

- Repairs are most satisfactory when made using natural rather than reconstituted stone. Defective stone should be removed completely or cut back square with joints to match the existing stonework. There are many patching materials available, based on cements, resins or other binders. They may be highly satisfactory in themselves and give good appearance at low cost but there is a risk in the long term that they may debond from parent stone due to differential movement. Resins can, however, be very useful in joining new stone to old.

- Replacement stones should be chosen for similar absorption and colour after weathering, assuming, as is often the case, that the original source is no longer available. Stones of different types may cause problems if mixed. Sandstone may, in particular, be affected by magnesium sulphate solution produced by the action of pollution on magnesian limestone or cast stone.

- Pointing mortars should be soft – for example, lime based – to attract moisture in preference to the stone. They can easily be replaced if they deteriorate.

- Softer stones accumulate dirt which helps retain moisture and, therefore, accelerates decay as well as looking unsightly.

- The safest cleaning method for removal of dirt is by water jets, with soaking as necessary to remove calcium sulphate deposits on limestone. Unfortunately, water may penetrate the structure so it is best to carry it out when in fine, warm weather to permit rapid drying. Abrasive blasting is possible on harder stones, though it may produce unacceptable dust levels and the attendant risk of silicosis to operators on prolonged exposure. Wet blasting solves the dust problem but creates an unpleasant slurry on and below the building face. Acid cleaners based on hydrofluoric acid are sometimes used. They will damage limestone or calcareous sandstones but are effective on softer, acid resistant stones that would be damaged by abrasion. They must be kept away from glazing. Adequate cleaning off is essential and they are hazardous to operators and any others in the vicinity of the work.

- A number of methods of protecting soft, porous stones from further deterioration have been tried. Treatments are commonly based on silicones which coat the pores of the stone, preventing water from entering in liquid form. Various formulations are available according to the stone type being treated (BS 6477). Maximum penetration is limited to about 3 mm. In some stones, the treatment may accelerate decay, since any water in the stone may then evaporate beneath the surface, possibly causing crystallisation problems and spalling off the treated layer. Careful consideration is necessary before using such treatments, especially if water may find other routes into the stonework. The Building Research Establishment has developed a treatment based on trialkoxyalkyl silane which penetrates to between 25 and 50 mm, depending on stone type. It stabilises softer stones, coats salts, rendering them inactive; and coats, but does not seal, pores. The treatment is expensive but is one of the very few methods which can be used to preserve stonework of special historic value.

Cladding

When stone is used for aesthetic purposes (and possibly for weatherproofing) and not for structural support, it is referred to as a cladding or facing (Figures 6.24 and 6.25). The thickness of stone can then be reduced, though it must be supported by the structural frame of the building and restrained laterally against wind loads. Many buildings have an ashlar cladding of thickness around 75 mm. Stone courses may be supported by concrete nibs or metal brackets at intervals, usually at each storey and restrained laterally by cramps at the edges. The joints should be pointed with a sand/lime mix. Fixings in such construction must not corrode and are best in materials such as phosphor bronze or stainless steel (Figure 6.25).

Where stone strength permits, the thickness can be reduced to about 25 mm, though each stone would then need both support and lateral restraint. Limestone facings have been cast into the wet mix during production of precast concrete cladding panels, relying on a chemical bond which develops between the two materials. Thin granite claddings are popular (Figure 6.22), though great care is necessary to support panels adequately and to allow for movement, while at the same time sealing joints to prevent moisture ingress.

Recent experience has indicated that the strength of very thin claddings can be significantly reduced by thermal effects, for example, sudden temperature changes caused by thermal radiation. Such changes would be absent in the parent rock from which stone is extracted and appear to affect grain interfaces, causing an irreversible expansion in limestones and marbles where a strength loss of as high as 50 per cent may occur. Granites may also lose strength but to a much smaller degree. When there is any doubt as to suitability, tests should be carried out and thin claddings used well within their stress capability, with adequate allowance for movement.

6.7 MASONRY MORTARS

Mortars may be defined as mixtures of sand, a binder such as lime or cement, and water. Their prime function in masonry is to take up tolerances between building units such as bricks or blocks, though they also have to satisfy some or all of the following requirements.

1 They should impart sufficient strength to the complete unit.
2 They should permit movement (unless this is negligible or joints are provided). When movement occurs within a well constructed masonry unit, it should take place in the form of microcracks within the mortar rather than cracking of the bricks or blocks.
3 They should be durable, that is, resist frost or other forms of environmental attack.
4 They should resist penetration of water through the unit.

To permit effective use, mortars should be workable, yet cohesive in the fresh state; bricks or blocks should need only minimal effort to bed them in the correct position.

The functions 1, 3 and 4 above are best provided by a strong mortar, though the wall as a whole may be up to five times stronger than the mortar on account of the very small mortar thicknesses used – typically 10 mm. The function 2 is best satisfied with weak mortars which crack readily if movement occurs. This leads to the working rule: *the mortar must not be stronger that the units it is bonding*.

Mortar mixes are classified in BS 5628 Part 3 (Table 6.9) For high strength units such as engineering bricks, type (i) can be used, the cement content being almost sufficient to produce a cohesive mortar without additional plasticiser. These mortars will give great strength and durability and can be used in situations of high exposure and high stress, though separate movement provision may be required. Some softer sands may not give adequate durability with type (i) mortars, since they are too fine and lead to high water requirements. Coarser sands to BS 1200 or finer concreting sands of BS 882 should be suitable.

Types (ii) and (iii) are suitable for general masonry work in winter. They have lower strength than type (i) but the lime acts as a plasticiser in the fresh material and helps bond the units and resist water penetration in the hardened material. (See the section on limes below.)

Types (iii), (iv) and (v) mortars contain progressively more lime to compensate for the reduced cement content in plasticising terms. Type (iv) mortars can be used for general purposes in summer, while type (v) are restricted to internal use because they have low durability. Note that when bonding weak products, such as lightweight blocks externally, low strength mortars are preferred – protection would normally be provided by a cladding or rendering. Where low strength stock facing bricks are used, a higher strength pointing can be used for protection, though such pointing is not recommended in highly stressed situations because it tends to transmit stress through the brick surfaces, being stiffer than the remaining material.

Instead of lime, liquid plasticisers may be used (Table 6.9), having an air entraining effect and, therefore, improving frost resistance, though mortars so produced are less cohesive in the fresh state and less impermeable in the hardened state. A further alternative is masonry cements which have properties intermediate between the above types (Table 6.9).

Table 6.9 BS 5628 Part 3 mortar types and chief properties. The range of sand contents is to allow for the effects of the differences in grading upon the properties of the mortar. In general, the lower proportion of sand applies to grade G of BS 1200 while the higher proportion applies to grade S of BS 1200

	Mortar designation	Type of mortar Cement:lime:sand Proportions by volume	Air entrained mixes	
			Masonry cement:sand Proportions by volume	Cement:sand with plasticiser Proportions by volume
	(i)	1:0 to $\frac{1}{4}$:3		
	(ii)	1:$\frac{1}{2}$:4 to $4\frac{1}{2}$	1:$2\frac{1}{2}$ to $3\frac{1}{2}$	1:3 to 4
	(iii)	1:1:5 to 6	1:4 to 5	1:5 to 6
	(iv)	1:2:8 to 9	1:$5\frac{1}{2}$ to $6\frac{1}{2}$	1:7 to 8
	(v)	1:3:10 to 12	1:$6\frac{1}{2}$ to 7	1:8

↑ Increasing strength and improving durability

Increasing ability to accommodate movements due to temperature and moisture changes →

Increasing resistance to frost attack during construction →

Improvement in adhesion and consequent resistance to rain penetration

←

Direction of change in properties is shown by arrows

As well as the functional requirements described above, mortars should contribute to the aesthetic quality of the walling. The commonest problem is associated with colour; to achieve the best appearance, uniformity of colour from batch to batch is essential. This requires consistency of colour of the sand, cement and, if used, pigment, together with accurate batching. Sand and cement should be obtained in adequate quantity to ensure uniformity during construction. Sand quality is often variable and although unwashed soft (building) sands are generally suitable, they should be checked for excessive clay content or fine material (BS 1200). To achieve repeatable mixes, the use of gauging boxes for batching is recommended. Machines are now available for automatic batch mixing of mortars and these should give successive batches of very uniform quality and colour. A further option is the use of ready mixed mortars in which case the supplier should guarantee consistency of colour. These mortars are set retarded so that they can be used up over a working day.

There is concern that, with steadily increasing demands for good thermal insulation, the mortar joints cause quite extensive cold bridging, especially in solid wall construction, increasing heat transmission by as much as 25 per cent. This can be overcome by the use of recently introduced lightweight mortars in which sand is replaced by a suitable crushed lightweight aggregate, such as perlite. It should be possible by careful mix design to satisfy the above functional requirements with the much lower heat transmission which is characteristic of lightweight materials. A further alternative is the use of thin layer mortars. These are designed to be used in thicknesses of between 1 and 3 mm and therefore reduce heat losses through the mortar component of blockwork. Effective use of such mortars requires somewhat improved dimensional tolerances than are obtained on standard lightweight blocks. While quite widely used in Europe, this new approach to blockwork has made little impact on the UK market at the present time.

As a final cautionary note, it should be appreciated that mortars exhibit very high shrinkage, often over 2000×10^{-6}. This is of little consequence when correctly used in small thicknesses (up to 15 mm) since a network of fine cracks results. Mortars should not, however, be used to fill large spaces, especially in one operation. Concrete or other materials with relatively low shrinkage would perform such functions much more satisfactorily.

6.8   USE OF LIMES IN BUILDING

There are vast quantities of limestone and chalk within the earth's crust, probably derived from marine life many millions of years ago. Some limestones contain magnesium carbonate caused by seepage of sea water through strata which causes exchange of the calcium ion for the more stable magnesium ion. They are known as dolomitic limestones. For general manufacturing purposes, pure limestones are preferred due to their wider applicability. During manufacture and use lime undergoes a complete chemical cycle involving calcination, slaking and carbonation as shown in Figure 6.26.

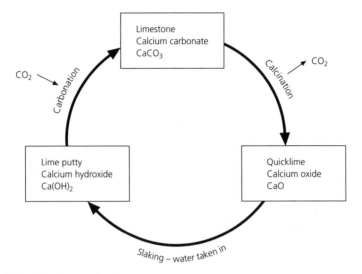

Figure 6.26 The lime cycle

The heating (calcination) of lime changes its chemical composition from calcium carbonate to calcium oxide:[1]

$$CaCO_3 \xrightarrow{\text{heat}} CaO + CO_2$$

The calcining reaction takes place at a temperature of under 1000 °C although in practice the kiln temperature may be higher – up to 1400 °C or higher according to the application. The raw material is also relatively pure, which together with gas firing produces lower pollution risk from the kiln. All reactions occur in the solid state since calcium carbonate does not melt until around 2000 °C. The energy input is similar to that of Portland cements made by the dry process, being about 3500 kJ/kg.

The lime may be marketed as granular oxide of particle size typically about 15 mm (in which case it will need to be kept in sealed containers or bags) or it may be slaked by the addition of water. This causes a reaction as the calcium oxide reacts to form calcium hydroxide:

$$CaO + H_2O \rightarrow Ca(OH)_2$$

---

[1] It might be noted that substantial amounts of carbon dioxide are emitted in this process. Considering atomic weights:

Ca 40
C 12
O 16

$CaCO_3$ therefore corresponds to an atomic weight of 100 and CO_2 to 44. Therefore 44 per cent of the mass of limestone is lost to the atmosphere on calcining.

The reaction is strongly exothermic.[2] The rate of this reaction depends on the temperature to which the lime was fired. As this temperature increases, the speed of reaction with water during slaking is reduced. The term *quicklime* refers to the less hard burnt materials which react very quickly. Hard burnt limes tend to be described as *calcium oxide* rather than quicklime. Where an excess of water is added, the process is known as *slaking* and the product is described as a lime putty. The attraction of lime putties is that they can be kept in open tubs for very long periods – of up to 10 years. They are said to *mature* as they age. Maturing has the advantage of ensuring that slaking is complete and produces a denser putty-like product. During the process water rises to the surface and this protects the lime from atmospheric hardening. However, the water must be prevented from freezing.

Alternatively the chemically correct quantity of water may be added. A powder is obtained and this is known as *hydrated lime*. Both slaked lime and hydrated lime are still widely available for construction purposes.

Pure (non-hydraulic) limes harden by very slow reaction with atmospheric carbon dioxide:

$$Ca(OH)_2 + CO_2 \rightarrow CaCO_3 + H_2O$$

Pure lime mortars harden gradually over a period of months to strength levels much lower than those of cement based products. The hardness of the lime mortar can be increased by adding a pozzolana such as brick dust which produces a hydrate (see 'Cements'). Proportions of 25–35 per cent by volume may be added. A possible advantage of non-hydraulic limes over Portland cement products is that during carbonation the substantial quantities of carbon dioxide that are emitted during the manufacturing process are recombined, reducing the environmental effect, unlike Portland cement products in which only a few mm at the surface become carbonated.

Alternatively hydraulic limes can be employed. They are manufactured from essentially impure limestone, containing silicates which react with the calcium oxide in the kiln. This produces a product which is really intermediate between lime and Portland cement. There are relatively few manufacturing plants for these limes in the UK, though hydraulic limes can also be imported from France or Italy, where they are more widely available.

To illustrate the relative strength properties of lime based and cement based mortars, Table 6.10 gives approximate values of 28 day compressive strengths of type 1 (1:3 cement:sand) mortars and three types of mortar with binder:sand ratio of 1:6. These are all based on curing in damp air. It will be noticed that inclusion of cement produces significant strengths even in ratios as small as 1:6 cement:sand. These mortars would be quite difficult to remove if bricks were to be recycled. The 1:1:6 cement:lime:sand is of somewhat lower strength than the pure cement equivalent, though the lime would confer other benefits as above. The remaining mortars have very low strengths. Both are friable in the hand and would be easily

---

[2] The emission of heat by limes on slaking is in fact the principle behind self-heating food cans. Puncturing a liner to allow water to contact the lime generates sufficient heat to warm the can.

Table 6.10 Comparative 28 day strengths of mortars made from cement and lime

Type	Type (i) 1:3 Cement:sand	1:6 Cement:sand	1:1:6 Cement: lime:sand	1:6 Hydr. lime:sand	1:6 Non-hydr. lime:sand
Approximate 28 day strength N/mm^2	16.0	4.2	3.5	0.4	0.2

removed, though the hydraulic mortar does have greater cohesive strength which would continue to increase with age. If obtainable, hydraulic limes might be considered to be a good substitute for cement mortars for normal cladding applications. They are specified by BS EN 459-1. A relatively rich mix of 3:1 sand:lime mortar of type HL 3.5 would reach a compressive strength of 3.5 N/mm^2 at one year of age.

Ready mixed lime based mortars are available for masonry and rendering applications and these have combinations of sharp or soft sand appropriate to the type of masonry and render/float/set coats. They are referred to as *coarse stuff*, *medium stuff* and *fine stuff* according to application. They offer the batching advantages of ready mixed mortars, long storage and the benefits of a traditional lime product in performance terms. When used for renderings limes exhibit much greater permeability than cement mortars so that there should be fewer problems of moisture build-up.

Lime may also be added to gypsum plaster to produce a smoother harder finish (though this is rare nowadays where premixed materials are preferred for their time saving properties).

A further use of limes is in limewash where a lime slurry is used for protection of more friable materials, such as earth structures, from the weather. In these applications a small proportion of linseed oil (added to the putty) or tallow (added during slaking so that the heat melts it) may be used to assist in producing water resistance in the coating.

6.9 GLASS AS A GLAZING MATERIAL

Modern scientific progress has brought with it new materials and methods but, as a glazing material, glass has remained unchallenged and seems likely to continue to dominate this field into the foreseeable future. Raw materials are plentiful and cheap, and glass combines relatively high stiffness with unrivalled abrasion resistance, light-transmission properties and resistance to weathering or chemical attack.

Ordinary glass is based on silica (silicon oxide) which, in crystalline form, exists in a tetrahedral array (Figure 6.27(a)). It has been explained that if silica is cooled at normal rates from the liquid, an amorphous structure results (Figure 6.27(b)). This may be regarded as a heavily and randomly distorted form of the crystalline structure. Although silica forms the basic network of glass, it is not used in the pure

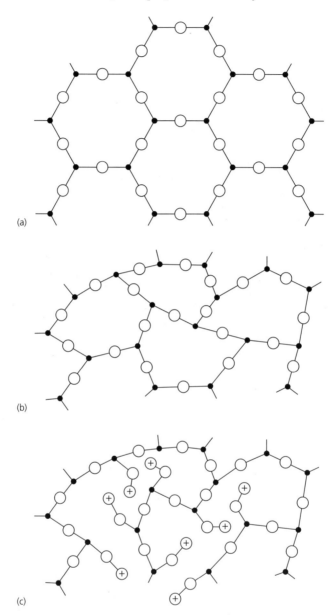

Figure 6.27   (a) Tetrahedral silica crystal. (b) Amorphous silica. (c) Silica, modified by incorporation of metal ions to reduce melting point. Note that in all cases silica is shown as trivalent to simplify sketches

form because its melting point is too high – in the region of 1700 °C. Instead, the silica network is modified by compounds such as sodium carbonate which, at high temperatures, decomposes to sodium oxide and then combines with part of the silica, forming sodium disilicate, thereby interrupting some of the rigid silicon–oxygen links. Hence, *soda glass*, as the material is known, melts at a much lower temperature, in the region of 800 °C (Figure 6.27(c)). Unfortunately, soda glass is water soluble and a further network modifying compound, calcium carbonate, is added to stabilise the glass. The approximate composition of the raw materials for a typical soda–lime glass is:

SiO_2	75 per cent
Na_2CO_3	15 per cent
$CaCO_3$	10 per cent

Smaller amounts of other materials may be added as follows:

■ Manganese dioxide to remove coloration due to iron in the sand.
■ Lead to improve surface lustre or, in larger quantities, to produce a high density glass resistant to X-rays.
■ Borax to produce glass having very low thermal movement – such glasses are highly resistant to thermal shock.

MANUFACTURE OF GLASS

The raw materials are mixed, in the correct proportions, with a quantity of scrap glass, *cullet*, and heated to about 1500 °C. The cullet melts first and permits reaction and fusion of the remaining ingredients at temperatures below the melting point of pure silica. The liquid is then cooled to a temperature of 1000–1200 °C, at which its viscosity is sufficiently high for forming. The most important processes are as follows.

The flat-drawn process

The glass is drawn upwards on a metal grille known as a *bait*, the sheet engaging with rollers which prevent its waisting. The glass is annealed to relieve cooling stresses and then cut to size. This type of glass contains slight ripples but is economical and was formerly used in dwellings, offices and factories. It is still used for agricultural applications but has now been largely replaced in the UK by float glass.

Rolled glass

The glass is drawn off in a horizontal ribbon on rollers and is then annealed. Such glasses do not give clear vision but can be given textured or patterned finishes, allowing light transmission but giving some privacy when used in glazed doors or partitions. Wire may be incorporated, producing *Georgian glass*, a material with higher fire resistance and increased safety against injury from impact. Wired glass is widely used in rooflights and low level glazing.

Float glass

This glass is optically flat and is produced by drawing it, while still soft, along the surface of molten tin in a bath. It is now used for general glazing purposes, as well as mirrors, shop windows and other situations where clear, undistorted vision is essential.

Strength of glass

The strength of any glass unit is determined largely by the effect of any surface imperfections it may contain. An indication of maximum tensile strength is obtained by tests on very thin glass fibres which are sensibly free of flaws and are found to withstand stresses of up to 3000 N/mm². The tensile strength of bulk glass, as measured by flexural tests, may vary between 20 and 200 N/mm², according to its surface condition. The highest strength is obtained immediately after manufacture, though strengths are still variable. Strength reduces on ageing, as surface imperfections increase, whether by chemical attack or simply mechanical abrasion, glass which has weathered for some years being much weaker than new glass. For this reason, old glass is more difficult to cut and more likely to fracture due to movement or fixing stresses.

Toughened glass

The surface flaws in glass can be removed chemically but toughening can be carried out more simply by heat treatment. Sheet glass is heated uniformly until just plastic and then cooled by air jets. The outer layers contract and solidify and then, as inner layers try to follow, they throw the outer layers into compression, tending to close the microscopic cracks. In this way, the overall strength of the glass can be increased several times and impact strength may increase seven-fold, a 4 mm thickness complying with BS 6206 class A (see below). Figure 6.28 shows the approximate stress distribution in a sheet of toughened glass. On bending, the

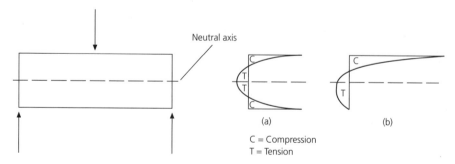

C = Compression
T = Tension

Figure 6.28   Simple representation of the stresses in a sheet of toughened glass subject to bending: (a) no load; (b) under load. Failure does not occur until the tensile strength at the lower surface is exceeded

compressive stress on one face reduces and failure occurs only when this has been reversed to the normal tensile limit. If the surface is broken, the stress distribution becomes unbalanced and the material shatters into small, relatively harmless fragments. Cutting or edge working are, therefore, not possible and it must be ordered to size. Toughened glass has widened the application of glass to include solid glass doors, large windows and suspended glass curtains. Very large areas of glass are more conveniently suspended from a structural frame since, when used in this way, the weight of glass itself contributes considerably towards stability. Large sheets may be joined by square metal plates at their corners, gaps being filled with transparent plastic material. Additional stability may be achieved by means of stiffening fins which may also be of glass. Such glass curtains are, themselves, an architectural feature and have been used as complete facades of two-storey buildings.

Laminated glass

This provides a high degree of resistance to injury from flying glass in case of impact. In its simplest form, two sheets of glass are bonded with a thin film of plastic such as polyvinyl butyrate, under pressure at a temperature of about 100 °C. The sandwiched plastic bonds well to the two glass surfaces and helps absorb energy in impacts but, most importantly, stops glass shattering and disintegrating if stressed to failure so that it often remains secure and weatherproof. Under severe blows, a hole may be punched through, the surrounding glass remaining intact. Higher levels of impact resistance can be produced by increasing the number of glass/plastic laminates; bullet- and missile-proof glazing are made in this way. The strength of laminated glass is rather less than that of toughened glass so that increased thicknesses may be required in a given situation. To achieve class A rating to BS 6206 (see below) a minimum thickness of 6.8 mm – two 3 mm sheets with a 0.8 mm interlayer – is required. It offers the advantage, however, that cutting is possible; two-ply glass is easily cut by scoring and breaking each side in turn. Re-order times should also be less since laminated glass can be cut from stock material.

SAFETY WITH GLASS

There are many potentially hazardous areas of domestic, commercial and industrial buildings, mainly in low level or overhead glazing, where risk of injury by broken glass must be recognised. Perhaps the greatest risk is associated with children, since they may run into glazing if unaware of its presence. Accordingly, BS 6206 defines three classes (A, B and C) of safety glass appropriate to risk. Glass is described as *safe* if, in a standard impact test, any one of the four conditions is met:

1 There is no breakage.
2 No shear or opening develops through which a 76 mm diameter sphere can be passed freely.

3 Disintegration occurs but the 10 largest crack-free particles remaining 3 minutes after impact together weigh no more than the mass equivalent to 6500 mm^2 of the original test piece.
4 Breakage results in several pieces but none of these present sharp edges which are pointed or dagger-like.

Unfortunately, a number of fatal accidents have occurred with imported toughened glass claiming to comply with BS 6206 by means of clause 3 above, since dagger-like splinters over 200 mm long may still comply with the area requirements of this clause. Where safety is an important consideration, it is recommended that great care is exercised in the choice and specification of glazing, and that compliance with BS 6206 is established by an accredited independent testing authority and clearly indicated on the glass.

Toughened and laminated glass are now used very widely for large scale commercial applications and where overhead glazing is employed it is particularly important that the highest standards of safety are achieved. There have been reports of failures in toughened glass, possibly due to nickel sulphide inclusions which may generate local stresses (for example due to heat build-up). In such applications greater safety can be achieved by use of laminated toughened glass, the toughened glass imparting the necessary strength while laminating reduces the risk associated with dramatic failure of the toughened material.

THICKNESS AND WEIGHT

Although the thickness of ordinary glass was traditionally measured in terms of its mass per unit area, thickness is now given directly, the common sizes being 3 and 4 mm thick. The thickness of glass for ordinary glazing should increase with wind load and glass area. Square sheets should be thicker in general than rectangular sheets since, for a given area, there is less restraint at the centre of a square sheet of glass. The thickness of toughened and laminated glass also depends on other factors, such as the impact loading they are likely to have to withstand. In many cases, it will, nevertheless, be more economic to supply glass of uniform thickness to withstand the most demanding situation in a given building, since this simplifies supply, installation and future replacement.

THERMAL PROPERTIES OF GLASS

These will be divided into two areas – solar gain and heat loss characteristics.

Solar heat gain

The aesthetic and light-transmitting properties of glass, combined with improved manufacturing technology, have encouraged the use of large glazed areas, offering adequate natural lighting to building occupants as well as the important psychological advantage of visual contact with the exterior environment. In the 1960s, many buildings were constructed to take advantage of this new freedom by use of the material, only to discover that thermal properties for occupants were very far from ideal on account of the greenhouse effect. This is largely caused by radiation transfer and is illustrated by Figure 6.29(a) which shows:

1 The radiation characteristics of the sun's rays received at the earth's surface.
2 The radiation emitted by a warm internal enclosure.

The sun's rays contain a mixture of ultraviolet, visible and infra-red radiation. All surfaces within a building also emit radiation, though this is of much lower intensity and longer wavelength (infra-red) on account of the much lower surface temperatures compared with that of the sun.

Glass transmits some ultraviolet light and is quite transparent to visible and infra-red light of wavelength up to about 3 microns. Hence, most energy in the sun's rays is transmitted by ordinary glass, causing warming of internal surfaces (Figure 6.29(b)). These radiate, according to their temperature, in the wavelength range 5–40 microns but glass is almost completely opaque to radiation of this wavelength and absorbs or reflects the heat re-radiated internally, effectively acting as a one-way heat filter – the greenhouse effect. The effect is most marked in south or west facing elevations especially when the sun is relatively low in the sky, when there is less reflection by vertical glazing. This can cause considerable discomfort in buildings with large areas of ordinary glazing. Caution is necessary when designing large glazed areas to the south or west, though there are several methods by which solar gain can be controlled:

■ Use tinted, heat absorbent types of glass (Figure 6.29(b)). These are available in a number of shades including grey and green, the glass itself absorbing as much as 50 per cent of the heat (and a similar proportion of light), though some heat will be re-radiated through the glass. The amount of heat absorbed depends on glass thickness and care is necessary in design with such sheets to prevent excessive temperature rise in the glass, with consequent overstressing, especially if glazing beads shield edges from the heat, producing a cool glass perimeter. Tinted glass permits a view of the interior of the building, rather than producing a 'mirror' effect.

■ Heat reflecting glass can be achieved by thin metallic surface coatings, usually applied to the inner face of the outer glass sheet in sealed double glazing, for protection, though they can also be used for single glazing. As much as 40 per cent of the heat and a similar proportion of light are reflected and, in this case, there is less heat absorption of the glass itself. These glasses have a metallic tint which may, typically, give a silver mirror-like appearance when viewed externally while there is a bronze tint to the external environment, when viewed from the inside. (Like tinted glass, the occupants tend only to be aware of tints when a window is opened.)

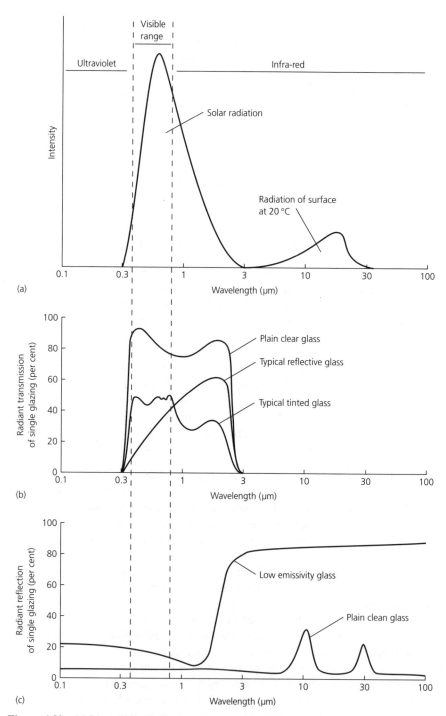

Figure 6.29 (a) Intensities of radiation due to sun and due to surfaces at room temperature. (b) Transmission characteristics of single glazing (various types). (c) Reflection characteristics of plain clean glass and low emissivity glass

Heat losses from glass

When solar radiation is reduced, as in winter, the heat flow in glass can reverse quite dramatically. Glass itself has a high thermal conductivity – around 1.0 W/m °C, so that it offers very little resistance to conducted heat flow. The main resistance of glazing to heat comes in fact from air layers at each surface which impede the convected heat flow from the warm interior air to the cooler glass surface. Double glazing operates by providing double the number of these glass/air interfaces, though a cavity width of 20 mm is the optimum for obtaining full advantage of the insulating qualities of the trapped air between panes. Sealing of the cavity is always recommended to prevent dirt and condensation problems and a further 10 per cent reduction in U value can be obtained by filling the cavity with an inert gas such as argon. Yet another possibility now being exploited is to reduce the absorption of the glass to internal radiant heat by low emissivity coatings. The situation is shown in Figure 6.29(c). Although ordinary glass does not transmit long wavelength infra-red radiation as explained above, it absorbs most of this radiation, becoming warm and, therefore, increasing heat loss by conduction/convection. If a low emissivity glass surface is used, normally on the cavity side of the interior sheet in a double glazing installation, the heat flow due to internally radiated room heat across the cavity can be greatly reduced, lowering the effective U value of the glazing. At the same time, useful solar heat can be admitted, since the low emissivity coating transmits the shorter wavelength solar energy. It will be evident that the precise U value of such a double glazing system will depend on window orientation and solar heat conditions prevailing at the time. In favourable conditions, a negative U value may be obtained; in other words, solar radiation may provide a net gain to the building in spite of a lower external temperature. More usually, reduced U values in the range 1.0–3.0 W/m² °C would be obtained. Where necessary, effective heat retention in winter months can be combined with solar control in summer months by use of double glazing systems incorporating solar control glass (Table 6.11).

BUILDING REGULATIONS IN RELATION TO GLAZING HEAT LOSSES

To save energy in dwellings, current *Building Regulations* restrict the size of glazed areas which normally have much higher losses than other elements of construction. The allowable glazing heat loss in any one enclosure is the amount which would be lost from a single glazed area of 12 per cent of the total perimeter wall area of that enclosure. This figure represents a typical average glazed area of traditional style housing in the UK, though larger glazed areas could be incorporated as follows:

- Use double glazing – this would permit about double the glazed area to be used on the basis of the value of 2.8 W/m² °C prescribed for double glazing, as against 5.7 W/m² °C for single glazing.
- Use treble glazing (*U* value 1.9 W/m² °C or less) – this would permit three times the 12 per cent figure indicated above.

Table 6.11 Solar control and thermal insulation properties of various glazing systems

Type	Function/thickness					Solar gain		Heat loss
	Plain glass	Solar control (bronze)		Insulation (double glazing)		Light transmission (per cent)	Total solar radiant heat transmission (per cent)	U value W/m^2 °C
		tinted	reflective	plain	low emissivity			
Plain single glass	4 mm	–	–	–	–	89	86	5.4
Plain double glass	4 mm	–	–	4 mm	–	80	75	2.8
Solar control single glazing	–	4 mm	–	–	–	61	70	5.4
	–	12 mm	–	–	–	27	47	5.2
	–	–	6 mm	–	–	10	23	4.0
Solar control double glazing	–	6 mm	–	6 mm	–	44	49	2.8
	–	–	6 mm	6 mm	–	9	16	2.3
Solar control low-emissivity double glazing	–	6 mm	–	–	6 mm	34	42	1.9
	–	–	6 mm	–	6 mm	7	13	1.8

■ If the external walls are of U value less than 0.6 W/m² °C, then additional glazing can be employed provided the total heat loss does not exceed that corresponding to an external U value of 0.6 W/m² °C and 12 per cent of single glazing.

The *Building Regulations* also have the flexibility to allow other glazing systems, provided the architect/services engineer can demonstrate compliance with the basic energy requirements as outlined above. It is likely that, in the near future, heat insulation regulations will be tightened to reduce further heat loss in domestic buildings.

FIRE PERFORMANCE

Ordinary glass has a poor performance in fire due to its tendency to shatter when heated. A period of 60 minutes stability and integrity to BS 476 Part 8 (1972) can, however, be achieved by use of wired glass or heat treated borosilicate glass, which can be laminated with float glass if high impact resistance is also required. An additional 30 minute insulation level under BS 476 Part 8 can be achieved with an intumescent layer which expands on heating to provide an insulating barrier, though transparency is lost in a fire. In all cases, a suitable frame is required, for example, hardwood glazing beads with intumescent seals located by steel inserts.

DURABILITY

Glass is unaffected by the atmosphere and by most acids, with the exception of hydrofluoric acid. Alkalis, which occur in cement or chemical paint strippers, attack glass and destroy the smooth surface and light transmission properties.

SOUND INSULATION

The sound reduction properties of glazing are typical of those of a thin panel or membrane. Resonances occur at low frequencies, the insulation improving towards high frequencies. High-frequency insulation is, however, reduced by the *coincidence effect*. In double glazing, the cavity size required for effective sound reduction is about 200 mm, much larger than that for best thermal insulation properties. To be satisfactory, it is extremely important that air paths through glazing should be prevented. The use of different thicknesses of glass in the two sheets and an absorbent material in the cavity increases the sound insulation. Typical sound reduction values at medium frequencies for opening windows are given in Table 6.12. Note that 50 per cent sound absorption leads to a sound level reduction of just 3 dB, hence, the 31 dB reduction of the system given in Table 6.12, corresponds to a transmission equal to less than $(0.5)^{10}$ or 0.1 per cent of sound energy.

Table 6.12   Sound level reduction in single and double glazing (opening windows)

Type and thickness of glazing	Reduction (dB)
3 mm single glazing	20
12 mm single glazing	22
3 mm and 4 mm double glazed window with absorbent material in 200 mm cavity	31

6.10   ASBESTOS

Asbestos exists in many different forms but is essentially a siliceous material with a fibrous molecular structure described earlier. It is, in fact, the only naturally occurring inorganic fibre. Being a mineral, it has extremely good durability and chemical resistance, and can be heated to high temperatures without melting or burning. Most uses derive from these properties, together with its high strength in the direction of the fibres. There are three main types of fibre:

- chrysotile
- crocidolite
- amosite

They are all of metamorphic origin, the type occurring in any one situation depending on the mineral balance; for example, sodium is needed to form crocidolite.

Chrysotile has the formula $Mg_3Si_2O_5(OH)_4$ and it contains small amounts of aluminium, iron and sodium. It has good heat resistance and, being resistant to attack by alkalis, is very suitable for use with Portland cements, as in asbestos cement products. The fibres are the longest, can be woven, and have a silky white appearance. It used to be by far the most widely used.

Crocidolite (blue asbestos) contains more iron than chrysotile, has greater strength and acid resistance and has a characteristic blue colour.

Amosite contains relatively long, stiff but brittle fibres and is very suitable for insulating boards.

All fibres are very fine, as low as 0.1 µm diameter; hence, they have very high aspect ratios and bond well to most matrices (see page 553). Embrittlement of asbestos fibres may begin at temperatures as low as 300 °C, due to loss of water on crystallisation, decomposition into simpler products occurring progressively up to about 1000 °C. Fusion is complete by about 1500 °C, depending on type. Part of the fire-proofing ability of asbestos products is that, rather like wood, the decomposed material provides effective insulation to underlying layers in a fire with the additional benefit of non-combustibility.

Asbestos fibres are milled before use to break them down into finer and hence more efficient forms. The effect of milling can be checked by measurement of specific surface, rather like cements. Values of 5000 m^2/kg are typical.

Health hazard

It is unfortunate that asbestos particles may seriously affect the health of those who are in prolonged contact with the material. Inhalation of fibres may produce disease of the lungs known as *asbestosis*, related lung cancer or a rare disease called *mesothelomia*. It is likely that crocidolite represents a particular hazard and, for this reason, the fibre has not been imported by manufacturers into the UK since 1970, though it is quite common in existing buildings and may still be imported in the form of manufactured goods. Stringent regulations now apply in areas where asbestos dust may be present. These require the provision of respirators, protective clothing and exhaust ventilation. Special storage and waste disposal facilities are necessary in particular situations. In general, hand tools produce less dust than power tools. Another hazard occurs in demolition work if asbestos has been used for insulation purposes. The most harmful type, crocidolite, may be recognised by its blue colour (chrysotile and amosite are generally white and brown, respectively) unless paint has been applied but great care is necessary when dealing with the material in any form.

It now also appears that there is some risk from asbestos products in use, mainly those which are not strongly bound, such as sprayed or low density products. Fibres can be released from these materials on ageing, so it is advisable to have all such materials correctly removed. Asbestos cement products are generally safe unless they are subject to abrasion.

Perhaps, ironically, fires, which many asbestos products are specifically designed to cope with, can cause a further hazard. Fibres are broken down in high temperatures but it is possible for unaffected fibres to be dispersed over quite large areas as a result of progressive destruction or collapse of a building in a serious fire.

Assessment of health risk

The risk associated with any mineral fibre is greatly influenced by its size, fibres of diameter below 2 μm and length in the range 5–100 μm generally being considered most harmful. Hence, when measuring asbestos levels, size should be taken into account, since larger fibres do not normally reach the sensitive lung tissues. Part of the reason for crocidolite fibres being more harmful is that they seem to reach the lungs more easily than other fibres. The method of assessment of risk has been the subject of much experiment, though currently, it is often specified in terms of the number of fibres in a given size range per millilitre of air. A typical safe level is considered to be 2 fibres per millilitre in working environments, with a reduction of 10 times for crocidolite. In making tests of this type, great care is necessary in sampling. A further note for caution is that, since fibre size is an important factor in determining risk, it is possible that other non-asbestos based inert fibres may also constitute some hazard.

Notwithstanding the above comments, asbestos still has important, though limited applications in construction where its excellent mechanical properties, combined with chemical and heat resistance, can be exploited without significant danger to health. The safest products are those in which the fibres are tightly bound by a hard, abrasion resistant matrix, for example, asbestos cement products.

Asbestos cement

This material is formed from a mixture of short fibres, Portland cement and water, built up in layers to form sheets which are moulded and cured. Alternatively, silica and lime may be added during manufacture, the mix then being steam-autoclaved. It has been used to form a wide variety of products, many of which are used externally, having a life of at least 40 years. During this time, however, impact strength decreases due to embrittlement (which also produces an increase in flexural strength). Some softening of the surface may also occur, due to the action of pollution on the cement. Life can be prolonged by painting, an alkali-resistant primer being essential. Since products are brittle, fixings must allow a certain amount of movement and must not cause localised stresses around them. Asbestos cement is still commonly found in rainwater goods, many types of pipe, cisterns, conduits and troughs. In sheet or tile form, asbestos cement exists in the fully compressed or semi-compressed states. Full compression results in high density – typically 1800 kg/m^3 – and high bending strength – typically 22.5 N/mm^2. Semi-compressed forms have densities of approximately 1450 kg/m^3 and bending strengths of 16 N/mm^2. Fully compressed products may have smooth finishes on both surfaces, while semi-compressed sheets have only one smooth surface. Pigments may be included for colouring, and treatments or coatings may be applied to smooth surfaces for purposes such as preventing fungus growth, providing attractive appearance or protection against acid attack. The resistance of roofing products to external fire is given as P 60, as defined in the current edition of BS 476 Part 3 1975 (corresponding to Ext. S. AA in BS 476, Part 3 1958), signifying that the material passes a preliminary ignition test and that fire does not penetrate in less than one hour. There is, however, a tendency for asbestos cement to shatter under the intense heat of an internal fire, so that roofing or cladding should be protected by lining sheets or panels in conjunction with insulation. Asbestos-free varieties of the above products are now available.

Low density insulating boards and wallboards

Insulating boards have a density of not more than 900 kg/m^3 and thermal conductivity not more than 0.175 W/m °C. They were traditionally used as surface membranes in roofs, walls, ceilings, partitions and ductings, or as protection to structural steel.

Wallboards are tougher than insulating boards, having densities in the range 900–1450 kg/m^3 and thermal conductivity not greater than 0.36 W/m °C. They were used where greater mechanical resistance was required, such as on doors or as overlays for floors. Boards of these types represent a health hazard and currently available boards now contain non-asbestos substitutes, such as glass, cellulose or polyvinyl alcohol fibres or rockwool.

Sprayed asbestos

Although sprayed asbestos has been extremely effective as a means of providing fire protection to structural steelwork and concrete, its use has now virtually ceased on account of the health hazard it poses. The former British Standard, BS 3590, has been withdrawn and protection is now provided by the preformed insulating boards and wallboards described above. It is advisable to seal or replace any existing sprayed asbestos coatings to minimise the dust hazard.

Other products using asbestos

Asbestos in the form of chrysotile has excellent weaving ability and has been formed into fire blankets, gloves, ropes, sleeving, etc., for heat/fire resisting applications. Although non-asbestos substitutes are now available, many of these are still in use. Further applications include vinyl floor tiles, roofing felts and dampproof membranes, the health risk from such products being generally small.

CHAPTER 7

CEMENTS, CONCRETE AND RELATED MATERIALS

The bonding of simple commonly available materials to form rigid structures has challenged the builder's ingenuity ever since the need for shelter was first recognised.

The basic problem is that if the raw materials to be bonded are durable and unaffected by weathering, abrasion and atmospheric exposure they are probably also not readily amenable to bonding in a chemical sense since they are likely to have a relatively stable surface structure. Such materials may show very little inherent bonding capability when pressed together, although at very low particle sizes there are secondary bonding forces which produce some degree of cohesion – this is the principle by which clays have some cohesive strength when reasonably dry.

Most inorganic cements take the form of liquid or plastic materials which enclose an inert solid 'filler' and then set hard, thereby providing a solid composite. This occurs largely by a mechanical keying action rather than by true chemical bonding. The keying action may be assisted if one of the materials to be bonded is porous so that the cement, while in liquid form, can penetrate the pores before setting.

Traditional cements are, in essence, inorganic materials which would be made to form a set mass usually by addition of water. Hydraulic cements are cements that set and harden by the action of water only, whereas non-hydraulic cements require some other agency, such as atmospheric exposure.

Traditional cements are obtained by very simple manufacturing processes, usually involving heating of commonly available materials. Examples are:

- building limes – based on calcium carbonate
- gypsum – based on calcium sulphate

Heating the raw materials produces chemical changes as a result of which the product is chemically attracted to water. During the process of this reaction a crystalline framework is obtained which corresponds to a 'set' product.

It is perhaps remarkable that the great majority of cements both in traditional and current use rely upon the presence of calcium in some form. This is normally in the form of calcium oxide, although other forms (such as calcium sulphate in gypsum plaster) are also significant. Calcium is reasonably abundant in the earth's crust and readily forms compounds with elements of similar atomic size, such as oxygen.

Materials based upon cements continue to play a key role in almost all forms of construction, often being manufactured *in situ*. A number of the problems and failures encountered in modern buildings are the consequence of mis-use of these materials and a comprehensive understanding of their behaviour, together with effective implementation of this knowledge, are essential for successful use of the wide range of products obtainable. Most reported failures could have been avoided by application of existing knowledge.

7.1 PORTLAND CEMENTS

A real advance was made when, in the early nineteenth century, it was discovered that if a mixture of clay and chalk was heated, a (hydraulic) cement was obtained which would develop excellent strength and durability in a reasonably short time without any atmospheric action being necessary. Such cements would set as well under water as they do in the air.

Portland cements are so named because concretes made from them have a colour similar to that of Portland stone. They remain by far the most important cements in current use.

MANUFACTURING PROCESS

The manufacturing process is, in essence, the same as that originally conceived, although by careful control cements of consistently high quality can now be achieved and various forms are obtainable according to application.

Cement chemists use specific (perhaps rather peculiar!) chemical shorthand to describe the compounds in cement. The chief component of chalk or limestone is calcium carbonate, $CaCO_3$, which on heating yields calcium oxide (quicklime) CaO. This is referred to as C.

The chief ingredients in clay are, in order of decreasing amount:

SiO_2 (silica)   abbreviated   S
Al_2O_3 (alumina)   abbreviated   A
Fe_2O_3 (ferrite)   abbreviated   F

These compounds, C, S, A and F, have little chemical interest in each other at normal temperatures but by heating to bright yellow heat (about 1500 °C) slow chemical combination occurs. All forms of chemical reaction take place much more readily when one of the constituents is liquid and in the case of cement manufacture the reaction begins when the metallic oxides, chiefly iron, fuse (melt). These materials are said to act as fluxes in the process. They combine with the calcium oxide to form liquid products and in this environment further reaction takes place with the silica. The latter slowly combines with calcium oxide to form two crystalline products based upon calcium silicate. The quantities of the raw materials are carefully chosen so that almost all the calcium oxide (derived from the chalk or limestone) is consumed chemically by the other ingredients. On cooling

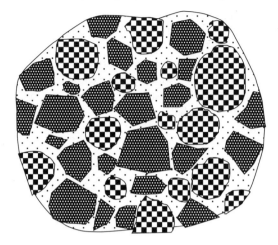

Figure 7.1   Schematic representation of a section through a grain of Portland cement

the liquid turns into a largely amorphous solid which encloses the crystalline calcium silicates forming a *clinker*. The crystalline materials formed at the high temperature are:

dicalcium silicate C_2S
tricalcium silicate C_3S

while the surrounding matrix contains:

tricalcium aluminate C_3A
tetracalcium alumino ferrite C_4AF

Figure 7.1 shows a schematic section of a grain of cement formed by crushing the clinker. The grain comprises angular crystals of C_3S together with more rounded crystals of C_2S in a non-crystalline background of C_3A and C_4AF. (In a typical cement grain of size 20 μm (0.02 mm) there would be many more crystals than shown). All four compounds are hydraulic though their behaviour is very different as summarised in Table 7.1. Their behaviour will be described in more detail later. It is important in the production of cement to control the relative quantities of each compound produced. Some aspects of the ratios can be controlled fairly easily, for example, the ratio of C_3S to C_2S which will affect the rate of hardening and heat output, can be regulated by adjusting the chalk/limestone:clay ratio (usually about 80:20). More of the former will tend to produce more C_3S which contains more calcium oxide. (Too much lime is, however, undesirable – it remains uncombined and produces *unsoundness*.)

Again, if iron oxide (F) is added specially as a separate ingredient, more C_4AF and therefore less C_3A will be produced, though manufacturers would prefer, if possible, to avoid the extra cost of additional ingredients to modify properties.

It will be apparent that both the composition of the raw materials and variations in the composition will affect the final product. In addition, the raw materials must

Table 7.1 Properties and typical proportions of compounds in ordinary Portland cement in current use

Name	Abbreviation	Approximate percentage in OPC	Properties	Heat of hydration J/g
Dicalcium silicate	C_2S	19	Slow strength gain – responsible for long term strength	260
Tricalcium silicate	C_3S	52	Rapid strength gain – responsible for early strength (e.g. 7 days)	500
Tricalcium aluminate	C_3A	10	Quick setting (controlled by gypsum) – susceptible to sulphate attack	865
Tetracalcium alumino ferrite	C_4AF	8	Little contribution to setting or strength – responsible for grey colour of OPC	420

be very finely ground in order to permit complete chemical reaction in the kiln – the core of larger particles would be unable to react with surrounding materials. The temperature of the kiln must also be carefully regulated.

There are two chief aspects of the manufacturing process: the first is to produce a finely divided mixture of the raw materials – chalk/limestone and clay/shale (maximum particle size approximately 100 μm). The second is to heat this mixture to produce chemical combination. Two main processes may be used – the wet and the dry process – and the one chosen will be dependent on locally available materials and conditions. However, since the water used in the wet process must all be evaporated, there is now an increasing trend towards using the dry process where possible, in order to save fuel costs in production.

Wet process

The raw materials are converted into a slurry with a water content of about 30 per cent which is pumped into tanks from which blends can be taken to obtain the precise ratio of calcium carbonate required. The slurry is fed into the upper end of a large rotating kiln, slightly inclined so that the material gradually passes down the kiln. The kiln is heated from the lower end by injection of powdered coal, oil or gas, the highest temperature of about 1550 °C being reached near the bottom end.

The energy consumption is about 5000 kJ/kg of cement, about half this being used to dry the slurry. The firing energy in a 50 kg bag of cement is hence about 250 MJ – the equivalent of 70 units of electricity.

Dry process

This requires non-cohesive materials, limestone and shale normally being used. They are ground and blended to a uniform free-flowing powder which is transferred to silos and then mixed to the desired composition. The powder is pre-heated by waste gases in a vertically arranged system of cyclones and then falls into a short kiln where nodules of clinker form.

A variant of this process, the semi-dry process, may be used for harder, drier ingredients. Nodules are formed by a water spray before firing. The dry and semi-dry processes require about 3000 kJ/kg of cement – only about 60 per cent that of the wet process.

Final processing of cement

The clinker is ground in large steel ball mills to particles mostly in the range 100 µm down to 10 µm to give satisfactory hardening properties, and gypsum is also added at this time to prevent the cement from flash setting when water is added. The proportions of gypsum used depend upon the composition of the cement, more being required for fine cements or those with high C_3A or C_3S contents. The proportion is usually about 5 per cent by weight. The energy required for grinding increases with the desired fineness. It is typically 100 kJ/kg of cement – much smaller than the energy used to produce the clinker – though this is electrical energy and therefore more expensive than the fossil fuel energy used to fire the kilns.

The final stage involves either bagging cement or transferral to tankers for transport in bulk. Bagged cement is ideal for use in small quantities but handling costs result in a price approximately 20 per cent higher than bulk cement.

HYDRATION OF PORTLAND CEMENT

Hydration is the process of chemical combination between the cement and water. It results first in setting (the cement becomes solid) and then hardening (increase of strength and stiffness). The processes are gradual and require the continuous presence of water. The reaction rate is in fact so slow that unground cement clinker can be stored outside for some days, or under cover for longer periods, only a very thin surface film becoming hydrated. The hydration rate depends on the surface area of clinker exposed, hence the fineness of grinding has a very significant effect, especially on the rate of hardening.

It must be emphasised that the atmosphere plays no part in hydration and the quite common belief that strengthening is associated with 'drying out' is erroneous, for drying out of concrete will bring the hydration process and therefore strength development to an end (though it may cause a small, short term, increase as explained in 'Shrinkage' on page 283).

It will be recalled that there are four chief compounds in Portland cement and that these react quite differently with water.

Immediately after adding water, the pH value of the mixture rises and laboratory tests show that the temperature of the mixture also rises rapidly. These effects are caused by reaction between the C_3A and water which is initially quite rapid since it takes some minutes for the gypsum to dissolve sufficiently to control the reaction of the C_3A. Thereafter, setting and gradual hardening take place by reaction of the C_3S and C_2S with water. In the early stages the C_3S is chiefly responsible for hydration. There is still some uncertainty as to the precise nature of the reaction between these compounds and water. They dissolve slowly and react chemically with the water to form a hydrate as follows (full chemical notation used here to enable equations to be balanced):

$$2(2CaO\ SiO_2) + 4H_2O \rightarrow 3CaO.2SiO_2.3H_2O + Ca(OH)_2$$

and

$$2(3CaO\ SiO_2) + 6H_2O \rightarrow 3CaO.2SiO_2.3H_2O + 3Ca(OH)_2$$

The hydration products shown here are the same in each case, though in practice a range of similar products may be obtained depending on the constituents, the presence of any impurities in them and the prevailing conditions. Evidence suggests that there may also be a bond between hydrated and unhydrated material – indicative of some additional surface reaction of the cement with water rather than simple dissolution.

Both the chemical and physical nature of these materials have a decisive effect on the behaviour of the cement product.

The calcium silicate hydrate takes the form of extremely small interlocking crystals which grow out slowly from the cement grains to occupy previously water-filled spaces. An indication of fineness is obtained when their specific surface (about 200 000 m^2/kg) is compared with that of unhydrated cement, about 300 m^2/kg. This microcrystalline material is responsible for strength in the hardened concrete and also for its susceptibility to moisture, for it absorbs water very strongly. The term *adsorption* is used. It is described as a *gel*, the nearest natural equivalent being *tobermorite gel*.

Equally important is the other hydration product, calcium hydroxide, which forms much larger crystals. These acts as fillers in the hardened cement but do not interlock and therefore do not contribute directly to strength. However if moisture is present in the concrete, the crystals partly dissolve to form an alkaline solution which is protective to any steel present so preventing corrosion of the metal. The pH value of saturated calcium hydroxide is approximately 12.5 though the pH of concrete pore water may be raised further to 13 or 14 by small quantities of sodium or potassium hydroxides which are present, especially in high alkali cements. Reinforced concrete depends on this alkalinity for its durability and it is unfortunate that calcium hydroxide at the surface of concrete is converted to calcium carbonate by atmospheric action, thereby becoming neutralised and ceasing to be protective to steel. To be durable the steel must be placed beneath this surface layer (see 'Carbonation of concrete', page 282).

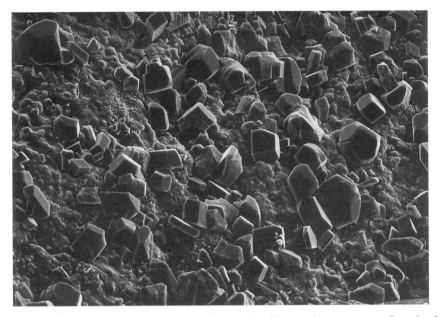

Figure 7.2   Scanning electron microscope photograph of hydrated cement paste. Crystals of calcium hydroxide are surrounded by a matrix of cement gel

The contrasting physical forms of cement gel and calcium hydroxide crystals can be seen in Figure 7.2 which is a scanning electron microscope photograph of the surface of hydrated cement.

The reaction of C_3A was mentioned earlier in that it tends to produce a flash set. Although vigorous, the reaction produces little strength. The gypsum, added to prevent a flash set from occurring, forms an intermediate compound, calcium sulphoaluminate (ettringite). Once the gypsum has been consumed, calcium aluminate monosulphate hydrate is formed. There should be sufficient gypsum for this reaction to occur fully – if gypsum becomes exhausted, the remaining C_3A may react quickly producing a sudden heat emission some hours after mixing. The monosulphate will revert to ettringite if the concentration of sulphates in solution rises as in sulphate attack. This will cause expansion and disruption of the hardened cement. It will be appreciated that the presence of C_3A, with its rapid setting, high heat emission and sulphate susceptibility, is undesirable in concrete. Alumina, the ingredient responsible, is found in most clays. It would be extremely difficult to remove it but the amount of C_3A in the cement can be reduced by addition of iron oxide as a raw material which combines with alumina in the kiln to form C_4AF instead. Some C_3A is helpful in manufacturing terms since it exists in the kiln as a liquid, facilitating silicate formation.

C_4AF makes little contribution to setting or strength but acts as a flux in the kiln and is responsible for the grey colour of cements; hence higher C_4AF cements will be darker.

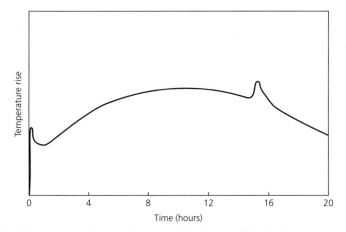

Figure 7.3 Temperature rise in early stages of hydration of Portland cements. Three distinct stages are apparent

The sequential stages in the early hydration of Portland cements are illustrated by a heat output curve for cement paste. Three distinct sections may be visible (Figure 7.3). The initial peak is due to uncontrolled C_3A hydration, followed by a much more gradual output as the C_3S and C_2S hydrate. A final short peak may occur if the mixture becomes short of gypsum, resulting in a sudden reaction of the remaining C_3A.

It should be added in conclusion to this section on the hydration of Portland cements, that the rate of hydration decreases continuously with age as the resistance to water penetration of unhydrated cement grains progressively rises. It is unlikely that hydration penetrates to a depth of greater than 4 μm even over long periods of time so grains of cement of size 10 μm or larger will always contain an unhydrated kernel.

WATER IN HYDRATED CEMENT

When it is considered that the volume of water in fresh concrete is normally greater than that of the cement (see page 247) it might be surprising that an impermeable product could ever be produced. The reason is that cement gel occupies over twice the space of the cement powder from which it is formed. An expansion of about 114 per cent occurs. For example, 1 g of cement (relative density 3.15) occupies

$$1/3.15 = 0.318 \text{ ml}$$

before hydration and, therefore, after hydration approximately:

$$0.318 + 0.318 \times 114/100 = 0.68 \text{ ml}$$

The increase of volume is $0.68 - 0.318 = 0.362$ ml.

Therefore if cement and water are mixed in the mass ratio 1:0.36, the volume of the hydrated material would be exactly the same as the volume of the original constituents. This water/cement ratio of 0.36 is one of the key values in understanding volume effects in cement products, though there are no sudden changes in performance at this value. There are two other key values of water/cement ratio which must be considered:

Water/cement ratio 0.23   This represents the water required for chemical combination as defined by chemical equations such as those on page 209.

Water/cement ratio 0.42   This is a more realistic value for full hydration of cement, the extra 0.19 being an allowance for water which will be absorbed into cement gel as it forms.

The effect of the water/cement ratio on hydration may therefore be related to the above figures:

Water/cement ratio below 0.23   Some hydration would occur but the hydrate formed initially would quickly adsorb any remaining water so that much of the cement would not be hydrated. The removal of the remaining water by cement gel is known as *self-desiccation* – it is noticeable that the inside of a well compacted cube of low water/cement ratio, crushed immediately after withdrawal from the curing tank, will often have a dry appearance, water being unable to penetrate to desiccated regions. There are practical difficulties with concretes of very low water/cement ratio in that compaction (air elimination) may be difficult. Nevertheless, if this can be achieved, for example, by a combination of pressure and vibration, very high strength products can be achieved. The porosity of such products is very low and the small amount of cement hydrate present is sufficient to 'glue together' the stronger unhydrated cement and aggregate.

Water/cement ratio 0.23–0.36   As the water/cement ratio increases in this range, the proportion of cement which hydrates will increase, especially if water can be 'imbibed', by curing to help offset desiccation. Hydration will however cease when all the free space around the cement/cement gel mixture, known as capillary pores, becomes filled with hydrate. Mixtures in this range of water/cement ratio will have very low porosity and permeability though strength will decrease with increasing water/cement ratio as the proportion of the weaker hydrate material increases.

Water/cement ratio 0.36–0.42   Figure 7.4 shows, schematically, the hydration of a cement paste of water/cement ratio 0.36. Once the value has increased to over 0.36 there will be capillary pores since the cement cannot now fill all the previously water-filled spaces. Hence, capillary porosity increases, as shown in Figure 7.5. Nevertheless, full hydration would still require extra water to be imbibed by curing. Strength continues to reduce with rising water/cement ratio – now due to the

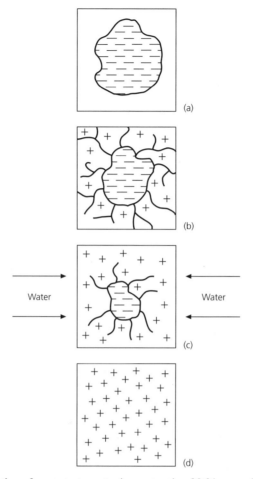

Figure 7.4 Hydration of cement at a water/cement ratio of 0.36, assuming extra water to be available during curing (the process would take some years to complete): (a) unhydrated cement and water; (b) hydrated cement containing capillary pores. Unhydrated cement kernel: (c) extra water required to fill capillary pores so that hydration can continue; (d) cement completely hydrated

increasing number of capillary pores as well as the increasing proportion of hydrate.

Water/cement ratio above 0.42 Hydration can now proceed to completion without the need for extra curing water, provided the initial water content is maintained by sealing against evaporation. This latter requirement becomes more acute as the water/cement ratios continue to rise since water escapes much more readily from such mixes in drying conditions. Strength continues to decrease with increasing water/cement ratio for the reasons given above.

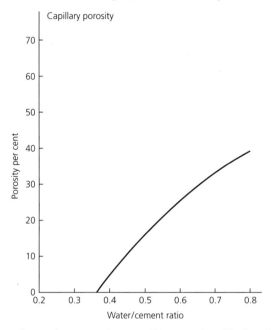

Figure 7.5   Effect of water/cement ratio on capillary porosity of hydrated cement paste

PERMEABILITY OF HARDENED CEMENT PASTE

The requirement that concrete be impermeable to water is of fundamental importance if it is to be durable since most aggressive agencies require penetration of water or aqueous solutions. Clearly the concrete can be no better than the cement paste matrix in this respect. Many specifications for durability attempt, usually indirectly, to control the permeability of this fraction. Cement paste becomes permeable, even if fully compacted, when capillary pores resulting from excess mixing water form continuous *canals* through the material. It is possible to have some capillary pores without leading to high permeability. However, the permeability of cement paste is highly sensitive to water/cement ratios in the practical range 0.4–0.7, since small increases of water in this range tend to lead to linked capillary pores. The effect is shown in Figure 7.6 which underlines the importance of controlling water content in concrete mixes if consistently good durability is to be achieved.

FACTORS AFFECTING THE PROPERTIES OF PORTLAND CEMENT

Modern construction methods often involve exacting specifications for concrete and it is therefore of prime importance that properties of the cement used be known and that they be as consistent as possible. The nature of the raw materials for cement (including fuel used for firing) is such that there is an inherent difficulty

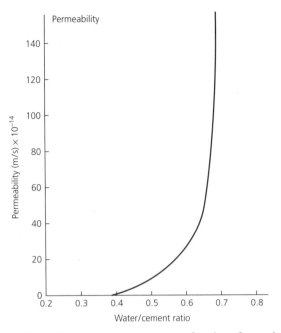

Figure 7.6   Permeability of hydrated cement paste as a function of water/cement ratio. Permeability is greatly increased at values of over about 0.5 since capillaries in such pastes remain connected

in producing totally uniform quality, since raw materials variations are bound to occur to some degree, from day to day as well as on a long term basis. However, a considerable amount can be done to minimise variations in cement properties and, when these do occur, it is usually possible to ascertain them by contacting the cement manufacturer on a day-to-day basis if necessary when very high levels of uniformity are required. The properties of cement are affected chiefly by its chemical composition and its fineness.

Chemical composition

The relative proportions of the four compounds already described depend on the relative proportions of the four constituent compounds, calcium oxide, silica, alumina and ferrite, which, in turn, depend on the relative proportions of clay and limestone, and the balance of minerals within the clay. Bogue produced equations which enable the quantities of the four compounds to be calculated:

$$\%C_4AF = 3.04\ (\%F)$$
$$\%C_3A\ = 2.65\ (\%A) - 1.69\ (\%F)$$
$$\%C_2S\ = 8.60\ (\%S) - 3.07\ (\%C) + 5.10\ (\%A) + 1.08\ (\%F)$$
$$\%C_3S\ = 4.07\ (\%C) - 7.60\ (\%S) - 1.43\ (\%F) - 6.72\ (\%A)$$

These equations might appear complex but they can be derived quite simply by consideration of the atomic weights of the component materials.

Note, for example, that C_4AF, the only compound containing iron, is determined completely by the percentage of F.

The percentage of C_3A is determined by the alumina content but with a deduction due to the alumina taken up by the C_4AF.

The C_2S and C_3S equations are more complex but note that, on adding the equations:

$$\%C_4AF + \%C_3A + \%C_2S + \%C_3S = \%F + \%A + \%S + \%C \quad \text{(approximately)}$$

The Bogue equations are useful, since the results of a normal chemical analysis of cement only give the percentages of the mineral constituents calcium oxide, silica, alumina and ferrite. The equations permit calculation of proportions of the active cementing ingredients and therefore of the likely properties of the cement. They also enable the effect of variations in the relative proportions of limestone and clay to be determined. For example, an increase in the calcium oxide content by 1 per cent would have the following effects:

silica   decrease by about 0.7%
alumina   decrease by 0.2%
ferrite   decrease by 0.1%

(based upon normal proportions in clay).

The change in C_2S is

$$8.60 \times (-0.7) - 3.07 \times (1) + 5.10 \times (-0.2) + 1.08 \times (-0.1) = -10.2\%$$

A similar increase of C_3S occurs and these changes will have a significant effect on the properties of the cement. Hence, it is essential that the balance between limestone and clay be carefully controlled.

Other methods of chemical analysis, such as microscopic examination of etched, polished sections of cement clinker, enable the compounds in cement to be measured directly.

British Standard 12 lays down limits for combustible or acid-soluble impurities, sulphates and chlorides. The amount of magnesium oxide in the clinker is also limited to 5 per cent because, like lime, it can cause unsoundness. The chemical composition of cement affects also the setting time and strength of cements but, since other factors also are involved, BS 12 measures these properties directly rather than in terms of the four main compounds described above.

(In any case, results using the Bogue equations, which are based on equilibrium between the four compounds in crystallised form, are subject to errors, since the compounds C_3A and C_4AF, which were formerly liquid, do not crystallise completely. There is always a proportion of 'glass' in the cement clinker and this affects the proportions of C_2S and C_3S.)

Soundness

If free lime in crystallised form in the cement clinker is present, this could lead to *unsoundness* – a slow expansion of hardened concrete caused by slaking. Since this is not easy to distinguish from the large quantity of chemically combined lime in cement, a separate soundness test for it is required (BS EN 196-3). The expansion of a pat of the hydrated cement after a period of boiling is measured, boiling being essential to accelerate hydration of the lime.

Fineness

The hydration of cement is a process which involves penetration of water into cement particles to produce a cement gel. Hence, a finer ground cement will hydrate more quickly and produce earlier strength. At the same time, more gypsum is essential to combat the extra C_3A revealed by the greater surface area of cement particles. Since, on drying, cement gel shrinks, finer cements will correspondingly exhibit greater initial drying shrinkage at an early age. Final strength and shrinkage values are, however, similar to those of ordinary Portland cement. Fineness is measured by the term *specific surface*, the average surface area of cement in m^2/kg. Fineness may be obtained by measurement of the permeability to air of a compacted cement bed of standard thickness, a finer cement being less permeable. Portland cements to BS 12 no longer have fineness requirements, since this is effectively controlled by other performance requirements, though cement of grade 42.5N can be expected to have a fineness in the region of 325 m^2/kg.

SETTING AND HARDENING PROPERTIES

Setting time

An initial and final setting time are defined empirically using a cement paste of standard consistency (BS EN 196-3). The initial setting time should not be less than 60 minutes for cements of grades 32.5 and 42.5 (a period related to the time after mixing required for placing, compaction and finishing of concrete). For higher grades the initial setting time is not less than 45 minutes. Final setting times are no longer specified in BS 12.

Compressive strength

The current BS EN 196-1 test for compressive strength of cements relates to samples of size $40 \times 40 \times 160$ mm, made from a 3:1 sand/cement mortar with water/cement ratio 0.5. The sample is first tested in flexure and then the broken pieces are tested in compression using a special 40×40 mm jig. The test employs a standard CEN sand of specified grading, the particle sizes being between 1.6 mm and 80 μm. A sand:cement ratio of 3:1 is employed with water/cement ratio of 0.4. Standard sands such as Leighton Buzzard sand should be checked against the BS EN 196 specification before being used. The mortar is prepared by mechanical mixing. Requirements for the compressive strength of mortar cubes are given in Table 7.2. Note that upper strength levels are required to guard against the use of cements of excessive fineness.

Table 7.2   Strength requirements of mortar cubes. Data taken from BS 12 (1996)

Class	Early strength (N/mm^2)		Standard strength (N/mm^2)	
	2 days	7 days	28 days	
32.5N	–	≥ 16	≥ 32.5	≤ 52.5
42.5N	≥ 10	–	≥ 42.5	≤ 62.5
52.5N	≥ 20	–	≥ 52.5	≤ 72.5
62.5N	≥ 20	–	≥ 62.5	–

An extra letter is added after the strength class to denote rate of early strength development. Most common cements are given the letter N, corresponding to normal development, while the letter R is used to denote a rapid rate of development.

OTHER TYPES OF PORTLAND CEMENT

Rapid hardening Portland cement

The above term is no longer used in Standards, though it is still in general use to refer to cements of grade 52.5N which is covered by BS 12 (Table 7.2). It is essentially different from ordinary Portland cement only in respect of its fineness and produces a strength approximately 50 per cent higher than ordinary Portland cement at three days, though long term strengths are similar. Accompanying the rapid early strength gain is a considerable evolution of heat so that rapid hardening Portland cement is often used in cold weather to assist in development of maturity. However, on the same account, it should not be used in mass concrete, where the heat concentration would cause a reduction in strength due to thermal stresses.

Ultra rapid hardening cements, which are very finely ground, are also available, the same arguments as above applying but to a greater degree. In addition, such a high degree of fineness in the cement tends to reduce workability and, therefore, there will be some loss of strength if the water content of the concrete is increased to compensate.

Sulphate-resisting Portland cement

Sulphate attack is due primarily to the effect of sulphates in solution of C_3A (see 'Hydration of cement'). Hence, the sulphate resistance of a cement will improve as the quantity of C_3A decreases; BS 4027 specifies a maximum content of 3.5 per cent. This can be achieved by the addition of iron ore to the raw materials so that more C_4AF is produced and the alumina is used up in this way instead of forming C_3A. In other respects, sulphate-resisting cement is similar to ordinary Portland cement. Note that even concrete made with sulphate-resisting Portland cement may be attacked physically by sulphates if porous, since, on drying, crystallisation will take place inside the concrete, causing expansion and disruption. Hence, the use of sulphate-resisting cement is no substitute for the production of dense, non-permeable concrete. Low alkali forms of sulphate-resisting Portland cement can be produced for use with aggregates which might pose a risk of alkali–silica reaction (see page 297).

Low heat Portland cement

This contains relatively small percentages of the compounds C_3S and C_3A which have the greatest heat evolution. BS 1370 requires a heat output of not more than 250 J/g by seven days and 290 J/g by 28 days. The rate of gain of strength is lower but this cement has an ultimate strength similar to that of ordinary Portland cement and is suitable for mass construction where heat concentration must be avoided. Concretes of high cement content must not, of course, be employed – they would negate the effect of the low heat cement. Its fineness, as required by BS 1370, must not be less than 275 m^2/kg in order to ensure satisfactory strength development. Note that, due to the small amount of C_3A, sulphate resistance of this cement is greater than that of ordinary Portland cement. Availability of the cement is limited because similar effects can be obtained with other cements described below.

Portland blast furnace cement

This consists of Portland cement clinker and gypsum, ground together, with up to 65 per cent ground granulated blast furnace slag, which contains lime, silica and alumina (BS 146). The hydration of the Portland cement initiates that of the latter (pozzolanic action), producing a strength progression similar to that of ordinary Portland cement, though early strengths may be lower, so that adequate curing is essential. Portland blast furnace cement has a higher sulphate resistance than ordinary Portland cement, hence its use in sulphate soils and for construction in or near the sea. Its use may also reduce the risk of damage due to the alkali–silica reaction (a slag content of at least 50 per cent is required for this purpose). A low heat form of Portland blast furnace cement is available and is covered by BS 4246. The heat output and strength development are similar to those of low heat Portland cement. Availability is limited to areas in which blast furnace slag is produced. Ground granulated blast furnace slag is also available separately giving greater flexibility in the proportion used; replacements of up to 90 per cent have been used in this way.

PFA cements

PFA – pulverised fuel ash – is a by-product from coal fired power stations and acts in a similar manner to blast furnace slag: it sets in the presence of lime released by hydrating Portland cement. It may be incorporated with Portland cement during manufacture with the advantage of accurate blending and ease of batching; or it may be used as an admixture, a minimum replacement proportion of 30 per cent being used to improve resistance to sulphate attack or ASR. Early strengths are lower than in corresponding OPC concretes so that adequate curing is essential. PFA cements are covered by BS 6588 [15–35% PFA] and BS 6610 [35–50% PFA].

White cement

This is made using china clay, which contains very little iron, the latter being responsible for the grey colour of ordinary cement. White Portland cement is one of the most expensive Portland cements, since there are few geographic locations where china clay is found, and modifications to the method of firing and grinding are required. In the absence of C_4AF, other fluxes may be used to assist in manufacture. The cement conforms to BS 12, though with relatively high C_3S contents the strength grade achieved is likely to be 62.5N (Table 7.2). White cement is used for the production of white or coloured concrete.

Hydrophobic Portland cement

This is ordinary Portland cement to which a small percentage of ground-in water repellent, such as oleic acid, is added. Such cements can be stored for a considerable time in damp atmospheres without subsequent deterioration. On mixing, the acid coating breaks down and behaves as an air entraining agent but mixing time should normally be about 25 per cent longer than for ordinary Portland cement. Early strength development of concretes using such cements is slightly reduced. The cement is only available to special order.

ORDINARY PORTLAND CEMENTS – A CURRENT PERSPECTIVE

Ordinary Portland cement is a high quality product designed to produce satisfactory setting and hardening properties at reasonable cost and without excessive heat of hydration and shrinkage. The basic production process remains unchanged but developments have nevertheless taken place, partly for commercial reasons in order to increase product competitiveness. Cements are finer than they were 50 years ago, typical fineness levels being around 325 m^2/kg for ordinary Portland cement. In composition terms, the main change has been in the C_3S level, typical percentages now being in the range 50–55 per cent compared with 40 per cent, 50 years ago. This results in more rapid hardening, a typical 0.6 w/c ratio concrete now having a 28 day cube strength of around 45 N/mm^2 compared with 30–35 N/mm^2 in the 1950s. Heat output in the first 24 hours is also increased, hence mass concrete in particular will become much warmer with consequent risk of damage in the case of modern cements. Of particular importance also, the strength/age relationship has altered, typical concretes reaching over 70 per cent of 28 day strength at seven days, compared with a little over 60 per cent 40 years ago. Increases from 28 days to one year have, by the same token, reduced from about 20 per cent to 15 per cent, giving less bonus over the 28 day strength design figure. The effect is shown in

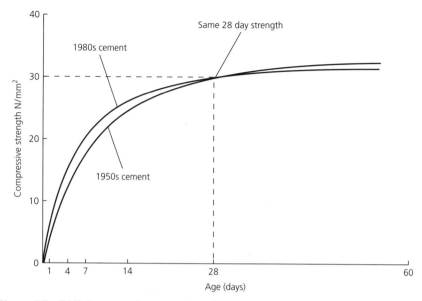

Figure 7.7   Differing strength progressions of 1980s cements compared with 1950s. If they are both used to produce a concrete of comparative strength 30 N/mm² at 28 days the 1980s cement would result in a concrete which is less strong in the long term

Figure 7.7 and it will be evident that concrete designed for a given 28 day strength using a modern cement would give lower long term strength than a corresponding mix using an older type. (The latter would require a lower water/cement ratio and hence more cement in the mix for a given workability.) A most important aspect of concrete performance is durability, known to be related to water/cement ratio. It is important, therefore, when taking advantage of the early age strength benefits of modern cements that economies are not made on cement content to the extent that durability is affected. The risk of problems will be further increased if curing is inadequate since higher water/cement ratio concretes lose water more quickly by evaporation when formwork is struck, especially if the concrete is warmed due to higher heat of hydration. As regards quality, there is little doubt that cements are now much more consistent and that variations in minor ingredients such as alkali levels, which could affect alkali-aggregate resistance, are smaller. Nevertheless, care is necessary when selecting cements which are now available from international as well as national sources and it is suggested that quality assured products should be specified, especially where strength or durability performance is of critical importance.

An alternative to ordinary Portland cement, *controlled fineness Portland cement*, has been in use for some years and is also covered by BS 12. Fineness values are a matter for agreement between supplier and purchaser though levels around those of the earlier (1978) standard, which requires a minimum fineness of 225 m²/kg, are likely. The cement gives a concrete which can be de-watered more easily, facilitating for example, production of pressed paving slabs, and the slower hardening makes it suitable for applications such as ready mixed or retarded mortars. It is not at present available in bagged form and is not similar to older cements in that it still has the higher C_3S contents characteristic of modern cements.

7.2 CEMENTS OTHER THAN PORTLAND CEMENTS

HIGH ALUMINA CEMENT

The cement was first developed in France, in response to the need for a concrete which would resist attack by sulphates. It is made by heating a mixture of limestone or chalk and bauxite (aluminium ore) in a special furnace, to a temperature of about 1600 °C. This causes complete fusion of materials. Cooling produces a clinker which is ground to give a cement with slightly lower specific surface than that of ordinary Portland cement.

The predominant compound in high alumina cements is CA (monocalcium aluminate), though compounds with other proportions of calcium oxide and alumina also exist, together with small proportions of iron, silicon, titanium and magnesium oxides.

On hydration of the cement, the chief compound produced is calcium aluminate decahydrate, CAH_{10} though some C_2AH is also produced. This means that, unlike Portland cements, there is no 'spare' calcium oxide since these compounds must be derived from CA as indicated above. Any excess of alumina produces an alumina hydrate gel. Setting is not rapid and BS 915 requires an initial setting time of not less than two hours compared with 45 or 60 minutes for Portland cements. However, once setting commences, strength development is much quicker than with Portland cements, the final set being not more than two hours after the initial set and the required strength of mortar cubes at 24 hours being 41 N/mm^2 minimum. Figure 7.8 shows comparative strength developments of a concrete mix with gravel aggregate, using various Portland cements and high alumina cement at a water/cement ratio of 0.6. This property of high alumina cement made it popular for precasting, allowing rapid turn-round of moulds. There is considerable early heat evolution (Table 7.3) so that water immersion or spray is necessary during the early stages to prevent excessive temperature rise. Alternatively, this heat could offset the effect of placing the concrete in subzero temperatures. The shrinkage of high alumina cement concrete is similar to that of ordinary Portland cement concrete.

Unlike Portland cements there is no calcium hydroxide in hydrated high alumina cement – indeed its sulphate resistance can at least in part be attributed to this. The question might then be asked as to how high alumina cement concrete protects steel from corrosion since it is largely the calcium hydroxide in Portland cements that affords this protection. It is found that the pH of high alumina cement concrete is quite high, in the region of 11–12. Although not as high as that of Portland cement based concretes this is sufficient for protection of steel. This alkalinity is afforded

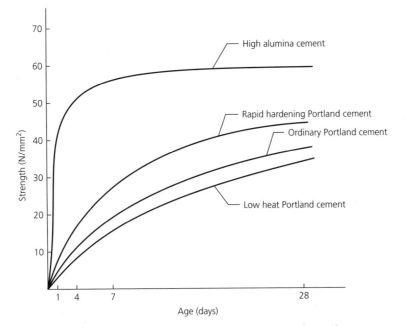

Figure 7.8 Strength development of gravel aggregate concretes using various cements at a water/cement ratio of 0.6

Table 7.3 Comparative heat outputs of common Portland cements and high alumina cement

Type	Approximate heat output in kJ/kg at normal temperatures			
	1 day	3 days	7 days	28 days
Grade 42.5N PC	240	280	310	380
Grade 52.5N PC	290	320	360	420
Low heat PC	160	180	220	260
Sulphate-resisting PC	180	210	240	300
High alumina cement	310	330	380	400

by the presence of hydrated alumina which is reasonably alkaline. Carbonation reactions (though not the same as those occurring in Portland cement concretes) do occur in which the CAH_{10} is converted into aluminium hydroxide and calcium carbonate. This leads to loss of alkalinity so that steel corrosion may occur as in Portland cement concretes if a permeable surface exists.

Conversion in HAC

If the temperature of high alumina cement hydrate is allowed to rise above about 25 °C in the presence of high humidity, the CAH_{10} becomes unstable, changing gradually into C_3AH_6 (tricalcium aluminate hexahydrate) and AH_3 (aluminium hydroxide), and this process may occur at any stage in its life:

$$3CAH_{10} \rightarrow CA_3H_6 + 2AH_3 + 18H$$

This change is accompanied by the formation of small pores, so that the resulting hydrate has consequently decreased strength and increased permeability. The process is known as conversion. It was originally thought that the strength reduction could be avoided if some unhydrated cement is still present during conditions of conversion, so that pores could fill with new hydrates as they are formed. It now seems that it is only at water/cement ratios of as low as 0.30 that strength loss is fully compensated. The loss of strength depends also on the rate of conversion, rapid conversion resulting in a more rapid loss. It is quite likely that concrete which is allowed to overheat during the first 24 hours could convert relatively rapidly at this stage. The conversion process occurs slowly even in normal environmental conditions and may reach completion after about 20 years especially where concrete is permanently wet – for example, below the water table or subject to leaks or condensation (Figure 7.9).

A further problem resulting from conversion is that high alumina cement concrete becomes more susceptible to chemical attack and, particularly, sulphates or alkalis, which might be leached out of neighbouring materials. The presence of chlorides also increases the rate of conversion. Note also from the equation above that conversion produces water which may dilute the pore fluid, reducing the pH value and therefore increasing the risk of steel corrosion.

Following the collapse of a school roof in 1974, high alumina cement, which for a number of years was widely used in precast units of all kinds, was withdrawn from structural use in the UK.

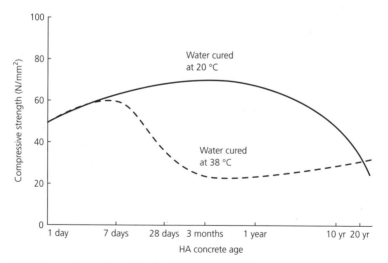

Figure 7.9 Strength variation with age of 0.4 W/C HAC concrete cured in water at temperatures of 20 °C and 38 °C. The strength at 20 years is similar in both cases. (Based on BRE Information Paper IP 8/88)

There are still many buildings employing high alumina cement elements, particularly floor and roof beams and the residual strength of these has been the subject of much research over the last 15 years. The degree of conversion can be estimated by differential thermal analysis (DTA) and similar techniques, while *in situ* strength can be obtained for core tests or pull-out testing. To satisfy the code of practice at the time (CP 116) a one day characteristic cube strength of about 50 N/mm^2 was recommended, implying a one day mean strength of over 60 N/mm^2.

Tests have shown that, on conversion, such concretes have a residual characteristic strength of about 21 N/mm^2, this figure providing a reasonable basis on which to check for structural safety of beams in use. Higher figures might be used where *in situ* tests justify this and there may be other cases where initial strengths did not reach CP 116 requirements so that the figure of 21 N/mm^2 may be too high. Many existing high alumina cement buildings are likely to remain in use for further long periods though continued appraisals will be called for as information on strengths at greater ages becomes available.

Other applications

A variety of other applications exists for high alumina cement, based predominantly upon the superior resistance of the material to chemicals such as sulphates. As a precaution against the effects of conversion, concretes of low water/cement ratio are, nevertheless, recommended, with careful attention being paid to compaction. Although high alumina cements require a water/cement ratio of about 0.5 for complete hydration, values below this do not lead to reduced strength, since there is chemical bonding between hydrated and unhydrated parts of cement grains.

A further important use of high alumina cement is as a refractory cement since, on desiccation by heat, a ceramic bond is developed which is retained at temperatures of up to 1300 °C. The cement is also used in lightweight concrete for flue linings.

Mixtures of high alumina cement and Portland cement

The setting time of such mixtures is reduced to a degree dependent on their proportions, a very rapid *flash set* taking place when between 30 per cent and 80 per cent high alumina cement is used. This is thought to be due to rapid setting of the Portland cement fraction owing to chemical combination of its retarder, gypsum, with calcium aluminate hydrates from the high alumina cement. The ultimate strength of such pastes increases with the percentage of high alumina cement but ultimate strengths are lower than those of ordinary Portland cement, unless at least 75 per cent high alumina cement is used. Mixtures may be useful when a very rapid set is required, for example, in sealing leaks and for making rapid fixings into masonry, but mixing of these cements is not generally recommended.

SUPER-SULPHATED CEMENT

This is made from granulated blast furnace slag, calcium sulphate (hence the name) and a small percentage of Portland cement clinker which behaves as an activator. The cement is highly resistant to sulphates and has a low heat output so that it may be used in mass concrete even in tropical conditions. The ultimate strength is similar to that of ordinary Portland cement. Water/cement ratios of less than 0.4 should not be used, since the cement has a high water requirement. Richer mixes than would be required with ordinary Portland cement are, therefore, essential for a given strength and workability. Super-sulphated cement is not suitable for steam curing since this retards its strength development. It must be stored dry. The cement is no longer manufactured in the UK, since sulphate-resisting Portland cement and other similar cements are more competitive in cost.

COMPARATIVE HEAT OUTPUT OF CEMENTS

It will be understood that the heat output of Portland cements depends principally upon the proportions of the constituents and the fineness of the material. Table 7.3 shows typical approximate values for several types of Portland cement together with high alumina cement for comparison. Note that the lower heat outputs of low heat and sulphate-resisting cements are due to lower C_3A contents. The long term heat output of high alumina cement is similar to that of ordinary Portland cement but its early heat emission is very much higher. Up until 6 hours after adding water there is little heat output but soon after the heat emission rises to about 3 times that of Portland cements with about 204 kJ/kg having been released by the age of 12 hours.

   These figures might be compared with the energy of manufacture of Portland cements of 3–6 MJ/kg. It might be argued that the energy released by conversion of cement back to a 'low energy' state by hydration should be comparable with that required to convert the raw materials to the 'high energy' state by manufacture. However even the dry process consumes about 10 times the energy released during setting and hardening processes. This is indicative of the large processing requirements, even of relatively improved production processes, and the fact that the materials must be heated to enable chemical change to be achieved.

7.3 AGGREGATES FOR CONCRETE

Aggregates may be described as clean, hard, inert material incorporated in concrete mixes. They usually serve the following main functions:

- They reduce the cost of concrete. Most aggregates are natural materials which require only extraction, washing and grading prior to transport to site.
- Well graded aggregates produce workable, yet cohesive, concrete.

- They reduce the heat of hydration of concrete since they are normally chemically inert and act as a heat sink for hydrating cement.
- They reduce the shrinkage of concrete since most aggregates are not affected by water and they restrain shrinkage of the hydrating cement.

Additionally, aggregates may serve the following purposes for specific applications:

- Control of surface hardness. Most aggregates have better abrasion resistance than hydrated cement. Where heavy abrasion is anticipated, hard, high strength aggregates such as granite or carborundum can be incorporated.
- Colour or light reflecting properties. Aggregates can be exposed to utilise their visual properties.
- Control of density. See below.
- Control of fire. See below.

In a typical medium strength concrete mix, the aggregates occupy about 75 per cent of the total volume and in lean (low cement content) mixes, the figure may be as high as 85 per cent.

TYPES OF AGGREGATE

Natural aggregates

The great majority of aggregates used for concrete are obtained from natural sources, either in the form of rock which is crushed to obtain the desired maximum size or gravel which is processed by crushing or screening oversized material. In any particular area, there is only likely to be one type of aggregate which is readily available and, due to high costs of transportation of aggregate, normal concrete will be made with this type.

Natural aggregates are now briefly classified petrologically, since there are some important properties which are specific to the various groups. The simplest means of classification is according to the mode of formation of the rocks concerned.

Rocks which were at one time molten are described as *igneous*. Examples are granite and basalt, and the flint group of aggregates is igneous in origin, the rock being broken up into gravel and often worn smooth by glacial action. When fine particles become cemented together over a period of time, the resultant rock is described as *sedimentary*, limestone and sandstone being of this type. Limestone has several important characteristic properties: it results in a concrete with a relatively low coefficient of thermal expansion and also excellent fire resistance. Abrasion resistance, conversely, is not generally as good as that of other aggregate types. Sandstones vary greatly according to source, some sandstones being of high strength but others being less strong and exhibiting significant moisture movement which is not found in aggregates generally. Where stones have been formed or modified by heat and pressure, a *metamorphic* rock is produced. The most important of these is marble, which is formed from limestone and used widely for its decorative effect, for example, in terrazzo flooring.

Artificial aggregates may be manufactured from natural or waste materials and their use, mainly in the form of lightweight aggregates, is rapidly increasing. They are described under the heading of 'Lightweight concrete'.

Recycled aggregates

Quite large quantities of crushed material with the potential for use as concreting aggregates are now available. The chief sources are:

■ demolition waste
■ crushed concrete
■ crushed brick (for example reject material)

It seems very likely that increasing levies will be imposed on aggregates obtained from land based sources and this would increase the commercial viability of aggregates obtained by recycling processes.

The chief characteristics of recycled materials are:

1 Densities are in general lower and more variable than natural aggregates.
2 Aggregate absorption is likely to be higher depending on type.
3 Material variability may be a problem.
4 The strength of the product may have a 'ceiling' value depending on source.
5 Shrinkage and moisture movement may be higher.

Provided a reliable source of material can be located (and availability is rapidly increasing) recycled aggregates may well provide an economical solution for a number of applications. Further details of performance levels are given on page 323.

AGGREGATE PROPERTIES AND THEIR EFFECT ON THE PROPERTIES OF CONCRETE

Density

Aggregate densities may vary within wide limits, according to the type of material. Some lightweight aggregates have densities below 500 kg/m^3, while others such as magnetite, have solid densities of over 7000 kg/m^3. In general, natural aggregates having solid densities in the region of 2600 kg/m^3 are most widely used, the former types finding application where low or high density concretes are required, respectively.

The solid density of aggregates is described by the term *relative density* (formerly, *specific gravity*), which is the ratio of the density of the aggregate to that of water. It is often quite important to have an accurate knowledge of aggregate density; for example, it affects the yield of a mix which is batched by weight and, in some types of concrete, the theoretical density of the compacted material must be calculated. Reference will, therefore, be made to the three definitions of relative density which arise according to the treatment of voids in the aggregate. An aggregate particle has

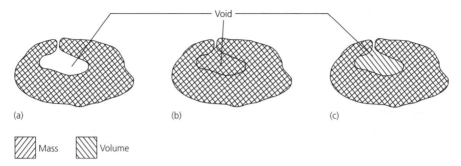

Figure 7.10 Representation of three methods of measuring relative density of aggregate: (a) apparent; (b) saturated, surface dry; (c) oven-dry

been drawn three times in Figure 7.10, the pores to which water has access being represented, for simplicity, as a single enclosure in the particle. The definitions are:

$$\text{apparent relative density} = \frac{\text{mass of dry aggregate}}{\text{solid volume of aggregate}}$$

$$\begin{aligned}\text{saturated surface dry} \\ \text{relative density (SSD)}\end{aligned} = \frac{\text{mass of aggregate + water in pores}}{\text{volume of aggregate including pores}}$$

$$\text{oven-dry relative density} = \frac{\text{mass of dry aggregate}}{\text{volume of aggregate including pores}}$$

The apparent relative density is highest, since permeable voids are, in effect, treated as external to the aggregate. The term is used for the calculation of theoretical dry density in tests for lean concrete for road bases. The SSD state is assumed for mix design purposes. The lowest value, resulting from the oven-dry formula, would be used to predict the density of the hardened concrete in the dry state.

Maximum size

Aggregates are classified as coarse aggregate if they are largely retained on a 5 mm mesh sieve and sand (formerly fine aggregate) if the majority passes a 5 mm sieve. The maximum size of aggregate to be used in concrete is governed mainly by the dimensions of the structure for which it is required. As a general rule, the largest size of aggregate should not exceed 25 per cent of the minimum dimensions in the structure, particular care being taken with the spacing of bars in heavily reinforced concrete. In most constructional engineering and building applications, the normal maximum aggregate size is 20 mm, while in roads it may be 40 mm, and larger still in mass construction, such as dams. In the latter, occasional large lumps of rock or masonry known as *plums* are included in addition to the mixed concrete, though these should not occupy more than about 30 per cent of the total volume of concrete. Since increasing the maximum size of aggregate reduces the surface area of aggregate to be bonded in a given volume, the cement and water requirement for a given strength may be reduced, producing a more economical mix. In mass concrete, this will reduce the heat production of the concrete, though above about 40 mm maximum size, higher strength concretes tend to become weaker, due to failure of the now heavily stressed bonds between aggregate and cement paste. This effect also tends to occur in very high strength concretes, for which relatively low maximum size aggregates may be preferred. The sand should also be matched to the maximum size of coarse aggregate: for example, a sand with fine grading combined with a large coarse aggregate may lead to a concrete of low cohesion.

Shape

This may vary between *rounded* (implying water-worn material) and *angular* (material with clearly defined edges, produced by crushing). Particles between these extremes would be classed as *irregular*, that is, having rounded edges. In normal concretes, angular material tends to produce concrete of lower workability but higher strength for a given water/cement ratio. In high strength concrete, workability is not affected in this way – some angular aggregates may produce higher workability than rounded aggregates. Other possible shapes include *flaky* and *elongated* but the non-isotropic nature and relatively high specific surface of these detracts from their value as aggregates for concrete.

Surface texture

The two extremes are *rough* or *honeycombed* textures, which will provide an extremely good key to cement, and *glassy* surfaces which do not form a strong bond with cement. Intermediate possibilities are *smooth* or *granular*. Although rougher surfaces will tend to reduce workability, they also result in increased strength – subject, of course, to the aggregate being itself satisfactory in other respects.

Crushing strength

There is little point in attempting to obtain high strength concrete using weak aggregates, since the latter, which generally have low E values, will take only a small stress under load, the cement paste being thus overstressed and failing at relatively low loads. Similarly, a high strength aggregate, such as crushed granite, would be best utilised where a high strength concrete is required. The extra cost of such materials would be of little benefit in the medium strength range. A guide to the suitability of a particular aggregate from the strength point of view may be obtained by inspection of crushed concrete cubes made with that aggregate. If a significant number of fractured aggregate particles is visible (more than, say, about 25 per cent), then it is likely that improved strength would be obtained with stronger aggregate. In practice, there is of course, the question of availability of stronger material and it may be more economical to provide more cement per unit volume in a mix than to transport stronger aggregates over considerable distances.

Grading

This term is used to describe the relative proportions of various particle sizes between the nominal maximum aggregate size and the smallest material present, which passes a 150 μm sieve. The object of grading aggregates is to assist in producing concrete with satisfactory plastic properties (workability, cohesion and resistance to bleeding), as well as satisfactory hardened properties (strength, durability and surface finish), using as little cement as possible. The need for grading arises from the requirements, first, that aggregates fill as much as possible of the total space and, secondly, that the aggregates and cement, being much denser than water, tend to settle, while the mixing water tends to rise. A well graded aggregate will ensure that there are no large volumes of cement paste and that settlement of solids is minimised by particle interference. By employing continuously graded material, voids between larger particles can be filled efficiently with slightly smaller ones, the process repeating down to and through the cement grain sizes which are themselves graded. Figure 7.11 is a photograph of a sawn section of concrete, giving an indication of the aggregate grading. Gradings are commonly represented graphically as the percentage passing a given BS sieve size

Figure 7.11   Section through a concrete cylinder showing grading of aggregate particles. There is perhaps an excess of smaller sizes in this section, together with some air voids – about 1 per cent – see BS 1881 Part 120

against size, sizes decreasing in ratios of about 1/2 from maximum size to 150 μm. A theoretical idealised curve is the Fuller curve in which:

$$\% \text{ passing a given sieve size} = \sqrt{\frac{\text{sieve size}}{\text{maximum size}}} \times 100$$

For example, taking a 20 mm aggregate, the proportion passing the 5 mm sieve would be

$$\sqrt{\frac{5}{20}} \times 100 = 50\%$$

A possible exception to this grading rule applies to sizes in the range 5–10 mm which can be omitted (gap grading) with some benefits on workability. Such mixes have been used for the special visual effects on exposing the aggregate or for applications such as sea defences where exposure of the larger stones helps resist abrasion of the sea. There is, however, an increased danger of segregation, especially of lean, wet mixes and the cost of the specially graded aggregates may be increased. Figure 7.12 shows typical grading curves, originally produced for

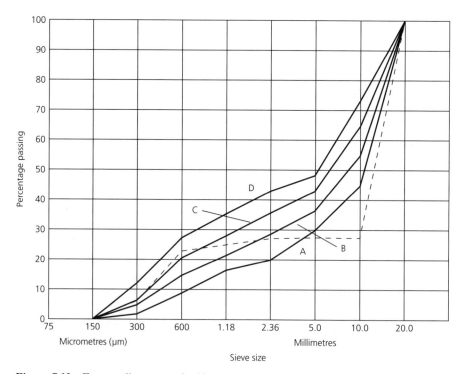

Figure 7.12   Four grading curves for 20 mm aggregates, given in Road Note 4, together with a gap-graded curve (indicated by the dotted line). (14 mm sieve requirement, introduced in 1992, not included)

the construction of concrete roads in the 1950s which have been found to produce satisfactory results, together with a *gap graded* curve. The coarser curves (A and B) would be satisfactory for higher cement content mixes while curves C and D (which approximate better to the Fuller curve for this aggregate size) would be suitable for leaner mixes. Having given these guidelines, experience has shown that with careful mix design good quality concrete can be produced with aggregates which do not conform to idealised grading curves.

Of central importance to any aggregate used for concrete of good quality and appearance is that aggregate gradings should be consistent from batch to batch and, on larger projects, throughout the construction period. If aggregates are stored in single stockpiles containing all fractions from maximum size to 150 μm there is a strong tendency to segregate, larger sizes tending to fall to the foot of the stockpile sides. This can be avoided if coarse aggregates (greater than 5 mm size) are stockpiled separately and then combined with sand by careful weigh-batching at the mixing stage. Better still, or if a large maximum size is employed, would be division of the coarse aggregate into two sizes, for example, 40–20 and 20–5 mm, though a cost penalty is obviously incurred. Sand has a fairly high natural cohesion provided, as is normally the case, it is damp. Segregation is therefore not likely.

Table 7.4 shows three alternative ways of obtaining aggregates for concrete requiring a 20 mm maximum size – the one in most common use – together with applications. BS 882 gives grading *envelopes* for these and other aggregate types. There should be little trouble in producing satisfactory concrete if aggregates fit the appropriate envelope. There are now three zones for sand, replacing the four more restrictive zones (zone 1, coarsest, to zone 4, finest) of the previous BS 882. Quite often, sands were borderline between zones, though with the new gradings this is less likely. Sands should be classified as coarse, medium or fine according to how their grading compares with figures given in Table 7.5. Some sands may fit in more than one zone, in which case they are given the classification of the finest zone into which they fit.

Table 7.4   Three alternative ways of batching graded aggregate of maximum size 20 mm

No. of stockpiles	Aggregate size			
	20 mm	10 mm	5 mm	150 μm
1		20 mm 'all in' aggregate (ballast)		
	←			→
2		20 mm graded course aggregate		sand
	←		→←	→
3	20 mm single size coarse aggregate	10 mm coarse aggregate		sand
	←	→←	→←	→

Table 7.5 BS 882 aggregate gradings (coarse, medium and fine) for sand. The 15 per cent figure for the 150 μm sieve can be increased to 20 per cent for crushed rock sands except when they are used for heavy duty floors

BS sieve size	Percentage passing BS sieve			
	Overall limits	Additional limits for grading		
		C	M	F
10 mm	100	–	–	–
5 mm	89–100	–	–	–
2.36 mm	60–100	60–100	65–100	80–100
1.18 mm	30–100	30–90	45–100	70–100
600 μm	15–100	15–54	25–80	55–100
300 μm	5–70	5–40	5–48	5–70
150 μm	0–15	–	–	–

Quality

This term manifests itself in a number of ways:

- Silt. In the context of aggregates this may be defined as material composed of particles passing a 75 μm sieve. (BS 882 refers to such materials as *fines*.) Owing to its high specific surface, its presence requires additional water for a given workability. Also, since such materials are often of a clayey nature, they decrease the bond between aggregate and cement, reducing the strength of concrete.
- Organic impurities. Such material, being acidic, reduces the alkalinity of cement paste which is essential for its hydration, thereby affecting setting time and strength. Impurities occur mainly in the form of vegetable matter such as top soil or leaves, usually resulting from poor stockpiling practice.

Other impurities are:

Soluble salts   These are present in marine aggregate and will leave a thin deposit on each particle unless washed in fresh water. Salts tend to accelerate the early hydration of cement but, more importantly, can cause corrosion of embedded metal. In reinforced or prestressed concrete, the total salt content must be carefully controlled.

Reactive or unsound inclusions   Some aggregates contain mineral particles which react to high alkali levels (see 'Alkali–aggregate reaction'). When there is any doubt, tests whould be made since this is a serious problem which is difficult to rectify in hardened concrete. A further, less serious, form of contamination is pyrites – a form of iron ore. Even quite small particles, on weathering, produce unsightly rust stains on the surface of affected concrete.

Moisture

Almost all aggregates contain some moisture, the important requirement for batching of aggregates being to establish whether this water will contribute to

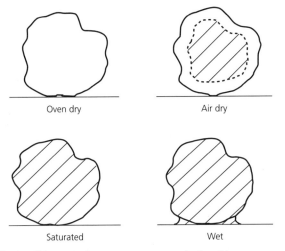

Figure 7.13   The possible states that an aggregate may be in with respect to moisture content

workability or whether water additional to the quantity based on the mix design will be necessary, due to absorption of aggregates. There are four possible states in which an aggregate may exist (Figure 7.13.)

■ Oven-dry: implying that, on heating to 105 °C, there would be no loss of weight. This state rarely occurs in practice.
■ Air-dry: there is no free moisture and surface layers of aggregate are dry. This state occurs in upper parts of aggregate stockpiles in dry weather.
■ Saturated, surface dry: this is the ideal state for an aggregate for concrete, since it requires no alteration to mixing water.
■ Wet: surplus moisture is present. This is the most common state.

The first two conditions will required extra water to be added at the mixer, especially in the case of lightweight aggregate concrete, since these aggregates are highly absorbent.

The last will require a deduction of water at the mixer equal to the total free moisture present in the batch of aggregate. Sand normally holds greater amounts of free moisture than coarse aggregate since, in a given mass, there are more points of contact between the larger numbers of particles involved. Water is held at these points by capillarity.

7.4   PRINCIPLES OF PROPORTIONING CONCRETE MIXES

There have been many attempts at producing rigorous design methods which would give accurate batch quantities for a wide range of performance and materials specifications, though the methods are usually unreasonably complex and difficult to operate. The principles of mix proportioning and design are quite simple in essence and satisfactory mixes can easily be produced provided some time is given to testing and modification of trial mixes. Hence, the emphasis here will be on the basics of mix proportioning and methods of adjusting trial mixes, should this be necessary.

The design and selection process must result in a concrete mix of appropriate proportions in the fresh and hardened states at lowest possible cost. These properties and their implications in mix terms must be examined.

FRESH PROPERTIES

There are two chief properties that must be considered: *workability* and *cohesion*.

Workability

This may be defined as the ease with which concrete may be placed, compacted and finished. Of these three operations, compaction (elimination of voids) is the most important. Workability is chosen according to the method of compaction to be used; high workabilities require little compaction while low workabilities require intensive vibration, sometimes together with applied pressure. Tests for workability are given below (page 252). The following are the chief factors affecting workability:

Water content   Water is the main lubricant in concrete mixes and small changes in water content produce marked changes in workability. An increase, for example, of 30 litres (or 30 kg) from 160 to 190 litres in one cubic metre of concrete containing over 2300 kg of material often alters the workability from low to high.

Aggregate type and size   Crushed or smaller maximum size aggregates have a larger surface area per unit mass (specific surface). They therefore increase the water demand in concrete. Other factors may be involved, for example, aggregate grading or cement fineness, though the effect of these is less pronounced than the above factors and they can be allowed for in trial mixes.

Cohesion (cohesiveness)

This is the ability of the concrete to resist segregation – separation of the constituents resulting from their differing densities. Cohesive mixes will also be easier to finish effectively than 'harsher' mixes. Cohesion is provided by the finer fractions of sand and by the cement. It follows that rich mixes of a given workability will tend to require less sand. Wet mixes are more prone to segregation and smaller aggregate sizes may also cause problems in this respect. Note also that a smaller maximum aggregate size increases sand requirements. It is important to appreciate that high cohesion (*fatty*) mixes require more water and would, therefore, require more cement, be more costly and would have higher heat of hydration and shrinkage for concrete of a given strength. Hence, the lowest satisfactory cohesion for a given application should be used.

Bleeding is a specific form of segregation involving separation of water from the concrete though cohesive mixes should not be affected by this problem.

HARDENED PROPERTIES

There are a great many properties of hardened concrete that might have implications in mix design terms but because many of these properties correlate to strength and in order to keep design methods as simple as possible, strength tends to be used as an index of all-round performance. This also applies to some degree to the question of durability though this will also be considered specifically in relation to mix design in view of its importance.

Strength

In the UK this is normally taken as compressive strength obtained from cube tests, though other strengths can be inferred from known relationships. Compressive strength is affected by the following factors:

- Cement type. Ordinary and rapid hardening Portland cements are normally taken for design purposes.
- Water/cement ratio. This implies the free water/cement ratio (by weight) at the time of setting and, therefore, does not include any water absorbed by aggregates. The effect of water/cement ratio is critical since it determines the capillary void content of the concrete.

  An approximate guide is:

water/cement ratio	strength
0.40	high strength
0.60	medium strength
0.8	low strength

  These values represent the normal working range though higher and lower values may be taken in special cases.
- Aggregate type. Crushed aggregates such as granite give increased strength at a given water/cement ratio particularly at high strengths where increases of about 7 N/mm² may be obtained.
- Age. The age of 28 days is most commonly used for specification purposes but strength at ages of 3, 7 and 91 days can also be used as the basis of design.

Durability

The best overall guide to resistance to water related problems such as frost damage and sulphate attack is permeability which is, in turn, closely related to the water/cement ratio as explained under 'Water in hydrated cement'. The figures of 0.4,

Table 7.6 BS 8110 requirements for durability of reinforced concrete. The mix details given in each category represent the minimum quality acceptable. Where higher concrete grades are used steel cover may be reduced. Air entrainment is recommended for very severe and most severe exposure conditions in which case lower concrete grades may be used

Exposure condition	Summary of exposure condition	Maximum free water/ cement ratio	Minimum cement content kg/m^3	Lowest grade of concrete	Minimum actual cover to steel for stated concrete grade mm
Mild	Protected against weather	0.65	275	C30	20
Moderate	No freezing whilst wet	0.60	300	C35	30
Severe	Saturation and occasional freezing	0.55	325	C40	35
Very severe	Saturation freezing and de-icing salts	0.55	325	C40	45
Most severe	Sea water tidal zone	0.45	400	C50	45

0.6 and 0.8 referred to above would result in good, medium and poor durability, respectively, assuming an impervious aggregate were used. However, with increasing emphasis being placed on the use of water reducing admixtures, some authorities may specify a minimum cement content to guard against the use of very low cement contents which might then be feasible. BS 8110 gives requirements for each durability category – maximum free water/cement ratio, minimum cement content and lowest concrete grade together with minimum cover to reinforcement appropriate to each grade since the concrete must protect the steel from corrosion (Table 7.6).

Note that while concrete grades give a general guide to durability, other factors must also be considered, for example, a granite aggregate concrete of a given strength would give a *less* durable concrete than a gravel aggregate of similar strength because it could be produced with a higher water/cement ratio.

SELECTION OF MATERIALS

This must be a compromise between the ideal for the particular purpose, the materials readily available and the relative costs. For example, there is a wide range of cements on the market and, although ordinary Portland cement is the cheapest, there may well be a reason for using a special cement, perhaps, if early or very early strength is required; where ordinary Portland cement would induce thermal stresses; or where extra heat would be advantageous, such as in winter concreting. It may, on the other hand, be worth considering the use of a modified ordinary Portland cement mix as an alternative to special cements. There is often little choice in selection of aggregates, since locally available materials are usually far cheaper than 'imported' aggregates. However, the use of lightweight aggregates is increasing and their use might be considered. Similarly, use of recycled aggregates might be considered for some applications assuming a supply is available. Admixtures can undoubtedly be of benefit in many situations, although their use constitutes another possible source of error, particularly in view of the very small dosages normally required.

PRELIMINARY TESTS ON MATERIALS

Cements

Provided cements comply with the appropriate British Standards, there should be no need for cement testing unless there are special requirements, for example, resistance to alkali–silica reaction. Cement tests are in any case not designed for site use and should be carried out by experienced personnel. Cement manufacturers are normally able to give detailed properties of cement on a regular basis if required. Concrete mixes may give higher strengths than design figures, which are usually quite conservative. Cement must, however, be fresh, any lumps present crumbling easily under slight pressure. Where surface finish is important, it is advisable to arrange for cement to come from a single source and to check that both the required volume and consistency of product can be reliably obtained.

Aggregates

The following tests are important, the first three acting as a guide to the overall suitability of a given aggregate and the fourth enabling the water content of the mix to be accurately calculated. The latter test should also be carried out regularly during production and the other tests at intervals or when there is cause for doubt. Relative densities of aggregates and bulk density of coarse aggregate may be required by some design methods. BS 812 describes methods for these.

Grading A grading curve for the aggregate is obtained using the BS standard sieves, ranging from the nominal maximum size of the aggregate to 150 μm.

(Note that in the 1992 edition of BS 882 a 14 mm sieve was introduced since, occasionally, 14 mm aggregates were found to comply with 20 mm grading curves. This sieve size is omitted below for simplicity.)

It is essential to obtain a representative sample of aggregate – a poorly produced stockpile may itself contain segregated sizes, larger sizes tending to be at the bottom. Several samples should be taken and mixed, and then divided, either by quartering, or by a riffle box to give a quantity suitable for sieving. The aggregate should be dry and too large a quantity will give false analysis due to 'blinding' of sieve apertures by particles. Table 7.7 shows gradings of a typical sand and 20 mm coarse aggregate, indicating the method for calculating the *fineness modulus* which summarises the aggregate grading in a single figure and can be useful for checking day-to-day variations in fineness (a high number means low fineness). The grading curves should be compared with BS 882 envelopes for sand and coarse aggregates,

Table 7.7 (a) Details of a sieve analysis of 0.5 kg of a sand. Comparison of the 'percentage passing' column with Table 7.5 shows that the sand just fits the type F zone. From below, the fineness modulus is 2.62. (b) Details of a sieve analysis of 5 kg of 20 mm coarse aggregate. The sum of percentages retained is equal to 260 and this together with the four smaller sieve sizes not used here makes the total percentage retained for fineness modulus calculations equal to 660. Hence the fineness modulus is 6.60 (14 mm sieve omitted for simplicity)

(a)

Sieve size	Amount retained g	Cumulative amount retained g	Percentage retained	Percentage passing
10.0 mm	0	0	0	100
5.0 mm	0	0	0	100
2.36 mm	85	85	17	83
1.18 mm	65	150	30	70
600 μm	65	215	43	57
300 μm	170	385	77	23
150 μm	90	475	95	5

Sum of percentages retained = 262.

(b)

Sieve size mm	Amount retained kg	Cumulative amount retained kg	Percentage retained	Percentage passing
37.5	0	0	0	100
20	0	0	0	100
10	3.25	3.25	65	35
5	1.50	4.75	95	5
2.36	0.25	5.00	100	0

the sand zone classification being obtained. The simplest method for combining gradings is illustrated in Table 7.8 using a 30:70 sand:coarse ratio, together with grading curve B from Figure 7.12 for comparison.

Silt ('fines') test   This is mainly a problem in relation to sand and a rapid assessment of silt content can be obtained by a field settling test (no longer a BS test). A sample of sand is shaken in a measuring cylinder containing salt solution. On allowing it to stand, silt and clay settle slowly as a distinct layer above the aggregate, the salt in solution helping to flocculate particles which otherwise might stay in suspension. If the result is greater than 8 per cent by volume, the material is suspect. BS 812 describes a more accurate method for determination of fines content by weight and this should be used if in doubt. Where excessive silt is present, the sand should be rejected or, if this is not possible, the cement content should be adjusted and trial mixes made to test suitability of the concrete. The cleanliness of coarse aggregate can be ascertained by visual inspection or, in cases of doubt, by a decantation method, again in BS 812, limits being given in BS 882.

Table 7.8   Calculation of combined grading of fine and coarse aggregates combined in the ratio 30:70 (14 mm sieve omitted for simplicity)

Sieve size	mm					μm		
	20	10	5	2.36	1.18	600	300	150
Coarse	100	35	5	0	0	0	0	0
Fine	100	100	100	83	70	57	23	5
70% coarse	70	24.5	3.5	0	0	0	0	0
30% fine	30	30	30	24.9	21.0	17.1	6.9	1.5
Obtained	100	54.5	33.5	25	21	17	7	1
Required (curve B, Fig. 7.12)	100	55	35	28	21	14	3	0

Organic impurities   Almost immediately after mixing, the calcium hydroxide liberated in solution in concrete raises its pH to a strongly alkaline value (12.40 or above, in the case of the standard mortar paste). Organic impurities, for example, topsoil, if present in significant quantities, will reduce the pH value of the concrete and, hence, affect setting and strength properties. When in doubt, concrete cubes should be made.

Moisture content   In the case of aggregates containing free moisture, it is the free moisture only which is required, since absorbed water will not affect the properties of the concrete. This is determined most commonly by one of the following methods:

■ Oven-drying method. This is a direct and simple way of determining moisture content. Results may be determined very quickly by use of microwave ovens. These would give total moisture content, since microwave ovens cause evaporation of absorbed, as well as surface, moisture.
■ Siphon can (BS 812). This is based on the principle that a given mass of aggregate containing free water will occupy a larger volume than the same mass of a dry aggregate. The volume of each is obtained by displacement of water in a specially designed can, suitable for site use. A sample of dried material is, however, necessary. If the free moisture content of the aggregate is required, this sample should be surface-dry only. If the total moisture content is required, it should be oven-dry. This test relies heavily upon constant relative density of material from sample to sample.
■ Chemical method. A rapid value can be obtained by mixing a sample of the damp aggregate with an excess of calcium carbide in a pressure vessel. The calcium carbide reacts with the moisture, producing a gas the pressure of which is proportional to the quantity of moisture present. The moisture content reading will correspond to a value somewhere between the free and total moisture content, depending on the degree of crushing of the material.

It is most important to note that moisture content often varies from one part of an aggregate stockpile to another, as well as with time. Great care should be taken in obtaining a representative sample, especially when the final sample is small, as in the chemical method.

In the case of absorptive aggregates, the absorption coefficient would also be required. BS 812 gives a method for this.

Quality of water

This is important, since contaminated water may lead to impaired performance of the hardened concrete. As a general rule, if water is suitable for drinking, then it is likely to be suitable for making concrete. If there is any doubt, it is advisable to carry out the setting time and strength test of BS EN 176, using samples of the water in question and also distilled water for comparison. BS 3148 indicates that a strength reduction up to 20 per cent in the former is acceptable.

Sea water is generally suitable for ordinary concrete, since the effect of salt in the normal concentrations found is slight, but it is not recommended for reinforced or prestressed concretes. A further problem of sea water is the increased likelihood of efflorescence resulting in unsightly salt deposits, particularly if drying is concentrated at certain positions in the structure. Many salts are also hygroscopic and, therefore, tend to perpetuate dampness at positions where they build up.

Examples of moisture content corrections

Non-absorptive aggregates   Calculate corrected batch masses of aggregate and water for a concrete mix if the given quantities are:

coarse aggregate	1300 kg
sand	500 kg
water	150 litres

and the sand and coarse aggregates contain 5 per cent and 3 per cent moisture, respectively, based on wet weight.

1300 kg of coarse aggregate contains $3/100 \times 1300 = 39$ litres of water and, therefore, only 1261 kg of coarse aggregate. Hence, use $1300 \times 1300/1261 = 1340$ kg of coarse aggregate.

500 kg of sand contains $5/100 \times 500 = 25$ litres water and, therefore, only 475 kg of sand.

Hence, use $500 (500/475) = 526$ kg of sand. These contain:

$$1340 \times 3/100 + 526 \times 5/100 = 40 + 26 = 66 \text{ litres water.}$$

Therefore, water to be added at the mixer is $150 - 66 = 84$ litres.

Absorptive aggreagates   Repeat the above example, assuming the aggregates have absorption coefficients of 5 per cent and 10 per cent by weight of dry material, sand and coarse, respectively. In this case, the given batch quantities of coarse and sand would be based on the saturated surface dry (SSD) condition. The corresponding batch quantities of dry and damp aggregates are:

If dry   *If damp*

coarse aggregate $1300 \times \dfrac{100}{110} = 1182$ kg $1182 \times \dfrac{100}{97} = 1219$ kg

sand $500 \times \dfrac{100}{105} = 476$ kg $476 \times \dfrac{100}{95} = 501$ kg

The water absorbed is based on the dry materials and is

$$10 \times \frac{1182}{100} + 5 \times \frac{476}{100} = 142 \text{ litres}$$

The water content is based on the wet materials and is

$$\frac{3}{100} \times 1219 + \frac{5}{100} \times 501 = 62 \text{ litres}$$

The corrected batch quantity of water is

$150 + 142 - 62 = 230$ litres

Note that the total mass of material is

$1219 + 501 + 230 = 1950$ kg

This equals the sum of the original batch quantities.

7.5   SPECIFICATION OF CONCRETE MIXES

The object of a specification is to ensure that the concrete is of quality and performance consistently appropriate to its application. There are four basic ways of specifying a concrete mix.

NOMINAL MIXES

This is the simplest and least rigorous method of specifying concrete. For oversite concrete or strip foundations it may be sufficient to specify ratios such as

1 part cement; 2 parts sand; 4 parts coarse aggregate

or even simpler:

1 part cement; 6 parts all-in aggregate (ballast)

In addition the aggregate size and workability should be stated.

It may be satisfactory to batch concrete by volume (for example using gauging boxes) for such mixes. They may produce perfectly adequate concrete, but offer little control over variations and so would not be acceptable where structural concrete is specified.

Table 7.9   Standard mixes based on grade 42.5N cement. Data taken from BS 5328

Standard mix	28 day characteristic strength (N/mm²)	Constituents	75 mm slump	125 mm slump
ST1	7.5	Cement (kg)	210	230
		Total aggregate (kg)	1940	1880
ST2	10.0	Cement (kg)	240	260
		Total aggregate (kg)	1920	1860
ST3	15.0	Cement (kg)	270	300
		Total aggregate (kg)	1890	1820
ST4	20.0	Cement (kg)	300	330
		Total aggregate (kg)	1860	1800
ST5	25.0	Cement (kg)	340	370
		Total aggregate (kg)	1830	1770
ST1 ST2 ST3		Sand (percentage by mass of total aggregate)	35–50	35–50
ST4 ST5		Sand (percentage by mass of total aggregate):		
		sand grading C	35–45	35–45
		sand grading M	30–40	30–40
		sand grading F	25–35	25–35

STANDARD MIXES

Examples of these can be found in BS 5328. Table 7.9 gives batch quantities for one cubic metre of concrete for 20 mm aggregate. The quantities shown will 'normally' give characteristic strengths stated in the table. Volume batching is permitted for mixes ST1, ST2 and ST3. This procedure permits a simple approach to specifying structural concrete, albeit of limited strength – maximum characteristic value 25 N/mm². Note the simple numerical differences between the figures in the table. Cube testing is not required but might be specified as a quality control tool.

DESIGNATED MIXES

These are recommended mixes for specific applications. BS 5328, for example, defines a GEN 1 concrete suitable for a range of applications, including

■ blinding and mass concrete fill
■ strip footings
■ mass concrete foundations
■ trench fill foundations
■ unreinforced house floors (under screed)
■ drainage applications

In all the above applications the standard mix ST2 is identified with workability appropriate to the application. Hence designated mixes provide a simple way of specifying concrete for common applications.

DESIGNED MIXES

These are examples of performance specifications: the purchaser specifies those aspects of the mix relevant to placing, for example, workability and maximum aggregate size, but gives a grade of concrete and/or other hardened performance requirements rather than mix proportions. The choice of mix proportions is left to the supplier who must ensure that concrete is of the correct performance, whether expressed in characteristic strength or other terms.

The Department of the Environment (DOE) method of mix design (1997) is an effective method of design and illustrates well the logic underlying design procedures. A full description of this method is given in the companion volume to this text, *Materials in Construction, an Introduction*. Attention here will be given only to the rationale of mix design used in that method and to the use of the DOE method for variations on the standard method. Very simple basic figures will be used throughout but these will be sufficient to indicate the important characteristics. All design procedures focus upon both the fresh and hardened states of concrete.

Fresh concrete

The *workability* of concrete is largely determined by its water content. For normal concretes with 20 mm uncrushed aggregates the 'starting point' for water content is 180 litres (or kg) per cubic metre of concrete. This is likely to lead to medium workability. The figure could be changed as follows:

Increase by 20 litres per cubic metre for higher workability or crushed aggregates.
Decrease by 20 litres per cubic metre for lower workability or uncrushed aggregates.

The cohesion requirement is addressed by consideration of fine material content. A starting point for mixes of average composition, strength and performance and uncrushed aggregate type of 20 mm maximum size is 33 per cent sand of total aggregate. This could be varied as follows:

Increase by 10 per cent (i.e., to 43 per cent) for high cohesion, coarse sand or lean
   concrete mix – say less than 200 kg of cement per cubic metre.
Decrease by 10 per cent (i.e., to 23 per cent) for low cohesion, fine sand or rich
   concrete mix – say more than 400 kg of cement per cubic metre.

All the changes suggested above can be interpolated or extrapolated according to the particular materials of characteristics required.

The density of most fresh concrete can be taken to be of the order of 2300 kg per cubic metre.

Hardened concrete

The key property in hardened concrete terms is compressive strength and this has already been described also as a key indicator to durability. Approximate mean strengths to be expected from hard uncrushed aggregates at 28 days using grade 42.5N portland cement are:

Water/cement ratio	Strength N/mm^2
0.4	60
0.6	40
0.8	20

Durability of these mixes would be excellent, moderate and poor, respectively.

Typical worked example

Produce batch quantities for 1 m^3 of concrete to have medium workability and mean 28 day compressive strength of 40 N/mm^2 using 20 mm crushed aggregates.

The figures given above indicate a water content of 180 litres per cubic metre of concrete for uncrushed aggregates. This would be increased by 20–200 litres for crushed aggregate. The water/cement ratio would be 0.4 for the given strength requirement from the above figures.

The cement content is obtained by dividing the water content by the water/cement ratio:

$$\text{cement content} = \frac{\text{water content}}{\text{water cement ratio}}$$

This gives 200/0.4 or 500 kg per cubic metre of concrete. (This figure is probably a little high since the crushed material may produce somewhat higher strengths than indicated by the very simple approach above.)

The aggregate content is obtained by subtracting cement figures from the density figure given above:

aggregate = 2300 − (200 + 500)
= 1600 kg total aggregate per cubic metre of concrete.

A fairly low sand content can be taken because we are dealing with a rich mix:

sand content = 23% × 1600 = (say) 370 kg.

The coarse aggregate content is 1600 − 370 = 1230 kg per cubic metre.

Note that the figures above are approximate but they illustrate the rationale of mix design.

CALCULATION OF CONCRETE FRESH DENSITY

An accurate estimate of concrete density is required for yield purposes. The technique for calculating fresh density from first principles is given here since an

Table 7.10 Calculation of fresh concrete density from densities of constituents

Material	Mass kg	Mass ratio per kg of cement	Relative density (SSD)	Volume litres
Water	200	0.4	1.00	0.400
Cement	500	1	3.15	0.317
Sand	370	0.74	2.65	0.279
Coarse aggregate	1230	2.46	2.65	0.928
Total	2300	4.60		1.924

understanding of the method will also enable the effect of modifications, for example, incorporation of air on the mix density, to be obtained.

Suppose, for example, the batch quantities per cubic metre are (worked example above):

water	200 litres
cement	500 kg
sand	370 kg
coarse aggregate	1230 kg

The relative densities of materials are required. Cement is assumed to be 3.15 and that of the aggregates 2.65.

The figures are tabulated as in Table 7.10 and, to simplify them, all figures are divided by the mass of cement, giving masses which would correspond to 1 kg of cement.

The density of the concrete is $4.60/1.924 = 2.39$ kg per litre or 2390 kg per cubic metre. The volume contributions of the components should be noted. The water volume is larger than the cement volume, even in this high cement content mix. The coarse aggregate is the chief filler, occupying almost as much space as the other three components combined.

If the presence of air is to be allowed for in calculating density, this is easily done by increasing the total volume of concrete, but not the mass, by the percentage of air which is expected. The same procedure would be used if air were to be entrained for frost resistance purposes.

TRIAL MIXES

These are a most important part of the mix design procedure. Initial trial mixes may be made in small quantities, for example, using a 20 kg sample of concrete. Water should be added until, by visual inspection, the workability is correct. The workability and cohesion should then be determined and adjustments made as necessary. The adjustments procedure will be given for a number of eventualities using the figures shown in the worked example above.

■ **Workability too low** Slump, say, 10 mm instead of 30–60 mm. There will be little point in proceeding with strength tests if it is found that extra water is required for workability, since this will decrease the strength of the concrete. Note the increase of water required and obtain the corresponding figure per cubic metre (say, 15 litres). Now increase the cement content to keep the water/cement ratio the same as previously.

For example, if water/cement = 0.40, the new cement content is

$$\frac{200 + 15}{0.4} = 537 \text{ kg} \quad \text{(formerly 500 kg)}$$

Find the new density of the material (with the higher water content) as above and hence adjust the mass of coarse aggregate. The proportion of sand should not need to be altered unless substantial alterations to the cement content are made.

- **Cohesion too low** If the cohesion is slightly low the sand proportion can be increased by say 5 per cent, affecting only the last stage of the design process. The workability of the adjusted mix may then decrease and, if this is the case, should be corrected as in the previous paragraph. More serious cohesion problems may require an inspection of the aggregate grading with consideration being given to the use of admixtures such as air entraining agents. Mixes with high workability, low cement content or small aggregate size are most likely to cause cohesion problems.
- **Concrete strength** In order to save time, 28 day concrete strength can be predicted by accelerated curing or by testing early and referring to strength/ age correlations. If the strength is incorrect it will be necessary to increase the cement content in order to reduce the water/cement ratio. A guide to the change necessary is that an increase in cement of 6 kg per cubic metre will increase the mean compressive strength of the material by 1 N/mm^2.

Concrete density

Once the concrete mix is shown to be correct, its fully compacted density should be accurately measured in order to permit accurate yield details to be calculated.

The suitability of the concrete is not finally proven until a full size mix has been produced and placed in a situation similar to that in which the concrete will be finally used, attention being given to placing, compaction and finishing techniques. Where appropriate, the quality of finish of the hardened product should also be checked.

MIX DESIGN METHOD FOR SPECIAL MIXES

Mix design methods such as the DOE method can be easily adapted for specific requirements or materials:

Sulphate resistance   For the particular class of sulphates encountered obtain the cement type, for example, from Table 7.14 on page 296 together with the maximum water/cement ratio and minimum cement content. Then carry out the procedure ensuring that these limits are complied with.

Air entrained concrete   Entrained air has two effects in mix design terms:

1 It is estimated that each 1 per cent of air reduces strength by 5.5 per cent for air contents in the range 3–7 per cent. Hence, if, say 4.5 per cent air is required, increase the target mean strength by the ratio

$$\frac{1}{\{1 - 0.055(4.5)\}} = 33 \text{ per cent approximately.}$$

For example, if the original target mean strength was 30 N/mm², the mix should be designed for target mean strength 1.33 × 30 = 40 N/mm²

2 The air has a lubricating effect which can be allowed for by selecting a workability level one category below that actually required. A small reduction in sand (say 5 per cent) may be allowed as a result of increased cohesion, permitting a further water reduction which will assist in achieving the lower water/cement ratio required without a large increase of cement content and hence cost. The final cement content will be in the region of 10 per cent higher than that of a similar non-air-entrained mix.

PFA concrete   PFA has two main effects in mix design terms:

1 It increases workability. As a guide, each 10 per cent of PFA (measured as a proportion of total binder) will lead to a water reduction of 3 per cent. Suppose 25 per cent PFA (of the total binder) is to be used; then a reduction of 25 × 3 = 7.5 per cent would follow. With a typical mix of 180 litres of water per m², this would be 13.5 litres, so that, on rounding, 165 litres would suffice.
2 It contributes to strength, though not to the same extent as cement. The DOE method uses a *cementing efficiency factor* which varies between 0.20 and 0.45 according to PFA and cement types, though an average figure of 0.3 may be used as a guide. Hence, if a water/cement ratio of 0.5 is required, this may be expressed in the terms:

$$W = \frac{0.5}{C + 0.3F}$$

where   W = free water content
   C = cement content
   F = PFA content

For the above water content of 165 litres, we obtain

$$\frac{165}{C + 0.3F} = 0.5$$

from which C + 0.3F = 330, or C = 300 − 0.3F.

Also

$$F = 0.25 \ (C + F) \ \text{(that is, 25 per cent of total binder)}$$

Substituting:

$$F = 0.25 \ (300 - 0.3F + F)$$
$$= 75 + 0.17F$$
$$0.83 \ F = 75$$
$$F = 90 \ \text{kg}$$
$$C = 330 - 0.3 \times 90$$
$$= 300 \ \text{kg}$$

Slag/microsilica concretes   Design of these concretes should be by trial mixes or reference to manufacturers' data. In the case of slag concretes, an efficiency factor can be used as with PFA though the value varies by quite large amounts according to the material used. There is, at present, an insufficiency of data regarding mix design with microsilica.

USE OF CHARTS IN CONCRETE PRODUCTION

The need for repeated mix design operations can be obviated in situations where concrete mixes of varying properties are to be produced using aggregates from a given source as in ready mixed concrete production. To obtain data for varying strength and given workability, the design process might be carried out at, perhaps, low, medium and high strengths with trial mixes as necessary. Then a range of mixes can be obtained by adjustment of cement and sand contents. Figure 7.14 shows such an example for 20 mm aggregate and medium workability. The cement content increases in 50 kg stages with the sand quantity reducing to maintain the same total volume. Hence, the reduction in sand is

$$50 \times \frac{\text{relative density of sand}}{\text{relative density of cement}}$$

Typically this would be

$$\frac{50}{3.15} \times 2.55 = 40 \ \text{kg}$$

hence, the 40 kg reduction in sand in Figure 7.14. Strength curves can be monitored and amended from test results and small adjustments made to aggregate quantities from experience of behaviour of mixes as they are produced, giving a convenient and finely tuned method for mix proportioning. Charts are now often computerised with materials being batched automatically to give mixes satisfying given performance specifications.

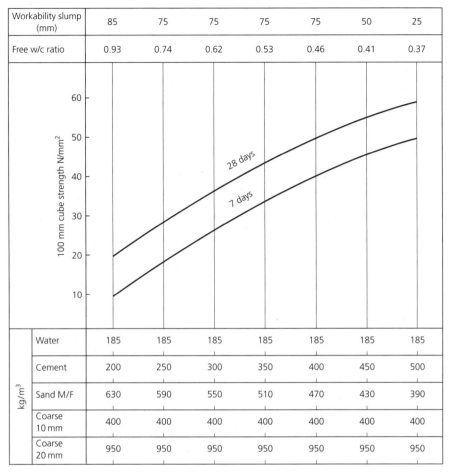

Workability slump (mm)	85	75	75	75	75	50	25
Free w/c ratio	0.93	0.74	0.62	0.53	0.46	0.41	0.37

kg/m³	Water	185	185	185	185	185	185	185
	Cement	200	250	300	350	400	450	500
	Sand M/F	630	590	550	510	470	430	390
	Coarse 10 mm	400	400	400	400	400	400	400
	Coarse 20 mm	950	950	950	950	950	950	950

Figure 7.14   Mix design chart by Owens for concrete of medium workability using 20 mm quartzite aggregate. Note that quantities of water, 10 mm and 20 mm aggregates are constant. As the cement content increases, the sand content decreases such that the total volume of materials remains constant. The strength graphs are obtained experimentally

7.6   MEASUREMENT OF PROPERTIES OF FRESH CONCRETE

It should be appreciated that the properties of hardened concrete are closely related to those of the fresh mixed, plastic material, so that it is of great importance to be able to measure and control the latter. Furthermore, if by means of tests the concrete is found to be unsuitable for its purpose, it is far easier to reject a mix before it has set and hardened, and easier still if the failure can be detected by tests before placing. Tests for workability are considered together with a method of analysing fresh concrete. In the case of tests covered in BS 1881 strict adherence to detailed procedures is essential if meaningful results are to be obtained.

WORKABILITY

Workability must be controlled within fine limits if concrete of consistent quality is to be produced. If the workability (and therefore the water content) is too high, the concrete will be more prone to segregation in the fresh state and, in the hardened state, the quality will be impaired. If the workability is too low compaction will be difficult, with increased risk of air voids. It should be noted that 5 per cent of entrapped air will lead to a strength reduction of about 30 per cent. The workability of concrete must be matched to the techniques of compaction, placing and finishing and should result in concrete of uniform composition and maximum density for the particular mix used. Since workability is a complex property involving the interplay between the quantity of water in the mix, the volumes of constituents and internal friction due to particle abrasion, its measurement must be made empirically and mixes of different proportions but of similar workability may give different results to a particular test. The workability categories, extremely low, very low, low, medium and high, can only act as a guide when specifying concrete and some reliance must be placed upon trial mixes. During production, however, workability tests provide a simple means of quality control. A further point is that, due to absorption of water by cement (and aggregates, if absorbent), workability may decrease rapidly after mixing. To compare results, therefore, tests should be carried out at a standard time after addition of the water and, preferably, just before placing.

Slump test

This is still the most widely used test, due to its simplicity and convenience. Concrete is placed and compacted in three layers by a tamping rod, in a firmly held slump cone (Figure 7.15). On removal of the cone, the difference in height

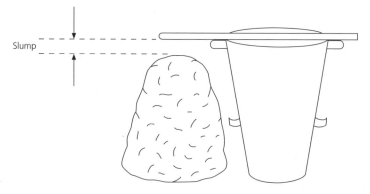

Figure 7.15   The slump test

between the uppermost part of the slumped concrete and the upturned cone is recorded in mm as the slump. Less cohesive or lean mixes tend to give a greater slump, may shear, or even give a collapsed slump, and aggregate type also affects the result at a given workability, so that allowances may have to be made. Dry mixes often give no slump at all and, in this case, another test should be used.

Compacting factor test

A sample of the mix is allowed to fall through two hoppers into a cylinder, thereby compacting itself to a degree dependent on its workability. The contents of the cylinder are weighed and then the same volume of fully compacted concrete is weighed. The compacting factor is calculated from:

$$\text{compacting factor} = \frac{\text{weight of partially compacted concrete}}{\text{weight of fully compacted concrete}}$$

A value near unity indicates a workable mix, while a value of, for example, 0.70 would indicate a dry mix. The compacting factor test is useful for drier mixes, although these tend to fall less freely through the hoppers. The problem of richness also applies, richer mixes compacting more easily for a given compacting factor. The apparatus is more bulky than the slump test apparatus and most forms require separate weighing equipment, so that it is not as commonly used as the slump test, though it is widely used in the production of concrete for road pavements.

Vebe consistometer

A slump test is first carried out, using a slump cone fixed inside a cylindrical container on a small vibration table. Then a weighted transparent plastic disc, held in a vertical guide, is allowed to rest on the surface of the slumped concrete. The vibrator, in the form of an electrically operated eccentric rotor under the table, is then operated and the time in seconds for the disc to fall, such that the concrete wets its whole circumference, is recorded. This is then the consistency of the concrete in *Vebe degrees*. The Vebe test is suitable for dry mixes and, particularly, those which are to be compacted by vibration, since the test itself involves vibration of the concrete. It requires an electric power supply (normally three phase) and is therefore restricted to use as a laboratory test and in precast concrete production.

Comparative results of the above workability tests are shown in Figure 7.16. The approximate form only is indicated, that actually obtained being dependent on the richness of the mix, and the type and grading of the aggregate. The graph, however, indicates the reasons for the suitability of each test as given above. In practice, having decided from the above arguments which type of test is to be used, it is normal to use that method only, since it is not easy to correlate results from

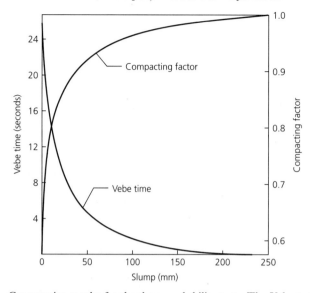

Figure 7.16   Comparative results for the three workability tests. The Vebe test is suitable for very low workability, the compacting factor test for low/medium workability and the slump test for medium/high workability

different methods and, in any case, these do not assist in quality control – an important use of workability tests.

Flow table test

This is designed for high workability concretes and has become more widely used since the introduction of super-plasticised (*flowing*) concrete. It consists of a 700 mm square board fixed by a hinge on one edge to a similar base board. The free edge of the upper board can be lifted to 40 mm height, when it reaches a stop. Concrete is poured into a small cone of similar shape to a slump cone mounted centrally on the dampened table, tamping lightly. The cone is removed and the board lifted and released 15 times, causing the concrete to spread out. The maximum spread in each direction is measured, by means of graduations, in mm.

COHESION

Although, perhaps, not of such importance as workability, there will be an optimum degree of cohesion for each concreting situation. The need for adequate cohesion has been explained but it should be re-emphasised that a concrete which is unnecessarily cohesive will also be unnecessarily expensive, since the extra fine material incorporated will increase the water and, hence, the cement requirement at a given strength level. There is no BS test specifically for the measurement of cohesion but the appearance of concrete after a slump test will give some indication. Alternatively, a simple test which may be carried out is by trowelling, say, 15 kg of concrete on a flat, non-absorbent surface. The degree of cohesion is assessed by the ease with which a smooth surface finish is obtained when trowelling to a thickness of about 75 mm. High cohesion is only necessary when the concrete may be dropped from some height during placing or when there are relatively large areas of finished concrete to be produced – as, for example, in thin slabs or slender columns. The flow table test also gives some indication of cohesion in high workability concretes.

ANALYSIS OF FRESH CONCRETE

An apparatus to analyse concrete mixes accurately and quickly – preferably within a few minutes – would be of great benefit in many types of concrete production. It would detect errors in batching equipment or in its operation and act as a further means of control, allowing action to be taken, if necessary, before placing. The main problem is that of obtaining sufficient accuracy quickly. The method described in BS 1881, for example, is slow and requires a knowledge of the relative densities of the aggregates.

Perhaps the most important aspect of the analysis of fresh concrete is that of measuring the cement content, since the free water content can usually be inferred with reasonable accuracy from the workability, especially if details of aggregate properties are known. A number of schemes have been developed in which cement is separated from the remainder of the concrete and either weighed or detected chemically. In one commercially operated system, for example, concrete is placed in a vertical *elutriator* column and the fine material is washed out by a controlled upward flow of water (Figure 7.17). A known proportion (say, 10 per cent) of the fine material is then washed through a 150 μm sieve and allowed to settle into a collecting vessel. After siphoning excess water to a predetermined level, the vessel and contents are weighed and the cement content obtained from a calibration graph. Once calibrated, a result can be obtained within five minutes although, in such a method, any silt in the aggregate will lead to an overestimated cement content. A correction can be made by repeating the test with a sample of the aggregate used without any cement. Such methods permit the measurement of cement content to within about 20 kg/m^3 of the correct value at a 95 per cent confidence level, provided concrete is tested within about two hours of mixing. Where delays are envisaged, the use of an appropriate quantity of retarder will delay hydration and, hence, increase accuracy.

Where an accurate and rapid indication of water content is required, nuclear techniques, involving the moderation of fast neutrons by water, have been employed but this technique is, at the present time, of a specialised nature. A further, though slow, alternative is oven drying.

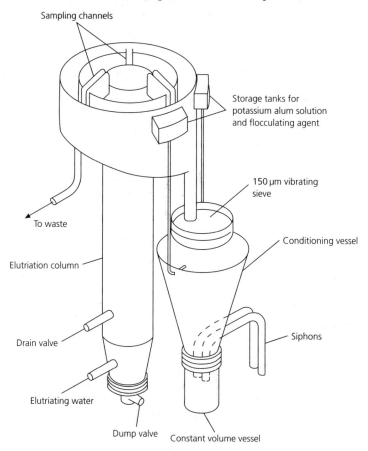

Figure 7.17   Rapid analysis machine for determining the cement content of fresh concrete. The five minute analysis can be operated automatically

7.7 COMPACTION OF CONCRETE

Although a detailed treatment of the methods of compaction is outside the scope of this book, a number of important points are summarised here.

The approximate effect on the strength of under-compaction has already been briefly stated, though it may be added that the relationship between strength S and porosity p is of approximately negative exponential form:

$$S = S_0 e^{-kp}$$

k being a constant and S_0 the strength at zero air voids. Constant k takes a value of about 8 in order that 30–35 per cent strength loss occurs in the 5 per cent voids. Note that the first 1 per cent of air voids leads to the largest percentage loss of strength – about 9 per cent. Hence the need to eliminate as much air as possible will be appreciated, although air contents in the region of 1 per cent are considered normal. Conversely, overcompaction will also lead to strength loss, particularly at surface layers, due to the effect of water rising to the surface of concrete, especially in lifts of some depth. What might be regarded as the exception to this effect is re-vibration, in which concrete is subjected to further compactive effort some time – perhaps in the region of two hours – after placing. If, for instance, external vibrators are operated at this stage, an increase of strength and improvement in surface finish might be obtained. The former is probably due to the extra consolidation achieved, together with removal of any plastic cracks, while the latter is due to removal of further air by the second application of vibration.

Delaying placing of concrete is in no way detrimental, provided full compaction is finally achieved. However, workability reduces steadily from the time of adding the mixing water and delays of over two hours may, especially in hot weather, make full compaction very difficult to achieve unless retarding admixtures are used. Any extra water added (other than the small amount to replace evaporation) will reduce strength according to the water/cement ratio rule.

7.8 CURING OF CONCRETE

This is the process in which, by means of moisture in the concrete, the material matures, increasing in strength and decreasing in porosity. Curing is in no way dependent on air so that, once setting is complete, ideal curing conditions are those in which the concrete is completely saturated. Temperature is also important. The ideal curing temperature for ordinary Portland cement concrete is in the region of 10 °C; higher temperatures lead to a more rapid hydration but a less favourable crystal structure. In the case of rapid hardening Portland cement, the optimum temperature is lower, in the region of 5 °C. However, temperatures of up to 25 °C do not significantly reduce long term strength. At temperatures lower than these respective values, strength development is slow, due to reduced hydration rates. Curing should continue until the material has sufficient strength to resist shrinkage cracking. Thin sections, in particular, tend to dry out very quickly as soon as protection is removed. High water/cement ratio concretes are permeable and will, therefore, be most seriously affected in this way, especially as they also exhibit high shrinkage. Even in mass concrete, a short curing period may cause problems since, on exposure, surface layers may dry and shrink, while inner layers continue to hydrate, resulting in surface cracking. It is clear from the above that the curing method and precise curing period will depend on the type of structure and its situation but ideally the concrete should be kept moist, if necessary by artificial means, for at least seven days. Minimum curing times are given in BS 8110 and these depend very much on ambient conditions. For example, at 10 °C and relative humidity between 50 and 80 per cent, a minimum period of three days is stated, increasing to four days when pozzolanic additions are included. In poor conditions, minimum periods of up to 10 days are required. In good curing conditions (greater than 80 per cent relative humidity) no requirement is given. Nevertheless, when there is any doubt, it is best to err on the side of a longer curing period if *in situ* strengths corresponding in any way to cube strengths are to be obtained.

Curing techniques depend on the situation but might typically be damp sacking or polythene sheeting for exposed concrete slabs, bitumen emulsion for concrete road bases or sprayed curing compounds for concrete road slabs. Formwork for structural concrete protects it from temperature changes and drying, and should be left in position as long as possible. Cements with a high early heat output present a problem, since temperatures within formwork may then rise greatly above the ambient temperature. The use of low heat cements is recommended where this is likely to happen but, in any case, if formwork is left in position as long as possible, sudden stresses, due to cooling and evaporation, will be minimised. There is often a temptation to spray with cold water a surface found to be warm on removal of formwork, but this is not advised due to the thermal shock and consequent surface stresses it produces in sections of thickness greater than approximately 450 mm.

In mass concrete, the best procedure is to insulate the sides and surface so that the structure as a whole heats and then cools. The difference between internal and external temperatures may be monitored and this should not exceed 20 °C if risk of cracking is to be avoided. Thermal stresses begin to be induced at the point of setting – usually about 12 hours after placing. In mass concrete, the peak temperature rise will be some time after this – between one and two days, depending on section size and ambient conditions. Hotter areas are thrown into compression as they react against colder areas which will be in tension. Since failure strains in tension are much lower than in compression, it is cold areas which are at risk at peak temperature.

Curing in hot weather

Hot weather will exaggerate some of the problems already mentioned. Owing to evaporation, the quantity of mixing water may have to be increased to produce the required workability at the time of placing, which may have to be carried out more quickly, due to more rapid hydration of cement and consequently decreased setting time. A warm, dry atmosphere may cause plastic cracking due to surface evaporation. Such cracks, which do not normally penetrate more than about 50 mm, may be retrowelled but steps must then be taken to conserve all possible moisture in the concrete. The use of a low heat cement or lean mixes will reduce the extra heat due to hydration. Aggregates form the main bulk of the concrete, so that if these are stored in large stockpiles, the diurnal temperature variation in them and, hence, the concrete, will be minimised. In general, concreting thin, exposed sections above about 30 °C in low humidities should be avoided.

Curing in cold weather

Freezing of concrete before it is fully cured may result in a substantial reduction in strength and durability and, if freezing occurs before it sets, concrete is rendered vitually useless, firstly due to damage produced as water expands to form ice and,

Table 7.11 Minimum pre-hardening times for frost resistance for various grades of concrete at curing temperatures in the range 5–20 °C

Characteristic strength at 28 days N/mm^2	Minimum prehardening time at stated curing temperature in hours			
	20 °C	15 °C	10 °C	5 °C
20	24	32	46	71
25	22	30	42	65
30	20	27	38	59
40	17	23	33	50

secondly, because ice has no chemical power to hydrate the cement. Concrete can be safely exposed to frost once it reaches a compressive strength of about 2 N/mm^2 (provided it is not saturated) and, once frost resistant, hydration will continue, though at a reduced rate, down to temperatures of approximately −10 °C, since water in small pores does not freeze until much lower temperatures. The rate of strength gain depends, in a complex way, on the grade of concrete, the temperature and the time after placing. The term *maturity* has been used previously where temperature, measured above the datum −10 °C, is multiplied by the time after placing in hours to give an index of the degree of hydration and, hence, strength development, but it is found that concretes of a given maturity so defined are weaker if cured at temperatures below 10 °C than those cured at higher temperatures. A 0 °C datum, conversely, leads to a low estimate of strength of concrete cured near freezing point. More complex formulae have been suggested but pre-hardening times are probably best expressed empirically, Table 7.11 indicating typical values of minimum pre-hardening times for various grades of concrete at curing temperatures in the range 5–20 °C. It is noticeable that pre-hardening times increase greatly for lower grades of concrete at lower temperatures. If concrete is likely to be saturated, larger values are necessary and concretes below grade 20, which have water/cement ratios over 0.7, may never be frost resistant. The periods may, in general, be reduced by about 25 per cent if rapid hardening Portland cement is used and by a further 25 per cent if calcium chloride is also included.

In all types of concrete, full compaction is necessary if maximum frost resistance is to be provided. When concreting is being carried out in cold weather, measurements of concrete temperature, using a thermometer in the concrete, protected by a metal sheath, should be made periodically, so that maturity can be estimated. The following steps will assist in obtaining the required maturity:

- Use a rapid hardening cement, an accelerating admixture or a higher strength mix requiring a lower pre-hardening time.
- Heat the aggregates and/or water to increase the initial temperature of the concrete.
- Use insulated formwork. The best insulator is a layer of trapped air. This may be obtained, for example, by a tarpaulin over an airspace on slabs, or formwork with a backing of an insulating material.

■ Enclose the structure in a temporary heated covering, for example, polythene sheeting on scaffolding.

If the concrete relies on its own heat, together with insulated formwork for protection, thin sections will be far more difficult to deal with satisfactorily since they are, in general, of greater surface area for unit mass. Such types of construction should be avoided in very cold weather unless satisfactory maturity before freezing can be ensured.

Steam curing

The object of steam curing may be to accelerate strength development or simply reduce the time that precast products need to remain in the mould. It can be carried out at low or high pressures.

Low pressure steam curing

Steam at a temperature of 55–80 °C is passed through chambers containing the units for a period of about 12 hours. Quite high rates of temperature rise are obtainable as the steam condenses on the concrete, releasing its latent heat, but temperatures over 50 °C should be avoided during the first two hours. A delay of two to five hours before commencement is advised, longer delays being needed for higher curing temperatures or weaker concrete mixes. The problem is probably due to the stresses caused by the rapidly rising vapour pressure of water as its temperature rises; rapid or early rises tend to cause porosity and thermal stress leading to microcracking. The rate of temperature rise should not exceed 20 °C per hour for this reason. The 12 hour strength of a steam-cured concrete may be three to four times that obtainable by normal curing and one-day strength is at least 50 per cent of long term strength so that large precast panels can be handled at this stage, allowing more rapid turnround of moulds. Although the chemical structure is similar to that obtained by normal curing, the less perfect gel structure results in a long term strength shortfall of 20–40 per cent compared with conventional curing. However, for a given strength, the concrete is stiffer (higher E value) and exhibits lower shrinkage which may be an advantage in some situations.

Low pressure steam curing is now widely used for obtaining estimates of the 28 day strength of concrete.

High pressure steam curing (autoclaving)

This is quite different from the method described above, the units being heated in pressure vessels, at a pressure of about eight atmospheres to approximately 180 °C. Ultimate strengths are less than those obtained by normal curing. The addition of finely ground silica restores much of this loss by a surface reaction similar to that which occurs in the manufacture of calcium silicate bricks, the lime being provided by the hydrating cement, in particular, tricalcium silicate. The optimum silica

Figure 7.18   Scanning electron microscope photograph within a pore of an aerated concrete block showing relatively coarse crystal structure

content is about 30 per cent of the total binder. As in low pressure steam curing, the rise of temperature should be gradual.

The specific surface of the resulting hardened paste is much lower than that of normally cured cement paste and the hydrate should be regarded as micro-crystalline rather than in gel form (Figure 7.18). As a result, drying shrinkage, moisture movements and creep are much smaller, due to reduced quantities of adsorbed water. The concrete is more durable than normally cured concrete and its resistance to sulphate attack is greater, since the hydrates are less reactive in the presence of sulphates. High pressure steam curing produces maximum strength in about one day but is not suitable for reinforced concrete, since bond strength is reduced. Precast concrete blocks are often cured in this way and particularly the lightweight aerated blocks now widely used to obtain low U values in walling.

Steam curing by either process cannot be carried out on high alumina cement which, in any case, develops strength rapidly under normal curing. The high cost of cement in the latter is balanced by the cost of steam curing apparatus and power in Portland cements.

7.9   STRENGTH PROPERTIES OF HARDENED CONCRETE

The strength of cement bound materials in compression, tension and shear follows the same general pattern as in the other ceramic type materials, such as stone or fire-clay products. Concrete contains many imperfections in the form of pores, voids and cracks which, combined with its heterogeneous nature, lead to stress concentrations at localised points in the material, in spite of a 'uniform' applied stress. The stresses in tension are amplified according to Griffith's theory (see Chapter 5) and, in consequence, tensile strength is less than 10 per cent of compressive strength. Indeed, it is likely that even compressive stress may eventually cause failure by crack propagation, if only due to Poisson's ratio strains, which occur in the plane at right angles to the direction of application of stress. The cracks may propagate at stresses which are relatively small compared to the ultimate stress, whether the concrete is loaded in tension or compression, but development of these cracks is normally arrested at low stresses and they are, therefore, referred to as *stable cracks*. At a later stage, and particularly in tension, the energy released by crack propagation becomes greater than the work required to form new crack surfaces and cracks propagate rapidly, leading to failure. In most concretes, the aggregate is harder and stronger than the cement paste and, in the case of a vertically applied stress for example, this would result in greater stresses above and below aggregate particles, the nature and variability of which have an important effect on the strength of the concrete as a whole.

As a result of its low tensile strength, concrete is generally reinforced in areas where tensile stresses arise, although some tensile or flexural strength is, nevertheless, assumed in such situations as unreinforced road slabs, ground floor slabs and foundations.

Factors affecting strength

Consideration has already been given in the section on the design of mixes to a number of factors affecting strength, though it may be appropriate to elaborate on some aspects of this information.

Water/cement ratio   As far as mix proportions are concerned, this is the most important factor affecting strength for given materials, lower water/cement ratios leading to higher strengths. The effect may be broadly considered as the same as that of compaction, higher water/cement ratios resulting in a more porous cement paste and, hence, lower strength. The situation is complicated, however, by the fact that strength continues to increase, subject to the achievement of full compaction, down to water/cement ratios as low as 0.1, compressive strengths in the region of 300 N/mm² having been achieved using cement *compacts*. The cement, in these cases, is largely unhydrated, the hydrated component merely filling voids and forming thin *glue lines* around each cement grain. The strength–water/cement ratio relationship is, in fact, approximately logarithmic in the normal strength range, the log of strength increasing uniformly with reduction in the water/cement ratio. Illustrating this point, the strength of concrete is increased by 25 per cent by reducing the water/cement ratio from 0.6 to 0.5 and further 25 per cent increases would be obtained by further reductions to 0.4 and then 0.3. Clearly, when the added advantage of high durability is considered, there would seem to be great benefit in producing powerful compaction methods for concretes of low workability

Table 7.12   Strength ratios for 0.6 water/cement ratio concrete at various ages (well cured)

Age days/months	1	3	7	28	2	3	6	12
Strength ratio	0.20	0.48	0.72	1.00	1.07	1.10	1.13	1.16

and low water/cement ratio, though these techniques are most suited to factory production. Though of comparatively minor importance, it is also found that at a given water/cement ratio, concretes of lower cement content (and, hence, lower water content) tend to be slightly stronger – presumably because they contain less of the relatively weak cement paste per unit volume. This further reinforces, on economic grounds, arguments already given for lean concrete mixes of very low workability.

Age   From an age of about 12 hours, the strength of concrete increases rapidly with time, the rate of hardening thereafter reducing, the strength approaching its long term value exponentially. Correlations between strength at different ages are important, since they often form the basis of 28 day, or later, strength prediction, by testing at early ages. Typical age factors are shown in Table 7.12. For 0.6 water/cement ratio concrete, continuously cured, the relationship between 7 and 28 day strengths has been the subject of particular interest, the traditional working guide being that 28 day compressive strength is 50 per cent greater than 7 day strength. This rule is no longer accurate for modern cements where the hydration process occurs more rapidly, the 28 day strength now showing a smaller percentage increase over the 7 day value. A rise of about 30 per cent is more likely. Situations causing even smaller increases would include the use of fine cements, high curing temperature and high strength concrete. Conversely, increases between the two ages may be greater when lean mixes are employed or when curing temperatures are low. Increases of strength after 28 days are now less than formerly for reasons given on page 221.

Curing   Curing may influence the strength of concrete both from the point of view of the time for which it is applied and the effectiveness of the method used. While it is accepted that continuous saturation represents the ideal condition for all concretes, the effect of loss through curing depends very much on the type of mix used and on the type of structure. In strength terms, for example, poor curing would be less damaging to mass concrete than to thin sections, which could dry out more quickly. When concrete is allowed to dry out, further bonds are formed in the cement paste fraction, so that the concrete strength can increase quite rapidly (Figure 7.19). Thereafter, since hydration can no longer proceed, strength development is arrested. On rewetting, there is a rapid strength loss caused by the absorption of water, followed by a recovery such that, in the long term, provided shrinkage damage has not occurred, the strength approaches the value that would be obtained by continuous curing. The effect of curing method depends on the water/cement ratio of the concrete; sealing moisture in, for example, by polyethylene

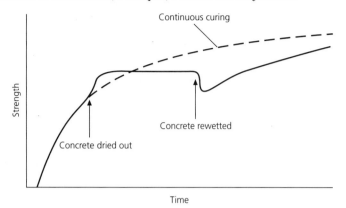

Strength

Continuous curing

Concrete rewetted

Concrete dried out

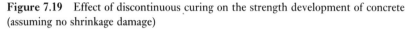

Time

Figure 7.19   Effect of discontinuous curing on the strength development of concrete (assuming no shrinkage damage)

sheeting, is adequate in concrete mixes of water/cement ratio over 0.5 – these contain ample water for hydration. Saturation is more important with concretes of low water/cement ratio, since these mixes may become self-desiccated. This is particularly the case in the early stages of curing: for example, capillary pores in a concrete of 0.4 water/cement ratio may be discontinuous by the age of three days, so that curing after this time may not greatly assist strength development, even if it does delay shrinkage. With weaker mixes, curing should be carried out for longer periods, since an impermeable matrix takes longer to form in these. The effect of temperature should also be considered – concrete cubes may now be *temperature matched* to the structure by means of temperature sensors and control equipment so that the effect on strength of varying temperatures can be accurately ascertained.

METHODS OF TESTING FOR STRENGTH

These may be classified as destructive and non-destructive, the former providing the basis for most design and production aspects of structural concrete, despite the fact that destructive testing, as well as non-destructive testing, must be regarded, in general, as an indirect way of ascertaining concrete strength. The basic problem of destructive tests is that it is rarely economical to carry out full scale *in situ* tests, while the problem of non-destructive tests is that they measure some property other than strength, thus requiring correlation information. Nevertheless, factors of safety in structural concrete design have decreased progressively in recent decades, reflecting the measure of confidence which can be attached to destructive testing in particular, together with increased understanding of correlations and improvements in the quality of concrete. There are also many situations in which non-destructive testing has made an important contribution, particularly when assessing the strength of concrete in connection with remedial treatment.

Destructive tests

Cube test (BS 1881 Part 116)   This is currently the most common type of destructive test for concrete in the UK, owing to the cheapness of the cube moulds and the comparative simplicity of manufacture and testing of cubes. Carefully obtained samples of the concrete mix are placed and compacted in accurately formed steel moulds, with machined inner surfaces. Bonding with the steel is prevented by coating with release agent. The surface of each cube is covered with an impermeable sheet, or the entire mould sealed. After 24 hours in a vibration-free place, the cube is removed and cured under water at about 20 °C, until tested. The cube is then placed centrally between the platens of a compression testing machine, trowelled face sideways, and the load is applied such that the stress increases at a given constant rate until failure. The maximum load is recorded. The compressive strength of concrete, as measured by the cube test, may be affected by the following factors.

The size of the cube   150 mm cubes, for example, fail at stresses approximately 5 per cent lower than those of 100 mm cubes. This would be consistent with the *weak link* theory, since the larger quantity of material present in a large cube is more likely to contain a very weak region. Whatever the explanation, the effect illustrates the earlier affirmation that the strength of concrete depends on the method of test. One possible advantage of large concrete cubes is that, since they contain a larger sample of concrete, there should be less variability in cube results and this effect is obtained in practice. The effect is more important when larger aggregates (say, 40 mm) are employed, for which 150 mm cubes are always recommended.

Moisture condition   Test results for cubes depend on their moisture condition, a cube dried just before testing giving a higher strength. Hence, since the wet state is the most reproducible condition, cubes should always be tested wet.

Loading rate   Decreasing the loading rate gives lower cube strengths, due to the increased contribution of creep to failure. While the effects of, for example, doubling or halving the BS loading rate of 15 N/mm^2/minute are probably less than 1 per cent, extremely high loading rates, for example, causing failure in 1 microsecond, could double the failure stress and very low loading rates would give significant strength reductions.

Stress concentrations   Low results will be obtained if there are stress concentrations at the surface of the cube. These may be due to particles of loose material on the cube surface, or irregularities on the surfaces of the loading platens, because of wear. The machine must stress all parts of the cube surface equally. To this end, it is most important that the upper platen of the testing machine, which is located on a ball seating, should lock on loading, so that the concrete cube is evenly

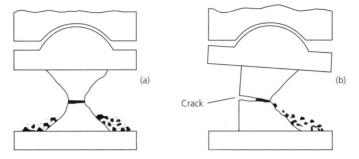

Figure 7.20   Possible modes of failure of a concrete cube: (a) correct; (b) incorrect

stressed, even if its modulus of elasticity varies from place to place or if the cube is not exactly centrally placed. Failure to lock would result in preliminary failure of the weaker or more heavily stressed areas of the concrete cube. It is found that best results are obtained if the ball seating of the testing machine is not lubricated with a high pressure lubricant. Correct and incorrect failure modes of concrete cubes are illustrated in Figure 7.20.

Test machine   The machine itself should also be checked for accuracy and stress distribution during loading; BS 1610 deals with load calibration but does not cover possible eccentricities in loading, such that a 'Grade A' machine may still be unsatisfactory in this respect. BS 1881 Part 115 covers other aspects such as flatness of platens and uniformity of stress distribution across cube sections. One effective way of checking the overall performance of a compression machine is by reference testing in which matched pairs of concrete cubes of various strengths are tested on the machine in question and on a carefully verified standard reference machine (BS 1881 Part 127). Any defect in the machine is then revealed by discrepancies in test results for each pair. If expensive in terms of pessimistic estimates of strength, it is, perhaps, fortunate that unsatisfactory machines tend to give low rather than high test results.

The 45° planes of Figure 7.20(a) would suggest a shear failure in the concrete, the shear stress being of maximum value in these planes when uniaxial stress is applied. However, the stresses in the cube are not uniaxial, since the platen surfaces themselves provide considerable frictional lateral restraint to the cubes, as illustrated in Figure 2.6. In a normal failure, the pyramid shapes could correspond to the regions of influence of the platen restraint, actual failure being in tension as a result of stresses induced in the remainder of the cube, according to the Poisson's ratio of the concrete. The effect can be simply demonstrated by taking two similar cubes, crushing one normally and loading the other between two rubber mats. The rubber reverses the Poisson's ratio effect since it squeezes out under pressure, producing tensile stress at the top and bottom surfaces of the cube. This leads to much earlier failure in the form of a vertical crack.

The observed compressive strength of concrete as obtained by cube testing has been estimated to be as much as double that in the actual structure for this and other reasons, for example, curing conditions. Hence, it is very difficult to correlate closely cube strength and structure strength and the main value of the cube test is not in attempting to obtain directly the strength of the structure, though the test may be useful in the following ways:

■ Cubes can be used for quality control. Assuming cubes are reliably made, cured and tested, cube variations will be indicative of variations in the concrete from which they are sampled. Hence, they provide a means of detection of changes in materials or errors in methods.
■ A minimum or *characteristic* cube strength is the normal way of specifying concrete strength, rather than attempting *in situ* measurements on the structure itself. The required cube strength for a given structure is decided upon by experience or from codes of practice. Appropriate factors, applied to the cube test result due to the above effect, are included as part of the design process.

Compressive tests on cylinders

These have some advantages and some disadvantages compared with the cube test. Since only the cylinder ends are loaded, the body of the mould need not be machined and can be formed from cheaper, non-reusable materials, such as plastic. On the other hand, the ends of the cylinder must be of accurate tolerance, requiring capping of one or both ends, dependent upon whether a machined base-plate is used. The capping process is slow and costly and the capping material itself must be correctly chosen to be of approximately the same stiffness as the body of the concrete. In particular, capping material which is weak relative to the concrete will induce lateral strain in the concrete, tending to cause splitting and a low failure load. The reverse effect, of a strong capping material on a weak concrete, is less important, though in all cases the capping should be no thicker than the minimum required at give a surface which is plane to a tolerance of 0.05 mm at right angles to the cylinder axis. A maximum thickness of 3 mm is normally allowed and the capping material may be based on Portland or high alumina cement, gypsum or sulphur.

The normal height-to-width ratio of a cylinder is 2:1, so that platen restraint is less than in a cube leading to lower apparent strengths. When correlating to cube strengths, a ratio of 1.25 is generally taken and a further factor will need to be included if the height diameter ratio is not 2:1 (see Figure 7.21). The Eurocode for Structural Concrete, EC 2, includes use of cylinders for compressive strength testing.

Cylinder testing has been in use for many years in the form of testing of cores cut from the concrete (BS 1881 Part 120). These allow visible examination and strength testing of the *in situ* material, though cutting them is expensive and, if reinforcement is also cut, there may be implications both in terms of the test result and the stability of the structure itself.

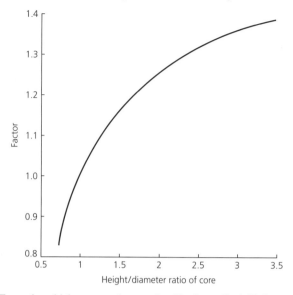

Figure 7.21   Factor by which measured strength of horizontally drilled cores should be multiplied to obtain equivalent cube strength (BS 1881 Part 120)

Direct tension testing

In situations where concrete is to be subjected to tensile stress, some form of tensile test is preferred to compression testing, since there is no unique relationship between compression and tensile test results, the effect of aggregate type, in particular, depending on whether tensile or compression tests are used. While a pure tensile test might be regarded as the ideal from which other types of tensile behaviour could be deduced, direct tension testing is, at present, rare since relatively elaborate means of gripping the test specimen are necessary. Methods involve the use of glued end pieces, the casting of concrete around a metal anchorage frame or the gripping of specimens by plates, relying to some extent on friction. Results obtained indicate that the ratio between uniaxial tensile strength and 100 mm cube strength varies from 0.08 at cube strengths in the region of 20 N/mm^2 to 0.05 at cube strengths of 60 N/mm^2. The absence of a British Standard test for direct tension testing reflects the difficulty in adopting this type of testing for routine commercial purposes.

Indirect tension test (cylinder splitting test) (BS 1881 Part 117)

In this test, cylinders which are typically 300 mm long and 150 mm in diameter, are loaded in a compression tester with their cylindrical axes horizontal, stress concentrations being avoided by use of hardboard or plywood strips about 12 mm wide (Figure 7.22). The successful operation of the test requires careful alignment of the cylinder (or use of a jig) and packing strips should be used once only to ensure

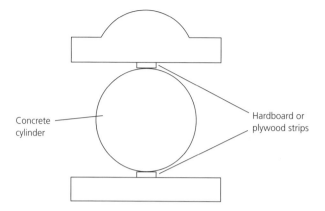

Concrete
cylinder

Hardboard or
plywood strips

Figure 7.22   The cylinder-splitting test for measurement of the tensile strength of concrete

uniform bedding, especially in the case of weak concretes, for which plywood is the
more suitable material. Except near the packing pieces, a tensile stress is induced by
concrete on the vertical plane and the tensile strength f_t at failure is given by:

$$f_t = \frac{2W}{\pi DL}$$

where W = load at failure
$\quad\quad\quad D$ = diameter of cylinder
$\quad\quad\quad L$ = length of cylinder

Note that, since the failure area is DL, the expression is the same as

$$f_t = \frac{\text{load}}{\text{failure area}} \times \frac{2}{\pi}$$

The test can be used for other specimen shapes, such as cubes, provided the
trowelled face is not loaded, the latter expression being used to calculate the
failure stress.

The strength measured in this way is similar to direct tensile strength for values
in the region of 2 N/mm² but, in the case of high strength concretes, indirect
tensile strength may be in the region of 25 per cent higher than direct tensile
strength.

A most convenient aspect of the indirect tension test is that it can be carried out
on a compression testing machine. Results are also more consistent than those from
other tension tests.

Flexural test (BS 1881 Part 118)

The test described in BS 1881 uses a two-point loading system on a 100×100 mm
or 150×150 mm beam, as shown in Figure 2.9 which produces a constant bending
moment between loading rollers.

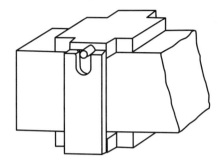

Figure 7.23   Equivalent cube test for measurement of cube strength of a broken part of a concrete beam

Assuming that normal bending theory applies, the extreme fibre stress at failure is given by

$$f_t = \frac{Wl}{bd^2}$$

where l = length of beam between supporting rollers
   W = load at failure
   b = breadth
   d = depth

provided failure occurs between the loading rollers.

The stress measured in this way corresponds to the weakest portion of concrete at the soffit in the central third of the span but stresses based on the flexural test are, nevertheless, approximately twice the failure stresses measured using direct tension tests on similar concrete. A contributory factor to this difference is probably that the neutral axis of the beam moves upwards during the test, increasing the proportion of the cross-section which carries tensile stress. BS 1881 describes an *equivalent cube test* which can be carried out on the broken ends of the beam after failure. These ends, which are not damaged in the flexural test, are loaded in compression between auxiliary platens 100 or 150 mm square, appropriate to the width of the beam and give an indication of the cube strength of the concrete (Figure 7.23). Results are normally about 5 per cent higher than a normal cube test result, due to the contribution to strength of the concrete outside the auxiliary platens.

Perhaps the main limitations of the flexural test stem from the fact that either a special machine or a compression machine attachment are necessary for its execution and the variability of results is somewhat greater than that produced by the indirect tension test.

NON-DESTRUCTIVE TESTS

Although three of the methods of non-destructive testing described here measure stiffness rather than strength, they are included in this section because their object is normally strength or quality determination. Further comments are made on the significance of elastic moduli obtained in the section which follows under that heading.

Schmidt rebound hammer (BS 1881 Part 202)

This is a small, portable instrument containing a spring loaded plunger. When pressed against a well-restrained concrete surface, the plunger is forced into the instrument, loading a spring to a point where a mass is released and, under the energy of the spring, hits the end of the plunger. The mass rebounds to a distance dependent on the hardness of the material against which the plunger rests. The rebound distance is recorded by a marker and may be related to strength for a given material by means of cube results. The distance of rebound depends on the inclination of the hammer, use on soffits, for example, giving higher readings due to the assistance of gravity on the rebound. Since the area of contact of the plunger with the surface is only a few square millimetres, a large number of readings (for example, 20) is essential to average out local variations in the concrete. Trowelled surfaces are best avoided and, to ensure adequate restraint and avoid the risk of damaging the concrete, readings should not be taken near to edges. The test is unreliable at ages greater than three months since, by this stage, progressive carbonation may have resulted in a hardened surface layer. The simplicity and convenience of the instrument make it a useful means of checking the strength progression or strength variations in concrete and models are now available that can be linked to a portable computer for data recording. However, it should not be regarded as a substitute for cube testing in new concrete or core testing in mature concrete. It is not recommended for high alumina cement concrete testing, owing to the influence of surface effects on results.

Electrodynamic method (BS 1881 Part 203)

This is based on the principle that the resonant frequency of a concrete beam depends on the velocity of compression waves through it, which, in turn, depends on the modulus of elasticity of the concrete. The simplest form of resonance, shown in Figure 7.24, occurs when the wavelength (λ) equals twice the length of the beam (l). If the frequency (f) of applied mechanical vibrations is varied until the pick-up response is a maximum, then the velocity (v) of the wave is given by:

$$v = f\lambda = 2fl$$

If the beam is assumed to approximate to an infinitely long rod, then v is related to E, the modulus of elasticity, by the equation:

$$v = \sqrt{\frac{E}{\rho}}$$

where ρ is the density of the concrete. Therefore,

$$E = 4f^2l^2\rho$$

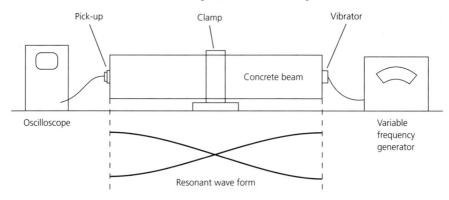

Figure 7.24   Determination of elastic modulus of concrete by an electrodynamic method

Care is needed when interpreting E values measured in this way; for example, the electrodynamic modulus (E_D) increases in saturated concrete since water filled voids transmit sound more effectively; there is no such effect in static tests. Since E_D is measured at very low stresses, results are higher than static (E_C) values. BS 8110 gives the relationship:

$$E_C = 1.25E_D - 19$$

In lightweight concretes where much lower E value aggregates are used, the difference is much less pronounced. For comments on the relationship between modulus of elasticity and strength, see page 277. The electrodynamic method can be used for strength prediction but is limited by the fact that it can only be used on small beams – it is not an *in situ* test. It is, nevertheless, useful for repeated tests on a single specimen in the laboratory to measure, for example, the effect of curing conditions on concrete properties.

Ultrasonic pulse velocity (BS 1881 Part 203)

Pulses of ultrasound, usually having frequencies in the region of 150 kHz, are passed through a concrete structure by means of transmitting and receiving transducers, and an accurate indication of the time taken is obtained using electronic circuitry. The equipment is extremely simple to use, the time in microseconds being displayed digitally. There are three possible modes of use, direct transmission, semi-indirect transmission and indirect transmission (Figure 7.25).

The most satisfactory method is the direct method, since the strongest signal is received in this way; it would normally be used to check beams and columns, giving an easily measured transmission distance. With floor slabs or encased columns, the indirect method must be used, the velocity being the gradient of a separation–time graph (Figure 7.26). Cracks can sometimes be detected in this way – there will be a sudden increase of transmission time if a transducer is moved over a crack. Greater care is necessary with the indirect method as the received signals are weaker.

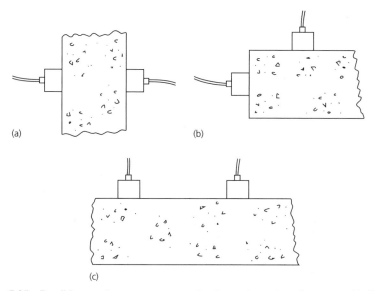

Figure 7.25 Possible transducer arrangements in ultrasonic testing of concrete: (a) direct; (b) semi-direct; (c) indirect

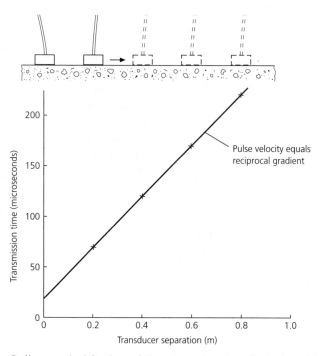

Figure 7.26 Indirect method for determining ultrasonic pulse velocity in a slab. Where there are variations in concrete quality or cracks a straight line will not be obtained

The pulse velocity, once obtained, can be used to find the modulus of elasticity of the concrete (E) from the formula:

$$E = v^2\rho\left[\frac{(1 + \sigma)(1 - 2\sigma)}{(1 - \sigma)}\right]$$

where v = pulse velocity

ρ = density of the concrete

σ = Poisson's ratio for the concrete

The use of the formula requires a knowledge of the Poisson's ratio for the concrete, so that there is an increasing tendency to interpret the pulse velocity directly.

The following aspects of concrete quality can be checked:

- Variations in concrete quality. *In situ* strength variations due to, for example, inconsistent batching or poorly graded aggregates can be easily located.
- Construction faults. Areas of honeycombing or excessive laitance (weak surface layers) can be detected. In columns, compaction planes may be detected, together with possible weak areas towards the top of lifts, where water contents are often higher.
- Concrete deterioration. Ultrasonic pulse velocities are affected by fire or environmental damage, or by progressive deterioration as occurs, for example, in high alumina cement concrete.

Precautions

The ultrasonic pulse velocity through steel is approximately 6 km/s, compared to 4 km/s for normal concrete, so that effective pulse velocities are increased when a significant quantity of steel is present in the transmission zone. It is preferable to avoid zones of influence when taking readings, though correction factors can be applied if reinforcement details are known. Reinforcement running at right angles to the direction of transmission has a relatively small effect and bars of diameter less than 10 mm can be ignored. Pulse velocity also increases with moisture content, so that standardisation of the curing and testing condition is important when producing correlations between cube strength and *in situ* pulse velocity. While pulse velocities are quite sensitive to changes of E in concrete, they are not highly sensitive to strength changes due to the less sensitive relation between E and strength. For dense aggregates, the range of velocities 3–4.5 km/s probably represents extremes of strength, from about 10 N/mm^2 to 60 N/mm^2. Hence, care is needed to obtain accurate results. In particular, good coupling between transducer and surface are required and this is best achieved by use of a coupling agent such as a thick grease unless surfaces are very flat.

Other tests

A number of other methods of assessing strength have been suggested, for example, pull-out tests on bolts cast into the concrete to a depth of 75 mm, or measurement of penetration of steel probes fired into the concrete. In each case, the property measured and strength can be correlated, surface effects not being so important as in, say, the rebound hammer test.

THE EXTENT OF TESTING OF CONCRETE

As with other aspects of quality control, the extent of testing of concrete depends on the quality and quantity of concrete to be produced. When small quantities of concrete only are produced, it may be considered better to provide an extra safety margin in terms of cement content rather than incur the expense of testing. Such practice is, however, open to grave errors which may pass unnoticed unless some check on quality is made, and in any case, many specifications require a certain strength level, so that cubes must be made. If a detailed check on strength development is required, E values by the electrodynamic or pulse velocity methods may be obtained at frequent intervals, being correlated with cube strengths periodically. Such methods may also be used to predict likely strengths at 7 and 28 days. Alternatively, accelerated curing techniques enable prediction of these strengths to be made within 24 hours of placing of the concrete.

Other tests described may be useful if, for any reason, the quality of the concrete· is in doubt. If, for example, cube test results are unsatisfactory, rebound hammer readings could be taken on the *in situ* concrete (provided, of course, the hammer is calibrated) and, if this also gives unsatisfactory results, a final decision on removal of concrete could be taken after cutting and testing cores.

COMPLIANCE TESTING

This is the procedure by which concrete, especially if ready mixed, is accepted or rejected on a batch by batch basis. Such schemes are essential in addition to quality control procedures, since they form a simple, clear basis on which to judge the material. (Reference should also be made to sections in Chapter 4 on quality control and acceptance testing.) In the case of concrete compressive strength, they relate to typical variabilities of the material, the figures below giving a guide to these:

cement (same works)	3 N/mm^2
cement (different works)	5 N/mm^2
concrete (excellent control)	4 N/mm^2
concrete (poor control)	8 N/mm^2

When good quality control standards are achieved, it will be possible to operate at lower mean strength without risk of failing compliance or characteristic strength requirements. The compressive strength requirements of BS 5328 are:

1 The average strength determined from any group of four consecutive test results exceeds the specified characteristic strength by:

    3 N/mm^2 for concretes of grade C20 and above
    2 N/mm^2 for concretes of grade C7.5 to C15

(C20 means characteristic 28 day cube strength of 20 N/mm² with 5 per cent failures allowed.)

2 The strength determined from any test result is not less than the specified characteristic strength minus:

3 N/mm² for concretes of grade C20 and above
2 N/mm² for concretes of grade C7.5 to C 15

If any individual result or mean of four fails a compliance requirement, all concrete represented by the test (including all batches involved in the mean of four requirement) is then 'at risk'. This implies that an engineering judgement must be made, based on:

- The strength shortfall in relation to the required result.
- The effect in structural or durability terms of the low strength material in the position used.
- Checks on the validity of the test.
- Other tests on the hardened *in situ* material.

Where such checks confirm non-suitability for the intended purpose, concrete should be replaced, though in many cases qualified acceptance will be possible.

The purchaser of concrete must decide on sampling rates, recommended values indicated in BS 5328 ranging from one sample for each 10 m³ for critical structures such as beams and columns to one sample for each 100 m³ for low risk applications such as mass construction. Lower sampling rates reduce testing costs but put more concrete at risk in the event of non-compliance.

Rejected concrete is very expensive both in terms of the cost of removal and replacement, as well as in terms of lost confidence in the product. Since concrete just complying with characteristic strength requirements has a risk of failing compliance regulations (the risk is statistically estimated to be about 13 per cent), most producers work slightly above the required standard. This could be achieved by aiming for, say, 1 per cent failures at the required characteristic strength instead of the permitted 5 per cent. With a standard deviation of 4 N/mm², this would require a mean strength increase of :

$$(2.33 - 1.64)4 = 3 \text{ N/mm}^2$$

(approximately) in mean strength (see Table 4.1 for these values). If the standard deviation were higher, a larger margin would be required to achieve the same effect – one of the penalties of poorer quality control.

7.10 PROPERTIES OTHER THAN STRENGTH

These are described under the headings of modulus of elasticity, creep, impact strength, fatigue resistance, abrasion resistance, permeability, carbonation, shrinkage, thermal cracking, thermal properties, fire resistance, frost resistance, sulphate resistance, resistance to alkali–silica reaction and resistance to acid attack.

MODULUS OF ELASTICITY

The modulus of elasticity (E) of concrete is important from two points of view. First, it determines the resistance to deflection of concrete structures. Since many structural components have to satisfy deflection criteria – concrete beams, for example – those manufactured from concrete of high E value would deflect less and would be preferred. Secondly, and conversely, high stiffness concretes generate higher stresses when movements, such as thermal movement or shrinkage, are restrained. From such standpoints, low stiffness concretes would be preferable, although these would naturally be less satisfactory from the first standpoint.

The modulus of elasticity of concrete is, of course, equal to the gradient of the stress/strain graph, although complications arise because the graph is non-linear (Figure 2.11). The departure from linearity for stresses of up to 50 per cent of ultimate is small but, thereafter, the modulus of elasticity decreases, due to combined effects of microcracking and creep. Commonly stated values of elastic modulus are initial tangent modulus and secant modulus based on 30 per cent of ultimate stress. Since the electrodynamic test described earlier is based on very small stresses, it approximates to the former, while the latter, which is usually 5–10 kN/mm^2 lower, is more realistic for predicting deflections in structural members. Even lower values may be appropriate where concrete is stressed locally to failure, as in shrinkage cracking. Stress/strain curves in tension and compression are essentially similar, except that failure values are much reduced in tension.

For a given type of aggregate, the elastic modulus and strength of concrete correlate well, although the relationship is not linear and it also depends on age. The modulus of elasticity of a concrete can be expressed in terms of the volume fractions and E values of the separate components, quite high values being obtained when volume fractions of aggregate are high (as, for example, in dry lean concrete) even when the cement paste fraction is of relatively high water/cement ratio and the concrete is, therefore, of low strength. Hence, decreasing porosity of the cement paste fraction, whether by increasing age, at least during the curing period, or by decreasing water/cement ratio, generally increases strength more than elastic modulus. The relationship employed in BS 8110 for normal weight concrete is, for example,

$$E_{t,28} = 20 + 0.2f_{cu,28}$$

where $E_{t,28}$ is the initial tangent static modulus at 28 days in kN/mm^2 and $f_{cu,28}$ is the characteristic cube strength in N/mm^2 at 28 days. Note that an increase of, say 20 per cent in cube strength from 30 to 36 N/mm^2 would cause a corresponding rise in modulus of elasticity from 26 to 27.2 kN/mm^2 (about 5 per cent). It will be evident that the failure strain of concretes will increase as design strength or maturity rise.

CREEP

This may be defined as time-dependent strain resulting from sustained stress. Such strains tend to build up over a period of months, and even years, in concrete under

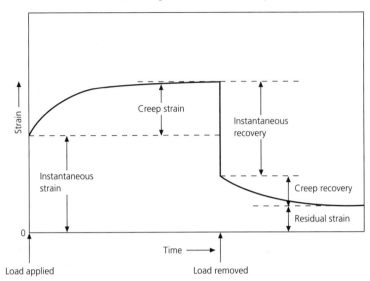

Figure 7.27   Creep and creep recovery in concrete

service stresses but they can also make a significant contribution to failure strain when samples are quickly loaded to failure, as in routine testing. A typical graph of creep against time is shown in Figure 7.27. It is noticeable that both creep strain and creep recovery strain tend to limiting values after a time, approximately 30 per cent of ultimate creep strain occurring in two weeks and 80 per cent after 30 months. The residual strain in the concrete is associated either with continued hydration or with permanent changes to the structure resulting from stress.

The origin of creep probably lies, to a large extent, in the water contained in the cement gel, since creep is much reduced in desiccated concretes. The exact mechanism has been the subject of much debate, theories including the effect of viscous shear in the cement gel, the movement of adsorbed water in the gel and microcracking. The effect in all cases is, however, a release of stress in the most heavily stressed areas – typically associated with larger aggregate particles in the cement paste. The term *specific creep* or creep per unit stress is used for comparison of different types of concrete. Specific creep is increased in weaker concrete, although actual creep in a concrete member manufactured from such a concrete would not be lower, due to the lower operating stress. Creep also increases as follows:

■ if finer cement is used
■ if the concrete is loaded at an earlier age
■ if the moisture content of the concrete is higher
■ if the load is applied under drying conditions

Creep increases with applied stress, being roughly proportional to stress at low stresses. At stresses above 75 per cent of ultimate, however, the concrete will eventually fail.

The effects of creep in concrete structures must be considered from two points of view. First, in any prestressed component, creep will reduce the prestress and hence reduce the load carrying capacity of the member, unless further prestress is applied after the bulk of creep has occurred. The creep in such situations is roughly equal to the shrinkage. Also, differential creep in massive structures may contribute to cracking; for example, in mass concrete when internal regions become warm due to cement hydration, the stresses may be relieved by creep. The temperature then reduces subsequently and the now more mature concrete is able to creep less under the tensile stresses caused by cooling, hence internal cracking may occur. Conversely, creep may, in many situations, relieve stress caused by, for instance, non-uniform loads and, therefore, reduce the danger of cracking.

BS 8110 gives a final (30 year) creep strain as:

$$\frac{\text{stress}}{\text{modulus of elasticity at time of loading}} \times \varphi$$

where φ is the creep coefficient which varies from 0.5 to 4.0, thinner sections, lower ages of loading and lower relative humidities tending to increase φ. For typical outdoor exposure loading at 28 days of age, the creep coefficient for a 300 mm section is approximately 1.5; hence the final creep strain is 1.5 times the instantaneous elastic strain.

IMPACT STRENGTH

Although the impact strength of concrete may be regarded as low, in the sense that small impact loads can cause considerable damage, the stresses in concrete at which failure occurs under impact are considerably higher than when failure is caused by low loading rates, this rule already having been given under testing of concrete for strength. In many situations, concrete has to withstand impact, whether by chance during service or by design, as in pile caps. Ability to withstand impact is determined by measuring the total energy absorbed at failure resulting from standard blows. After high initial energy absorption caused by local damage, the energy imparted to the concrete during impacts decreases and then, prior to failure, energy absorbed rises as cracks multiply and propagate. Impact strength generally correlates well to tensile strength, stronger concretes absorbing less energy per blow, since they behave more elastically, but requiring considerably more impacts to cause failure.

FATIGUE

In some structures such as concrete machine bases, floors and roads, stresses fluctuate continuously, with the result that failure is often attributable to fatigue. Damage is progressive, a network of cracks building up in the concrete, and the deterioration is readily monitored by ultrasonic equipment. There is no well defined endurance ratio such that resistance can be guaranteed below a certain stress irrespective of the number of stress reversals applied. For this reason, endurance ratios based on 10^7 stress reversals are often given. In addition, the amount of variation of stress is an important factor, higher stress variations causing earlier failure. Taking a concrete slab, for example, the number of cycles to failure for loads varying between zero stress and half ultimate static strength would be in the region of 10^6. If the variation of stress were reduced (for example, when a proportion of the stress comprises the steady stress imposed by self-weight) this number would, however, be substantially increased. It will be appreciated that, since factors of safety generally above 2 are used, very large numbers of stress reversals are normally required for fatigue failure in structures unless, for some reason, stresses exceed the design level.

ABRASION RESISTANCE

This may be defined as resistance to wear due to friction and in the context of concrete refers almost exclusively to floors, usually without a surface topping or coating. Such surfaces are commonly used in industrial environments where solid or even steel tyred vehicles can cause rapid deterioration. Abrasion resistance relates to the hardness and strength of surface layers, mainly the top few mm. Hence, the 'skin' of the concrete needs to be very hard for an abrasion resistant surface, and it must be well bonded to underlying layers to avoid detachment under frictional stresses. Lack of abrasion resistance causes dusting in early stages followed by progressive roughening as aggregate particles of increasing size are lost. Abrasion resistance can be measured in terms of wear in mm caused by loaded wheels, rotating about a vertical axis. It is a function of mix proportions, finishing technique and curing technique. Figure 7.28 shows that high strength (low water/cement ratio) mixes are best, though the effect is not pronounced. More important are the finishing and curing techniques. Repeated power trowelling closes pores left by evaporation and has a marked effect on abrasion resistance. Hand trowelling is ineffective in this respect. Curing is also important, sprayed curing membranes being best, while air curing gives poor results. *Dry shakes* are sands applied prior to trowelling and the metallic type enhance abrasion resistance: they improve surface hardness. Finally, if all else fails, an improvement can be obtained by applied hardeners after curing, the in-surface seals based on polymers in organic solvents giving best results.

PERMEABILITY OF CONCRETE

The permeability of concrete is the most important factor in determining resistance to environmental attack since impermeable concrete resists admission of water or aqueous solutions, restricting deterioration to the surface. The effect of water/cement ratio has already been described (Figure 7.6) but further steps by which maximum impermeability can be obtained include:

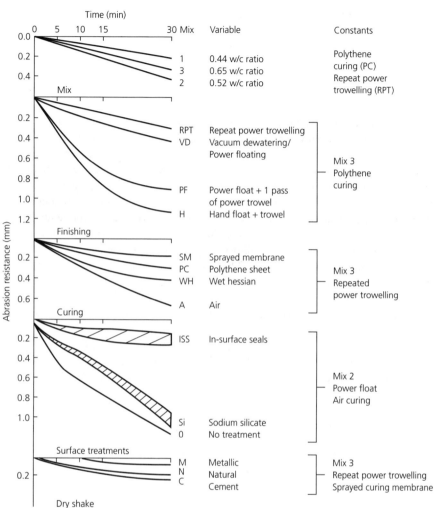

Figure 7.28 Factors affecting the abrasion resistance of concrete (Source: *Concrete Society Journal*, Vol. 20, No. 11, Nov. 1986; reproduced with permission of The Concrete Society)

- ensuring that full compaction is achieved
- adequate curing
- where aggressive agencies are concerned, use of pozzolanas, such as PFA, GGBFS or microsilica (see 'Cement replacement materials')

A rapid assessment of the permeability of concrete can be made by spraying with water when dry. Water should run off an impermeable surface and it should dry quickly. Where the water soaks in and the surface remains visibly damp for some time, high permeability is indicated. A useful check on permeability is provided by the initial surface absorption test (ISAT) covered by BS 1881 Part 208.

CARBONATION OF CONCRETE

Carbonation is the result of interaction between atmospheric carbon dioxide and the alkaline constituents of hydrated cement, chiefly calcium hydroxide. The reaction requires the presence of water but may be summarised:

$$CO_2 + Ca(OH)_2 \rightarrow CaCO_3 + H_2O$$

Carbonation requires penetration of carbon dioxide into the concrete and therefore proceeds usually very slowly, though at a rate dependent on the environmental conditions and the permeability of the concrete to the gas. Carbonation occurs most rapidly at relative humidities in the range 55–75 per cent and it is important to appreciate that it occurs, therefore, in interior as well as exterior environments. In exterior environments, carbonation is often greater in drier, less exposed situations – rates are very slow in saturated environments.

The permeability of concrete is a most important factor: Figure 7.6 indicates that weak mixes may have permeabilities hundreds of times higher than low, well compacted water/cement concretes. The rate of carbonation falls off exponentially with time in the absence of cracking. In good quality concrete, ultimate penetration may be as little as 3 mm but in permeable concrete it may reach depths of 20–30 mm, especially if permeable aggregates, such as some lightweight aggregates, are used.

The depth of carbonation of Portland cement concretes can be easily checked by breaking a piece of concrete away and immediately spraying with a 1 per cent aqueous solution of phenolphthalein indicator. Deeper, uncarbonated areas turn purple, while surface, carbonated areas remain colourless (Figure 7.29).

Figure 7.29   Phenolphthalein test for carbonation of concrete. The phenolphthalein must be sprayed onto a freshly broken surface

Carbonation is not itself harmful to concrete – indeed, there may be a strength benefit in some cases where internal stresses are relieved by recrystallisation and pore blocking by the calcium carbonate formed reduces permeability. Some shrinkage often accompanies carbonation (see below) but the chief effect is that the durability of steel in carbonated concrete is seriously impaired. This important topic is examined under the heading 'Reinforced and prestressed concrete', page 329.

SHRINKAGE

In general usage, the term 'shrinkage' denotes any contraction, irrespective of cause, but in the context of concrete 'shrinkage' is used to denote contraction due to loss of moisture. Concrete, effectively cured, has a tendency to expand slightly as it matures. The expansion is, however, small – long term values of 100×10^{-6} being typical – and, on drying, it is quickly reversed, a shrinkage level of over 2000×10^{-6} being possible in cement mortars. When hardened concrete is dried for the first time, a particularly large contraction occurs, followed by smaller expansions and contractions on subsequent wetting and drying cycles, respectively. The effect is shown in Figure 7.30 from which it is evident that subsequent movements reduce in magnitude. The origin of these movements lies in the cement gel component of the concrete. The expansion on wet curing is caused by the absorption of water into the gel pores, which are extremely small, about 0.15 μm in diameter. The pressure of this water swells the gel and hence the concrete. When the concrete dries for the first time, some of the gel water is removed and consequently the gel contracts, new bonds tending to form, as gel surfaces which were previously separated by water

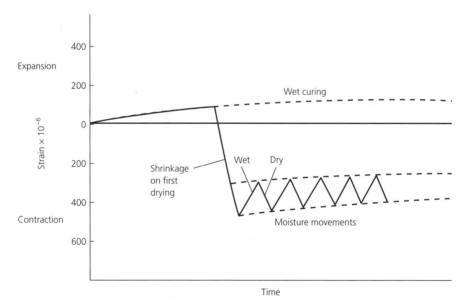

Figure 7.30 Initial expansion, subsequent shrinkage, and finally moisture movements in concrete

Table 7.13 Approximate long term shrinkages of concrete as a function of water/cement ratio and cement content per cubic metre

Cement content kg/m³	Long term shrinkage × 10⁻⁶		
	w/c ratio		
	0.4	0.6	0.8
450	500	1000	segregates
300	300	500	700
200	200	300	500

come together. Hence, shrinkage results in a short term increase in strength, provided restraint does not lead to cracking. If the concrete is rewetted, water access is reduced and, therefore, not all the shrinkage strain can be reversed. Repeated wetting and drying cycles then produce progressively less movement, as the cement gel gradually stiffens and water access becomes more limited. These movements are referred to as *moisture movements* to distinguish them from the initial larger movement for which the term *shrinkage* is reserved. Since the cement paste fraction is the active component of the concrete in respect of shrinkage, those concretes containing larger proportions of cement paste, particularly of high water/cement ratio, shrink most. Table 7.13 shows approximate figures for long term shrinkage using inert aggregates.

These figures may be compared with the failure strain in tension of concrete which is in the order of 100×10^{-6}, depending on strength. Hence, all concretes represented in the table would fail if shrinkage in them were completely restrained though complete restraint is rare in practice. The rate of shrinkage depends on the porosity of the cement gel but, in general, approximately half occurs within one month of casting. The aggregate size may also affect shrinkage, concretes employing smaller aggregates shrinking more because water requirements are increased. Also, whereas most aggregates restrain movements of the cement paste, some are moisture susceptible – certain sandstones, for example. Shrinkage of concretes employing such aggregates can in consequence be increased.

In order to limit shrinkage damage, the following guidelines should be followed:

- Do not use excessive cement and water content. BS 8110, for example, specifies a maximum of 550 kg of cement per cubic metre of concrete, to avoid possible shrinkage problems in thin sections.
- Where concrete is restrained, it is particularly important to provide contraction joints.
- Cure concrete until its tensile strain capacity is sufficient to accommodate shrinkage strains.
- Avoid rapid drying out, which may cause differential shrinkage problems.
- Surface cracking may be controlled by secondary reinforcement such as steel mesh or use of fibres.

Figure 7.31   Plastic settlement cracking of a reinforced concrete ground beam.
A reinforcement cage of the type used has been placed over the beam. The crack pattern is
seen to follow this closely. Note that cracks close at the sides of the beam. They could have
been closed while plastic by tamping or trowelling

Plastic cracking

This occurs within a few hours of placing and may be caused by plastic shrinkage
or settlement. Shrinkage results due to moisture loss in conditions of severe drying,
for example, hot, windy weather. Settlement cracks are caused by excessive
bleeding together with restraint, often caused by steel reinforcement (Figure 7.31).
In both cases the cracks close up at the edge of the concrete and do not penetrate
to great depths. They are easily removed by revibration/finishing as the concrete
stiffens, provided they are noticed in time.

THERMAL CRACKING

The most critical stage in avoiding thermal cracking is within one day or so of
casting the concrete. Typical figures for heat of hydration are shown in Table 7.3.
Taking a 24 hour figure of 300 J/g or 300 kJ/kg of cement, this would translate
into approximately $300/6 = 50$ kJ/kg of concrete on the assumption that the mass
of concrete is 6 times the mass of cement. The specific heat capacity of concrete is

around 950 J/kg°C so the temperature rise obtained if no heat were to escape (adiabatic conditions) would be:

$$\frac{50\,000}{950}$$

or 53 °C. This could be considered to be the maximum possible temperature rise and values approaching this would not be obtained in practice unless the concrete were of very substantial thickness – perhaps with a minimum dimension of 10 m or more. Nevertheless in mass concrete, of thickness perhaps 2 m or more, significant temperature rises, in the region of 20 °C, might occur at the core.

It should be emphasised that temperature rises are not likely to be the main problem in concrete; it is temperature *differentials* between different parts of the same structure that cause problems. For example, if the concrete at the centre of a mass concrete structure becomes hotter than the outer layers, early hardening will occur with the concrete in a thermally expanded state compared with outer layers. When cooling occurs the core material will contract, but this movement will be restrained by the cooler outer material. Hence the inner layers will be thrown into tension. This is rather similar to the process of toughening glass except that much larger thicknesses of material are being considered. The maximum tensile strain concrete is likely to withstand is in the region of 100×10^{-6} and given typical thermal expansion coefficients of 10×10^{-6} per °C, this would correspond to a temperature fall of 10 °C. Given that the problem is due to differentials within the concrete, a temperature differential of 20 °C would actually be needed between core and outer layers in order to generate a compressive strain in outer layers of 100×10^{-6} and a tensile strain of 100×10^{-6} in inner layers. It will be appreciated from this simplified analysis that thermal cracking is certainly possible in mass concrete if heat loss from outer layers leads to a substantial temperature differential across the structure and that the greatest risk of cracking is at the core of the material. Furthermore, if the differential of temperature across mass concrete can be reduced, the risk of damage will also be reduced. One way of achieving this is by insulation of outer layers which reduces early heat losses and therefore the harmful differentials that might arise between core and surface.

THERMAL PROPERTIES

It need hardly be mentioned that, owing to moisture effects in concrete, it is not possible to quote a coefficient of thermal expansion, as would be possible, for example, for steel. When concrete is heated, water diffuses from gel pores into capillary pores and the concrete as a whole tends to lose weight. Hence, an initial expansion may be offset by the consequent moisture change, which may take some considerable time to complete. A further effect which influences expansion is the decrease of surface tension of water with increase of temperature. This results in a further expansion of concrete due to the reduction of the compressive effect of water in capillary pores. This capillary effect is greatest at humidities of about

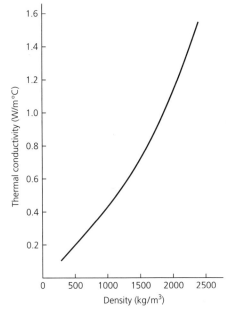

Figure 7.32 Approximate relationship between the density and thermal conductivity of dry concrete. The exact relationship depends on aggregate type

50 per cent since, at high humidities, capillaries are full of water and there are fewer water/air interfaces while, at low humidities, they are almost empty. The expansion coefficient of concrete also depends on that of the aggregate used. As far as it is possible to quote an average value, normal concretes have coefficients of thermal movement around $11 \times 10^{-6}/°C$ which, fortunately, is similar to that of steel, so that relative movement in reinforced concrete is usually insufficient to destroy the steel/concrete bond. Some aggregates, for example quartzite, may, however, have coefficients of $13 \times 10^{-6}/°C$ or more, while that of limestone concrete is in the region of $8 \times 10^{-6}/°C$.

The thermal conductivity of concrete is dependent on its density (Figure 7.32), though it is also affected by aggregate type and moisture content. Generally, dry, dense concrete has a thermal conductivity of about 1.4 W/m°C – similar to that of clay brickwork – so that neither of these materials is effective as an insulating material unless used in great thickness. When saturated, thermal conductivity values may be up to 15 per cent higher.

A valuable property of normal dense concrete that is being increasingly used for control of the internal environment is its high thermal inertia. This was defined (on page 44) as $k \rho C_p$ where k is the thermal conductivity, ρ the density and C_p the specific heat capacity of the material. Dense concrete has quite high values of k and ρ, giving it high thermal inertia and the ability to absorb heat during the daytime in hot summer weather (if not insulated by plaster coatings), the heat being re-emitted during the cooler night time period.

FIRE RESISTANCE

In some respects, concrete offers advantages over other materials in relation to fire, for example, it is non-combustible, it does not emit harmful gases on heating and, perhaps most important for escape, concrete surfaces have a high thermal inertia so that they absorb heat in a fire, delaying the onset of flashover. Indeed, concrete has been widely used for the protection of structural steel.

As a structural material, concrete is affected both physically and chemically by fire though by careful design, satisfactory performance can usually be obtained at reasonable cost.

The prime physical effect is due to thermal expansion, the main problem being spalling. Some types of aggregates, notably siliceous materials, disintegrate on heating due to changes in their crystal structure, though this effect is usually confined to the surface of the concrete, individual particles breaking away. More seriously, surface layers of concrete may break as a result of their thermal expansion and the added effect of pressures due to trapped water vapour beneath the surface. The problem is again mainly encountered with siliceous aggregates and will be increased when the moisture content of the concrete is high. Reinforcement often produces a plane of weakness so that spalling may occur at this depth, exposing reinforcement directly to the fire. Cement paste shows pronounced shrinkage on heating to over 100 °C as the gel and calcium hydroxide components become dehydrated so it is hardly surprising that aggregates which expand least when heated tend to produce the most fire resistant concrete. Figure 7.33 shows the thermal expansion characteristics of concrete containing three aggregate types.

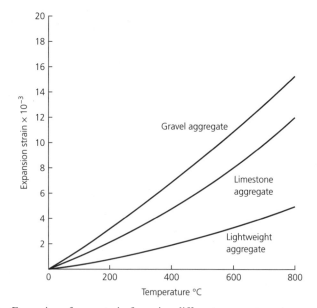

Figure 7.33 Expansion of concrete in fire using different aggregates

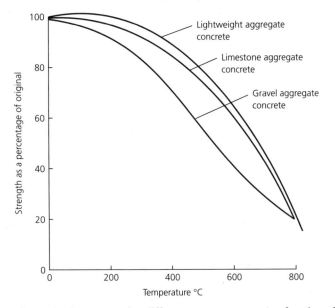

Figure 7.34 Strength of concrete using different aggregate types as a function of temperature. Note that limestone, and especially lightweight aggregates, are also slower to heat in a fire than gravel aggregates

Figure 7.34 shows how strength is affected over the same range. It is noticeable that concretes containing lightweight or limestone aggregates perform best, the latter aggregate owing to its small thermal coefficient up to about 400 °C together with the fact that calcination (conversion to calcium oxide) is endothermic – it absorbs heat. Lightweight aggregates perform well due to their low expansion, and because they are heated during manufacture and are therefore usually chemically stable up to at least 800 °C. By the time the temperature has reached 800 °C, all concrete has virtually failed due to breakdown of the cement paste. Prediction of precise behaviour of concrete in fire can be quite difficult in view of the fact that the loss of strength depends additionally on mix parameters, the actual temperature regime and the amount of load during the fire; concrete under compressive load weakens at a slower rate than if not loaded.

Design for fire resistance

In reinforced or prestressed concrete beams or floors, the critical property in fire is normally the tensile steel strength, since this steel is normally near the soffit and, therefore, heats more readily than the concrete section as a whole. There is a consequent risk of steel yielding. Factors affecting performance in this situation are as follows.

Concrete type

Lightweight concretes are best in fire for reasons given above and because they transmit heat more slowly to the steel than dense concrete (lower diffusivity).

Steel type

High grade steels are more susceptible to high temperatures (see 'Fire resistance of steel').

Cover to steel

For example, to obtain two hours fire resistance the cover needed for a simply supported reinforced concrete beam will be 40 and 50 mm for lightweight and dense aggregate concretes respectively. This means, in practical terms, that there is a minimum width for such beams; for example, for two hours fire resistance it is about 200 mm while in waffle floors the minimum rib width is about 125 mm. For higher fire resistance, extra cover is needed and if this is to be in the form of concrete (rather than, say, gypsum plaster), additional *secondary* reinforcement may be needed. The use of stirrups is now preferred to mesh for this purpose, since the latter tends to cause placing problems. Cover to this steel must also be adequate to eliminate corrosion risks.

In the design of columns for fire resistance, account must be taken of the effect of the fire on the bearing capacity of both concrete and steel as well as the fact that with a reducing concrete section there is an increased risk of buckling. Cover to steel again has an important effect and it is not surprising that larger sections which take longer to heat up in fire have better fire resistance. In most cases, for two hour resistance it will be a question of *checking* for compliance rather than *designing* for compliance, especially with lightweight concretes. For design purposes, tables based upon standard depth/temperature/time curves in dense and lightweight concrete together with standard curves for the effect on strength properties of concrete and steel are normally used.

It will be appreciated that fire resistance considerations may be more important where high strength concretes are concerned since smaller section sizes may become feasible, resulting in higher rates of temperature rise. In addition, since stronger concretes tend to be more impermeable, pressure from heated water vapour may result in increased risk of spalling. It is possible that use of a small percentage (say 0.5 per cent by volume) of polypropylene fibres may improve fire resistance by allowing dispersion of water vapour and therefore reducing the risk of surface concrete spalling to expose the steel. There would be a small strength penalty if using such fibres but significant benefits could emerge, especially in concrete, with greater risk of spalling such as those employing siliceous aggregates.

Remedial treatment

The main emphasis in fire protection is always on escape and prevention of spreading, with little attention either in design or during fire fighting being given to preserving the structure itself; quite often the quenching effect of water in a serious fire adds greatly to the damage to the structure, especially in the case of concrete. Nevertheless, remedial treatments may be feasible. The first step is to establish

damage to remaining concrete; some indication of this may be obtained from the
colour of the concrete, provided the sand or coarse aggregate contains traces of iron:

300 °C	600 °C	800 °C
pink	grey	buff/yellow

(Some aggregates, such as granites or limestones, may not show this effect.)
A pink coloured surface indicates significant strength loss, further strength
reduction being progressive and by 800 °C the concrete is likely to be virtually
useless. In most fires, damage is limited to spalling, even when the period of fire
equals or exceeds the design value. Underlying material should be removed until
undamaged concrete is reached – the state of this can be checked by tests such as
ultrasonic pulse velocity, internal fracture tests, the Schmidt hammer, or cores.
Minor damage can be made good with epoxy mortar while, for larger areas, guniting
(sprayed concrete) is suitable. Note that the repair must be effective in future fires
and epoxy resin repairs must themselves be protected for fire resistance purposes.
It may be possible to seal cracks caused by movement with resins provided cracks
are stable. The situation is much more complex where main steel reinforcement is
exposed, particularly if this has led to excessive deflection of beams. High yield
steels may have reverted to mild steel and may, therefore, be no longer structurally
adequate, hence it is quite important to ascertain the maximum temperature
reached by affected steel. In a minority of cases, where significant deflections or
damage to steel have occurred, replacement or structural alterations to reduce load
bearing requirements of damaged concrete may be called for.

FROST RESISTANCE

The risk and occurrence of frost damage to concrete varies greatly, reflecting
the range of concrete types and mixes used and the range of service conditions.
Provided the seriousness of exposure to frost is adequately predicted at the design
stage, there is no reason why concrete should not perform successfully even in cases
of very severe exposure.

The mechanism of frost attack on porous materials generally has already been
covered (see page 34) and so it remains to examine how the principles can be
applied to concrete. A saturated horizontal surface of air entrained concrete will be
taken by way of example (Figure 7.35). At or slightly below 0 °C, water will tend to
be drawn from capillary pores or cracks towards the concrete surface where freezing
will occur. Once a continuous layer has formed at the surface, an ice front will
advance downwards into the concrete. Water in gel pores does not freeze, since
surface forces prevent this – it is estimated that temperatures of as low as -70 °C
would be required to cause freezing of gel pore water. Instead, water tends to be
drawn from smaller pores into capillaries or cracks where it freezes, adding to the
volume of ice already formed at these positions. This movement, together with
the expansion of water on freezing, leads to hydrostatic pressure (Figure 7.35).
In regions adjacent to entrained air bubbles, the pressure can be relieved because

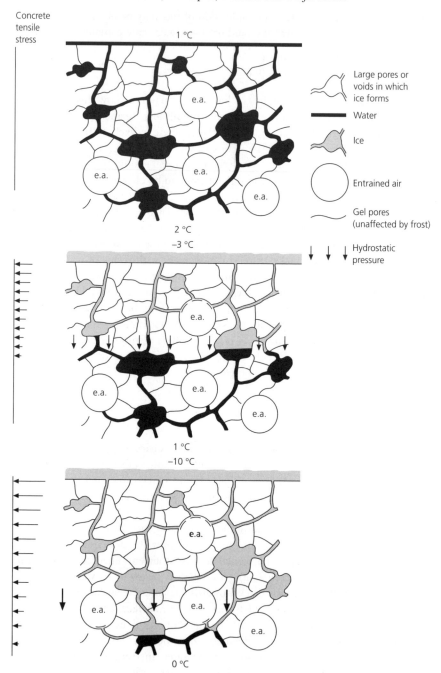

Figure 7.35 Schematic representation of ice front advancing into concrete as the surface temperature falls. Entrained air relieves local pressure as the ice pushes water deeper into the concrete

water can flow from pores into these bubbles (which do not normally fill with water) where freezing can occur. The extent of stress relief in the concrete depends on the proximity of all parts of the paste to entrained air bubbles – if the air content is inadequate, then stress relief only occurs around each bubble. As the temperature falls further, the ice front travels further into the concrete, generating higher pressure as it pushes water in front of it. This pressure, combined with corresponding tensile stresses in the frozen material, may lead to swelling and, ultimately, tensile failure.

Progressively lower temperatures increase the risk of damage because:

- in frozen areas, water from smaller pores continues to feed growing ice crystals in larger pores, and
- the area of stressed material grows larger

A further problem is caused by osmosis – if there is any salt in the pore water, it tends to be rejected on freezing so that there is a build-up of salt in solution at the point of maximum hydraulic pressure (indicated by the vertical arrows in Figure 7.35). Water tends to flow by osmosis towards these regions to equalise concentrations thereby further increasing the pressure just below the advancing ice front.

Damage produced by frost is progressive, each freezing cycle causing further break-up, surface layers being worst affected. It has already been stated that air entrainment, if adequate, can largely dissipate the pressures in freezing concrete and the extent to which this occurs is illustrated by the fact that there may be a net contraction in frozen air entrained concrete caused by shrinkage, resulting from the removal of water from pores, similar to that of drying shrinkage. In fact, an assessment of the resistance of concrete to frost can be made by checking length changes of a sample on repeated freezing. Where expansion occurs, there is risk of damage in the long term.

In practical terms, the main points to be considered when designing for frost resistance are as follows.

- The risk depends very much on the degree of exposure and saturation of the material concerned. The risk is greatly increased for external horizontal surfaces which do not shed water effectively and which are subject to rapid radiation heat losses. If de-icing salts are used, the risk is further increased by crystallisation/osmotic effects.
- Low permeability concrete is less seriously affected because it admits less water. Low water/cement ratio concretes have greatly increased frost resistance. As an indication of the relative rates of deterioration, 50 freeze/thaw cycles would reduce the compressive strength of concretes of 0.4, 0.6 and 0.8 water/cement ratio by approximately 10 per cent, 20 per cent and 60 per cent, respectively. 100 cycles would cause 30 per cent, 70 per cent and 100 per cent reductions, respectively. It will be appreciated that fully exposed concrete, even in the UK, must be manufactured using low water/cement ratios if long service is to be achieved.

- Full compaction is vital to keep capillary pores or voids to a minimum.
- The use of entrained air provides a major benefit, even when low water/cement ratio concrete is used.

SULPHATE ATTACK

Sulphates occur widely in the ground, especially calcium sulphate (gypsum), sodium sulphate (Glauber's salt) and magnesium sulphate (Epsom salt). Chemical (as distinct from crystallisation) activity can only take place in the dissolved state. A key guide to chemical activity is therefore reflected by their solubilities, which are quite different:

calcium sulphate	1 g/litre
sodium sulphate	400 g/litre
magnesium sulphate	300 g/litre

Sodium and magnesium sulphates are, in consequence, much more aggressive than calcium sulphate. Magnesium sulphate is generally the more common, occurring in sea water and groundwater, especially in conjunction with clay soils, where they tend to be leached by rain to lower levels. Sulphates may reveal themselves in newly excavated ground as white crystals or as a white deposit on trench walls in drying conditions. Landfill sites may also be a hazard, especially where industrial processes, such as coal or gas production, have been involved. Concentrations may vary from place to place and may be altered when the ground is disturbed, for example, exposure of lower levels of soil may lead to concentration by evaporation or induced water flow.

Chemical effects of sulphates

The main reactions involve tricalcium aluminate and calcium hydroxide in the concrete. Take, for example, sodium sulphate:

calcium aluminate + sodium sulphate → calcium sulphoaluminate (ettringite)
+ aluminium hydroxide
+ sodium hydroxide

(schematic equation given for simplicity), and:

$$Ca(OH)_2 + Na_2SO_4.10H_2O \rightarrow CaSO_4.2H_2O + 2NaOH + 8H_2O$$
gypsum

The respective reaction products, ettringite and gypsum, each occupy increased space and lead to eventual disruption of the concrete, if formed in sufficient quantities. However, the reaction rate depends on the type of sulphate present.

Calcium sulphate, with its low solubility, reacts only slowly with calcium aluminate hydrate and not at all with calcium hydroxide, while the reaction rates with sodium sulphate are slowed down by the fact that one of the reaction products, sodium hydroxide, is soluble and concentrates in solution (unless washed out). This 'congests' the solution, which is the reaction medium, reducing the rate of attack. The effect of magnesium sulphate is, however, more serious because magnesium hydroxide has low solubility and is precipitated out, leaving the solution clear for the reaction to proceed. Magnesium sulphate has the additional effect of reacting with calcium silicate hydrates and its presence, therefore, poses a serious risk to concrete.

Avoidance of sulphate attack

The first step is to estimate the concentration of sulphates by soil tests. Either the soil may be tested or, if groundwater seeps into excavations, this may be tested. In either case, careful sampling is necessary. When soils are tested, the sulphates may be extracted with water – commonly a 2:1 or 1:1 water:soil ratio by weight – or by dissolving in hydrochloric acid. The latter method dissolves calcium sulphate and may produce misleadingly high sulphate contents when a substantial proportion of sulphates are in the form of calcium sulphate. Table 7.14 shows typical concentrations, together with resulting classifications. The table also gives details of cements and mix specifications for concrete using 20 mm aggregates in each situation. When designing/producing concrete to resist sulphates, the following points should also be considered.

- Replacement of cement by PFA, ground granulated blast furnace slag or microsilica can give improved sulphate resistance, though tests may be necessary to check performance in given situations.
- Special care is necessary when significant quantities of magnesium sulphate are present.
- The risk of attack depends upon the ground conditions. Significant water movement, high or fluctuating water tables all increase the risk of attack since higher levels of sulphates in solution become likely.
- In class 5 conditions, concrete should be protected by an efficient, impermeable membrane.
- Risk of serious structural damage is greater in slender structures where a given degree of penetration will represent a greater proportion of the whole section.
- The need for full compaction to produce impermeable concrete cannot be overemphasised.
- Sulphate attack appears to be less serious in sea water, since the reaction products, ettringite and gypsum, tend to dissolve in the sea water, rather than crystallising and causing damage.
- Drying effects can concentrate salts and, therefore, increase the rate of attack both chemically and by crystallisation of salts. Areas subject to wetting and drying, for example, sea walls around high tide mark and foundations just above ground level, are particularly vulnerable.

Table 7.14 Classification of sulphates together with recommended concrete mix details. Instead of PPFAC, OPC or RHPC may be used with: (i) 70–90 per cent ground granulated blast furnace slag or (ii) 25–40 per cent pulverised fuel ash

Class	Sulphate concentration			20 mm aggregate concrete		
	In soil		Groundwater g/litre	Cement type	Maximum free water/cement ratio	Minimum cement content kg/m³
	Total per cent	2:1 water:soil extract g/litre				
1	<0.2	–	<0.3	OPC, RHPC, PBFC, PPFAC	0.60	300
				SRPC	0.55	280
				PPFAC	0.55	310
2	0.2–0.5	–	0.3–1.2	OPC, RHPC, PBFC	0.50	330
				SRPC	0.50	330
3	0.5–1.0	1.9–3.1	1.2–2.5	PPFAC	0.45	380
4	1.0–2.0	3.1–5.6	2.5–5.0	SRPC	0.45	370
5	>2.0	>5.6	>5.0	SRPC	0.45	370

OPC   Ordinary Portland cement
RHPC   Rapid hardening Portland cement
PBFC   Portland blast furnace cement
PPFAC   Portland pulverised fuel ash cement
SRPC   Sulphate resisting Portland cement

Thaumasite

The thaumasite form of sulphate attack (TSA) is of relatively recent discovery. The composition of thaumasite is $CaSiO_3.CaSO_4.CaCO_3.15H_2O$. Clearly, formation of this material requires the presence of carbonates as well as sulphates, these being present when aggregates contain a significant quantity of calcium carbonate, such as limestone. Where groundwater contains high sulphate levels, the calcium silicate component of the hydrated cement is attacked and this can convert affected areas of concrete into a pulp. Unfortunately, the use of sulphate-resisting Portland cement does not prevent attack since it is the calcium silicates that are affected. Because high sulphate concentrations together with wet conditions are needed, the problem is found mainly underground. One cause has been found to be exposure of iron pyrites (which contains sulphur) in clay by excavation of clay soils. This oxidises, producing high sulphate concentrations on backfilling.

There have been cases where cement mortars have been attacked. The origin of the carbonate appears in this case to be acid rain. In exposed conditions the mortar is again converted to a pulp. Avoidance of the problem is a matter for continuing investigation, though clearly where damp conditions prevail there will be a need to avoid high levels of sulphate contamination. For example in exposed brickwork, the risk of problems will be reduced by use of low salt content bricks.

Mundic

This is yet another form of sulphate attack in which pyrites in mining wastes, used to form building blocks, has been attacked. As above, the pyrites oxidises over time, producing sulphates which attack the concrete from within if damp conditions prevail. Fortunately, the problem is confined mainly to parts of Cornwall in which the mining waste was used as a cheap aggregate for building blocks.

ALKALI–SILICA (ALKALI–AGGREGATE) REACTION (ASR)

Although still relatively uncommon, the problem of alkali–aggregate reaction has evoked strong general concern, possibly because, first, the total extent of the problem may be difficult to estimate and, second, there are no straightforward remedies for structures, once seriously affected.

The most common form of attack results from the presence of reactive silica in certain types of rock, such as opal and tridymite. These are attacked, in the presence of water, by the hydroxides of sodium and potassium which, though present in only small quantities in hydrated cement, may increase the pH value of the pore solution by 1 or more above the normal value of about 12 found in uncarbonated concrete. The resulting alkali–silica gel attracts water and expands gradually within the framework of the hardened concrete, causing unsoundness in the form of spalling or *pop-outs*, or, in more serious cases, map cracking – an array

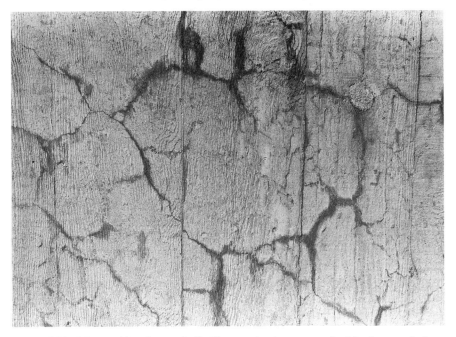

Figure 7.36   Map cracking due to alkali–silica reaction in concrete. In this photograph the cracks are highlighted by dampness

of cracks which may affect the whole structure. Figure 7.36 shows an example of map cracking which is found on plane surfaces away from the influence of reinforcement. Because many of the cracks originate from a point at angles of about 120°, it is common to find them meeting in groups of three, another name being, consequently, *Isle of Man cracking*. (Note that such cracking is more related to the general expansive movement of the substrate than to the chemical origin of the problem, and may also be found on occasions where stresses due to sulphate attack or excessive shrinkage arise.) Where reinforcement restrains movement, cracks may run parallel to reinforcing bars. The reaction process tends to be self-perpetuating, since cracks readily admit the water needed to continue the attack. The speed of attack depends on the fineness of the reactive aggregate – in very fine aggregates, the reaction may occur within days or weeks and is less likely to cause damage. In coarser aggregates, water takes much longer to penetrate and the problem may take up to 10 years to appear, cracking becoming progressively more serious over a further period of years. An additional difficulty lies in the fact that quite small proportions of reactive aggregate – 1 per cent or less – may cause the worst damage; indeed, there is a 'pessimum' reactive aggregate level, above which the aggregate swamps the alkali in the cement, overcoming the problem. Flint – a major source of aggregates in the UK – may be reactive but where more than 60 per cent of the aggregate is flint material, the risk is considered negligible.

Prediction of risk

ASR will only occur in the presence of a reactive aggregate, a high-alkali cement and water. The first step is to consider the aggregate used both mineralogically and from the point of view of service history. If there is any risk (and this could occur with quite a small proportion of reactive material), then steps should be taken to control either of the other two factors – namely, the alkali content of the cement or the degree of exposure to water.

The reactive alkali content of Portland cement is normally measured as sodium oxide (Na_2O) equivalent:

$$\% \text{ equivalent } Na_2O = \% \ Na_2O + 0.658 \times (\% \ K_2O)$$

(the factor applied to potassium oxide takes account of its different molecular weight).

It should be noted that sodium chloride appears to generate additional alkalinity in concrete so that its presence in aggregates or admixtures should be taken into account. Exposure to salt laden on-shore winds near the sea may increase attack for the same reason. It is found that an equivalent amount of alkali corresponding to the chloride ion content of aggregates is:

$$\% \text{ equivalent } Na_2O \text{ (aggregate)} = 0.76 \times \% \ Cl^-$$

BRE Digest 330 classifies aggregates as of low, normal or high reactivity – it should be noted that most aggregates in the UK are classified as normal reactivity. Table 7.15 gives recommended limits for the alkali contents of concretes according to the aggregate reactivity class and cement alkali content. The limits given are based upon the assumption that the alkali input from other sources such as aggregates or admixtures does not exceed 0.2 kg of Na_2O equivalent per cubic metre of concrete.

One way of keeping to the limiting values would be to partially replace cement by ground granulated blast furnace slag (GGBS) or pulverised fuel ash (PFA) as

Table 7.15   Recommended limits for alkali contents of concretes in situations of risk of ASR

Aggregate type or combination	Alkali content of a Portland cement or the BS 12 component of a combination with GGBS or PFA		
	Low alkali cement [% Na_2O < 0.60 eq guaranteed on spot samples]	Mod alkali cement [% Na_2O < 0.75 eq]	High alkali cement [% Na_2O > 0.75 eq]
Low reactivity	Self-limiting	Self-limiting	Limit < 5.0 kg Na_2O eq/m^3
Normal reactivity	Self-limiting	Limit < 3.5 kg Na_2O eq/m^3	Limit < 3.0 kg Na_2O eq/m^3
High reactivity		Limit < 2.5 kg Na_2O eq/m^3	

these have been found to be highly effective in the control of ASR. However, each of these materials contains some alkali which must be allowed for, so that limits are not exceeded. A 40 per cent minimum replacement level is recommended for GGBS and at least 25 per cent replacement level for PFA. BRE Digest 330 should be consulted for further details. Microsilica has been found to be effective in controlling ASR, though experience of use is somewhat limited at present. In some situations, it may be possible to control attack by preventing exposure to water. The practicality of this mode of protection obviously depends very much on the situation and waterproof coatings should only be used where their serviceability can be guaranteed. Even when protection from rain is achieved, ASR can occur wherever relative humidities (once cured) of 75 per cent or over are maintained so that careful consideration is necessary. In general, structures which are highly exposed to water, such as bridges or water retaining structures exposed to high vapour levels, such as swimming pools, or are of critical importance structurally or in safety terms, should not rely on the exclusion of water as a protection method.

Remedial treatment

When ASR is suspected, it is important to arrive at a firm diagnosis, since similar symptoms may be caused by other agencies, such as shrinkage, sulphate attack or frost damage. Where cracking patterns suggest ASR, cores should be taken and examined microscopically to check for signs of gel and associated damage to aggregate particles. Where exudations are present, these may be checked for composition. Remaining reactivity in cores can be checked by recording length changes at 38 °C in a 100 per cent humidity environment, though samples should be saturated first to separate from moisture movement effects. Figure 7.37 shows possible effects. The behaviour of cores and that of the actual structure (which should be monitored) may reveal that ASR becomes exhausted as reactive aggregate produces stable products. If this is the case and cracking has not affected reinforcement, it may be possible to apply a protective coating to prevent further damage. Note that protective coatings often require specialist application and may,

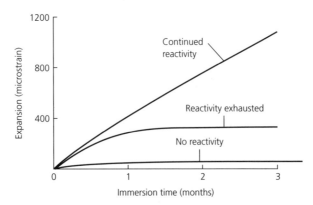

Figure 7.37   Behaviour of core saturated at 38 °C to check for continued ASR reactivity

Figure 7.38 Additional structural support plus rain protection provided for a viaduct affected by ASR, near the sea

if incorrectly used, serve to trap moisture already in the structure rather than prevent ingress. In the viaduct illustrated in Figure 7.38 additional structural support was provided, plus protection from on-shore winds – the structure is near the sea. Where cracking is serious, impairing structural performance or causing risk of corrosion of reinforcement, the only option will probably be removal and replacement of affected sections.

ACID ATTACK

The calcium hydroxide in hydrated cement is readily attacked by dilute acids, most acids producing salts which are sufficiently soluble to be leached out, allowing continued attack. The aggressiveness of acids can be measured in pH terms, the extent of damage being as follows:

pH > 6.5   little effect
pH 5.5–6.5   slight damage
pH 4.5–5.5   medium damage
pH < 4.5   severe damage

Flowing water is always more aggressive, since it removes reaction products more effectively as well as providing fresh supplies of hydrogen ions.

One of the commonest sources of problems is carbon dioxide which dissolves in water and can result in pH values as low as 5 in peaty soils. The net effect is rather different from that of carbonation, in that calcium bicarbonate, rather than calcium carbonate, is formed which, being soluble, can be easily leached out by flowing water, permitting continued access of the acid.

Acid attack can also occur in sewage pipes, sulphur in the sewage being converted by bacteria first into hydrogen sulphide. Some of this escapes as a gas and is converted into sulphuric acid if, for example, condensation is present on surfaces above the water line. The sulphuric acid then attacks the concrete with damage usually being concentrated just above the water line. The problem can be reduced by improving ventilation so that condensation levels are lowered. Acid attack is also quite common in unlined brickwork or blockwork flues where the sulphur given off by fossil fuels is oxidised to sulphurous or sulphuric acid which attacks the cement mortar.

The net effect of acids is to produce a spongy, weak material, though, unless the concrete is highly permeable, the rate of attack should decrease with time as penetration distances to unreacted calcium hydroxide increase.

Good quality impermeable concrete has reasonable resistance to acid attack but durability can be further improved if the surface is allowed to carbonate before exposure, pores then being blocked by the insoluble material. A similar effect is obtained by treating the surface with solutions of sodium silicates or silicofluorides.

7.11   ADMIXTURES FOR CONCRETE

These may be defined as materials, other then cement, aggregates and water, which are added at the mixer. The term *additive* normally denotes a material incorporated with the cement during manufacture.

Admixtures have been used to modify the properties of concrete since time immemorial – indeed, bull's blood was used for air entrainment as far back as Roman times. Specifiers are sometimes wary of admixtures in the absence of an understanding of their principles of operation and especially the effect of overdoses, together with a lack of standard specifications. It is probably fair to indicate that the latter problems are now largely overcome due to increasing appreciation of their chemical behaviour and the development of standards to control quality. Hence, admixtures now have a sound technological basis and, provided precautions are taken, particularly with dispensing procedures, they can make a real contribution to good quality concrete. They will be described roughly in the order of decreasing use. Note that addition rates are normally given as a proportion by weight of cement.

WATER REDUCERS/PLASTICISERS

These are based on the same compounds incorporated for different purposes – water reducers are designed to reduce water content at a given workability, or increase

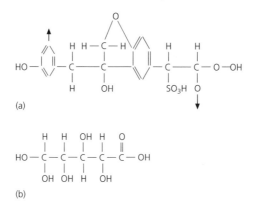

(a)

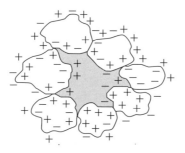

(b)

Figure 7.39 (a) Lignosulphonic acid repeat unit. The COOH acid group is at the right hand end and units join up at the arrows. (b) Gluconic acid – a typical hydrocarboxylic acid used as a water reducer

Figure 7.40 Cement particles, on account of surface charges, form into flocs, trapping water and reducing workability

workability at constant water content. The chief compounds used are lignosulphates, a by-product of the paper-making industry, and hydroxylated carboxylic acids (Figure 7.39). These molecules are quite large, containing two essential parts. They have a long chain-like 'body' and at one end a 'head' comprising a chemically active acid group (–COOH) which tends to dissociate in aqueous solution, giving positive and negative charges, the acid end group becoming negatively charged. In addition, the compounds described have electric charges (polar groups) along the body of the molecule. Cement grains also contain charges – normally slightly positive – which tend to lead them to attract each other (van der Waal's forces) causing *flocculation* – that is, the formation of groups which trap water between them (Figure 7.40). The admixture molecules become electrically attracted to the cement grains giving them a considerable negative charge and causing them to repel each other, giving great fluidity to the mix and releasing water which was trapped in flocs. A secondary effect is that some admixtures, such as lignosulphonates, tend to stabilise tiny air bubbles, the chain-like body of each molecule (which is hydrophobic, repels water) aligning itself inside a bubble while the charged end is in the water (Figure 7.41). This is a disadvantage of normal plasticisers, together with the fact that they tend

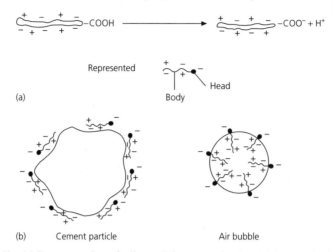

Figure 7.41   (a) Representation of a lignosulphonate molecule with its negatively charged acid group which is hydrophilic (attracted to water) and long body which is hydrophobic (repelled by water). (b) The body of each molecule attaches to the surface of cement particles or aligns inside air bubbles – away from water. Both cement particles and air bubbles become negatively charged and repel one another. Aggregate particles may also be charged negatively

to retard hydration due to the presence of hydroxyl ions (OH^-) on the chains. Since all components of the mix become negatively charged, repelling each other, plasticisers are said to *disperse* the mix, reducing surface tension and tending to cause bleeding, though air entrainment offsets this to some degree.

Dose rates of plasticisers must be small if excessive side effects are to be avoided – typically, 0.1 per cent by weight of cement. They can be used to:

1 increase the slump by 50–70 mm at constant water content, or
2 reduce water content by 5–10 per cent at constant workability, or
3 save 15–30 kg of cement per cubic metre of concrete at constant water/cement ratio

The latter mode of use illustrates the strong commercial reasons for using plasticisers, especially in the ready mixed concrete industry. In terms of overall performance of concrete, the second is perhaps more relevant if advantage is taken of lower water/cement ratios which will lead to improved strength and durability characteristics at reasonable cost. Mortar plasticisers work on similar principles, though they are not so highly refined and could therefore lead to unacceptably high air entrainment and retardation if used in concrete.

AIR ENTRAINING AGENTS

These are typically based on neutralised vinsol resins which have an effect rather similar to that of plasticisers. The main difference is that the end of the chain is more active than the body, the latter having fewer charges – it may be a simple hydrocarbon chain as in oleic acid. Air bubbles are stabilised and become negatively

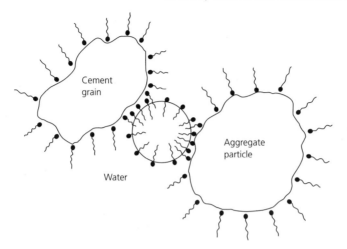

Figure 7.42 Action of air entraining agents in stabilising air bubbles and producing a cohesive but workable mix. Note that the heads, not the bodies, of the molecules attach to the solid surfaces

charged as before but action on solid surfaces is more one of neutralisation as the negative ends (not the bodies) of chains attach themselves to cement grains and aggregate particles. The charged air bubbles are attracted to the solid surfaces, adhering to them by means of the hydrophobic bodies of each molecule and leading to a cohesive effect (Figure 7.42). The air bubbles act both as a lubricant and as 'ball bearings', producing a workable but cohesive mix, unlike water reducers which tend to have a dispersive effect.

The amount of air entrained for a given quantity of admixture depends on mix proportions and, in particular, the amount of water available. Hence, wetter mixes tend to result in higher air contents while mixes rich in sand entrain less air as fine material competes for available water; hence, the sand content should be reduced if air entrainment is to be used, the air itself behaving as a sand. Air entraining agents can be used to compensate for bleeding effects in harsh mixes, or simply to produce a cohesive mix with a resulting high quality surface finish. Typical air contents would be in the region of 5 per cent for 20 mm aggregate and this leads to a strength reduction of around 20 per cent, though some of this loss can be regained due to the lower water contents which follow.

Perhaps the most important application of air entraining agents is their ability to improve resistance to frost and de-icing salts. The air bubbles resulting are able to relieve the hydraulic pressure caused by an advancing ice front in the concrete (Figure 7.35). Bubbles produced by air entrainment are quite different from entrapped air because:

- They are sealed and would not be filled with water during normal saturation of the concrete.
- They are very small and well distributed.

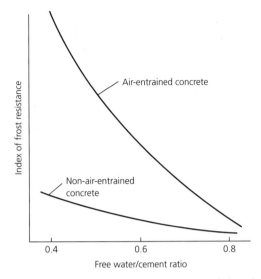

Figure 7.43 The effect of air entrainment on frost resistance of air and non-air-entrained concrete of variable free water/cement ratio

Studies have indicated that, to be effective, bubbles should not be separated by more than about 200 μm (0.2 mm), so that no point in the cement mortar is more than 100 μm from an air bubble. To achieve this requires very small bubbles – sizes usually being in the range 50–100 μm, leading to an air content in the cement mortar itself of about 10 per cent. Taking the concrete as a whole, the required air content will decrease as the aggregate volume fraction increases, ranging from about 4 per cent for 40 mm aggregate to 7 per cent for 10 mm aggregate. Provided there is sufficient air, the improvement in frost resistance is spectacular, even at low water/cement ratios (Figure 7.43). Not surprisingly, many specifications for concrete to be exposed, especially in the form of slabs such as roads, require that the surface layers be air entrained and their use must be seen as a sensible precaution in any situation where there is risk of frost attack. Dosage rates are, like plasticisers, small – in the region of 0.1 per cent by weight of cement.

SUPER-PLASTICISERS

These are of more recent introduction, their basic action being similar to that of ordinary plasticisers but with the following differences:

■ Whereas cross-linking in lignosulphonates tends to cause bunching of chains, in super-plasticisers the chains are relatively straight so that they operate more efficiently.
■ The hydroxyl (–OH) groups which cause retardation in ordinary plasticisers are absent in super-plasticisers so that larger doses can be used without significant retardation.

■ They have higher molecular weights than ordinary plasticers – typically 5000, obtained by polymerisation using formaldehyde groups (as in many thermosetting plastics).

These molecules, on adsorption to cement grains, cause considerable cement grain repulsion. The polymerisation process unfortunately, however, also increases the cost.

Two main types are available – sulphonated melamine formaldehyde and sulphonated naphthalene formaldehyde condensates.

Super-plasticisers can be added in proportion up to 2 per cent of cement content and, if added to a normal slump mix, will produce *flowing concrete* which requires no compaction and is, therefore, very easy to place, though concreting on slopes of over 2° should be avoided with such mixes.

Alternatively, water reductions of 30 per cent can be achieved without loss of workability and very high strengths can be obtained in this way. In each case, the effect of the super-plasticisers is quite short lived, perhaps due to cement hydration. Higher temperatures or C_3A contents accelerate the stiffening process. Concrete should be placed within 30–45 minutes of adding to the mix. For this reason, the admixture is often added to the concrete just before placing in the case of ready mixed concrete, followed by a second short mixing period.

Hardened properties, such as strength and shrinkage, of flowing concrete appear very similar to those of the normal material.

Super-plasticisers are likely to increase the cost of concrete by about 10 per cent, hence they are only used where either the benefit of a free-flowing mix or very high strength are required.

A further development in this field is synthetic polymer based *hyper-plasticisers* which permit very low water contents to be used while retaining workability. Such high-rate addition materials have opened up the possibility of producing concretes with much higher fines contents than would previously been possible for structural applications. This in turn offers the opportunity to develop new placing techniques. For example *self-compacting concretes* (SCC) with a maximum aggregate size of around 10 mm can be used almost like grouts, having been used to produce columns by bottom injection, the material passing upwards around reinforcement to produce a uniform product of adequate structural performance without any compaction. The advantage of injection of concrete from the base of a mould is that air does not have to travel upwards through the concrete as happens during normal compaction. The concrete merely displaces the air upwards as its level rises. The essential composition of self-compacting concretes is high filler content to produce very high cohesion, together with high plasticiser dosages to increase fluidity. Their use is likely to increase in the future.

RETARDERS

Retarders are used to delay stiffening of concrete and are, therefore, designed to interfere with hydration in the first 12 hours or so after mixing. The active components in most retarders are hydroxyl (–OH) groups attached to carboxylates or aldehydes, which are able to combine with initial hydration products to form very thin, but stable, layers on cement grains. They act mainly on C_3S but also on C_3A in the cement grains. Their effect is eventually overcome as ion concentrations in solution increase so that hydration proceeds as before. In the long term, strength is not reduced by use of retarders – in fact, it is often higher because:

- An improved gel structure results from the slower initial hydration.
- Retarders are usually blended to give water-reducing properties as well – this has little effect on cost, giving mixes with the added benefits of reduced water content.

Care is necessary with retarders, since overdoses may severely retard concrete mixes – indeed, retarders are often carried by ready mixed concrete vehicles in order to prevent the concrete setting in the drum in the event of breakdown. However, even in cases of large overdoses, resulting in delays to setting of over one day, experience shows that concrete, on eventual setting, will hydrate to reach normal long term strength.

Retarders may be used in the following situations:

- When there are delivery or placing difficulties – particularly with ready mixed concrete.
- To offset the accelerating effect on setting of placing concrete in hot weather.
- To ensure that a monolithic structure is formed in large pours which often take some hours to complete.

ACCELERATORS

Accelerators are normally designed to accelerate the hardening process rather than the stiffening process, although the two are both the result of the same process – cement hydration – and show a degree of interdependency. Most soluble salts accelerate hydration – acid salts by enhancing dissolution of calcium oxide and basic salts by aiding dissolution of silica and alumina. The chloride ion is particularly effective because of its small size, calcium chloride being traditionally the most commonly used on account of its low cost. The admixture affects the hydration of C_3A and C_3S, although it is the action on the latter which accelerates strength development. The action appears to be catalytic, that is, the calcium chloride does not become chemically involved but rather the chloride ions assist in the dissolution of cement and passage of ions through the hydrated layer around each grain of cement. The more rapid hydration generates extra heat which in turn further accelerates strength development and an excess of the admixture may cause a flash set – this has been used for leak plugging purposes. The activity of calcium chloride is increased by use of rich mixes or finer cements. Substantial improvements of strength are obtainable between the ages of one and seven days, especially in cold weather concreting, one of the main applications of the admixture. After three days, the strength is doubled by the addition of 2 per cent of calcium chloride if cured at

Table 7.16 BS 8110 limits for chloride content of concrete

Type or use of concrete	Maximum chloride content expressed as a percentage of chloride ion by mass of cement (inclusive of PFA or GGBFS if used)
Prestressed concrete; Heat cured concrete containing embedded metal	0.1
Concrete made with: Sulphate resisting Portland cement (BS 4027) Super-sulphated cement (BS 4248)	0.2
Concrete containing embedded metal and made with: Ordinary Portland cement (BS 12) Portland blast furnace cement (BS 146) Low heat Portland cement (BS 1370) Low heat Portland Blast furnace cement (BS 4246)	0.4

2 °C, this effect being assisted by slight plasticising action. Long term strengths are unaffected, although at low temperatures it may take ordinary concrete up to 1 year to 'catch up' with accelerated concrete.

The chief problem of calcium chloride is its effect on durability. The concrete has increased shrinkage creep and permeability, making it more vulnerable to frost, sulphate attack and alkali–silica reaction. The most important side effect is increased risk of corrosion to steel reinforcement, since the admixture depresses the pH value of cement, increases its electrical conductivity and, in sufficient quantities, destroys the passivating effect of the −OH ions present. Opinions vary as to what concentrations cause problems but BS 8110 restricts total chloride ion contents to the values shown in Table 7.16. These contents include any chlorides in aggregates or mixing water.

Note that contents are given in chloride ion terms and that 1 g of flake ($CaCl_2 . 2H_2O$) contains 0.73 g of anhydrous $CaCl_2$ which contains 0.47 of chloride ions. Hence, care is necessary when dispensing the admixture, especially if solid (it should also be dissolved first in mixing water).

It is now generally accepted that the occurrence of any chlorides in concrete may pose a risk, perhaps because, like many other salts, migration can occur in wetting and drying cycles, causing concentrations at some positions. In view of the risk of corrosion due to calcium chloride, non–chloride based accelerators are being increasingly used, although these are more expensive and, in most cases, much larger doses are required. Examples include calcium formate, calcium nitrate and calcium nitrite. The catalytic action of these appears to be similar to that of calcium chloride, except that the passivation effect of steel is not destroyed so that corrosion risk is greatly reduced.

In practice, many accelerators are blended to give maximum efficiency at minimum cost, types being classified chiefly as chloride based or non-chloride based, in view of corrosion implications. As with retarders, accelerators are also commonly combined with water reducers to give the additional advantages of plasticising action.

WATERPROOFING AIDS

It is worthwhile re-emphasising that well designed concrete mixes of low water/ cement ratio have very low impermeability if they are fully compacted. Admixtures will only be necessary therefore when for some reason, perhaps to reduce cost, heat evolution or shrinkage, such mixes are not appropriate. Waterproofing admixtures may work in two ways.

- They may act as pore fillers. These can be inert materials such as sand, limestone or bentonite. They may be helpful if cement contents are low but may only increase water requirements of richer mixes tending to offset any pore blocking action by leading to increased water/cement ratios. It is preferable if pore filling admixtures are to some degree reactive, since they then become bonded within the concrete and contribute to strength. Examples are microsilica and pulverised fuel ash; these are each described under separate headings.
- They may deposit water repellant substances on pore surfaces, thereby reducing water penetration even in permeable concrete. Most are based on stearates of calcium, sodium or aluminium, though butyl stearates are also used. A hydrophobic layer of calcium stearate is produced, though there may be some strength reduction due to air entrained by these surface active materials. Since pores are not normally blocked by such materials, the permeability of concrete to water under pressure is not greatly reduced.

PUMPING AIDS

Cement has two opposing effects on the flow properties of concrete: first, being hydrophilic, it imparts cohesion to the mix, thereby preventing water separation; second, especially at higher cement contents, it increases particle friction, thereby making pumping more difficult. For pumping purposes, the ideal cement content should be rather more than that required to fill the voids left by the aggregate. In medium cement content mixes ($200-350$ kg/m^3), this is normally the case, although air entraining plasticisers may be used to produce good cohesion, especially if there are problems with or variations in sand grading. At low cement contents (below 200 kg/m^3), thickening agents can be used to prevent water separation which is possible under high pipeline pressures. Examples of these are polyethylene oxides, cellulose ethers and alginates. Polyethylene oxide is particularly effective because it does not entrain air and it bonds to water surfaces forming a skin which assists pumping and helps prevent water loss during curing. At high cement contents, the basic need is for increased fluidity in the mix and this is normally achieved by use of ordinary plasticisers, leading to increased slump at constant water content.

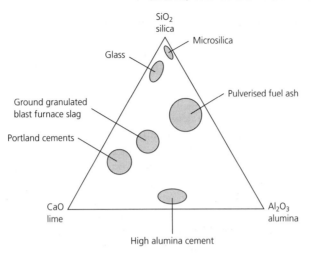

Figure 7.44 Compositional relationships between cements and cement replacement materials. There are more than 10 possible chemicals containing various combinations of C, S and A

Table 7.17 Typical properties of PFA, GGBFS and microsilica

Type	Silica content (per cent)	Lime content (per cent)	Relative density	Fineness	Particle shape
PFA	50	3	2.3	6% retained on 45 μm sieve	spherical
GGBFS	38	40	2.9	specific surface 400 m²/g	irregular
Microsilica	92	0.1	2.2	specific surface 15 000 m²/g	hollow or solid spheres

7.12 USE OF CEMENT REPLACEMENT MATERIALS

There are several siliceous materials, formerly regarded as waste products, which can have beneficial effects on concrete. The chief ones are pulverised fuel ash (PFA), ground granulated blast furnace slag (GGBS) and microsilica (silica fume). The compositional relationships between these are shown in Figure 7.44 and Table 7.17.

PULVERISED FUEL ASH (PFA)
(see also PFA cement)

This is an example of a pozzolanic material. It has, in itself, no inherent cementing properties but in the presence of lime released by hydrating cement, it displays hydraulic action, thereby contributing to strength. PFA obtained from power stations is processed by cyclones to remove unwanted coarse material until the residue on a 45 μm sieve is about 5 per cent (12.5 per cent maximum – BS 3892). The main constituents are silica and alumina. Particles are spherical so that they have a lubricating effect on concrete mixes and, since PFA has a low relative density (approximately 2.3), a given mass produces a larger volume of material than the same mass of cement (relative density 3.15). Hydration is slow and therefore requires increased curing times, though heat of hydration is reduced. If use is made of its plasticising action, PFA increases long term strength and reduces the shrinkage and permeability of concrete. The effects on strength and permeability are partly caused by a pore-blocking action of the resultant gel. A further reason for low permeability may be the relatively low lime content of the hydrated material so that there is less risk of lime being leached out. Resistance to dilute acids is increased for the same reason. PFA forms an economic method of controlling thermal cracking and enhancing durability and has been used on a number of large scale projects, particularly in civil engineering. Resistance to alkali–aggregate reaction and sulphate attack is also improved. PFA has been used as a pumping aid, especially in lightweight concrete.

GROUND GRANULATED BLAST FURNACE SLAG (GGBS)
(see also Portland blast furnace cement)

Slag from blast furnaces is tapped off whilst red hot and cooled rapidly by water jets to give a glassy solid. Close control over the raw materials results in consistency in both the iron and slag produced. The slag is then ground to a powder of similar fineness to ordinary Portland cement (minimum 275 m^2/kg – BS 6699). GGBS contains mainly lime, silica and alumina (see Figure 7.44) and is described as a *latent hydraulic binder*, the hydraulic action requiring activation by alkalis or sulphates, both of which are present in Portland cements. Particles are irregular with a relative density of 2.9 and the material produces a light creamy finish to concrete when incorporated.

GGBS has a slight plasticising action and when used as a cement replacement, will reduce permeability and enhance resistance to sulphates and alkali–silica reaction, provided it is well cured. In view of its inherent hydraulic properties, quite high replacement levels – up to 85 per cent – have been used. Applications are based on these properties, together with lower heat of hydration of GGBS cement blends.

MICROSILICA (SILICA FUME)

This is a by-product from electric arc furnaces used to produce silicon and ferrosilicon, most material being imported from Scandinavia. Microsilica contains at least 85 per cent silica in the form of microscopic spherical particles of mean size 0.1–0.2 μm, giving a specific surface of about 15 000 m^2/kg – many times that of cement. The relative density is about 2.2 – lower than that for silica normally because some spheres are hollow. It is available either in fine powder form of very low bulk density (200–300 kg/m^3) or it can be blended to form a slurry which makes handling easier, especially on site. The powder form can, however, be *densitised* or *pelletised* to assist site handling.

Typical use involves about 10 per cent replacement of cement which gives a cohesive, workable concrete with improved pumping and finishing properties.

Two actions occur in the hardened concrete: the very fine material effectively blocks pores and, in addition, there is pronounced pozzolanic action resulting in the formation of calcium silicate hydrates. Compressive strengths can be considerably increased while early heat of hydration is reduced in cement replacement applications. The low permeability and free lime contents in concrete lead to improved resistance to alkali–silica reaction. Sulphate resistance, frost resistance and protection to steel are also improved.

Although relatively new, these promising materials have become quite widely used. Some are available with Agrément certificates and future use looks likely to increase considerably.

7.13 LIGHTWEIGHT CONCRETES

These may have dry densities between 400 kg/m^3 (aerated) and 2000 kg/m^3 (structural lightweight concretes) compared with 2000–2600 kg/m^3 for normal concrete. They may have the following advantages:

- They produce lower foundation loads and are particularly useful in upper storeys of tall buildings.
- They may be placed in higher lifts than dense concrete on account of lower formwork pressures.
- They improve the thermal performance of buildings by reducing their thermal inertia.
- They have better fire resistance than dense concrete.
- Lightweight aggregates are often produced from waste products, hence they are reasonably cheap.
- Fixings may be made more easily than with dense concrete; for example, good fixings can be made into some types of lightweight concrete with cut nails.

As concrete density decreases, strength follows the pattern based on the law given earlier, relating voids to strength. However, the strength can vary considerably for a given density, according to both the structure and surface characteristics of the aggregate particles. The modulus of elasticity may be as low as 50 per cent of that of dense concrete having a similar strength, especially in the very low density concretes, reflecting the reduced stiffness of lightweight aggregates. Shrinkage and creep are, on this basis and on account of higher cement paste contents for a given strength, also increased. Lightweight concrete may be broadly classified into three types – no-fines, lightweight aggregate and aerated concretes.

NO-FINES CONCRETE

As the term implies, this type of concrete contains only coarse aggregate normally graded between 20 and 10 mm. The material produced has an open texture such that, when used in walling, a good key is provided for plastering internally. Externally, no-fines concrete is normally protected from rain by rendering, although the material has inherent low permeability, since pores are too large to permit capillarity. The upper strength limit is about 15 N/mm^2 but shrinkage and moisture movement are considerably less than those of normal concrete, due to the discontinuous nature of the cement paste fraction. No-fines concrete may be made using natural or lightweight aggregates, the former being stronger but denser. Correct batch quantities are best obtained by trial mixes, the water content being so chosen that each particle of aggregate is well coated with cement grout. Too little water reduces cohesion and too much causes the cement grout to segregate at the base. Wetting aggregates before use is the best way of obtaining a consistent water content. No-fines concrete was formerly widely used for *in situ* walls in low rise housing, though it is doubtful whether it would satisfy the now more stringent *U* value requirements. It is still, nevertheless, used as a lightweight roofing screed.

LIGHTWEIGHT AGGREGATE CONCRETE

Although naturally occurring lightweight aggregates have been used (e.g., volcanic cinders and sawdust), the majority of aggregates are manufactured from denser materials, such as clay or slate. They are covered by BS 3797. Some examples are as follows.

Foamed slag

This is blast furnace slag, cooled quickly by using water. On crushing, an angular, rough material is produced, giving concrete of strength up to 40 N/mm^2 at 28 days.

Expanded clay

This process is based on the fact, already mentioned under 'Clay bricks', that rapid heating of clay causes bloating, due to expansion of trapped gases. In one process, the clay is heated in the form of small rolled lumps, the basic shape being retained after firing. Alternatively, a mixture of clay and colliery shale is heated, ignition taking place and producing a fused clinker. Crushing the clinker produces an angular material. The aggregates may be used to give strength up to about 35 N/mm^2 at 28 days.

Expanded pulverised fuel ash (PFA)

The ash, which is obtained as a waste product from power stations, is mixed with water and powdered coal to form nodules. Sintering causes ignition and the nodules expand into hard, spherical particles. The material produces concretes of high strength/density ratio with strength up to 65 N/mm^2, together with fairly low shrinkage. Ready availability has led to widespread use, including in ready mixed concrete. It is also used in small precast members, such as prestressed lintels, to increase ease of handling. It is one of the most promising of the lightweight aggregates.

Design of lightweight aggregate concrete mixes

Mix design is complicated by the high absorption of aggregates, though for a given aggregate type, there is a fairly well defined relation between water/cement ratio and strength. Ratios are often quoted in volumes for lightweight materials, masses for weigh-batching being obtained from bulk densities. Water added at the mixer should be corrected according to:

- the free water in aggregates (deduct value)
- the absorption of the aggregates (add value)

There is more justification in lightweight than dense concrete for final correction of water at the mixer to produce correct workability, since the above corrections are substantial and may vary considerably from batch to batch. It is also found that additions to maintain workability do not appear to reduce strength as they are later absorbed by the aggregate, though there is advantage in delaying final compaction until the extra water has been absorbed. When lightweight concrete is pumped, there is a tendency for the hydrostatic pressure to drive water into the aggregate, blocking the line. This can be overcome by use of extra mixing water or by pumping aids. Natural sand used in place of lightweight sand increases strength and workability and reduces shrinkage but also increases the density of concrete.

Precautions

In view of the increased shrinkage of lightweight concrete, it is important to allow adequate contraction and movement joints in structural concrete. A further most important consideration from the point of view of durability is that carbonation in lightweight concrete often reaches depths of 15 mm or more, so that reinforcing steel must have substantially increased cover. It has been demonstrated, however, that by appropriate modifications to design and practice, both reinforced and prestressed lightweight concrete can be successfully employed.

Higher strength lightweight concrete mixes

If the water/cement ratio is reduced to 0.4 or below, quite high strengths can be obtained using stiffer aggregates such as expanded pulverised fuel ash. There is, however, a tendency for large aggregate particles to act as stress raisers in strong concrete since they are less stiff than the mortar and deform preferentially. Reduced aggregate sizes, for example, of about 15 mm, may help in such cases. Plasticisers or super-plasticisers may also help to produce acceptable workability at low water/cement ratios. Precast roof and bridge structures have been produced economically in this way, strengths of 80 N/mm^2 or more being obtainable.

A further possibility is to incorporate materials such as microsilica, which is particularly effective if the units are autoclaved. Strengths of over 100 N/mm^2 can then be obtained with a density less than 2000 kg/m^3. Such high strength mixes are suited to precast items where the benefits of intensive vibration and strict quality control can be obtained. Shrinkage is of little importance in such items and carbonation levels should be acceptably low.

AERATED CONCRETE

A newer range of extra high insulation blocks has been developed with k values in the region of 0.11 W/m^2 °C, so that U values below the current required value of 0.6 W/m^2 °C can easily be obtained by a single skin 225 mm solid wall. (Dampproofing would be obtained by external cladding or rendering.) Such construction eliminates the risk associated with poorly constructed cavities and can also reduce building costs in domestic housing. For use below the DPC, the more conventional aerated blocks should be used.

Aerated blocks have surprisingly good resistance to frost to which they sometimes may be exposed during construction in winter. Blocks should, nevertheless, be kept as dry as possible in order to prevent the large mass increase which occurs on wetting and to avoid shrinkage problems on drying out.

7.14 HIGH STRENGTH CONCRETE

Increasing use is now being made of high strength concretes – concretes of compressive 28 day strength over 60 N/mm^2. The following advantages may ensue:

- Rates of strength gain are increased, permitting faster construction.
- High strength concretes usually have a high strength/density ratio, hence where the self-weight of structures is important, foundation loads are decreased and section sizes can be reduced. When used as beams larger clear spans become possible since the self-weight of sections is reduced. (Best performance will be obtained by using prestressed concrete.)
- Other mechanical properties improve with strength, for example, stiffness, impact resistance and resistance to abrasion.
- Permeability and carbonation rates are reduced so durability should be enhanced.
- High strength concretes often contain some form of pozzolana and it is found, particularly if two or more types are blended, that chloride ion penetration is greatly reduced. This may be because the chlorides become chemically bound as they penetrate the concrete. These concretes are therefore more suited to hostile marine environments.

Figure 7.45   Production of high rise structures using high strength concrete as an *in situ* core

Table 7.18   Mix details for three grades of high strength concrete

Concrete grade N/mm²	70–100	100–150	Greater than 150
Water/cement ratio	0.4–0.3	0.3–0.25	Less than 0.25
Aggregate type	Gravel	Crushed granite	Synthetic
Max. aggregate size	Any	10 mm	5 mm
Additions	Plasticiser/ super-plasticiser	Super-plasticiser/ Microsilica	Microsilica/PFA blend/ Hyper-plasticiser

■ If strengths are achieved by use of very dry mixes (rather than high cement contents) there should be environmental advantages due to lower quantities of raw materials for a given load bearing capacity.

High strength concrete is suited to construction projects where good quality control is practised and is at present most likely to be found in high rise buildings with concrete frames (Figure 7.45) and in precast components. Compressive strengths of around 70–80 N/mm² can be obtained quite easily with conventional mix design methods and with modifications to materials and manufacturing methods, strengths of over 100 N/mm² are obtainable, often at quite early ages.

Table 7.18 summarises techniques by which three levels of high strength performance can be achieved.

Attention should be given to the following areas when contemplating use of high strength concretes.

Aggregates

As design strengths increase, the need for strong aggregates that bond well to the cement paste becomes more critical. Crushed granite aggregates are ideal in this respect, though their advantages must be weighed against the cost of transport, if not locally available. Gravel aggregates can be used to obtain strengths of about 100 N/mm^2, though relatively high cement contents would be necessary. It has already been explained that stresses tend to concentrate around normal dense aggregate particles since they are stiffer than the cement paste fraction. It is therefore beneficial to restrict maximum aggregate size, to say around 10 mm for strengths up to 150 N/mm^2 or 5 mm for even higher strengths in order to minimise such effects. For highest strengths (above 150 N/mm^2) synthetic aggregates such as calcined bauxite would be essential. Since failure in high strength concretes may be initiated at the aggregate/cement interface, a further method of producing very high strength is to use graded cement clinker as aggregate, since the whole unit then becomes bonded into an integral mass. High alumina cement clinker has been used in this way, though the high cost detracts from its use except in specialist applications.

Having regard to the fact that high strength concrete is most meaningful in relation to density, the possibility of use of lightweight aggregates in this context might also be considered for strengths up to 100 N/mm^2.

Water/cement ratio

This is the most important parameter in the production of high strength concrete. If cement contents are limited to a maximum of, perhaps, 500 kg/m^3 in order to avoid excessive heat of hydration and shrinkage, then water contents must be reduced to the minimum feasible to obtain very high strengths. Hence, very low workabilities are essential, entailing intensive compaction methods – the best means of compaction is a combination of pressure and vibration. Alternatively, in some applications, dewatering may be applied. Units such as pressed paving slabs have been made in this way for many years, excess water being pressed out after moulding.

The advent of super and hyper-plasticisers has greatly assisted high strength concrete production, highest strengths being obtainable when they are used in extremely low water content mixes with intensive compaction. Water/cement ratios of around 0.25 can be obtained in this way, giving compressive strengths of 150 N/mm^2 or higher. Alternatively, with low or medium workability mixes, super-plasticisers can be used for *in situ* concrete to give strengths of around 100 N/mm^2.

Additions/curing

The use of some form of addition for high strength concrete is very widespread. Many of these (especially microsilica) have very small particle sizes and it may be that they operate by reducing capillary porosity, which according to Griffith's principles are the cause of weakness. Best results appear to be obtained when microsilica is blended with another pozzolana such as PFA.

One of the problems associated with high strength concretes is that of self-desiccation – internal regions cannot fully hydrate because curing water does not have access through the impermeable material. For precast products in this situation, autoclaving can be beneficial since water is then introduced under pressure. Autoclaving of concretes with microsilica additions can produce very high strengths within 24 hours of casting.

PRECAUTIONS WITH HIGH STRENGTH CONCRETE

Although the stiffness (E value) of high strength concrete is higher than that of normal strength material, it does not increase proportionately with strength, hence working strains such as deflections in beams will be larger at the working stress. Crack widths, for instance, need careful monitoring in tension zones. Shrinkage of high strength concrete is also high, though not excessive. Attention may need to be given to strength development with age. The strength/time curve for high strength mixes is much steeper initially with less strength 'bonus' after the 28 day value which normally forms the basis of structural design.

7.15 VERY HIGH STRENGTH CEMENT PRODUCTS – MDF CEMENTS

Macrodefect-free (MDF) cements probably represent the ultimate in cement performance and are the result of studies into the origins of weakness in hydrated cements. These may be regarded as twofold:

- The strength of cement hydrates is limited by porosity, which is still considerable, even at very low water/cement ratios. The minimum porosity achievable by normal means is about 0.2, due to the presence of gel pores.
- Cement products contain *macrodefects*; these are flaws, such as pores, which affect strength according to Griffith's principles, the critical feature being their length (see page 133).

Tests have indicated that, in conventional concrete, they vary in size from a few microns only up to several mm, the larger ones having a critical effect on breaking strength. The origin of the larger pores or defects appears to be packing problems in cement particles caused by friction between them. It is now possible to eliminate these in certain special Portland and high alumina cements by addition of water soluble polymers, such as polyvinyl alcohol/acetate. These reduce particle friction,

Figure 7.46 Some simple artefacts made from macrodefect-free cement

allowing defects to be removed by rolling low water/cement ratio mixes into a dough-like material until macrodefects are no larger than 0.1 mm. Ordinary plasticisers and super-plasticisers cannot be used for this process because they do not appear to reduce particle friction in such mixes.

Residual porosity can be removed by heating the moulded material to 80 °C for a short time. This causes a shrinkage of almost 10 per cent in the material as it hydrates, corresponding to a large reduction in porosity. The final product has a flexural strength of around 150 N/mm^2 with a modulus of elasticity of about 50 kN/mm^2, indicating a strain capability of around 3000×10^{-6}, rather larger than the yield strain of mild steel.

The chief problems of the material are its cost (though it could be used in quite thin sections or sheets), brittleness and susceptibility to water – the polymeric component swells and loses strength on wetting. A suitably protected fibre reinforced form of MDF could, nevertheless, find application in future building, for example, lightweight claddings. Figure 7.46 shows some small artefacts produced from MDF.

7.16 LEAN CONCRETE

Sometimes known as dry lean concrete, this material may contain as little as 5 per cent cement by weight. It is normally used for road bases, being compacted by heavy or vibrating rollers, so that workability may be much lower than is normally possible for concrete. The correct water content of lean mix concretes is obtained by compaction tests at various values as for soils, the optimum moisture content (i.e., moisture content at which density is a maximum) being determined. For gravel aggregates the value is approximately 6 per cent total by weight of dry materials, corresponding typically to approximately 4.5 per cent free water, assuming an aggregate absorption of 1.5 per cent. Trial mixes may be designed using the strength:free water/cement ratio relationships given in the 'Mix design' section, using values at the top end of the range – for example 0.8–1.0. If, for example, the strength requirements results in a free water/cement ratio of 0.86, the cement content (by weight of dry materials) would be:

$$\frac{4.5}{0.86} = 5.2 \text{ per cent}$$

The aggregate content would, therefore, be $100 - 5.2 = 94.8$ per cent (oven-dry). The quantities are hence:

cement	aggregate	free water
5.2	94.8	4.5

The aggregate and water batch quantities will need to be adjusted according to the absorption and moisture content of the particular aggregates used. The sand content employed is normally between 35 and 40 per cent of the total aggregate. This is rather higher than in normal concrete, in order to avoid the possibility of under-sanded parts of the mix, with consequent segregation, and to obtain a sealed surface. The maximum aggregate size may be either 20 or 40 mm, depending upon local availability. Compressive strengths of lean concrete can be surprisingly high, typically 10–20 N/mm^2, depending on cement content. High strengths are not necessarily an advantage; they may lead to widely spaced cracks in the hardened material which tend to 'reflect' through the bituminous surfacing. Finer, more closely spaced cracks are preferable. To ensure full compaction, there is normally an additional requirement that the dry density of the *in situ* material, as measured by a sand replacement method, is at least 95 per cent of the theoretical dry density calculated from the relative densities of the constituent materials.

Considerable use is now made of rolled lean concrete in the construction of dams, on account of the low heat of hydration, low shrinkage and potentially rapid construction that is possible with such material. The lean concrete acts as a structural fill in conjunction with an impermeable upstream face and may offer significant overall cost advantages compared with conventional gravity structures.

7.17 CONCRETE ROAD PAVEMENTS

Concrete offers the following attraction when used as pavements for roads:

- High strength and abrasion resistance.
- Resistance to rutting which often occurs in flexible surfaces constructed from bituminous materials.
- Long life provided roads are well designed and constructed.

Possible problems experienced with concrete in this context include:

1 Need for joints which can be quite difficult to incorporate successfully and affect riding quality.
2 Difficulty of access to underlying services.
3 Need for good weather conditions during construction.
4 High noise levels in some forms of finish such as brushed surfacings.

Considerable research has taken place in developing materials and techniques which exploit the advantages given above, while overcoming some of the disadvantages. One example of a product which appears to offer good prospects for use is *Whispercrete*. The key features of this material are:

■ Continuously reinforced sections, eliminating the need for joints.
■ Provision of an exposed aggregate finish, achieved by application of a retarder to the fresh material and then later removal of the laitance layer by brushing.

Figure 7.47 shows the surface finish of a typical whisper concrete road. An aggregate maximum size of 10 mm may be preferred to give best overall results. It is clearly very important that weather conditions are suitable for such processes and the timing of the brushing to remove the laitance layer also calls for careful

Figure 7.47　Whisper concrete, obtained by use of a surface retarder. An idea of scale is given by the hard shoulder marking at the base

judgement. The results can be impressive, noise levels being reduced compared with hot rolled asphalts and when dark coloured aggregates are employed for the surface layer, road users may be unaware that they are driving on concrete.

Use of concrete roads is normally confined to areas in which services will not be incorporated beneath the surface, but they can provide good performance for 40 years or more, eliminating the distortion problems often associated with flexible roads subject to heavy traffic.

7.18 USE OF RECYCLED MATERIALS IN CONCRETE

Reference has already been made under 'Aggregates' to recycled materials which might be considered for incorporation into concrete. Use of such materials not only eliminates the cost of landfill disposal but also the levy on virgin aggregates which is likely to increase as environmental constraints are tightened. To be of concreting quality, recycled material, and especially demolition waste must be carefully sorted before use. Removal of each of the following undesirable components requires a separate stage:

large timber pieces
steel
non-ferrous metal
asphalt and tar
wood, plastic and paper

These processes are being increasingly mechanised to reduce costs.

Probably the most satisfactory approach to use of recycled materials is to use them to replace the coarse aggregate content (only) of concrete since continued crushing adds to costs and tends to produce sand of unpredictable and often excessive dust content.

To reduce variability, large stockpiles are necessary and plant should be capable of producing a uniform blend of material. The ideal is probably to obtain sufficient material for any one application at the same time.

It may be prudent to regard any recycled aggregates as of 'high reactivity' in relation to possible alkali–silica reaction (see page 299).

Three classes of recycled aggregate (RCA) are defined in BRE Digest 433 according to their brick content as follows:

Classification	% brick by weight
RCA (i)	0–100%
RCA (ii)	0–10%
RCA (iii)	0–50%

Essentially:

- type (i) is of low quality
- type (ii) is of high quality, and
- type (iii) is a mixture of type (i) and natural aggregates

This is of course on the assumption that the brick component forms a 'weak link' in the material and this depends partly upon the type of brick – the strength of clay bricks can vary between 5 and 80 N/mm^2. Invariably the stronger bricks such as engineering bricks will produce better results. The Digest also recommends maximum levels of the impurities identified above, allowing, for example up to 0.5 per cent of wood and up to 1 per cent of all other foreign materials in a type (ii) aggregate. Permitted values increase for types (ii) and (iii).

Since many aggregates can be quite absorbent, it is recommended to soak them before use and then to allow draining so that they are batched in the SSD form. This reduces uncertainties in allowing for absorption of aggregates.

Tests should be carried out to ascertain strength levels obtainable. These will in general be lower than the material from which the aggregates originate since the bond between the new and old cement mortar components is not as effective as that between cement mortar and natural aggregate. The water/cement ratio rule applies as would be expected but strengths achievable depend upon the average crushing strength of the recycled aggregate. A good indication of performance of the concrete should be obtainable from the water/cement ratio and its fresh density (Figure 7.48). There is little point in attempting to achieve high strength levels from material which includes significant proportions of softer components such as crushed brick. Such materials have low density and act as weak points in the concrete, initiating failure. For this reason use of 150 mm cubes for testing purposes is preferred, especially where softer materials are included in the aggregate. This should give greater consistency in cube results.

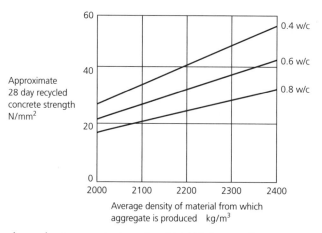

Figure 7.48 Approximate concrete strengths obtainable by recycling aggregates

With care it should nevertheless be possible to obtain concrete of consistent strength in the range 20–40 N/mm^2 suitable for a variety of applications. Where exposure to weather is likely an air entraining agent could be used to produce frost resistance. Recycled aggregate concretes may have inferior abrasion resistance to normal concretes and should not in general be used for flooring unless a covering is added. However one application which would certainly form a valuable outlet for recycled aggregates is for blockwork in (prestressed) beam and block floors since the blocks in these situations have to withstand neither weathering nor abrasion.

7.19 HIGH QUALITY SURFACE FINISHES TO CONCRETE

Inspection of existing concrete structures suggests that the achievement of a lasting high quality surface finish to concrete is not an easy task, either in design or constructional terms. In many cases, initially attractive finishes are spoiled by the effects of dirt and uneven weathering, many buildings having a grim appearance by the age of 20 years or so – well within the design life of most structures. A number of points are given here, which have some bearing on this difficult and complex subject.

Structural form

Experience shows that relatively simple structures tend to resist the effects of weather better because they contain fewer crevices where dirt can lodge (Figure 7.49). Areas sheltered from rain invariably become darker with time. Window openings and sills very commonly become unevenly weathered.

Materials/quality control

The consistent achievement of high quality finishes requires careful attention to materials. Aggregates must be of consistently reliable quality and grading, especially where an exposed finish is to be used. Some siliceous aggregates contain small amounts of pyrites, based upon iron oxide, which even in small quantities can cause unsightly rust stains on the surface of otherwise satisfactory concrete. Sources should be carefully checked for presence of such impurities. Cements vary in colour, especially if the source changes. This should be checked. Very careful batching is necessary so as to achieve the same mix every time – samples of concrete should be taken from successive batches to check for colour variation. Air entraining agents tend to improve surface finish by reducing bleeding. Weaker mixes are more permeable and therefore much more likely to attract dirt. Use of low water/cement ratio concretes will minimise dirt absorption. Precasting offers major attractions as regards quality because site problems such as synchronisation of diverse activities and problems relating to weather are avoided.

Figure 7.49   The simple structural form of this building helps to produce uniform weathering

Figure 7.50 shows results that can be achieved by precasting. The surface profile in this example was achieved by casting against GRP mouldings.

Formwork

The concrete finish can be no better than the surface finish of the forms used. Great skill is necessary to obtain the best finishes from timber forms, where joints tend to leak as a result of poor tolerances or moisture movement. It is not possible to hide joints completely, so it is best to make these features of the construction. Release agents are essential to avoid concrete bonding to the forms but some types lead to blowholes, especially on smooth form face materials, such as steel or hardboard. Release agents containing surface active agents can reduce this problem but trial panels should be constructed to check acceptibility.

Surface texture

The surface texture may vary greatly according to the materials/technique used. Very smooth textures, such as obtained from steel, plastic or hardboard form linings, are not generally attractive because blowholes and other small defects are often apparent and the concentration of cement mortar at the smooth surface encourages surface crazing. Plywood forms – one of the most common types –

Figure 7.50 High quality finishes achievable by precasting (courtesy of Tarmac Precast Concrete, Ltd)

produce a slight texture which generally has a 'utility' appearance. All smooth surfaces can be greatly improved by bush hammering in which surface laitence is removed with a hand-held percussion tool. Although expensive in time terms, this treatment, if skilfully carried out, will remove surface blemishes and produce a finish which weathers well. Quite common in existing buildings is the sawn board finish which was widely employed in the 1970s. Solid timber boards with a coarse surface texture, usually of width 150–200 mm, form the casting surface. Great attention to detail is necessary and the process considerably increases the cost. Textures as coarse as this also tend to attract dirt and so weathering can produce unsightly areas, especially where the initial colour is very light.

A further possibility is to use exposed aggregate finishes (Figure 7.51), a pleasing effect being obtained if an attractive aggregate, such as red granite, is used. This

Figure 7.51   Use of exposed aggregate panels. These panels at the Roman Catholic Cathedral, Bristol, remain attractive after 20 years

finish can be obtained in *in situ* concrete by coating the formwork with a thick, uniform layer of a concrete retarder which prevents hydration of the concrete surface. On form removal, the laitence is then removed with a water jet.

Alternative techniques include:

■ For precast panels: aggregate transfer in which a thin layer of selected aggregate is arranged on the base of the panel mould and ordinary concrete placed over it.
■ For precast panels: cover the *in situ* concrete with a workable but cohesive surface layer of mortar and produce, while still plastic, deep textured designs in this layer. These can be very effective, though this is a skilled craft trade.
■ Use moulded formwork linings (Figure 7.52).

With all the above techniques, the success depends largely on the skill and experience of the concrete manufacturer/producer.

Figure 7.52   Decorative facing achieved by moulded formwork lining, Cwmbran

7.20   REINFORCED AND PRESTRESSED CONCRETE

While a detailed treatment of these topics is beyond the scope of this book, a brief comparative study will be made, since the precise modes of operation of these materials are often not clearly understood.

Since the tensile strength of concrete is normally less than 5 N/mm², elements under tensile or flexural load will normally require some form of reinforcement, although in some situations of limited stress, for example, ground floor slabs, concrete can perform satisfactorily without any reinforcement.

Both reinforced and prestressed concrete might be regarded as environmentally attractive materials compared to, say, structural steel frames because the use of the high energy material, steel, is confined to the regions in which tensile stresses are to be withstood. In each case a volume of steel of between 1 and 2 per cent of total volume is normally sufficient to provide adequate performance. Perhaps slightly countering this argument is the fact that separation of steel and concrete for re-use at the time of demolition can be difficult.

REINFORCED CONCRETE

To produce reinforced concrete, steel is first positioned in the mould and then fixed
to prevent movement under forces arising from placing and compaction. The steel
is passive at this point – it carries no load. Hence, when, after hardening of the
concrete, the element is stressed, load must be communicated to the steel through
the concrete. This requires the concrete to undergo movement or strain. Operating
tensile strains in steel, if used efficiently, may be up to 1500×10^{-6} (depending on
type) and so strains in the concrete must also be of this magnitude if the steel is
to carry a full load. Unfortunately, concrete fails in tension at a strain of about
100×10^{-6}, so clearly must be in a cracked state in tension zones before the steel is
fully utilised. The nature of the cracking depends on the type and arrangement of
the steel used. Hooked ends are common in simple beams to prevent steel 'pulling
out' under stress. However, if there were no local bond, there would be a strong
tendency for a single large crack to form at the position of maximum bending
movement (Figure 7.53a). To overcome this, there should be bonding between

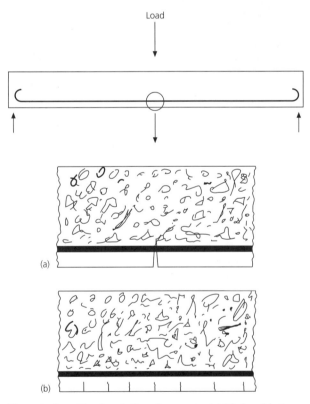

Figure 7.53   Control of cracking in reinforced concrete. (a) No bond between steel and
concrete – a single large crack tends to form. (b) Effective bond between steel and concrete
– a distribution of fine cracks is obtained

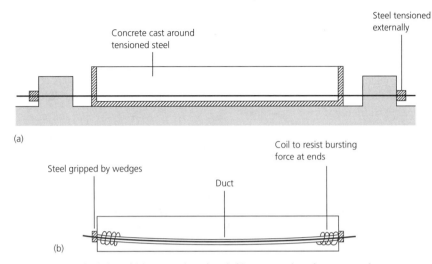

Figure 7.54   Principles of (a) pretensioned and (b) post-tensioned prestressed concrete

concrete and steel all the way along the bar which would then produce a fine distribution of cracks (Figure 7.53b). As the operating stress of the steel increases, local bonding requirements also increase and high performance steel must have bonding characteristics to match their strength. Deformed bars fulfil this need. It will be recognised that some cracking is inevitable in tensile zones of reinforced concrete, though, by controlling the size of these cracks at the surface, concrete of good durability can be achieved. Opinions vary as to what crack widths are acceptable, though in BS 8110, surface of cracks of up to 0.3 mm width are considered acceptable.

PRESTRESSED CONCRETE

In prestressed concrete, the steel is tensioned by jacks and then released so that it imparts a compressive stress to those areas of concrete which, in service, would be in tension. There are two basic processes as illustrated in Figure 7.54. In pretensioning, the steel is tensioned before the concrete is placed and then, when the concrete is mature, the ends are released and the concrete prestressed by local bond. The initial stressing operation requires anchor plates fixed to the floor and is therefore a factory process. In post-tensioning, ducts are cast into the concrete, usually in a curved profile to give increased bending at the centre and to help overcome shear stresses at the ends of the beams. When the concrete is mature, tendons are inserted and stressed against the specially strengthened ends of the beam. They are locked in place and the ends released. Grouting of the duct prevents corrosion. The essential feature of prestressed concrete is that tensile cracking can be completely eliminated (indeed, if a unit is over-tensioned, it can fail upwards due to excessive tensile stress in the top of the beam). Even if a prestressed unit is overloaded, resulting in cracks at the soffit, these cracks should close completely once the overload is removed. It will be evident that risk of corrosion of reinforcement due to tensile cracking in prestressed concrete will normally be much less than that in reinforced concrete.

One problem of prestressed concrete is that demolition can be hazardous on account of the elastic energy stored in the tendons even under zero load, which can be released suddenly if care is not taken.

7.21   CORROSION OF STEEL IN CONCRETE

There is little doubt that corrosion of reinforcement in concrete is the largest single cause of distress or failure in reinforced concrete structures and has contributed greatly to the less favourable perception of the material which emerged in the 1980s. There is certainly considerable risk when employing steel in concrete on the basis that:

- Unprotected steel corrodes in the presence of quite small amounts of water unless passivated by an alkaline environment.
- Carbonation of concrete results in progressive loss of alkalinity in surface layers – the regions where steel acts most efficiently in a structural sense.
- Tolerances in positioning of steel in reinforced concrete elements are sometimes poor so that cover to steel may vary.
- When corrosion to steel occurs, the damage resulting from consequent expansion (due to the corrosion product) will, if not treated in time, lead to eventual structural failure.
- Remedial treatment of affected structures is often difficult and expensive.

A graphic illustration of the problem is shown in Figure 7.55. The cladding panels were precast, the exposed aggregate finish being applied to a preformed concrete backing using a cement mortar, with reinforcement in the mortar. The mortar soon carbonated and the steel in over 100 panels was soon severely corroded.

The perception of the problem is, perhaps, reflected in current attitudes to housing with prefabricated reinforced concrete (PRC) frames (Figure 7.29). There are about 300 000 such buildings in the UK built to meet the post-war housing needs. Many are affected by corrosion of reinforcement, although there are few cases of advanced deterioration. The Housing Defects Act (1985) provided for 90 per cent grants towards repair of these properties when bought from local authorities by tenants. However, a 30 year minimum life was required by the Act which, together with warranty requests from building societies, led to the formation of PRC Homes Ltd., a subsidiary of the National House Building Council, in order to set repair standards. PRC Homes gave licences for such repairs and stipulated that all load bearing elements of the PRC dwellings must be made redundant by construction of a separate structural frame – usually in brickwork.

Perhaps a double irony is the fact that reinforcement corrosion problems in many types of building, including PRC, appeared to increase in the late 1950s, as a result

Figure 7.55 Serious damage to precast cladding panels on a 12 storey block of flats

of experience of mix design using the then current Road Note 4. Test results showed that modern cements performed much better in strength terms than those upon which the design method was based. Hence, in some cases, cement reductions were made, with durability implications. At the same time, there are examples of buildings over 100 years old in which carbonation depths are very small and steel reinforcement is in good condition.

The technology of steel protection is now well established: Table 7.6 gives cover requirements of BS 8110 for various concrete mixes and environments. It should be appreciated that these are minimum cover requirements so that, in practice, cover should be such that the minimum figure is obtained even in the worst cases.

An important development in terms of steel cover is that actual cover can now be checked with much greater accuracy than previously using the *covermeter*. Modern versions of this instrument give bar size and depth up to depths as high as 150 mm. Covermeters should comply with BS 1881 Part 204.

Table 7.19   Corrosion resistant steels and their relative costs

Type	Approx. cost ratio
[High-yield steel	1]
Galvanised steel	1.6
Epoxy coated steel	2.2
Grade 316 stainless steel	8.7

Difficulties of obtaining required cover in practice have led to investigations into other ways of preventing steel corrosion. High risk situations may include:

- construction in difficult climates, for example, hot climates
- inadequately skilled labour force
- chloride contamination

There are two main methods of reducing the risk of reinforcement corrosion.

Corrosion resistant or corrosion protected steel

Examples are given in Table 7.19, together with approximate relative costs.

Of those indicated galvanised steel is one of the best established. Its main limitation is probably in aggressive environments containing, for example, localised concentrations of chloride ions which might cause rapid loss of zinc. Galvanising is, however, suitable as an added protection to steel in normal environments and may permit a reduction in size of precast goods by reducing cover required. Note that:

- Bending is best carried out before galvanising, since cracking or flaking may result in regions of tight bends.
- Defects due to cutting or welding should be made good with a zinc/resin coating.
- Chromating of the finished surface may be advised if cements to be used have very low chromate composition. Non-chromated bars may react chemically with the cement, producing hydrogen which reduces the surface bond.

Epoxy coated steels are relatively new – they are produced by electrostatically coating cleaned, heated bars with epoxy powder particles which are then gelled and cured by heat. A coating of 0.17–0.30 mm is required (BS ISO 14654) and it is very important that an even thickness in this range is achieved. The standard refers to *holidays* which are non-visible defects in the coating. These may be detected electrically and are limited to a maximum of 4 per linear metre. Bars can be bent without damage, using special plastic faced mandrels provided a flexible type of coating (type A) is used. It is vitally important that any imperfections in the coating, such as at cut ends, must then be made good to avoid corrosion spreading under the film, especially if chlorides are present. Although information is limited at present it appears that bond strengths to concrete of epoxy coated bars may be slightly reduced. Most applications have involved marine or water retaining structures but other applications might include situations in the vicinity of cathodically protected underground pipes – the electrically insulating coating would prevent corrosion due to stray currents in such instances.

Stainless steels are likely to give the greatest durability, especially in more aggressive environments, though their cost is such that applications are likely to be limited to prestige types of work, for example, restoration of important buildings.

Austenitic types (for example, type 316) are preferred, especially where chlorides are present. Note:

- If used in conjunction with ordinary steel, great care must be taken to ensure that the latter is out of reach of carbonation, otherwise, rapid electrolytic corrosion of the ordinary steel could occur.
- Deformed bar bond strengths may be slightly lower than those of ordinary steel.

Concrete surface treatment

These can be used to reduce carbonation rates by reduction of the permeability of the concrete to carbon dioxide. At the same time, there should be some degree of microporosity to the smaller water vapour molecules to prevent build-up of moisture underneath the coating (see page 33). They also offer protection to carbonated concrete since they reduce rates of water admission. The following points are relevant when considering the type of material to use:

- Several coats are preferable – successive coats cover imperfections in underlying coats.
- Coatings must be sufficiently flexible to accommodate substrate movement.
- Thicker films are better – they have lower permeability to carbon dioxide.
- Use of pigmented coatings improves performance since larger thicknesses can be obtained and the durability of the coating itself is also improved.

Surface coatings may be used in new or remedial work and there are many products available, most products being based on synthetic resins. Treatment is probably best left to the specialist who may be prepared to guarantee performance.

Monitoring of corrosion rate in reinforced concrete

It will be explained (see Chapter 8) that the susceptibility of a metal to corrosion depends on its effective electrode potential in a given environment and, therefore, will vary from place to place according to the chemical make-up of the surrounding medium. In buildings where some corrosion has been detected (normally by localised spalling), some means of assessing the risk of further problems is a distinct advantage. For example where tests suggest that the entire structure is at serious risk, the possibility of demolition might have to be considered, whereas if the problem is localised, surface treatment of the remaining parts may suffice.

There are several techniques by which the extent of potential corrosion problems can be estimated. These hinge upon measurement of electrical activity within the concrete and especially at the concrete/steel interface.

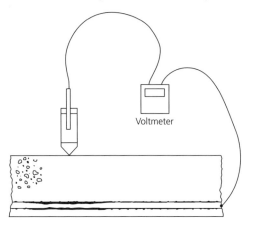

Figure 7.56   Measurement of half-cell potentials to determine corrosion susceptibility of steel

Half-cell potential

The electrode potential of steel relative to concrete cannot be measured directly since another electrode must be introduced to measure voltage. However, by use of a second 'half cell', the electrode potential of metal at any place can be inferred. Common arrangements are copper in a copper sulphate solution or silver in a silver chloride solution. Such half cells are described as *reversible* – the current can flow either way in them. The reinforcement must be exposed in one position and the circuit is completed by means of a damp sponge (Figure 7.56). When the steel has a high corrosion risk (as in carbonated concrete or when salts are present) it will give rise to a higher negative potential, as for the reactive metals in Table 8.3 in chapter 8. A typical risk assessment would be:

Voltage	Risk
less than −200 mV	low
−200 to −300 m V	medium
more negative than −350 mV	high

If readings at any position vary, then the concrete surface should be wetted by a 0.5 per cent detergent solution (as should the sponge) to give readings steady to within ± 20 mV over a 5 minute period. If this state cannot be achieved, the half-cell method is probably inappropriate.

Resistivity measurements

These can give additional information to the half-cell technique because they measure the ability of the concrete to conduct ions between the anodes and

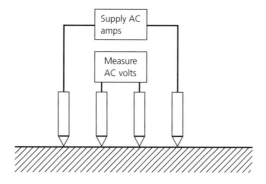

Figure 7.57 Schematic representation of resistivity measurement in reinforced concrete

cathodes of corrosion sites. The term 'resistivity' applies to a *material* and is defined such that:

electrical resistance of a given current path

$$= \text{resistivity} \times \frac{\text{path length}}{\text{area of conducting material}}$$

(Longer path lengths *increase* resistance while larger conducting areas *decrease* resistance.)

It will be noted from the above that the units of resistivity are resistance times length (usually measured in kΩ cm). The method of measuring resistivity involves applying an alternating current between two points on the concrete surface, the voltage between two intermediate points being measured (Figure 7.57). Alternating current must be used to avoid effects of polarisation that would occur with a direct voltage. Resistivity of the concrete will depend upon its moisture content, the presence of chlorides, and oxygen levels, oxygen being a key part of the corrosion process. The corrosion rate corresponding to a given resistivity will depend upon the half-cell potential as obtained above though an approximate guide is that resistivities less than 50 kΩ cm signify a risk of problems.

Recording and interpretation of data

It will be appreciated that there may be a lot of work in assessing the corrosion risk in a whole structure. Ideally the above techniques should be used in combination with each other, the half-cell technique providing information on the potential problems caused by, say, carbonation, while the resistivity technique measures the ability of the concrete to transmit ions. Measurements may take the form of spot checks, though where a significant risk is detected it may be worthwhile to map the whole structure. The process can be expedited by use of a *potential wheel* in which the half-cell electrolyte is connected to a saturated tyre which is run along the

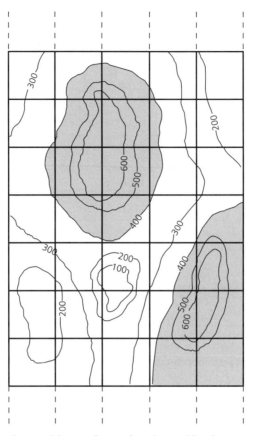

Figure 7.58   Half-cell potential map of part of a column. All voltages are negative (mV). High risk areas are shaded

concrete surface. Figure 7.58 shows the result of a survey of a column involving the half-cell technique. The chief risk in the section shown is in the shaded areas. Computerisation of results can lead to rapid production of corrosion risk contours.

There are also more recently introduced techniques in which both current and voltage are monitored by the same process, giving a more accurate guide in one survey. It must be appreciated that in all cases the readings obtained depend on conditions, for example weather and humidity conditions, and it may be necessary to monitor structures over a period of time.

REPAIRS TO REINFORCED CONCRETE

The effective repair of concrete structures affected by corrosion of reinforcement is a complex subject and best undertaken by specialists, though some principles of treatment are briefly covered here.

Diagnosis

It is important to establish clearly both the nature and extent of the problem. This would normally involve cleaning of the structure to facilitate inspection followed by tests to indicate areas affected. Very often, initial indications are but the tip of the iceberg, and the full extent of the problem may only be ascertained during remedial work, hence accurate Bills of Quantities may be very difficult to produce. While reinforcement corrosion is a very common cause of distress in reinforced concrete structures, it is by no means the only one and the possibility of other causes such as shrinkage cracking, alkali–silica reaction, frost attack or sulphate attack should also be considered – in some cases several agencies may be involved.

Repair specification

This must address the functions of the repair which may be to:

1 Replace damaged material with a durable substitute.
2 Restore structural performance, if affected.
3 Prevent deterioration in the remaining structure.
4 Provide acceptable finished appearance.

   To achieve these functions, attention must be given to procedure and workmanship. Considering the above headings in turn:

Functions (1) and (2)

The damage to steel must be assessed. If there is significant loss of section, or large areas of structural steel in situations of high stress have been exposed by concrete spalling, it may be simpler to replace the whole section, though replacement or additional steel can be used in some situations. It should be remembered that to simply replace damaged concrete will not in general assist the structural performance of the affected member if loss of concrete has led to increased deflections. Replacement concrete will help protect against further deterioration but will not normally become involved in a structural sense.

   Any damaged concrete must be replaced, ideally with a material having similar elastic and movement properties to the material it is replacing, though clearly it must provide adequate protection to the reinforcement. A number of materials might be considered.

Conventional concrete

In terms of elastic and movement properties, this is ideal, though its use is mainly confined to larger areas, since it is not a good patching material. Careful attention must be given to bonding between new and old concrete, and to the effects of shrinkage of the replacement material which may lead to interface cracking. The rules relating to shrinkage of concrete apply, that is, shrinkage increases with richer, wetter mixes. Hence, use of plain cement mortars, which fall into this category, to patch large, especially deep, areas of damage is certain to lead to severe shrinkage cracking.

Polymer modified concrete/mortars

These are cement based, containing resins such as styrene butadiene rubber (SBR) or styrene acrylate in latex form. They improve water resistance, reduce carbonation and help to control, if not reduce, shrinkage. Hence they can be used to patch concrete which is damaged due to inadequate cover. They also bond better to existing concrete than unmodified mixes, a small amount of dampness being acceptable on account of their hydraulic nature. They protect steel by passivation due to alkalinity provided by the cement binder. With suitable aggregate grading they can be used for up to about 50 mm thickness, though water content must be carefully controlled to avoid shrinkage. Pre-packed versions of composition suited to specific repair depths are frequently used for convenience.

Resin concrete/mortar

In these materials, the cement binder is replaced by resin, hence there will be no passivation of steel; protection relies upon impermeability. The advantage of resins, especially epoxy, is that they have excellent adhesion to steel and concrete combined with rapid setting and low shrinkage. However, since their modulus of elasticity is low and thermal movement high compared with conventional concrete, they must be carefully blended with filler/aggregate to reduce these effects while still providing satisfactory working properties. They are especially suited to smaller patch repairs; indeed, their high cost may limit use to such applications. See page 344 for further details.

Functions (3) and (4) – protection and appearance

Nearly all repairs to concrete are localised, reflecting variations from place to place in concrete quality, cover to steel and exposure conditions. Inevitably, therefore, many repairs take the form of patching which is difficult to match to existing material, even when the patching material is similar to the original. In most cases, cosmetic treatments are applied to disguise the repair. These may take the form of conventional renderings or masonry paints, though there is an increasing trend towards using coatings which are both decorative and which reduce carbonation rates, thereby decreasing the risk of deterioration of the structure as a whole. These coatings are described on page 335.

Procedure and workmanship

The need for the highest quality of workmanship cannot be overemphasised if successful repairs are to be achieved. Matters for attention include:

- Carrying out work in suitable weather conditions, or provision of protection.
- Removing all carbonated concrete areas around corroded steel. These often extend beyond visibly cracked areas.
- Cleaning steel by grit blasting until clean, bright surfaces are obtained. This also increases the steel surface area available for bonding by creating a textured surface.
- Application of bonding agents to steel surface and concrete.
- Thorough mixing of components. Repair materials containing polymers and resins are best mixed from accurately pre-batched containers.
- Timing of successive operations, for example, some resin systems should be applied while bonding coats are tacky.
- Production of a fully compacted repair material.

ALTERNATIVE REMEDIAL TREATMENTS

There may be situations in which treatments other than conventional repairs might be considered. For example localised problems may have resulted in surveys which reveal relatively large areas of potential problems, but without significant visible damage at the time of inspection. It might be possible to arrest corrosion processes, or at least retard them, without the inconvenience of large scale repairs. Such treatments might well be carried out where the structure concerned could not easily be replaced or cease operation during repair. Road bridges form a typical example. Buildings which are important for architectural or other reasons might be candidates for such treatments. The following processes might be considered.

Cathodic protection

This method appears to be successful in situations in which patch repairs are ineffective, for example, when substantial amounts of chlorides are present in the concrete. Localised repairs are carried out as necessary, then, typically, electrically conductive paint is applied to the whole structure followed by a decorative finish. Reinforcing steel to be protected must be connected to the DC power supply with the reinforcement negative. Monitoring probes may be included if required. Current densities of between 10 and 100 mA per square metre may be necessary to generate the small voltage necessary to stop corrosion. Installation of the system is expensive and additional charges will be incurred for the continuous supply of electricity to the system. If correctly installed, however, it should provide a permanent solution.

Chloride removal

The structure in question is surrounded by a water jacket and a direct current supplied between a steel grid in the jacket and the reinforcement (Figure 7.59).

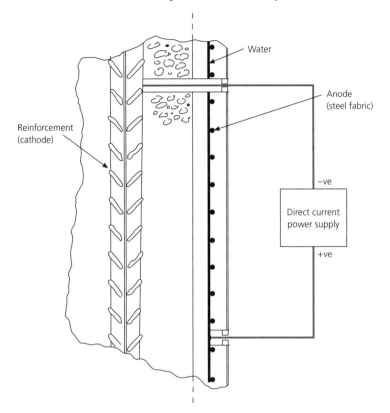

Figure 7.59   Chloride removal by application of a DC voltage between the reinforcement and the concrete surface

The negative chloride ions are drawn towards the jacket, where they can be removed from the water. Quite high current densities (1 A per square metre) are used and since the reinforcement is electrically active in the process, most of the chlorides removed tend to be between the bar and the surface, although up to 40 mm of concrete behind the bars can be improved also. Not all chlorides can be removed but the treatment is effective in reducing the corrosion risk. The process may take some weeks to complete.

Realkalisation

This is equivalent to a reversal of the carbonation process. The arrangement is similar to that of Figure 7.59 for chloride removal except that a solution of sodium carbonate surrounds the concrete. Sodium ions diffuse inwards towards the reinforcement, which together with hydroxyl ions generated due to cathodic behaviour of the steel, raise the pH value of the concrete. (See 'Electrolytic corrosion'.)

7.22 CONCRETES CONTAINING POLYMERS

There is now a wide range of concretes or mortars which contain a proportion of polymers. When the polymer is present in small quantities, the properties approximate to those of normal concrete, though with important improvements. When larger proportions of polymers are included, the properties approach those of the polymer (and, of course, the price rises so that the material becomes of a more specialist nature). In some cases, the binder may include both polymeric and traditional cement hydrate components.

In selecting a material appropriate to purpose from the wide selection available, the performance criteria of the composite must be identified, preferably in order of importance. Performance parameters may include (no specific order):

impermeability to specific liquids or gases
protection to steel
stiffness
tensile/flexural strength
toughness
abrasion resistance
temperature/fire resistance
creep resistance
low shrinkage

Note that compressive strength is not normally an important requirement, since conventional concretes are satisfactory in this respect. A number of fairly common polymeric materials are now considered.

POLYMER MODIFIED CONCRETE/MORTARS

These are based on concrete/mortar mixes and contain an emulsified thermoplastic polymer, such as styrene-butadiene rubber (SBR). Styrene is a hard monomer which, when combined with butadiene, a soft monomer, gives a copolymer with the good film forming characteristics required by emulsions (page 490). Since both are hydrophobic, water resistance is also good. Other polymers, such as styrene acrylate and PVA, may also be used, though the latter are not so good in wet conditions. Proportions of polymer in the region of 15 per cent by weight of cement are typical, though since emulsions are about 50 per cent water, the proportion of emulsion would be about twice this figure. A major advantage of polymer modified concretes is that curing times are greatly reduced, 24 hours usually being sufficient – by this time the material is completely waterproof.

The emulsions break to form thin films or strands in microcracks or pores with effects similar to those obtained by incorporation of soft fibres such as polypropylene, though some are claimed to interact chemically with the cement. Permeability is reduced and flexural and impact strength are improved, partly on account of plasticising action, allowing reduced water content. A further advantage of these composites is that of adhesion to other materials, especially smooth surfaces; the bond is better than either the concrete or polymer individually. Many applications are based on this property combined with impermeability. They may be used in:

- Mortars for renderings or screed – thicknesses as low as 6 mm are possible in the latter.
- Mortars for use with engineering bricks, brick slips, glass blocks and ceramic tiles.
- Concrete repair, thin concrete toppings on roads and pavements.

Most varieties are marketed as liquids but emulsions can be converted into powders by *spray drying* which may have advantages in certain applications.

POLYMER CEMENT CONCRETE

This description is given to concretes in which both polymer and cement contribute to binding action. Thermosets are used for this purpose and should be present in sufficient quantity to form a continuous matrix in the hardened concrete. The material hardens rapidly to give high compressive strength and high tensile strength, characteristic of the resin binder, though creep tends to be a problem. A variety of resins can be used and a polyester resin/cement mixture is available as a powder which is activated simply by adding water. Polymer cement concretes/ mortars are expensive on account of their high resin content but are used where their rapid hardening, together with improved strength, toughness and durability justify this extra cost. Applications include thin floor toppings, concreting onto metal substrates, high speed repairs and bedding mortars.

POLYMER OR RESIN CONCRETE

In this material, Portland cement is completely replaced by resin which, therefore, acts as the lubricant in the fresh material and the binder in the hardened material. The chief problem of such concretes is cost, since the resins, polyester, acrylic or epoxy, may cost £2000 per tonne or more. The relative merits of these three resin systems are shown in Table 7.20. The fire hazard associated with acrylic resins should be emphasised; monomers may have flash point temperatures as low as 10 °C. The effect of resin content on compressive strength of resin concrete with 20 mm graded aggregate is shown in Figure 7.60, the modulus of elasticity showing a similar relationship. At low resin contents, both properties fall off sharply due to incomplete compaction, while at high resin contents performance is reduced as the low stiffness resin has to act increasingly as a filler rather than an adhesive. The optimum resin content depends on aggregate characteristics. With a 20 mm well graded aggregate and intensive compaction, a resin content of about 6 per cent gives highest strength and this would result in an epoxy concrete with a cost of about £600 per tonne (compared with approximately £50 per tonne for normal concrete). If a more workable concrete or smaller aggregate were employed, the resin content

Table 7.20   Comparative properties of epoxy, polyester and acrylic resins for resin concrete

Resin type	Viscosity	Shrinkage	Relative rate of strength development	Safety	Cost
Epoxy	Medium	Low	Low/Med		High
Polyester	Medium	Med/High	High		Medium
Acrylic	Low/Med	Medium	Medium	Fire hazard	Med/High

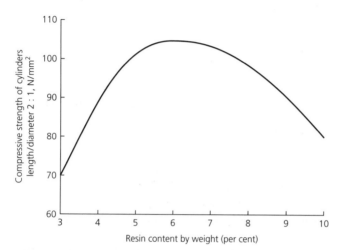

Figure 7.60   Compressive strength of epoxy resin concrete as a function of resin content. The exact relationship depends on compaction method

would increase, being typically 15 per cent by weight for a patching mortar. The cost would, therefore, increase proportionately, although polymer concretes/mortars can achieve performance quite unobtainable using any other binder and are only normally required in small quantities.

Particular advantages include high tensile strength, abrasion resistance, impermeability and durability. Use as a repair material is widespread and resin concretes/mortars are being increasingly used for production of small, high performance precast products, where the cost of resin is not of such great importance and can be offset by use of smaller sizes obtainable. A typical application is in simulated roofing slates; the high cost of splitting traditional slates is avoided by crushing the rock and bonding with polyester resin, to produce a 'slate' which is very similar in appearance to natural slate. There are many uses for resin mortars in new as well as repair work. They can be combined with fibres to produce a buttery consistency for use as bedding mortars in a wide range of applications, such as bonding of steel plates for strengthening structures (in conjunction with mechanical fixings).

POLYMER IMPREGNATION

Polymers can be used to impregnate conventional concrete after hardening, filling fine cracks, pores and voids. Very low viscosity resins, such as acrylics, can be applied, allowing them to soak in, or full impregnation can be achieved by pressure/vacuum. The method appears to work well for reducing surface permeability but where significant cracking is present there is the risk that if cracks are not sealed they can absorb water which subsequently becomes trapped. Pressure methods are used mainly for precast products when large increases in strength can be achieved. There is a high risk associated with use of acrylic resins, especially on site, due to their flammability.

7.23 RENDERINGS

These may be defined as cement based mortar coatings for external or other surfaces which must be water resistant. They have the following chief functions:

- to provide an aesthetically pleasing easily maintained surface finish
- to resist rain or damp penetration

To achieve these functions, renderings must adhere well to the background, be reasonably impermeable and free of cracks.

The mortars used for rendering are basically the same as those for brickwork, though careful attention should be paid to sands (see BS 1199). Excessively coarse sands will lead to mortars with poor adhesion while very fine sands or sands containing clay require more water and lead to shrinkage problems. Sharp sand is generally recommended, but where the local sand is too coarse, a blend of sands might prove satisfactory. As with jointing mortars, the mortar should not be stronger than the background. Lime is considered a valuable material, especially in weaker mortars.

Correct technique for rendering involves application of at least two coats. The first coat is designed to level the background and to even out its suction or water absorption; this may vary considerably from place to place, being lower in the case of lintels, mortar joints and dense bricks, and higher with lightweight blocks. Treatment of the latter with PVA emulsion may be necessary to reduce suction to acceptable levels. Saturation with water is not recommended; it leads to later shrinkage. Raking of mortar joints is essential for good adhesion if the render is to be applied to a smooth background such as common bricks. The ideal thickness for the undercoat is about 15 mm.

Adequate trowelling in the undercoat is very important; this compacts the mortar, closing cracks as the water is drawn into the background. Trowelling would normally be completed within about one hour of application. The undercoat is then 'scratched' to assist in producing a uniform distribution of small cracks and to provide a key for the final coat. Curing for at least three days is necessary; a weak friable product results if there is inadequate water for cement hydration. Premature drying is easy to spot since the material becomes much lighter as it dries. After curing, drying should be allowed in order to permit a network of fine cracks to from.

The final coat should be applied after the first coat has cured and dried, usually about one week. It is important to plan work for the final coat so that plain flat surfaces can be completed in one operation. This will obviate unsightly joints. In large areas a team of operatives may be required so that a 'wet edge' can be maintained throughout the process. The final coat may take several forms:

Smooth rendering A smooth finish is obtained by trowelling, with a wooden float to about 3 mm thickness. Metal trowelling produces a very smooth finish more likely to be affected by crazing problems. Smooth rendering can be shaped and ruled to produce the appearance of natural stone at a fraction of the cost.

Pebble dashing Pebbles of size 3–10 mm are dashed into the final coating while plastic.

Rough cast Mortar is dashed against the wall producing a textured finish. The coarseness of texture increases with the maximum aggregate size.

Some cracking invariably occurs in the final coat but is obscured in coarse textured finishes. In smooth rendered finishes, a very fine crack network results which should be invisible if the final coating is only a few mm thick. Smooth renderings are normally finally decorated with a masonry paint which will fill and obscure such fine cracks. Curing of the final coat is essential in all cases for best results.

The production of good quality renderings requires skill and experience as well as organisation of work and flexibility so that, for example, changing weather conditions can be accommodated.

CHAPTER 8

METALS

Metals display a considerable number of properties not found in any other major group of materials, for example, generally high tensile and compressive strength and density as well as the ability to deform plastically without damage; surface oxidation in the atmosphere; and good heat and electrical conduction properties. With their range of types and versatility, the future of metals in the construction industry is assured, though with high production costs there is likely to be increasing emphasis on 'high tech' forms to give maximum performance with minimum metal used. Metallic properties such as conduction of heat and electricity are easily explained by reference to the nature of the metallic bond but other properties require a more detailed examination of metal structure if they are to be understood.

8.1 METALLIC CRYSTALS

Crystals have been defined as very large, regular arrays of atoms conforming to a given pattern. The basic repeat unit is known as the *unit cell*. Virtually all pure elements, when in solid form, pack in a crystalline manner and metals are no exception. The unit cells of metals are quite simple since the metallic bond is largely non-directional in character. It results in close packing of metallic ions such that attraction due to bonding is balanced by ion–ion repulsion. There is an equilibrium distance corresponding to any one metal ion at which neighbouring ions will try to position themselves.

The shape of the unit cell produced may be predicted easily by studies of close packing of spheres. The maximum number of spheres that can be made to touch a single sphere of equal size is twelve and this can be obtained in two ways, the sphere in each case being surrounded by a hexagon of spheres with all spheres in the same plane (Figure 8.1(a)). The other six occur in two groups of three above and below this hexagon, fitting in the spaces between those in the original hexagon. In one case the upper triplet is directly above the lower triplet, and in the other case, each upper sphere is above a gap in the lower triplet (see Figure 8.1(b) and (c) respectively). The former is known as a hexagonal close packed lattice (HCP) and the latter as a face-centred cubic lattice (FCC). In the latter, the corners and centre

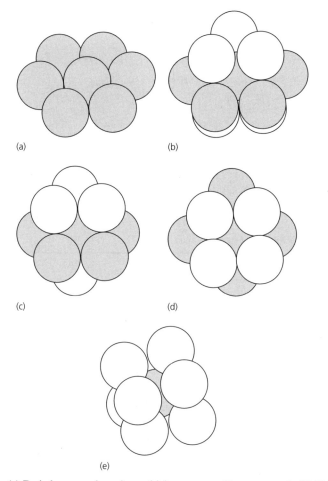

Figure 8.1 (a) Basic hexagon shape into which many metallic atoms pack. (b) Triplets of adjacent hexagon sheets directly above one another (shading for clarity). (c) Upper triplet above gaps in lower triplet. (d) As (c) but shaded to show face-centred cubic structure. (e) Body-centred cubic lattice. All atoms are identical – shading is for clarity only

atom of the face-centred cube have been shaded (Figure 8.1(d)), in case it is difficult to see the relationship between the hexagonal structure and the cubic unit cell. These unit cells are more commonly represented as in Figure 8.2. Also included is the body-centred cubic (BCC) unit cell (Figure 8.1(e) and Figure 8.2(b)) in which the atoms of some metals (for example, iron at room temperature) pack. This structure is less close packed, each atom having eight near neighbours. The reason for this is that iron at room temperature has a degree of covalency in its bonding, with the BCC structure in this case of lower energy than either of the above forms and, therefore, more stable, owing to the partially directional quality of bonds which results. The FCC and HCP structures are also of different energy

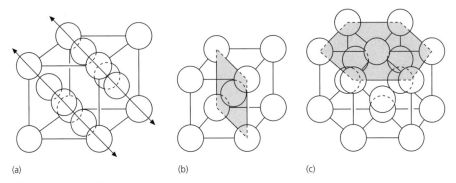

Figure 8.2 Metallic crystals: (a) face-centred cubic; (b) body-centred cubic; (c) hexagonal close packed. Typical densely packed planes are indicated in each case

Table 8.1 Crystalline form of some common metals

Crystalline structure FCC	HCP	BCC
Aluminium	Zinc	Iron (below 910 °C)
Nickel	Magnesium	Niobium
Copper	Titanium	Molybdenum
Lead		Vanadium
Iron (910–1390 °C)		Chromium
		Tungsten

and it is normal for metal atoms to pack in one or other form in given conditions. Table 8.1 shows the classification of some common metals.

The deformation properties of pure metals in particular groups are similar and they depend on the symmetry of atoms within the crystals. If a crystal could be observed under the microscope, planes of atoms would be immediately apparent, rather like lines of plants in a geometrically planted array, only in three dimensions. The mechanical properties depend on the population and spacing of these planes and they are influenced by certain imperfections which occur in virtually all metallic crystals.

SLIP PLANES IN METAL CRYSTALS

It has already been explained in Chapter 5 that plastic distortion in metals takes place as a result of shear or slip in crystal planes, the bonds between ions continuously breaking and re-forming as distortion occurs. The ability of metals to withstand plastic flow without damage is largely due to the non-directional character of the metallic bond, combined with the fact that, in a pure metal, all ions are of identical size.

Crystal slip does not take place haphazardly; it occurs in certain planes between which the metallic bond is more easily broken. These planes are normally the most

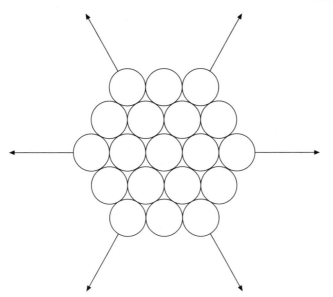

Figure 8.3 Close packed directions in a hexagonal array. These are the directions of slip in face-centred cubic and hexagonal close packed crystals

densely packed planes in the crystal, since such planes are more widely separated than any others and the bonding between atoms in adjacent planes is, therefore, weaker. A further requirement for crystal slip is that the atoms involved should move in such a direction that interatomic spacings remain as large as possible in order to minimise ion–ion repulsion. This results in slip taking place only in specific directions, which are also close packed directions (Figure 8.3).

It will be appreciated that in real situations the applied shear stress would not, in general, be in the exact orientation of any one slip plane or direction. Therefore, the more non-parallel, densely packed planes a crystal contains and the more close packed directions within each plane, the more likely slip would be to occur under a given stress. Hence, there are three factors affecting the ability of a crystal to flow under stress:

(a) The closeness of packing of crystal planes. The HCP and FCC structures each contain the hexagonal array with its very dense packing and, hence, would be most ductile from this point of view.

(b) The number of non-parallel close packed planes in the crystal.

(c) The number of close packed directions in each close packed plane.

The product of the numbers of (b) and (c) is called the number of slip systems for that crystal. Inspection of an FCC crystal, for example, would reveal four non-parallel sets of hexagonal planes seen as lines of atoms. Figure 8.2(a) has been drawn in such a way as to illustrate one such set of planes. There are three close packed directions in each plane (the three non-parallel sides of each hexagon) and therefore $4 \times 3 = 12$ slip systems. The FCC metals (Table 8.1) are widely used on account of their ductility, though this results in yield stresses which are relatively low compared to metals of other crystal groups.

There is only one set of hexagonal planes in the HCP structure (Figure 8.2(c)) and, again, three slip directions, as in the FCC crystal, giving three slip systems. Hence, HCP metals are generally less ductile with higher yield stresses than FCC metals.

The BCC group is rather more complicated. There are six densely packed planes and these join diagonally opposite edges of the unit cell (Figure 8.2(b)). In each plane, there are two close packed directions (body diagonals), giving 12 slip systems. There are, however, other planes of lower density packing along which slip can take place, each set including the close packed body diagonal direction. In fact, there are 48 slip systems in total. BCC metals are, nevertheless, generally less ductile than FCC metals because of the lower density packing within slip planes which, therefore, slip less easily.

IMPERFECTIONS IN CRYSTALS

The crystals in all bulk metals contain imperfections in their structure. Some of these, such as impurities remaining after extraction from the ore, are normally unwanted but difficult to remove. Other 'impurities' may be included intentionally to modify the properties of the metal, as in alloying. Imperfections may be classified as point imperfections, line imperfections and surface imperfections, and the presence of each has an important influence on the properties of the bulk metal. The diagrams, Figures 8.4 to 8.9, which are used to represent imperfections are

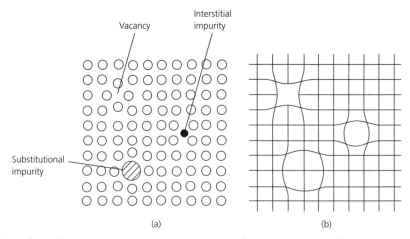

(a)   (b)

Figure 8.4 Point imperfections: (a) representation of a vacancy, a large substitutional impurity and an interstitial impurity; (b) the lattice lines which indicate the stresses resulting from the imperfections in (a). Where lattice lines are close together, there are compression zones. Where they are further apart, there are tensile zones

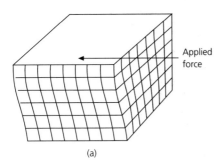

(a)

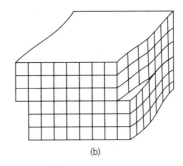

(b)

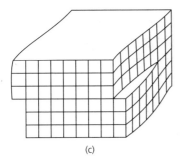

(c)

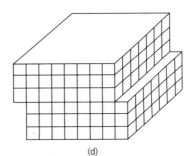

(d)

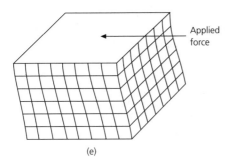

(e)

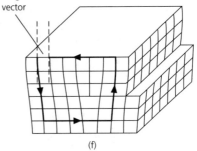

(f)

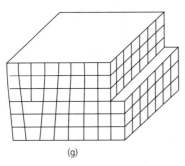

(g)

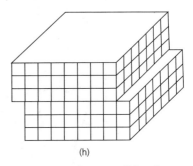

(h)

Figure 8.5 (a) to (d) Crystal slip by production and movement of a screw dislocation; (e) to (h) crystal slip by production and movement of an edge dislocation

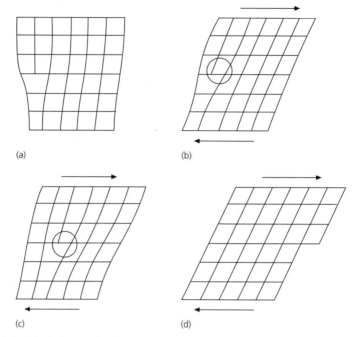

(a)   (b)

(c)   (d)

Figure 8.6 Plastic flow assisted by an edge dislocation: (a) zero stress; (b) elastic deformation under shear stress (unstable area circled); (c) extra plane having moved one spacing (new unstable area circled); (d) dislocation moved to edge of crystal

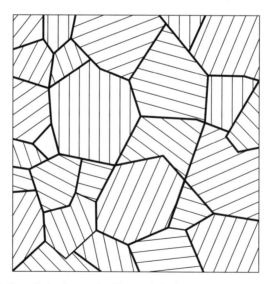

Figure 8.7 Grain boundaries in metals. The straight lines represent atomic planes

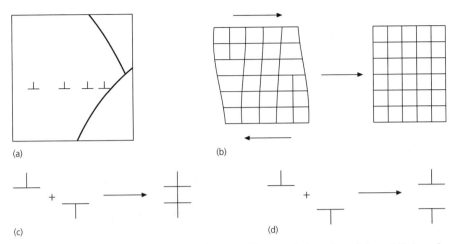

(a)   (b)

(c)   (d)

Figure 8.8 (a) Pile-up of edge dislocations caused by a grain boundary; (b) annihilation of dislocations by plastic flow; (c) simple representation of the situation in (b); (d) combination of dislocations to form a vacancy

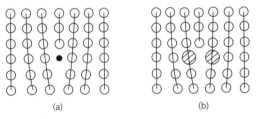

(a)   (b)

Figure 8.9 Relief of the tensile stress at the end of an edge dislocation by: (a) an interstitial impurity atom; (b) two substitutional impurity atoms

drawn using simple cubic arrays for simplicity. It should be appreciated that, although they serve the purpose of illustrating imperfections, they are not, in general, actual slip planes, since they are not densely packed and, indeed, the FCC and HCP crystal types do not contain such simple cubic planes.

Point imperfections

Vacancy A lattice site is unoccupied. This defect arises if the speed at which the crystal is grown is too high for perfect packing to take place. Slower cooled metals contain fewer vacancies. There is a tendency for atoms to move towards the gap as if the vacancy were causing a vacuum at that point and this is illustrated in Figure 8.4(b) by distortion of the lattice lines.

Substitutional impurities These may occur if a foreign material is present in the metal. Provided the atoms are of similar size (for example, zinc and copper, the atomic diameters of which are in the ratio 1:1.04), the impurity ions may fit into lattice sites of the host element. If the ionic diameters are quite different then the impurity material cannot 'dissolve' in the parent metal in this way and must form a separate compound or phase. Metals of the same crystal classification dissolve more readily in one another. Copper and nickel, for example, which are both FCC, dissolve in any ratio in one another. Substitutional impurities form the basis of most types of alloy production.

Interstitial impurities   If foreign atoms which are much smaller than the host ions are present, they may occupy space between lattice sites, causing distortion of the lattice. Carbon in iron, forming steel, is an example of this type of impurity. There will be a limit to the amount of such impurities which can be dissolved in the metal. In the case of iron at room temperature, this limit is less than 0.01 per cent by weight of carbon, since its presence causes considerable distortion of the BCC iron lattice.

Line imperfections (dislocations)

These are already present in large numbers in unstressed metals and these, like vacancies, are the result of imperfect crystallisation caused by rapid cooling. Their nature may be understood by considering the effect of stresses on a metal block consisting of a large, perfect crystal (Figure 8.5).

If the block is fixed at the base (Figure 8.5(a)), the application of force in the position and direction shown would cause a distortion of the lattice such that, in the most heavily stressed areas of the crystal, upper atoms will tend to occupy positions almost over atoms adjacent to those they were previously above. On increasing the stress, the crystal will finally slip along a plane by one atom spacing, producing a 'screw dislocation' as in Figure 8.5(b). Screw dislocations are represented by the symbol ⌒ or ⌒, since they may have a clockwise or anticlockwise sense. Note that, apart from the distortion of the lattice, it is perfect all round the dislocation which extends as a line through the crystal: hence, the term 'line imperfection'. If the stress is maintained, a further plane of atoms slips and the dislocation moves at right angles to the stress direction until finally the entire upper half of the crystal has moved one atom spacing to the left (Figure 8.5(d)).

If the stress is applied centrally to the block, then the distortion would be as in Figure 8.5(e). The dislocation would, in this case, be an extra plane of atoms in the upper part of the crystal, in a plane at right angles to the stress direction, known as an 'edge dislocation' and represented as in Figure 8.5(f). The dislocation will move parallel to the stress direction until the entire upper half of the crystal has slipped one atom spacing. Note that the slipped crystal finishes up with a perfect lattice of slightly different shape as a result of each kind of dislocation movement. This process may continue further by repeated dislocation formation and movement, causing eventual failure of the crystal, unless the dislocations are obstructed in some way.

Burger's vector

This is obtained by counting equal numbers of lattice spacings around the dislocation. For example, if, for the edge dislocation in Figure 8.5(f), five spacings are travelled, downwards, to the right, upwards and then to the left, a gap is found to remain and this is the Burger's vector of the dislocation. This is equal to the distance which the crystal will slip as a result of the dislocation movement and, since a larger distance will require more energy, dislocations tend to operate in planes which are close packed, these planes having smaller values of Burger's vector.

The effect of dislocations

The importance of dislocations is due to the fact that they allow the slip process to take place in small steps, in lines of atoms, instead of whole planes, so that the deformation occurs at a stress much lower than that which would be required to cause the upper part of the block to slip bodily by simultaneous disruption of all bonds in the slip plane.

The situation is analogous to moving a carpet by forming a ruck in it; only one part moves at a time although eventually, with much less effort than would be needed to move it all at once, the whole carpet moves. The movement of caterpillars is similar. Another illustration is zip fasteners. They are opened and closed one link at a time and therefore open and close easily while the fastener would be very difficult to open just by pulling when closed.

Dislocations act as a 'low gear', producing the same net effect as slip of complete planes but at much lower stresses. Although the example of stressing a crystal, given above, was used to explain the origin of dislocations, these defects are present in large numbers in unstressed materials. They may also occur in closed loops in the metal, the dislocations in places then being partially edge and partially screw dislocations. When materials are stressed, plastic flow takes place by movement and multiplication of dislocations already present, rather than by creation of dislocations as above since lower stresses are needed to move an existing dislocation than to form a new one. Figure 8.6 shows, for example, the effect of shear stress on a crystal containing an edge dislocation. In this case, the extra plane and the associated dislocation move to the right. This is because the plane to the right of the extra plane is distorted and the extra plane eventually joins the lower half of this distorted plane, leaving the upper half as the new 'extra' plane. Since the dislocation was already present, this effect occurs at much lower stresses than the plastic flow illustrated in Figure 8.5.

Surface imperfections

These arise due to the fact that, on cooling from liquid, metal crystals begin to form on a multitude of nuclei simultaneously rather than by gradual growth of single crystals. The crystals grow to a point where they meet and then, since the orientations of different crystals will be different, there will be narrow bands of semi-ordered structure known as grain boundaries (Figure 8.7). The size of crystal formed depends on the cooling rate of the metal; if cooling is slow, as in annealed metals, crystals tend to be large, as they form fewer nuclei and have more time to form. Rapid cooling reduces crystal size and very rapid cooling (for example, quenching) may completely prevent crystal formation. If metals are 'soaked' at temperatures near to their melting points, especially if under stress, then crystals can reform to relieve internal stress and they may also grow, small crystals being absorbed into larger crystals.

8.2 DEFORMATION OF METALS

The following general effects occur when load is applied. On first stressing the metal beyond its elastic limit, crystal planes will begin to slip by movement of dislocations already present. If the metal crystals are large and the metal is pure (for example, wrought iron), considerable plastic distortion will be possible without weakening, although the material will gradually harden as dislocations arrive at grain boundaries, where they must stop. (Adjacent crystals, having different orientations, cannot normally allow continued movement.) Hence, dislocations will tend to pile up behind one another if they are of the same type (Figure 8.8(a)). This process is known as work hardening and is a disadvantage in many manufacturing processes, since decreased ductility results in increased power consumption in shaping processes, greater wear on machinery and increased likelihood of damage to the component being made. Heating of the component at an intermediate stage in the shaping process (annealing) allows crystals to redistribute themselves so that softening occurs and the original properties of the metal will be restored. Some metals, notably lead, recrystallise at room temperature and, therefore, do not work harden (see Table 8.2). They have, in consequence, excellent forming properties. Conversely, those metals which work harden can be 'cold worked' to improve yield stress. Metals such as steel and copper are often used in work hardened form.

To some degree, dislocations may, on movement, cancel one another. For example, if an extra plane of atoms in the upper layers of the metal travelling one way encounters a similar plane in lower layers travelling in the opposite direction, the two dislocations will cancel (Figure 8.8(b), (c)). If the upper plane were one spacing above the lower, a vacancy would be formed (Figure 8.8(d)).

The behaviour of dislocations may be changed considerably by reduction of temperature. In BCC crystals, in particular, dislocations move less easily, so that, on stressing below a certain temperature, stress concentrations arise around dislocations but plastic flow does not occur. This may lead to cracks if the energy needed to

Table 8.2   Approximate recrystallisation temperatures of some common metals

Metal	Recrystallisation temp. (°C)
Iron	400–500
Copper	100–200
Aluminium	150–250
Lead	Below room temp.
Zinc	Below room temp.
Tin	Below room temp.

propagate microcracks is less than that needed to yield the crystal and, in such cases, failure within the crystal may occur. Failures of this type which occur suddenly and without plastic flow are called brittle fractures. Since yield is associated with shear stresses, brittle fracture can sometimes occur when applied stresses are such that there is no shear component – biaxial tensile stress is an example. When this type of stress is present, great care is necessary to avoid sudden, brittle failure.

Grain size can affect a number of properties. For example, in smaller crystals (finer grain structure), there is less scope for dislocation movement, so that yield stress is increased. It might be expected that the poor bonding at grain boundaries would cause premature failure at these positions but, fortunately, grain boundary failure stresses are normally considerably greater than yield stresses of ordinary bulk metals at normal temperatures, so that grain boundary failure without yielding is not a common phenomenon. Failure at grain boundaries may, nevertheless, play a significant role in overall failure, especially at high temperatures when grain boundaries are quite weak. This does not, however, imply that high temperature fracture would be brittle, since plastic flow in the crystal occurs more easily at high temperatures and flow may also be possible in boundaries before failure if the temperature is high enough. Hence, the amount of plastic flow before fracture generally increases as temperature rises.

If a metal contains impurities, these will, in general, tend to make it harder, since they interact with dislocations, blocking their paths. For example, an interstitial impurity at the end of an extra plane of atoms will cause the dislocation to 'lock' at this point (Figure 8.9(a)). If the foreign metal is present in a quantity above its solubility, it may form separate compounds at grain boundaries. It is for this reason than carbon is included in steel: it has a strengthening and hardening effect on iron. Another means of strengthening metals is by substitutional impurities. For example, two substitutional atoms slightly larger than the host atoms, just below a dislocation plane, would cause the dislocation to stop, unless a higher stress were applied (Figure 8.9(b)).

Where dislocations are locked at each end of the line, it is possible, on continued increase of stress, for the centre part of it to move further, bowing out, and eventually completely enclosing the original obstruction, giving rise to a new dislocation (Figure 8.10). This is known as a Frank Read source. There are several other ways in which dislocations can overcome obstructions.

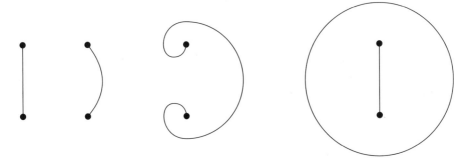

Figure 8.10   Stages in the distortion by stress of a locked dislocation to form a new dislocation

8.3 ALLOYS

An alloy consists of a metal mixed intimately with one or more metals (or, in some cases, non-metals) of a different type. The other metal may be dispersed in the first, possibly in the form of a 'solid solution', though metals do not have to be soluble in one another to form alloys. For one metal to 'dissolve' in another, the atoms of the former must be accepted into the lattice of the other and this may be possible in two ways: interstitially or substitutionally. In each case, the solubility of the foreign material will depend on its size; in the former, atoms should be small, while in the latter, they should be within about 15 per cent of the size of those of the host materials. Certain valency requirements must also be satisfied. Substitutional alloys are the most common. There is a very large range of these and many metals used in the construction industry today are alloyed in this way to some degree. If atomic diameters are very similar, then solubility may be complete so that, on cooling any proportion of two metals in liquid form, a single crystalline material is obtained. This is not to say that the two metals are evenly mixed in all crystals in the solid. One property of alloys, whether or not solubility is complete, is that different crystals have different proportions of the constituent metals, unless, after formation, the structure has been able to diffuse to a uniform composition.

The basic principle of alloy formation can be considered by reference to temperature–composition diagrams. Consider two metals A and B which are completely soluble in one another, having melting points as shown in Figure 8.11. On cooling a mixture, for example, of composition 60:40, A:B (indicated by the vertical line), a solid will first begin to form at the temperature represented by points a and it will be of composition corresponding to point a_s. As the temperature falls to a value corresponding to points b, a solid with b_s per cent of A exists so that the remaining liquid will be richer in metal B and will, in fact, be of composition corresponding to point b_L. (The lower and upper lines are on this account known as the solidus and liquidus lines respectively. At any given composition, the alloy cannot be all solid above the solidus line and it cannot be all liquid below the liquidus line at that composition.) Eventually, on reaching points c, the average composition of the solid will be denoted by c_s and, since this is that of the original liquid, the metal must now be completely solid. Between temperatures corresponding to a and c, solid and liquid exist together in equilibrium. Alloys therefore become increasingly plastic as they solidify rather than simply changing from liquid to solid.

This property in itself has been most useful, with application, for instance, in wiping joints using lead pipes. The plastic material can be moulded around the pipe but is not so fluid that it runs off it. The resultant solid in this case is known as a single phase solid, since all crystals are of one type though with varying composition, the core of each crystal being rich in metal A and the outer parts rich in metal B. In fact, if thermal energy were sufficient, diffusion of each metal would take place from richer to leaner areas, ultimately causing even distribution.

Alloys in which the two metals are completely soluble do exist, for example, copper and nickel, but their mechanical properties are very similar to those of the

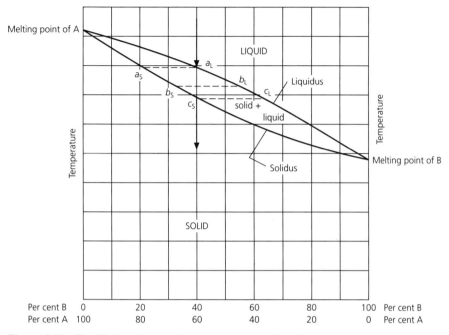

Figure 8.11 Equilibrium diagram for two metals A and B which are completely soluble in one another

pure metals and there may not always be much advantage in using them in alloy form. Most commonly used alloys consist of metals which are only partially soluble in one another, for example, zinc and copper, producing brass. When copper contains up to 36 per cent zinc, an FCC crystalline form known as *alpha brass* results, as in pure copper. Between 36 and 46 per cent zinc, a second type of crystal appears, known as *beta brass*, which is of BCC structure. Finally, above 46 per cent zinc, a complex phase known as *gamma brass* is produced.

The simplest general form of temperature–composition diagrams for metals which are partially soluble in one another is shown in Figure 8.12. The material (phase) produced when B dissolves in A is known as the alpha phase and when A dissolves in B as the beta phase. Note that these solubilities, denoted by the sloping lines at the left and right of the figure at the base, increase with temperature, as do all solubilities. The liquidus line this time reaches a minimum value at a point known as the eutectic point, signifying that the melting point of two-phase metal alloys is, for most proportions of its constituent metals, lower than the melting points of each of the pure metals. The products which result on cooling a mixture of these metals are as follows (see Figure 8.12).

1. Percentage of B below the percentage solubility S_B at room temperature The alloy will begin to solidify at the temperature where the vertical composition line intersects the liquidus and will finish solidifying where

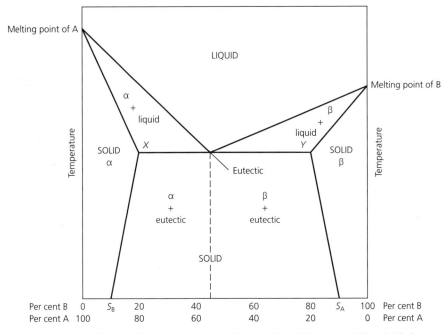

Figure 8.12   Equilibrium diagram for two metals A and B which are partially soluble in one another

this line intersects the solidus. The properties of the resultant product will be similar to those of pure metal A.

A similar argument applies to percentages of A below the solubility limit S_A dissolved in B.

2. Percentage of B below the high temperature solubility level X but above the room temperature level   The alloy will solidify as in (1) except that, when the temperature occurs at which the alpha phase becomes saturated, the pure metal B containing some of A (i.e., the beta phase) must begin to separate. If cooling is slow, the beta phase will form in grain boundaries but if cooling is rapid, the metal will be trapped, producing quite different properties in the alloy.

3. Percentage of metal B higher than that in (2) but below the eutectic composition   The liquid will commence to solidify, producing alpha phase material at the point given by the intersection of the composition line with the liquidus, as before. Solidification will continue until the temperature has fallen to the solidus line, the remaining liquid at this temperature having eutectic composition. However, because, owing to the limited solubility of each metal in the other, formation of a single phase of this composition is not possible, the solid which solidifies must contain proportions of alpha (B in A) and beta (A in B) phases at their respective solubility limits X and Y. Hence, the resulting alloy contains some alpha crystals and some of the alpha + beta mixture. As the alpha + beta mixture cools, argument (2) applies, owing to the decreasing solubilities. A similar argument applies if the composition of the liquid is between the eutectic value and that corresponding to point Y.

The nature of the alpha + beta mixture (eutectic) is such that the alloys containing it are much less ductile than pure phases. Alloys containing some eutectic may, therefore, be cast rather than cold worked. Cast iron is a typical example.

8.4 MECHANICAL PROPERTIES OF METALS

These may be described under the headings of tensile strength, creep, impact strength, fatigue and hardness.

TENSILE STRENGTH

The elastic nature of the deformation of materials has been described in Chapter 5 and this is obviously of prime importance, since nearly all metal components are designed to undergo elastic movement only. But to estimate factors of safety, it is essential to know at what point elastic movement changes into plastic flow. Also, since plastic movement during manufacture is responsible for considerable modification of metal properties, an understanding of both the elastic and plastic portions of stress–strain curves is of advantage. Although many components in service may be under compressive or shear stresses as well as tensile stresses, tensile tests are the standard way of assessing strength and this is perhaps appropriate since, for example, steel for reinforcement is principally under tension, as is the lower flange in a simply supported beam. Provided the specimen can be gripped, the results of tensile tests are easier to interpret than those of compression or shear tests.

Tensile tests are carried out by loading uniformly to destruction either bars or wires (as in the case of reinforcement or prestressing wire) or carefully cut strips of the metal component.

The exact nature and significance of stress–strain relationships produced by tensile testing depends on the nature and formation process of the metal and the differences are best understood by consideration of the behaviour of some simplified structures.

If a single perfect crystal were stressed, the yield point stress would, in the absence of dislocations, be very high. The stress would be dependent, however, on the orientation of the applied stress with respect to that of close packed planes – it could vary quite widely, especially in a HCP crystal in which there is only one set of close packed planes. Once dislocations have formed and yielding has occurred, further plastic flow would take place with no further stress increase – in fact, depending on test-machine characteristics, a yield point drop may be obtained

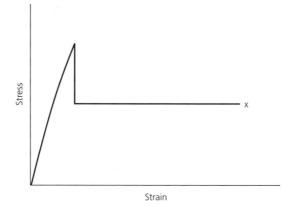

Figure 8.13 Stress–strain diagram for a single, perfect crystal. The yield point would depend on the crystal orientation

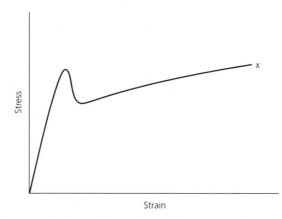

Figure 8.14 Stress–strain diagram for a polycrystalline metal (perfect crystals)

since dislocations, once formed, travel easily in the absence of interference from impurities or other obstructions. Hence, in this case, the yield stress would be equal to the ultimate stress (Figure 8.13).

If the metal were polycrystalline but each crystal were perfect, the yield stress would correspond to an average value, taking into account the random orientation of grains. There may, again, be a yield point drop, as newly formed dislocations travel easily through crystals but, unlike the single crystal, dislocations cannot now proceed unhindered from one grain to the next (assuming recrystallisation does not occur) and so there is a further progressive rise of stress with strain as work hardening occurs (Figure 8.14). It should be emphasised that, since dislocation-free crystals are never obtained in practice in bulk metals, these stress–strain relationships do not normally occur; they have been described in order that the effects of the various crystal defects be understood.

When stress is applied to a polycrystalline metal containing dislocations, yielding occurs much earlier, since there is now no need to generate dislocations, the sole requirement being to cause the existing ones to move. In most metals, the presence of dislocations, together with other impurities and associated internal stresses, often causes plastic flow at very low stresses. In such metals, there is no clearly defined yield point, since the stress at which individual dislocations begin to move depends on their position. Stress–strain graphs of the form shown in Figure 2.2 (see page 14) result and this is exhibited by most non-ferrous metals. Working stresses in such metals are normally based on 0.2 *per cent proof stress*. To obtain this stress, a line is drawn on the stress–strain graph, parallel to the curve at the origin but commencing at 0.2 per cent strain (Figure 2.2). The stress at which this intersects the curve is the stress required. The addition of substitutional impurities to give single phase alloys does not fundamentally alter this behaviour, although yield stress may be increased due to the resulting distortions within crystal planes and consequently increased resistance to dislocation flow.

The case is quite different when interstitial impurities are present, as in ferrous metals. The dislocations in the crystals then tend to gather congregations (*atmospheres*) of interstitial impurities around them, as in Figure 8.9(a). Light elements, such as carbon and nitrogen, dissolve in this way. On stressing the material, these atmospheres restrain the movement of dislocations, causing an artificially high yield point. As a consequence, when the yield point is reached and the dislocations move clear of the impurities, they become more mobile and the stress in the metal suddenly drops. An alternative possible cause could be the creation of new dislocations just before yielding when they multiply rapidly, moving more slowly as they multiply, assuming the strain rate to be constant. This multiplication of dislocations on yielding allows straining to continue at a reduced stress. The form of a stress–strain curve is, therefore, as in Figure 8.15.

The reduction in stress is known as yield point drop (point A). It cannot occur in substitutional alloys because the larger substitutional impurities have much lower mobility than interstitial impurities and cannot move to dislocation sites. On removal of the load at point X, the dotted line is followed as the stress reduces, being retraced on reloading so that the material is now effectively stronger than originally, in the sense that its yield point is now higher, though it is now permanently deformed. Note also that there will be no yield point drop on retesting. If, on the other hand, the metal is left for some days, the interstitial atoms will diffuse back to disclocation sites so that, on loading, a yield point drop again occurs. This process is known as strain ageing and the metal becomes effectively stronger by the action of dislocation pinning. The yield point drop may be restored quite quickly by heating the metal to about 200 °C or higher. Note that yield must not be allowed in service because, although no damage is done to the metal, the corresponding distortions to the component would be quite unacceptable.

Where a yield point drop occurs, design codes are normally based on the lower yield stress, since the upper point is difficult to measure and varies from specimen to specimen, though the latter might be considered more relevant to normal design.

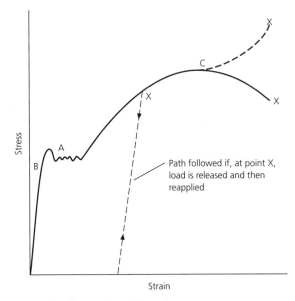

Figure 8.15   Stress–strain diagram for mild steel

Figure 8.15 is, in fact, typical of the form of the stress–strain curve for mild steel. In addition to the above features, the following points are worthy of note:

■ There is normally a slight deviation from Hooke's law (i.e., stress proportional to strain) before the elastic limit is reached – point B.

■ As plastic strains increase after yielding up to point C the metal will become thinner, so that the true stress in the metal is actually higher than that obtained by calculations based upon the original cross-sectional area.

■ From point C onwards, deformation continues at a single point on the bar only, due to stress concentration at this point. This might perhaps be caused originally by a slight dent in the bar, plastic movement then being sufficient to cause a local reduction in cross-sectional area, known as *necking*. The apparent stress–strain curve appears to have a negative gradient from this point but, in fact, the true stress in the necked region is still increasing (dotted line). When carrying out tensile tests on mild steel, the onset of necking corresponds to an observable falling of the load and signifies that the end of the test is near.

■ Since failure occurs after significant plastic flow, it is normally referred to as *ductile*. After extensive plastic deformation, there is a tendency for small voids (*inclusions*) to form in grains, possibly started where second-phase particles separate from the matrix. These grow and, on the principles of Griffith's theory, concentrate the stress in the remaining parts of the crystal. The crystal subsequently fails in tension (*cleavage*), forming a fracture surface in the most heavily stressed central region. The external annulus fails finally in shear at 45° to the tensile axis, leading to a *cup and cone* fracture (Figure 8.16).

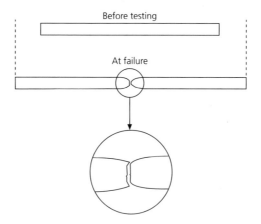

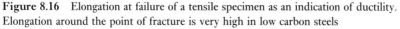

Figure 8.16 Elongation at failure of a tensile specimen as an indication of ductility. Elongation around the point of fracture is very high in low carbon steels

It should be noted that, when considerable work hardening is carried out as, for example, in cold-drawn prestressing wire or cold-worked steel reinforcing bars, the strain ageing effect mentioned above is small compared to the hardening effect obtained by dislocation interaction. Hence, yielding of these materials is similar to that of non-ferrous metals, as in Figure 2.2, and proof stresses are again employed. Figure 2.2 also compares stress–strain curves for a number of metals commonly used in construction.

CREEP

This is defined as time-dependent strain which occurs when a steady stress is maintained. The phenomenon has been described in concrete but, although the same effect occurs in metals, the cause is quite different.

Creep in metals can be satisfactorily explained by reference to the dislocations they contain. All dislocations have a certain amount of thermal energy and this will fluctuate, according to statistical laws, from atom to atom in the metal and with time. If occasions arise when a dislocation, previously locked by other dislocations or defects, has sufficient thermal energy, then it will, when the metal is under stress, tend to climb over the obstacle and thereby allow some plastic movement in the metal. A typical strain–time diagram for a metal undergoing creep is shown in Figure 8.17. During the first stage, known as primary creep, strain increases rapidly from the initial instantaneous value, the rate falling off as dislocations become trapped. During the secondary stage, diffusion of atoms leads to a gradual increase of strain as dislocations interact and annihilate one another. The process is known as *recovery*. Finally, in the third stage, shear between grain boundaries produces cracks (Figure 8.18). Vacancies may also accumulate at grain boundaries to produce voids and there may be some necking, these effects combining to produce failure at much lower stresses than would occur in short term tests. It is clear that increasing

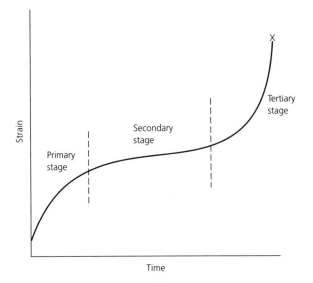

Figure 8.17   Stages in the fracture of a metal, due to creep

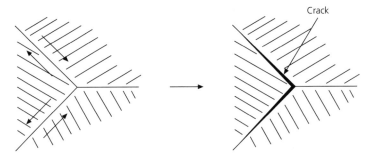

Figure 8.18   Effect of shear stresses on grain boundaries leading to crack formation

stress and increasing temperature will both contribute to creep strain so that the scale of the time axis will depend on the stress, temperature and metal used; it may vary between minutes and years. Reducing the temperature or the stress may reduce the creep rate to the extent that the secondary and/or tertiary stages of Figure 8.17 are not reached.

It would seem likely, also, that some creep could occur even at low temperatures, since even then there would be some statistical probability of dislocation movement. Although creep does occur at low temperatures, significant strains do not normally occur at temperatures below about 40 per cent of the melting point temperature of the metal in K (degrees K = °C + 273). Lead, for example, has a melting point of 327 °C = 600 K. Room temperature (20 °C) equals 293 K and this is 49 per cent of 600 K, so that lead would be expected to creep at normal temperatures. This fact is well known and adequate restraint of lead components, which are stressed due to their weight, is essential in order to avoid distortion with time. The melting point of iron is, on the other hand, 1539 °C, so that creep would not be expected in ferrous metals at normal temperatures.

A related phenomenon does, however, occur in cold-worked ferrous metals on stressing. The distortion of the crystals on cold-working leads to stresses in steels which tend to reduce by dislocation movement when a component is stressed in service. This is known as *relaxation* and allowances must be made, for example, for the loss of prestress that occurs in prestressed concrete, particularly when the working stress is near the characteristic strength of the material. Relaxation does not, however, lead to ultimate failure of the component, even under high stress, and it can be removed by initially stressing the wire beyond its in-service stress. Aluminium and its alloys may be subject to creep at temperatures over about 100 °C. If a metal operates at a temperature at which creep is likely, its performance will be improved with increase of grain size (and, therefore, reduction in the number of grain boundaries), since grain boundaries at such temperatures tend to be receptive to dislocations, thus permitting plastic flow. Work hardening increases creep resistance, provided recrystallisation does not occur. In lead, for example, a small amount of cold working is beneficial but extensive cold working would initiate recrystallisation reducing grain size. Therefore, the improvement in creep resistance would be lost, especially at higher operating temperatures. Creep resistance can be improved by alloying, although metals which dissolve readily in the parent metal are not, unfortunately, very effective in pinning dislocations at high operating temperatures. Various methods of producing precipitates or dispersions of alloying elements have been developed where high temperature creep resistance is required.

IMPACT STRENGTH – BRITTLE FRACTURE

The energy absorbed by impact of a metal is measured by means of a small notched specimen of the metal. The impact is produced by a pendulum and the energy absorbed by the specimen is proportional to the difference in heights of the pendulum when at rest before and after impact, a further indication of brittleness being given by the proportion of the cross-section of the specimen which has failed in a brittle manner (Figure 2.12). Brittle fracture gives a bright crystalline surface whereas a ductile fracture gives a duller, dimpled surface. The energy absorbed is affected by the following factors:

Temperature The impact strength of metals depends on the degree to which plastic flow can take place in the small interval of time between initial impact and fracture. If the yield point of the metal is increased, as occurs on cooling, thermal energy of dislocations is reduced and, therefore, plastic flow will be more difficult. As a result, the energy absorbed during impact will decrease. In the case of some metals, notably BCC structures such as ferrous metals, the impact strength may drop very rapidly with temperature (Figure 8.19). This may be attributed to the fact that, on cooling, a temperature is reached at which the yield point stress

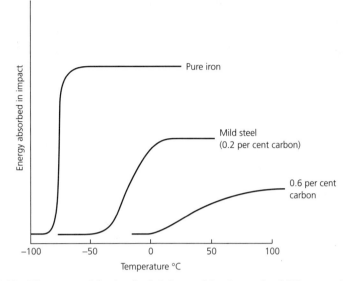

Figure 8.19 The nature of the ductile–brittle transition for steels of different carbon content

exceeds the fracture strength as determined by Griffith's principles so that virtually no plastic movement occurs. This transition is known as the ductile–brittle transition and, for low carbon steels, it takes place between 0 and −50 °C, dependent on the type of impact. The ductile–brittle transition is more gradual for higher carbon steels but note that the impact strength of medium/high carbon steels is quite sensitive to temperature changes at normal temperatures. The effect of lowering temperature on HCP metals such as titanium and zinc is similar to BCC metals but, in the case of FCC metals such as aluminium, copper and lead, with their large number of slip systems, there is no such transition and ductile behaviour is exhibited even at very low temperatures.

The geometry of the notch at which failure occurs   A smaller radius at the root of the notch produces a larger stress at this point, so that cracks which originate here form more quickly. The impact strength of metals in service depends, similarly, on surface defects and stress concentrations in them. The ductile–brittle transition temperature is also raised by any stress-concentrating mechanism.

Grain size   Coarser grains correspond to reduced yield stresses and might, therefore, be expected to increase resistance to impact failure. A more important effect, however, is that slip planes are longer in coarse-grained crystals, so that there is a greater tendency for dislocations to form clusters, giving rise to stress concentrations and, subsequently, cracks. Grain boundaries, in addition, tend to stop any cracks from being transmitted. Hence, by grain refining, as occurs in normalising or by addition of some alloying metals (such as niobium or manganese in steel), impact strength may be increased.

The straining rate caused by the impact The more suddenly the impact occurs, the less time will be available for absorption of energy by yielding. It is known that yield stresses tend to be higher when loading is more rapid, because dislocations take a finite time to move in order to cause yielding. Hence, a very severe impact may cause metals that are normally ductile to behave in a brittle manner. Again, this effect is particularly marked in BCC metals, due to dislocation pinning and, in HCP metals, due to their inherent low ductility.

FATIGUE

Fatigue damage normally starts at a fault in the metal, such as a weld fault, sharp concave corner or void. Each of these causes a local concentration of stress, resulting in yielding at that point.

Over a period of time work hardening occurs, reducing the ability of the metal to flow under stress and, eventually, leading to the formation of microcracks. These further concentrate the stress and, therefore, propagate until the stress in the remaining material is sufficient to cause failure. There is, very often, virtually no visible distortion before failure and the presence of fatigue cracking can usually be ascertained only by close inspection or non-destructive testing techniques. In situations where stress reversals are encountered, great care may, therefore, be necessary to avoid catastrophic failures.

Fatigue performance of metals may be investigated by rotation of a bar supported in bearings at both ends and loaded at its centre point. As the bar rotates about its own axis, the load causes fluctuation of the bending moment. Alternatively, in the case of larger components, for example, steel beams, a fluctuating load may be applied to the mid-point of a simply supported span. The number of reversals N to cause fracture at a given stress S is measured and results are plotted in the form of an S–N curve (Figure 8.20).

Curves are of two types. Metals in which strain ageing occurs show a certain stress below which fatigue does not occur. This is because, for example, in the case of steel, the carbon 'atmospheres' prevent dislocation movement. The stress is, in the case of mild steel, about 0.4 times the ultimate tensile strength (known as the fatigue limit). The term *endurance ratio* is also used. This is equal to the stress, expressed as a proportion of normal yield stress at which the metal would withstand a given number of cycles – say 10^7. In metals which do not exhibit strain ageing (for example, aluminium), failure may occur at very low stresses on continued stress reversal.

Fatigue strengths increase with tensile strength or decrease of temperature, and decrease in the presence of impurities, surface defects and corrosive environments. Tensile stresses tend to cause cracks to propagate so that they result in lower fatigue strengths than compressive stresses. If a compressive stress can be imparted to the surface as, for example, by carburising or nitriding, fatigue strength will be increased. Refining grain structure also increases fatigue resistance, since, as mentioned under 'Impact strength', smaller grains tend to reduce stress

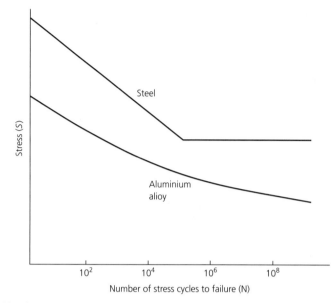

Figure 8.20 *S–N* curves for steel and an aluminium alloy

concentrations and grain boundaries are able to inhibit the spread of microcracks. Where a steady stress is superimposed on a reversing stress, as for example in bridges due to 'dead load', stress variations of a given amplitude will produce failure much more quickly, especially if the steady stress is a high proportion of the yield stress value.

HARDNESS

The hardness, as such, of metals for constructional purposes is not normally an important criterion. Hardness testing, however, has been used for quality control purposes during the manufacture of, for example, rolled steel joists, since local deformation of metal is involved and it is possible, for a given metal, to correlate hardness and yield strength. Hardness testing can also be used to assess fire damage to steel. *Brinell hardness* is measured by pressing a hardened steel ball into the surface of the metal under a load appropriate to the softness of the metal, see Figure 2.15. In the case of steel, the load is normally about 30 kN with a 10 mm diameter ball. The diameter of the dent is measured microscopically and the Brinell hardness is then equal to

$$\frac{\text{load}}{\text{surface area of the indentation}}$$

In some metals, hardness may be limited to a maximum value, since it is indicative of a brittle structure. For example, BS 4622 for grey cast iron pipes includes the use of the hardness test for this purpose.

8.5 METALLIC CORROSION

This may be described under the headings *oxidation, electrolytic corrosion* and *acidic corrosion*.

OXIDATION

This has important implications as regards choice of metals for high temperature applications. It also affects many forming, heat treatment and fusion processes, though oxidation is not a major cause of metallic corrosion in buildings under normal conditions.

Although the metallic bond has shown itself to be such that metals often have considerable strength and toughness in a mechanical sense, almost all metals are intrinsically unstable in an oxygen-containing atmosphere at normal temperatures. This is due to the fact that it is not possible for a crystal lattice to be perfect at the surface of the metal grains. Therefore, the metal atoms at the surface tend to be highly reactive and, in the case of most metals, combine very quickly with oxygen to form a more stable arrangement. Oxygen molecules in the atmosphere split into atoms, each of which contains six electrons in its outer shell, thus requiring two electrons to make a stable octet. These electrons are supplied by atoms at the surface of the metal so that ionic bonding occurs and a metallic oxide is formed (Figure 8.21). The precise behaviour of a particular metal with respect to its oxide coating depends on a number of factors, as follows:

The relative stability of the metal and its oxide   This varies from metal to metal. Gold is, in fact, more stable than its oxide in normal conditions, although a surface layer of oxygen atoms does exist to satisfy the bonding requirements of surface atoms. Other metals, such as silver and copper, have a relatively small heat

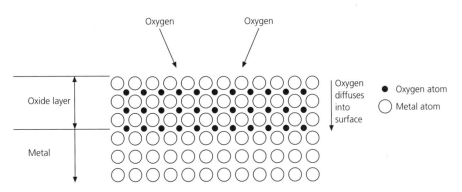

Figure 8.21   The combination of oxygen with the surface film of a metal to form an oxide layer. In this case, the oxide lattice is coherent with the metallic lattice and the oxide coating is, therefore, tenacious and impermeable

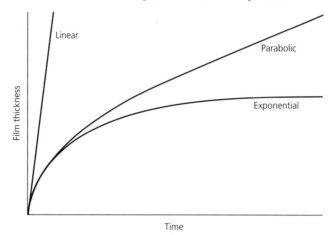

Figure 8.22   Relationships between oxide film thickness and time, all starting at the same oxidation rate

of formation with respect to their oxides, that is, the energy release when the oxides are formed is small so that the oxides are not extremely stable and the metals are corrosion resistant in clean atmospheres. Increasing oxygen pressure always increases stability of oxides in the same way that an increase of humidity in the atmosphere increases the tendency for water to exist in liquid form.

The physical properties of the oxide layer   In order for the oxidation process to continue, oxygen must have access to metallic ions below the surface of the existing oxide layer. The rate at which this can happen depends on the permeability of the oxide coating to metallic ions and/or oxygen ions. In the case of some metals, notably those whose atoms have only one or two electrons in their outer shell, the metallic ion is smaller than that of the metal from which it was formed. Thus, shrinkage of the metal occurs on oxidation, resulting in stresses in the oxide film and, possibly, cracking. The film will, therefore, be porous and will allow further oxygen to penetrate it so that oxidation continues. Such behaviour is shown by light metals such as magnesium. The oxidation rate of magnesium is therefore linear with respect to time (Figure 8.22), that is, the film thickness is proportional to time. Such metals may burn if heated, for the same reason.

In most heavier metals, ionic sizes may be comparable to, or larger than, pure metal atoms so that the oxide film is impervious and, possibly, in compression. Iron and copper are examples of this and since, in this case, ions must diffuse through the oxide coating, oxidation rates will decrease as the film thickness increases, though oxidation will never cease completely. This may be expressed in the form:

$$\frac{\mathrm{d}x}{\mathrm{d}t} = \frac{k}{x}$$

where x = film thickness,
          t = time,
          k = a constant

Integrating gives:

$x^2 = kt + C$ ($C = 0$ if $x = 0$ when $t = 0$)

or $x = \sqrt{kt}$.

The equation represents a parabolic relationship between film thickness and time (Figure 8.22).

In some cases, the oxide itself may be very tightly bound to the metal or it may be not electrically conductive so that it will not permit diffusion of ions or electrons through it to allow further oxidation. In this case, the growth rate of the film will decrease exponentially and, after a time, oxidation will cease (Figure 8.22). Zinc, chromium, lead and aluminium exhibit this behaviour, each having wide use in building as a result. Aluminium goods are often oxidised (anodised) after manufacture to give a uniform protective coating. The above arguments apply to dry non-polluted oxygen-containing atmospheres. Pollutants or dampness may give rise to other reactions, accelerating corrosion.

Temperature   As with any chemical reaction, oxidation proceeds by chance encounters of oxygen and metallic ions and, since temperature increases thermal energy and diffusion rates of ions, it will inevitably increase oxidation rates. Hence, for example, metals obeying the exponential law at room temperature might obey the parabolic law at higher temperatures. Oxide layers formed at room temperature are in general too thin to be visible, but films produced at higher temperatures become visible initially by interfering with reflected light, as occurs when steel is *tempered*. Steel ingots, before hot rolling, may form considerable thicknesses of *mill scale*, a grey coating, in quite short times and most metals in the earth's crust exist in the form of oxides (*ores*), which is indicative of high temperatures at some stage in their history.

The effect of alloying elements   Alloying elements are often added to metals in order to produce a protective film on their surfaces. Restrictions are, however, placed on the use of alloying elements, since they may also alter the mechanical properties of the parent metal. For example, aluminium, when alloyed with iron, forms an effective oxide coating but affects its forming properties. Hence, use in percentages required for protection is confined to treatments applied to the finished article. Perhaps the most important alloying elements for iron are chromium and nickel. Chromium forms a protective oxide layer while nickel has the effect of preserving the austenitic state (see 'Stainless steel'). Alloying elements added for other reasons may, by oxidation, affect the properties of the parent metal. Carbon in steel, for example, oxidises more quickly than iron and is lost as a gas, so that the surfaces of carbon steel components tend to soften on ageing.

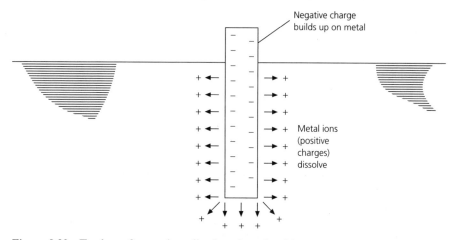

Figure 8.23   Tendency for metals to dissolve when placed in water

ELECTROLYTIC AND ACIDIC CORROSION

These corrosion mechanisms have very important implications for the performance of metals in virtually all environments in buildings.

They are caused in the first instance by the tendency for metals to dissolve (ionise) in aqueous solutions (Figure 8.23). If the surface of a metal becomes moist, metallic ions enter the water, leaving behind a negative charge on the remaining solid metal.

$$M \leftrightarrow M^{(n+)} + n(e^-) \qquad n \text{ equals the valency}$$
$$\text{Metal} \quad \text{Metal ion} \quad \text{electron(s)}$$
$$\text{(remain on metal)}$$

The arrow shown indicates that the reaction may proceed in either direction. This reaction may, theoretically, take place to some degree in any solid material in an attempt to balance the concentration of ions in the solid and adjacent liquid but, in the case of metals, it is particularly important due to the effect of the negative potential which builds up on the remaining metal with respect to the solution. Metallic ions continue to form until the electrons which are left behind have set up a sufficient negative voltage to oppose further release of positive ions. Even then, the equilibrium should be regarded as dynamic rather than static, that is, ionisation still occurs but is balanced by the opposite process – *deposition* or *plating out* of the metal.

The magnitude of the negative voltage which arises depends on the type of metal, the temperature and other factors. In practice, there is the problem of measurement of this potential, since the other *reference* electrode which must be used for measuring the difference in voltage between the metal and the solution will itself produce an electrode potential, so that a relative reading only is obtained. The problem is overcome by use of a standard *hydrogen electrode* which is used due

Table 8.3 Standard electrode potentials of pure metals

Metal	Electrode potential (volts)
Magnesium	−2.4
Aluminium	−1.7
Zinc	−0.76
Chromium	−0.65
Iron (ferrous)	−0.44
Nickel	−0.23
Tin	−0.14
Lead	−0.12
Hydrogen (reference)	0.00
Copper (cupric)	+0.34
Silver	+0.80
Gold	+1.4

Note that the most reactive metals are the alkali and alkaline earth metals which hold their electrons most loosely

to the significance of hydrogen in acidic corrosion. Table 8.3 gives the standard electrode potentials of some common metals in solutions of their ions, with respect to the hydrogen reference electrode. Metals which are more reactive than the hydrogen electrode produce negative voltages, while those that are less reactive produce positive voltages. The former are known as anodic metals and the latter as cathodic or noble metals. Electrode potentials are also affected by the state of the surface of the metal. Imperfections in the surface, such as grain boundaries or points of intersection of dislocation lines with the surface, will result in weaker bonding of atoms, easier dissolution and higher effective electrode potentials at these points. It should be emphasised that all metals have some tendency to dissolve, including those below hydrogen in the table.

So far, although the basic mechanism of ionisation and electrode potentials has been explained, the causes of continued corrosion, which occurs when ionisation proceeds with little or no hindrance over long periods of time, is not yet apparent. In the above examples, corrosion stops when the electrode potential is reached. Corrosion can only continue if the actual potential of the metal changes for some reason, so that more metal must dissolve in attempting to restore its original value, that is, the electrons which are the cause of the potential must somehow 'drain away' from the metal. This electron-consuming reaction may occur in two ways.

In the presence of acids – acidic corrosion

Acids contain free hydrogen ions which, on reaction with electrons, produce hydrogen gas:

$$2H^+ + 2e^- \rightarrow H_2\uparrow$$
in solution   from metal   gas

The supply of electrons required for this reaction is obtainable from metals which are above hydrogen in the table of electrode potentials. Hence, these metals dissolve in many acids. The more concentrated the acid is, the more rapidly will a metal corrode, since such acids contain greater concentrations of hydrogen ions. Metals below hydrogen in the table do not corrode in normal acids because hydrogen itself, which may be regarded as a 'metal', tends to ionise to a greater degree than these metals. Therefore, hydrogen, rather than the metal, will remain in ion form. The corrosion of the noble metals requires an oxidising as well as an acidic effect. This is, however, provided by certain concentrated acids or mixtures of acids.

Different metals in contact – electrolytic corrosion

Consider, for example, strips of copper and zinc immersed in the same solution and joined by an electrical conductor (Figure 8.24). The zinc would normally reach a voltage of -0.76 V on the hydrogen scale and the copper, $+0.34$ V. Therefore, on joining them, there must be a potential difference of 1.1 V between the zinc and copper. Electrons will flow through the wire from the zinc to the copper (corresponding to a flow of conventional current in the opposite direction) and the zinc will continue to dissolve, in an attempt to replace them, since the process will cause a reduction in its negative electric potential. The copper, on the other hand, will donate electrons to positive ions in solution in an attempt to maintain its former voltage of $+0.34$ V. If the solution contains hydrogen ions, they will collect electrons from the copper, forming hydrogen gas:

$$2H^+ \quad + \quad 2e^- \quad \rightarrow H_2\uparrow$$
in solution   from copper   gas

The zinc ions which dissolve will form a zinc salt by combination with negative ions in solution. If, for example, the solution contained sulphate ions, then zinc sulphate would form.

The corroding metal (in this case, the zinc) is known as the anode and the protected metal (in this case, the copper) as the cathode. Electron flow is always

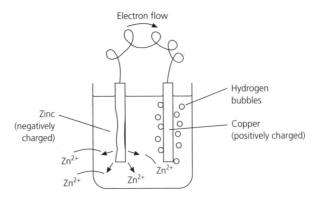

Figure 8.24 Electrolytic corrosion resulting from zinc and copper strips immersed in an aqueous solution while in electrical contact

Table 8.4 Some common situations in which electrolyte corrosion occurs

Situation	Metal which corrodes	Remedy
Galvanised water cistern or cylinder with copper pipes; traces of copper deposited due to water flow	Galvanised film corrodes electrolytically at point of contact with the copper particles. Film is destroyed, then steel corrodes similarly	Use a sacrificial anode in cistern. Otherwise use a plastic cistern or copper cylinder
Brass plumbing fittings in certain types of water	Zinc – dezincification	Use low zinc content brass or gunmetal fittings
Copper ballcock soldered to brass arm	Corrosion of solder occurs in the damp atmosphere resulting in fracture of joint	Use plastic ball on ballcock
Copper flashing secured by steel nails	Steel corrodes rapidly	Use copper tacks for securing copper sheeting
Iron or steel railings set in stone plinth using lead	Steel corrodes near the base	Ensure that steel is effectively protected by paint
Steel radiators with copper pipes	Steel radiators corrode	Corrosion can be reduced by means of inhibitors

from the anode to the cathode. Note that the above arrangement is the principle of the Daniell cell, in which the electromotive force of 1.1 V is used for the supply of small quantities of electric power. As with the Daniell cell, however, the electrolytic action does not necessarily proceed unhindered until the zinc is destroyed. Many other corrosion reactions are similarly slowed down, due to the resistance to ion movement. This may be caused, for example, by formation of insoluble films on the anode. Aluminium, for this reason, normally behaves cathodically with respect to zinc. Chromium is corrosion resistant in the same way. Alternatively, hydrogen gas bubbles evolved at the cathode may tend to form a 'third electrode' and also resist the flow of further positive ions to the cathode. This is known as polarisation.

ALTERNATIVE CAUSES OF ELECTROLYTIC CORROSION

There are a number of instances where commonly used combinations of different metals in building or constructional engineering result in electrolytic corrosion; examples are given in Table 8.4. Just as important, however, and less obvious, are instances in which electrolytic corrosion cells may form within a single metal in the presence of moisture. Table 8.5 gives a number of possible causes. The effect of grains and electrolyte concentration in causing corrosion will be readily appreciated, as these both cause changes of electrode potential but the effects of oxygen and stresses require some amplification, as both occur commonly and steps must be taken to prevent consequent deterioration of metals.

Table 8.5   Situations in which electrolyte corrosion of a single metal may occur

Cause	Anode	Examples	Remedy
Grain structure of metals	Grain boundary	Any steel component subject to dampness	Keep steel dry
Variations in concentration of electrolyte	Low concentration areas	All types of soil	Cathodic protection
Differential aeration of a metal surface	Oxygen-remote area	Improperly protected underground steel pipes	Cathodic protection
Dirt or scale	Dirty area (oxygen remote)	Exposure of some types of stainless steel to atmospheric dirt	Use more resistant quality or keep the surface clean
Stressed areas	Most heavily stressed region	Steel rivets	Protect from dampness

THE ROLE OF OXYGEN IN ELECTROLYTIC CORROSION

Oxygen plays a part in many corrosion processes but the case of steel is particularly important and, therefore, by way of example, the effects of oxygen on steel are discussed. If steel is immersed in fairly pure water, there is only a very slight chemical reaction. Rusting of steel is chiefly electrolytic, the anodes and the cathodes being different parts of the same piece of steel. The anodes may be stressed regions, for example, the head of a nail, or grain boundaries. Although the electrode potential of steel varies slightly from place to place on this basis, corrosion of the metal in pure water is very slow. Ions form at anodes:

$$Fe \quad \rightarrow \quad Fe^{2+} \quad + \quad 2e^-$$
metal   ferrous ion   electrons remain
     in solution   on metal

in the water:

$$H_2O \rightarrow H^+ + OH^-$$

at the cathode:

$$2H^+ \quad + \quad 2e^- \quad \rightarrow H_2\uparrow$$
from the   electrons   gas
  water   from cathodes

and in the water:

$$Fe^{2+} + 2(OH)^- \rightarrow \quad Fe(OH)_2\downarrow$$
ferrous hydroxide
precipitate (green)

Since water is only slightly ionised, the reaction is very slow and a steel nail immersed in boiled or distilled water may not show any visible corrosion product for some time.

Steel corrodes quite rapidly in ordinary water, however, and this is due to the presence of oxygen which reacts with electrons from cathodes to form the hydroxyl ions required by the above equation, to cause rusting:

$$2H_2O + O_2 + \quad 4e^- \quad \rightarrow \quad 4(OH)^-$$

from   hydroxyl

cathode   ions

Iron exhibits two valencies – of two and three – and rust is, in fact, ferric hydroxide, $Fe(OH)_3$, obtained when ferrous hydroxide is further oxidised by air, although the product will be black – anhydrous magnetite, Fe_3O_4 – if the air supply is limited, as in many closed heating circuits. (It follows that changing the water in central heating circuits will, initially, increase the corrosion rate, since a new supply of oxygen is injected with fresh water.)

When plenty of water is present, corrosion products form in solution rather than at anodes or cathodes and the exact position depends on the diffusion rates of the positive and negative ions. In the case of steel, Fe^{2+} ions are smaller and, therefore, more mobile than $(OH)^-$ ions, so that they meet near the cathode. Hence, unlike atmospheric corrosion, the corrosion product may be incapable of protecting the metal, since it does not form on the corroding part, the anode. Again, in a central heating system, the corrosion product tends to form in the water and offers no protection to remaining metal.

PITTING CORROSION

This occurs because ferric oxide (mill scale) itself behaves cathodically with respect to iron which may be exposed by a scratch in the oxide coating, so that the steel continues to corrode, causing cavities, whilst rust builds up on cathodes until it covers the entire surface (Figure 8.25). Another factor contributing to pitting is the

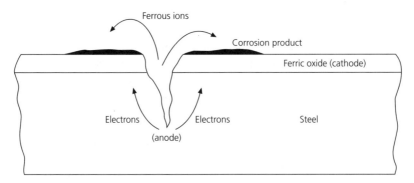

Figure 8.25 Pitting of steel originating from a crack in the oxide coating

fact that the anode, being normally much smaller in area than the cathode, corrodes at a faster rate and perforation may render components useless when the total metal loss may be as little as 5 per cent of the original (even so, the total loss of weight caused by pitting will be less than that caused by uniform corrosion of the surface under similar conditions, since the small anodic area acts as a 'bottle neck' in the corrosion process). Careful cleaning of metal components prior to application of surface coatings discourages pitting.

DIFFERENTIAL AERATION

This may occur when different parts of the same piece of metal are exposed in different degrees to oxygen. Oxygenated areas tend to form cathodes due to the production of hydroxyl ions, as explained above. Steel posts, for example, tend to corrode just below ground level, the corrosion cell consisting of an anode just below the ground where the oxygen level is relatively low and a cathode just above the ground where the oxygen level is higher. Underground steel pipes are sometimes seriously corroded by the action of bacteria in anaerobic (that is, oxygen-free) soils such as clays, which contain sulphates and organic matter. The bacteria cause the sulphates to react with and remove the hydrogen which would normally prevent corrosion by causing polarisation at cathodes.

EFFECT OF PH VALUES

Yet another factor affecting corrosion of metals is the pH value of the mixture with which they are in contact. Just as a reaction involving electrons such as:

$$Fe \rightarrow Fe^{2+} + 2e^-$$

depends on the e.m.f. of the metal with respect to solution, so a reaction such as:

$$Fe^{2+} + 2(OH)^- \rightarrow Fe(OH)_2$$

will be dependent on pH, since it is affected by the balance of hydroxyl ions in solution. The equation:

$$2H_2O + O_2 + 4e^- \rightarrow 4(OH)^-$$

will be affected both by the electrode potential of the metal and the pH of the solution. Diagrams relating the relative stability of different states of a metal to their electric potential and the pH of the solution are known as Pourbaix diagrams. A simplified form of that for iron is shown in Figure 8.26.

The boundary lines between Fe and Fe^{2+} ions and between Fe^{2+} and Fe^{3+} ions are horizontal since, as explained above, electrons only are involved in the change: pH has no affect. Note, however, that at higher pH values – that is, alkaline environments – ferric oxide, Fe_2O_3, is, within a range of voltages, more stable than the metal and forms a protective coating on it. This process is known as passivation and it prevents electrolytic corrosion.

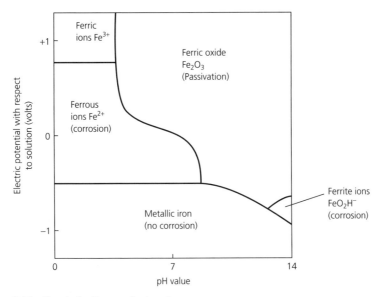

Figure 8.26 Pourbaix diagram for iron/water

The passivating effect of alkaline environments in steel is responsible for the protection of steel in reinforced concrete and is the basis of some types of corrosion inhibitor for boiler systems. The effects of alkaline and acidic environments on steel are also well illustrated by the fact that a bricklayer's trowel, left uncleaned, will not corrode for some time, whereas a plasterer's float, if left in this way, will develop a layer of rust in a few hours, due to the acidic solution resulting from gypsum plasters. Corrosion of steel, similarly, is greatly accelerated if saline solutions are present, as in sea water or in concrete containing chlorides.

Although Pourbaix diagrams are extremely useful in showing how metals react to varying conditions, they should not be taken as a final indication of their corrosion properties, since factors already mentioned, such as polarisation, the nature of the corrosive environment and impurities in the metal also play a part.

POLARISATION DIAGRAMS

The driving force (electromotive force, e.m.f.) behind corrosion currents is the difference between electrode potentials based on Table 8.3 for the two metals concerned but in addition to this the actual current flowing (and therefore the corrosion damage) depends on the sum of the following resistances:

1 resistance at the cathode surface
2 resistance at the anode surface
3 resistance of the corrosion medium

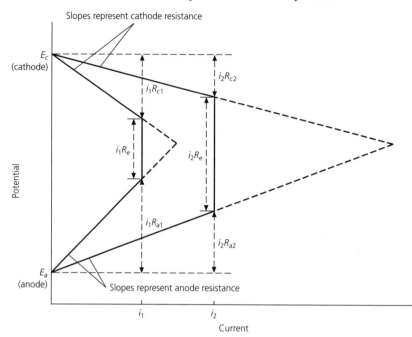

Figure 8.27   Polarisation diagram representing the parts played by anode, cathode and electrolyte resistance

Resistances (1) and (2) will be increased by

■ any corrosion products deposited on the surface
■ bubbles of gas if formed – for example hydrogen at cathodes
■ surface coatings such as paint or oxides
■ small anode or cathode – though as stated small anodes lead to rapid localised attack

Resistance (3) will be increased if

■ salt contents in solution are small
■ anode and cathode are well separated
■ the temperature is low – in freezing conditions corrosion would be minimal

The situation may be represented as in Figure 8.27. E_a and E_c represent the electrode potentials of anode and cathode hence $E_a - E_c$ is the e.m.f. of the corrosion cell. The sloping lines represent the anode and cathode resistances (R_a and R_c) – as soon as current flows the voltage available to drive the current through the solution is reduced. A steep slope means high polarisation voltage loss resulting in a low corrosion current i_1R_e. The anode, cathode and electrolyte resistances are in proportion to the lengths of vertical dotted lines at current i_1. If the polarisation loss is small (gently sloping lines) more voltage is available to drive the current through the solution – hence corrosion is more rapid – current i_2.

In Figure 8.27 the electrolyte resistance has been maintained at the same value (R_e) so the potential fall across the electrolyte increases as the current rises from i_1 to i_2.

Clearly all resistances should be maximised to reduce corrosion rates and in particular the larger the surface voltage drop at either or both electrodes, the lower the total corrosion rate will be. Protection systems achieve this end by various means.

STRESS CORROSION CRACKING

Some alloys, when under a tensile stress and in corrosive conditions simultaneously, have exhibited cracking, leading to sudden failure. This may occur in certain environments well below the yield point of the metal and may have disastrous consequences where considerable loads or pressures are sustained.

Stress corrosion may be caused either because a tensile stress at a point assists dissolution of metal, effectively altering its electrode potential, or simply due to corrosion of parts of a metal which would otherwise prevent spreading of a crack. The latter occurs in grain boundaries which would normally bond grains effectively together but which, in corrosive environments, become anodic, producing hairline cracks.

Stress corrosion may be regarded in some cases as an extension of pitting, as occurs in steel. As a pit becomes deeper, the stress at the root becomes higher, causing increased corrosion at this point, so that a crack propagates through the metal. This type of cracking has led to explosions of riveted boilers. Seepage of treated boiler water past rivets, and subsequent evaporation, leads to concentration of alkali around them (the alkali is used in water treatments for passivation of the metal). As the Pourbaix diagram indicates (Figure 8.26), large concentrations of alkali which produce high pH values may lead to local corrosion of the steel rivets, causing pitting and eventual failure as above. Riveted connections are particularly vulnerable to this type of attack, since some are inevitably stressed more than others, but stress corrosion may also occur in welded connections, which are more common today, if the process results in residual stresses in the metal. Chloride solutions increase the severity of such corrosion, although its incidence can be avoided by annealing after welding.

ALLOY CORROSION

Some alloys tend to exhibit corrosion due to electrochemical differences between the grain boundaries and the grains themselves, particularly if grain boundaries contain a different phase, as in the case of annealed steels. In certain stainless steels, in particular, chromium carbide forms at grain boundaries when it is heated, as in welding, so that, in certain environments, electrolytic corrosion will occur with chromium carbide as cathodes and neighbouring chromium depleted areas as anodes. This leads to cracking and eventual fracture under stress. High strength aluminium alloys are similarly affected.

8.6 PROTECTION OF METALS AGAINST CORROSION

Wherever possible, the choice of metal for a given situation should be such that electrolytic corrosion does not arise, since effective protection methods involve, firstly, the initial cost of treatment and, in many cases, the subsequent cost of maintenance.

It will be appreciated that all electrolytic and acidic corrosion processes require the presence of water which forms the electrolyte. In design terms therefore, steps should be taken to exclude water completely if possible, but at the very least ensure that water is not allowed to collect or become trapped by capillarity adjacent to metal surfaces. Some simple illustrations are:

- When steel components, or their connections, result in the possibility of water ponding, provide effective drain holes and take steps to ensure that blocking of such holes by debris is prevented.
- Where metals are joined by overlapping, apply a non-hardening mastic to the joint before assembly so that water cannot penetrate. Putty is not really suitable for this function since it embrittles with age and can absorb very little movement.
- Where access to metals is difficult, tapes impregnated with non-hardening mastics can be used to build up substantial coatings immediately, obviating the need to wait for successive coats of paint to dry.

In damp atmospheres, metals which form impervious corrosion films are most suitable and the use of different metals in contact in these situations should, if possible, be avoided. In particular, small anode areas will corrode rapidly when in contact with large cathodes, for example, copper sheet fixed with steel nails. Extra care is necessary in polluted atmospheres or corrosive environments which will accelerate corrosion by reducing electrolyte resistance. A number of metals are very prone to certain types of chemicals in solution; these are described under the headings of individual metals. The task of protection of metals in corrosive situations may be tackled from several standpoints; a metal may be protected by:

- coating with an impervious coating
- reversal of the corrosion reaction making it into a cathode (cathodic protection)
- by passivation or by means of inhibitors

PROTECTION BY IMPERVIOUS COATING

This is a very common method of protection of metals, since it is cheap and, in many cases, the only feasible method, for example, protection of structural steels, though some types of paint possess an inhibitive property in addition to their waterproofing qualities. The provision of a waterproof coating on metals depends on a number of factors relating to the paint and its application. The most important are outlined below:

- The best time for painting is immediately after manufacture, before water has had a chance to penetrate into seams or cavities.
- The film must be unbroken for effective protection, unless it is an inhibitive film. For example, although structural steel is often protected by a prefabrication primer, this will be damaged during fabrication and should be made good afterwards. Otherwise, corrosion will occur by differential aeration at discontinuities in the protective coating.
- The surface of the metal must be suitably prepared. For example, mill scale should be removed from steel before priming, since it has a different coefficient of thermal movement from steel and will tend to flake off in time. In factory processes, scale is normally removed by acids, as in pickling or phosphating. These methods are, however, not suitable for site application: shot- or grit-blasting is more effective. Wire-brushing achieves little if substantial thicknesses of rust are present.
- Since no paint is perfectly waterproof, the protection afforded depends on the thickness of the film. Sharp edges often protrude through a single coat of paint, due to its surface-tension effect, so that if such edges cannot be rounded off before painting, extra coats of paint should be applied at these points. Thixotropic versions of paints have the advantage that greater thickness can be obtained. For the same reason, polyvinyl chloride (PVC) coatings, factory applied, give excellent protection, provided they are not punctured, and are now commonly used for protecting a wide variety of components. A number of metals are often applied to steel to give impermeable coatings. Examples are tin, lead, nickel and aluminium. The latter is especially useful for structural steel, since it can be sprayed *in situ* after erection.

CATHODIC PROTECTION

If the electric potential of a metal in aqueous solution can be maintained at its equilibrium value, then corrosion will not take place since, as explained, the metal only ionises in order to try to establish this potential. If the electric potential is held by some means at a value more negative than the electrode potential, then the metal will behave cathodically and, although hydrogen may be generated from solution by the excess electrons in the metal, corrosion cannot occur. This is known as cathodic protection and is illustrated by the lower part of the Pourbaix diagram in Figure 8.26. It would be misleading to suggest that the exact value of the standard electrode potential of the metal must be maintained for full protection. If, for example, in the case of steel pipes in hard waters, a protective 'scale' of a magnesium or calcium salt forms, protection will be achieved at a voltage which would, in normal conditions, cause the steel to behave anodically. On the other hand, the concentration of ions in solution may be reduced as, for example, when there is flowing water. This constant diffusion of ions away from the metal will increase the negative potential needed to be applied to prevent corrosion. The corrosion rate is, in all cases, proportional to the difference between the actual potential of the metal and its effective electrode potential, under the given conditions.

In order to protect a metal from corrosion, electric current must be so supplied to it that the draining away of electrons from it is counterbalanced. Conventional current will, therefore, be away from the metal to be protected. This current may be provided by means of a second cell, either in the form of another more anodic metal, known as a sacrificial anode, or by a direct current source connected into the corrosion cell.

Sacrificial anodes

The use of these in protection of steel is very widespread. To be effective, the metals used must be higher in the electrochemical series than iron. It might be considered that protection would automatically be achieved with any such metal but this is not the case since, initially, the corroding metal will be some margin below its standard electrode potential and the sacrificial anode must be of sufficiently negative potential to increase the negative potential of the corroding metal by this margin. The impressed e.m.f. required depends on the resistance of the solution. If it is low, then a metal near to iron in the table will be satisfactory, for example, zinc. High resistance solutions will require a metal higher in the series, such as magnesium, if the steel is to be fully protected. The situation is illustrated schematically in Figure 8.28.

The metal which was formerly corroding now behaves as a cathode with respect to the sacrificial anode, which must therefore be near to the protected metal. Large areas of steel will need a number of sacrificial anodes, so positioned that all parts are maintained at the requisite negative voltage with respect to solution. The ideal form of sacrificial anode is, therefore, a uniform and continuous coating, zinc being most commonly used for the purpose (Figure 8.29), since it behaves anodically to steel but has reasonable durability in normal atmospheres. The following coatings offer protection to steel in this way.

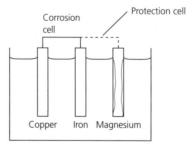

Figure 8.28 Cathodic protection using a sacrificial anode. The protection cell must be sufficiently active to restore iron to its standard electrode potential in spite of the influence of the copper

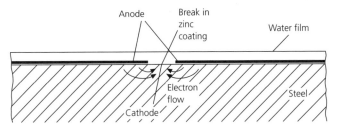

Figure 8.29 Protection provided by galvanising at breaks in the zinc coating

Hot-dip galvanising

Steel is prepared by degreasing, pickling and flux application, the latter ensuring metal-to-metal contact between steel and zinc. It is important that crevices – as for example caused by overlapping sheets – are avoided as they would trap acid. The component is then dipped in molten zinc at a temperature of about 450 °C. Initially zinc/iron alloys form on the surface of the steel but these are covered, on continued immersion, by a layer of pure zinc.

The thickness obtained depends on the steel composition, particularly the silicon content, and the immersion time. Values chosen in practice range from 50 μm to 200 μm according to application, 100 μm being a common thickness. Most mechanical properties should not be significantly affected though a work hardened steel may be softened by the process. Residual weld stresses can be reduced by galvanising, improving strength, though some distortion may occur. Thicker galvanised coatings may crack at sharp bends. The approximate life of a 100 μm thick coating varies as follows:

indoor atmosphere	infinite
inland, rural atmosphere	100 years
city atmosphere	50 years
industrial atmosphere	25 years
fresh water (hard)	30 years
fresh water (soft)	10 years
sea water	8 years

The life times of coatings of other thicknesses will be roughly in proportion to thickness. Components are often used without painting, the bright zinc surface weathering to dull grey. Paint can, however, be applied (with suitable surface preparation) and can add to protection in very aggressive environments as well as providing choice for the final colour. Since paint and galvanising protect each other, the life of a painted galvanised surface may be twice that of the sum of their lives if used individually.

Zinc spraying

In this process the zinc is not chemically bonded to the steel, so the prepared surface must be roughened to give mechanical grip. Molten zinc is atomised and projected at the steel surface, forming a slightly rough layer, which is porous initially, becoming impervious when the zinc reacts with moisture. Alternatively sealers may be used into which pigments or metallic aluminium may be incorporated. Since the steel is not heated there is no distortion. The method is more expensive than galvanising but can be used on large fabricated structures and thicknesses of over 100 μm of zinc can be obtained, giving excellent durability.

Electroplating

This produces an even zinc coating of about 30 μm thickness, the process being limited to articles which are small enough to fit into the bath. The process is also used to plate wire or sheet, since the coating has good flexibility.

Sherardizing

Prepared articles are tumbled in fine zinc dust just below its melting point – at about 380 °C. A good bond forms, the zinc having a thickness of about 30 μm which, in the case of nuts and bolts, for example, does not impair the action of the thread. Durability is clearly less than that obtained from greater thicknesses.

Zinc-dust paints

These comprise finely divided zinc particles in either an organic or an inorganic drying medium. The degree of sacrificial protection depends on the electrical conductivity of the hardened film and generally requires a large proportion of metallic zinc. These coatings are often used as prefabrication primers, though adequate surface preparation such as grit-blasting is essential to provide a clean, slightly rough surface to grip the paint.

Impressed current method

Protection of underground steel pipes may be successfully achieved by this method. By connection of the negative terminal of a direct current source to the steel, the positive terminal being connected to a further electrode in the soil, electrons are supplied to the steel, maintaining it at a sufficiently negative voltage to avoid corrosion (Figure 8.30). As with sacrificial anodes, the supply of electrons at any one part of the pipe depends on a completed cell being formed at that point, with the DC source. Hence, due to soil resistance, electrodes must be positioned at intervals along the pipe, the actual interval depending on the resistance of the soil. The current consumed will also depend on the soil resistance, being higher in more corrosive conditions. The other electrode will be at a positive potential and will, therefore, corrode unless made of a noble metal. Steel electrodes, for example, will require replacement, whereas more expensive platinum or carbon electrodes would be corrosion resistant. Impressed currents used for protection of underground steel pipes may adversely affect neighbouring buried metals. Also, direct currents in the

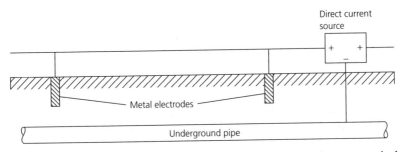

Figure 8.30 Protection of an underground steel pipe by the impressed current method

ground in urban areas may interfere with the pipes themselves; such currents were at one time produced by electric railway power supplies. Alternating currents are now more common but even these may adversely affect some buried metals, such as aluminium, the oxide of which has a rectifying action. Neither the above impressed current nor the sacrificial anode method is intended to obviate the need for painting of underground pipes. Complete protection by these methods alone would be very costly. They are merely intended to supplement the protection given by traditional waterproof coatings, such as bitumen, which may not be 100 per cent effective due to, for example, incomplete coverage or damage during laying of the pipe. (See also 'Cathodic protection of steel in concrete').

PASSIVATION AND USE OF INHIBITORS

Passivation is based on Pourbaix diagrams. If the pH value of the environment in which the metal is to be used, and the potential of the metal, are such that it is covered by a stable, impervious corrosion film, then corrosion will cease. For example, steel in an alkaline environment is passivated, though aggressive ions, such as those of chlorine, will destroy the protective film, allowing corrosion to continue. Hence, steel in uncarbonated regions of reinforced or prestressed concrete is protected, though the use of calcium chloride in such concretes is not recommended. By charging positively, steel can be passivated even in acidic environments (Figure 8.26), since ferric oxide is formed. The method relies, however, on careful control of the applied voltage, which depends, in turn, on the concentration of the acid concerned. The method is used for the protection of steel storage vessels. It is known as anodic protection.

Inhibition (anodic–cathodic)

Inhibitors are added to corrosive media to resist corrosion of metals immersed in them. They may retard either the anodic or cathodic reactions of a system. For example, if steel anodes can be coated with a layer of ferric oxide, Fe_2O_3, corrosion will not be possible. This requires an oxidising agent, for example, sodium chromate, and a slightly alkaline environment, to ensure that the ferric oxide is stable. Sodium chromate will work in a de-aerated solution but other inhibitors, such as sodium carbonate or sodium silicate, require an external supply of oxygen. Hence, these inhibitors would not be effective in, for example, boiler circuits where oxygen contents are low. The above might seem contradictory to earlier statements about the accelerating affect of oxygen on corrosion. The accelerating effect is still theoretically present but, in this case, the oxygen in combined form with the metal produces a situation in which corrosion cannot occur, in spite of the fact that the oxygen in solution behaves as a depolariser. The Pourbaix diagram for iron shows that oxide coatings are not stable at low pH values so that other methods must, in these cases, be used.

An example of inhibition by the interference of salts with cathodic products occurs in hard natural water. Calcium carbonate and magnesium hydroxide, which form a scale on the metal, prevent the penetration of oxygen to act as a depolariser. Unfortunately, such scales tend to accumulate, increasing the frictional resistance of pipes and adversely affecting their heat conduction properties. Other inhibitors, such as polyphosphates, retard cathodic and anodic reactions, though, as in the case of all inhibitors, the presence of aggressive ions such as nitrates, sulphates and chlorides, tends to reduce their efficiency. Soft waters, which often contain carbon dioxide dissolved to form carbonic acid, may be improved by addition of alkalis such as sodium carbonate (washing soda), which will precipitate calcium carbonate from solution. Excess use of alkalis is, however, not recommended; it can lead to localised corrosion cells.

8.7   IRON AND STEEL

As a metal forming element iron is second only to aluminium in abundance. It would be difficult to imagine how modern construction could exist without steel, an alloy of iron and carbon. Its use is almost universal in virtually all forms of construction, whether in steel frames, reinforcement or prestressing for concrete, fixings for timber frame construction, general fixings or simply sheet metal for claddings. Use of steel is central to the construction industry and there is no reason to suggest the situation might change in the future. Apart from its abundance steel has a number of other very important attributes:

- It is relatively easy to manufacture, the temperatures required being within reach of ordinary furnaces.
- Steel combines high working stresses and high stiffness at relatively low cost.
- Its dual crystal structure (BCC at room temperature and FCC at elevated temperatures) permits relatively easy hot working.
- Its dual crystal structure offers benefits in heat treatment terms.
- Steel alloys readily with many other elements.
- Large sections do not deteriorate rapidly by corrosion, so that substantial proportions can be recycled.
- Its magnetic properties assist handling and sorting from other materials.

Production of pig iron from its ore

Iron, like many metals, occurs in the form of an ore, generally iron oxide, together
with earthy material, such as silica and alumina. The first stage in the production
of iron involves the blast furnace, in which the iron oxide is reduced to iron by
carbon monoxide produced from ignition of coal and limestone, combustible gases
providing the necessary heat. The impurities and calcium oxide from the limestone
collect as a slag on top of the liquid iron, which is run off. The iron itself is run
into moulds forming *pigs* which contain between 2 and 4 per cent carbon and
quantities of silicon, sulphur, phosphorous and manganese.

Products based on iron

The impurities above have a marked effect on the microstructure of iron and,
by modification of pig iron to different degrees, materials are produced having,
apparently, quite different properties.

Iron is allotropic – it exists in several different crystal forms.

■ Up to a temperature of 910 °C, iron is body-centred cubic (BCC) and dissolves
very little carbon – less that 0.01 per cent at room temperature, increasing to
0.04 per cent at 700 °C. (This relatively low solubility in a crystal structure,
which is not as closely packed as FCC structures, is due to the fact that the
FCC crystal nevertheless supplies the largest interstitial sites – the centre of each
cube.) BCC iron with its small percentage of carbon is called alpha iron or ferrite
and is soft, ductile and readily cold worked.
■ At 910 °C, the BCC crystal structure changes to FCC and this lattice can
accommodate up to 1.7 per cent carbon by weight interstitially (or 9 carbon
atoms per 100 iron atoms), the material being known as gamma iron or austenite.
■ At approximately 1390 °C and before melting, the crystal structure reverts to
BCC. Pure iron melts at 1537 °C.

Liquid iron can dissolve substantial quantities of carbon, since it is non-
crystalline, but a eutectic point occurs at a temperature of 1130 °C, with a carbon
composition of 4.3 per cent.

The various forms of iron–carbon alloy are based on the above phases and their
properties depend greatly on the quantity of carbon in the original molten metal.
They are considered in order of increasing carbon content.

Wrought iron

This is the purest form of iron, containing about 0.02 per cent carbon, which is
almost completely dissolved interstitially in the iron lattice. The metal is tough
and ductile with a yield point of about 210 N/mm^2 and a tensile strength of about
350 N/mm^2. It is produced from pig iron by melting in a reverberatory furnace,
in which impurities are oxidised into slag. As the iron becomes purer, its melting
point rises and it therefore becomes 'pasty'. Balls of iron are withdrawn, and the

slag, which is more fluid than the iron, is forced out by steam hammering. The iron is then rolled into bars or the process repeated to obtain greater purity.

Wrought iron was formerly used for structural members subject to tensile stresses (compression members being made from cast iron). It is now produced on a very limited scale but is still occasionally used for ornamental ironwork, chains and hooks, since it has admirable working properties, and great toughness and corrosion resistance. More general use of the metal has, however, given way to steel, since the latter is much cheaper to produce.

Steel

Steel may be defined as an alloy of iron and carbon, in which the carbon is fully dissolved in the high temperature FCC gamma iron form. Steels may, therefore, contain up to 1.7 per cent carbon. Austenitic steel, as it is known, is, at this high temperature, soft, ductile and suitable for forging and rolling – commonly referred to as *hot working*. Heavy steel sections are, therefore, invariably heated to temperatures over 910 °C for forming purposes.

It will be clear that, on cooling, carbon must be rejected from the crystals as they revert to the BCC form. Pure carbon does not, however, form on cooling but rather iron carbide, Fe_3C, which contains 6.67 per cent carbon and is known as *cementite*. Cementite has a different crystal structure from ferrite, with properties more like a ceramic than a metal, since the high percentage of carbon atoms (6.67 per cent by weight corresponds to 25 per cent numerically) increases strength and prevents plastic deformation. The form in which the cementite is produced depends on the carbon content of the steel and the formation of the various compounds is best understood by reference to iron–carbon equilibrium diagrams.

Figure 8.31 shows the portion of the iron–carbon equilibrium diagram which is concerned with the various types of steel formation. When austenitic steel cools slowly, so that alpha iron is produced, the resulting steel must contain suitably balanced proportions of ferrite and cementite, and the mechanism of changes is rather similar to that by which a two phase alloy solidifies from a liquid mixture:

1 If the carbon content corresponds to the area to the left of line A, pure ferrite is formed, crystals growing gradually from the former austenite crystals, as the temperature is decreased from 910 °C, until a completely new crystal structure is produced (Figure 8.32(a)).
2 If the carbon content corresponds to the area between lines A and B, ferrite crystals form initially, as above, but since the solubility of carbon decreases with reducing temperature, cementite crystals must begin to form, occupying a small space in grain boundaries.
3 If the carbon content corresponds to the area between lines B and C, ferrite grains begin to form at 910 °C, carbon being rejected into the remaining austenite. The austenite composition follows the upper critical line, while that of the ferrite follows the lower critical line. Figure 8.33 follows stages in the ferrite crystal growth. At 723 °C the austenite contains 0.83 per cent carbon and

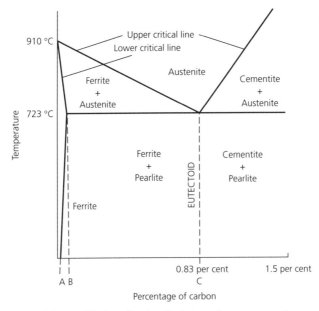

Figure 8.31   Part of the equilibrium diagram for iron-carbon compounds

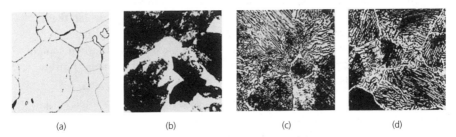

(a)   (b)   (c)   (d)

Figure 8.32   Magnified photographs of polished and etched sections of steel having various carbon contents: (a) Carbon content almost zero. Grain boundaries are clearly visible. (b) Carbon content 0.5 per cent. Ferrite grains are light and pearlite grains dark. (c) Carbon content 0.83 per cent, eutectoid steel. In this case, the ferrite/cementite layers are clearly visible. (d) Carbon content 1.2 per cent, hypo-eutectoid steel. Cementite appears as light-coloured layers around the pearlite grain boundaries

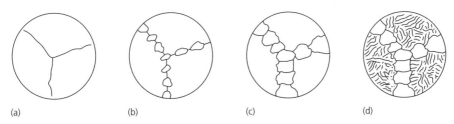

(a)   (b)   (c)   (d)

Figure 8.33   Stages in the cooling of a hypo-eutectoid steel: (a) over 910 °C austenite grains; (b) 850 °C ferrite grains begin to grow at boundaries; (c) 800 °C ferrite grains continue to grow; (d) 723 °C remaining austenite changes to pearlite

cannot cool further without changing its crystal structure. At this point, known as the eutectoid (as distinct from eutectic, which would be used for a liquid/solid equilibrium diagram), an intimately mixed, layered combination of ferrite and cementite will form, this combination being known as *pearlite* since, under the microscope, polished and etched specimens may have a pearly appearance. Steels containing a carbon content in this range are known as *hypo-eutectoid steels* and consist of a mixture of ferrite crystals grown at the higher temperature and pearlite crystals formed at 723 °C. Steels with a carbon content at the lower end of this range, for example, 0.1 per cent carbon, would comprise mainly ferrite. In Figure 8.32(b), which shows a section of steel with a 0.5 per cent carbon content, there is a slight preponderance of pearlite (the darker material).

4 If a steel containing 0.83 per cent carbon is cooled, no crystal change occurs until the temperature of 723 °C is reached and then a transformation to pure pearlite occurs, resulting in a microscopic structure, as shown in Figure 8.32(c).

5 When austenite of carbon content above 0.83 per cent (hyper-eutectoid) is cooled, cementite begins to separate from austenite grains at a temperature indicated by the upper critical line of Figure 8.31, the carbon content of remaining austenite being reduced progressively until, at 723 °C, the eutectoid composition of 0.83 per cent is reached. Since cementite contains 6.67 per cent carbon and it is rare for steels to contain much over 1 per cent carbon, these cementite layers are usually quite thin. At 723 °C, the austenite changes into pearlite, as previously described. Figure 8.32(d) shows a steel of carbon content approximately 1.2 per cent, the cementite layers being visible around pearlite grains. Figure 8.34 shows the stages in the process.

The above transformations will only take place as stated if cooling is very slow, allowing diffusion of carbon from within the austenite crystals to form ferrite and cementite products. The mechanical properties of ferrite are similar to those of pure iron, while cementite is intensely hard but brittle. Pearlite, a mixture of these two in the proportions 88 per cent ferrite and 12 per cent cementite, is, as would be expected, intermediate in properties. The material is much harder than ferrite but less hard than cementite, plastic deformation being possible along the ferrite layers. Pearlite is stronger than ferrite but weaker than cementite.

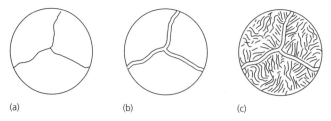

(a)   (b)   (c)

Figure 8.34   Stages in the cooling of a hyper-eutectoid steel: (a) over 900 °C austenite grains; (b) 800 °C cementite layers forming at grain boundaries; (c) 723 °C remaining austenite changes to pearlite

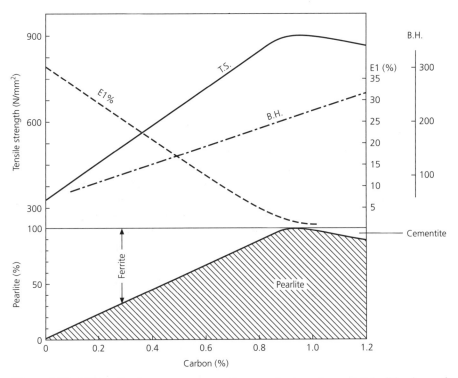

Figure 8.35 Effect of carbon content on tensile strength, elongation and Brinell hardness of annealed steel (source: Rollason, *Metallurgy for Engineers*, Butterworth-Heinemann)

Steels may be divided in respect of carbon content into:

- Low carbon steels, containing up to 0.15 per cent carbon.
- Mild steels, containing 0.15–0.25 per cent carbon.
- Medium carbon steels, containing 0.2–0.5 per cent carbon.
- High carbon steels, containing 0.5–1.4 per cent carbon.

It will be appreciated that the major factor affecting the properties of steels is their carbon content. The situation is represented in Figure 8.35 which shows how elongation at failure, ultimate tensile strength and Brinell hardness of annealed steels depend on their carbon content. Low carbon and mild steels exhibit excellent forming properties, are easily welded and have moderate strength. These steels are most widely used in the construction industry. The tensile strength of medium carbon steels is higher than that of mild steel, and, equally important, such steels benefit to a greater extent from the heat treatment processes described later. Impact resistance, weldability and ductility are, however, adversely affected. High carbon steels, which include the eutectoid composition, are highly susceptible to heat treatment and may be hardened to a very high degree. Applications are, however, limited by the very low elongation at failure of such steels.

8.8   STEEL PROPERTIES

Part of the attraction of steel is its enormous versatility, though this derives from the complexity of the metallurgy of steel. The properties are affected by a range of factors as follows.

METHOD OF FORMING THE METAL

The basic grain structure of a steel depends on the history of the product while in the austenitic state. Within the austenitic range of temperatures, thermal energy is high and the crystals tend to grow continuously, smaller crystals merging into larger ones, so that products maintained in the austenitic state for some time tend to have a coarse-grain structure, leading to an unsatisfactory structure at lower temperatures. A fine-grain structure is generally preferable, since it results in increased yield strength, toughness and ease of heat treatment. There are, of course, practical limitations to the size of grain which can be obtained. Ingots allowed to cool in air will tend to have a finer grain structure in external layers than internal layers which cool more slowly. Hot rolling breaks down the grains, destroying the grain structure of the original ingot and resulting in a more even size distribution. Recrystallisation takes place continuously during hot rolling. Cold rolling distorts or breaks down the grains and leaves the structure in a stressed, though harder and stronger state, since cold rolling is carried out at a temperature which is too low to allow recrystallisation to occur. Larger steel components, for example, steel joists, would be very difficult to cold roll because the metal is too hard, hence, they are hot rolled. Smaller products, for example, wire, can be cold worked or drawn, however, and this process results in smaller grain size and higher strength.

EFFECT OF COOLING RATE

Treatment by heat may produce a change of phase, crystal redistribution or passage of a constituent into, or out of, solid solution. The phase diagram given in Figure 8.31 only applies to steels cooled so slowly that carbon is able to diffuse from austenite grains to form equilibrium products. When the cooling rate is faster than this, diffusion to equilibrium states cannot complete and a frozen condition of non-equilibrium is obtained. In order of decreasing rapidity, the main processes are:

- quenching (cooling by immersion in water, oil or iced brine)
- normalising (controlled cooling in still air)
- annealing, in which, by slow cooling, steel is brought more or less to equilibrium conditions

Quenching/tempering

Quenching of a medium or high carbon steel completely prevents diffusion of carbon atoms so that they are held in an unstable condition in a body-centred

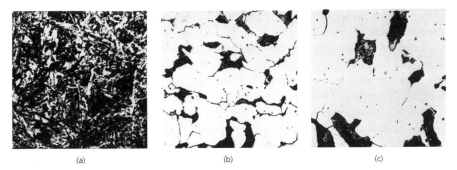

(a)   (b)   (c)

Figure 8.36 The effect of heat treatment on a 0.13 per cent carbon steel: (a) Martensite, obtained by quenching; (b) Fine-grain structure, obtained by normalising; (c) Coarse-grain structure, obtained by annealing

tetragonal lattice (Figure 8.36(a)). Quenching in iced brine or water is particularly drastic, since very high heat removal rates are obtained by vaporisation of water at the metal surface. In some cases, distortion or cracking of the article may result. The resultant material, which is called *martensite*, is intensely hard and brittle, and internal stresses are so high that further heat treatment is invariably carried out to improve toughness. The process is known as *tempering* and involves heating to a temperature which is sufficient to allow some diffusion of carbon to form small cementite crystals. Temperatures of about 200–400 °C are normally used; higher temperatures cause greater loss of tensile strength and hardness but increased toughness, as increased quantities of carbon are rejected and ferrite begins to form. This structure is known as a *sorbitic* structure. The degree of tempering is often measured by the temper colour resulting from the oxide layer, as explained under 'Oxidation'. Colours change from straw (at approximately 230 °C), to brown, then purple and finally blue (at approximately 300 °C). The effectiveness of quenching depends upon the physical size of the component. Products of small section such as wires can be cooled very quickly because they have little thermal inertia. Quenching of a rolled steel joist would be ineffective because surrounding water would simply turn to steam and the cooling rate achieved would be very low.

Normalising

This involves heating the steel to a temperature slightly above the upper critical line and 'soaking' for a time to allow small austenite grains to form. This is followed by cooling in air, resulting in a fine ferrite–pearlite grain structure in hypo-eutectoid steels (Figure 8.36(b)). Such steels are slightly harder and stronger than annealed steels, though heavy sections cannot be normalised so easily because cooling rates will be correspondingly lower.

Annealing

This relieves internal stresses resulting from cold working, forging or welding, and may serve to refine an excessively coarse-grain structure. It involves heating the steel as in normalising so that austenite grains form. The essential part of annealing is that cooling is very slow – often carried out in the furnace so that equilibrium products form, as given in Figure 8.31. Such steels will exhibit higher ductility and are easier to machine, though they have lower yield strength than normalised steels. It is quite possible to over-heat during annealing, leading to austenite grains of excessive size such that, on cooling, ferrite crystals form within austenite grains as well as at boundaries (*Widmanstatten structure*), leading to weakness and brittleness. Conversely, in under-annealing, austenite may only partially form, some areas remaining unrefined – this may occur, for example, near to welds. Figure 8.36(c) shows a section of an annealed steel.

In a further type of annealing, known as *subcritical annealing*, the steel is heated to about 600 °C, which is sufficient to allow regrowth of ferrite grains distorted by cold working. Subcritical annealing may be carried out several times during extensive cold working of steel products.

Time–temperature transformations (TTT diagrams)

While the above processes give a broad indication of the effect of cooling rate, the exact transformations to which steel samples are subject on cooling from the austenitic state are found to depend on the temperature at which they are finally held, even when relatively high finishing temperatures are used. Studies of these effects can be carried out by quenching thin disc shaped samples to a selected temperature and then holding at this temperature, quenching again after specified time intervals so that any unchanged austenite changes to martensite. Typical TTT or S curves so obtained are shown in Figure 8.37. Note that there is a start and

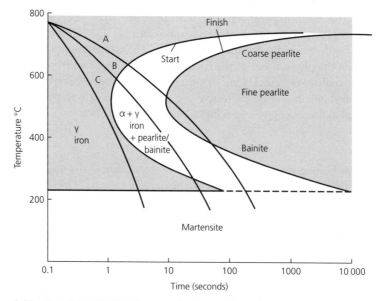

Figure 8.37   Principle of TTT diagrams for determining phase composition of steels at varying cooling rates

finish time at each temperature, the loci of these points being the curved S lines. It will be seen that holding a carbon steel at temperatures in the range 200–400 °C produces *bainite* (rather than pearlite) – a finely divided mixture of ferrite and cementite of feathery structure at high temperatures and needle-like structure at lower temperatures.

Once S curves are established for a given alloy, metallurgists can 'tailor' the cooling curve to obtain desired proportions of pearlite, bainite and martensite. Figure 8.37 shows three such curves, resulting in quite different final compositions for the same original austenite composition.

A – typical of normalising – will produce 100% pearlite.
B produces some pearlite/bainite together with some martensite.
C – typical of water quenching – produces 100% martensite.

It should be appreciated that in larger sections internal cooling rates will be slower than external cooling rates giving correspondingly different cooling curves and therefore differing final compositions.

EFFECT OF ALLOYING ELEMENTS

Steels may be classified as *plain carbon steels* or *alloy steels*, according to the quantities of alloying elements they contain, though even plain carbon steels often contain small quantities of other elements which have significant effects on their properties. The most common elements found in plain carbon steels are:

Sulphur and phosphorus   These both increase the brittleness of steels, so that very small percentages only are allowable (for example, 0.05 per cent).

Manganese   This may be used for deoxidation purposes, in which case it would be removed as a slag, but mild steels still benefit from its presence since it 'dissolves' in ferrite and refines the grain structure, increasing strength and hardness. High yield steels normally contain more manganese than mild steel, which usually contains about 0.5 per cent.

Silicon   This has an effect similar to that of manganese, only to a lesser degree.

Niobium   This element forms a carbide within the ferrite lattice, refining the grain structure and increasing the yield strength of steels. It also tends to increase the temperature of the ductile–brittle transition but the problem may be overcome by normalising, if high impact strength is required.

EFFECT OF DEGREE OF DEOXIDATION OF THE STEEL

Iron oxide forms on the surface of molten steel very quickly, due to its high temperature, and then combines with carbon in the steel, forming iron and carbon monoxide gas. The gas dissolves in the steel but tends to form bubbles on cooling, as its solubility decreases. To avoid porosity in the ingots, therefore, the evolution of gas must be controlled. The addition of manganese, silicon or aluminium prevents gas formation, a slag being formed instead. A steel in which no gas forms is known as a *killed steel*. Killed steels also have the advantage that added ingredients act as nuclei for austenite grain formation during cooling from the liquid. The fine-grain structure is preserved throughout the cooling process leading to high yield strength. Shrinkage in these ingots produces, however, a waste area at the top, so that killed steels are more expensive than the forms described below.

Semi-killed steels contain a proportion of gas which compensates for shrinkage and permits more efficient use of ingots. They are also known as *balanced steels*.

Rimming steels (non-deoxidised) contain large quantities of gas in the form of blow-holes. Although this might seem unsatisfactory, rimming steels have the advantage that the surface layers are usually very pure, since the gases sweep away impurities from the ingot surface, hence these steels are suitable for drawing and pressing. Working of the metal closes and welds the blow-holes, provided they are deep seated, so that they do not adversely affect mechanical properties.

8.9   STEEL FOR STRUCTURAL USES

Steel for structural uses may be divided into hot- or cold-rolled steel sections for structural steelwork, on the one hand, and bars or wire for reinforced and prestressed concrete, on the other.

STRUCTURAL STEEL SECTIONS

The 1990s have seen a marked upturn in the use of structural steel sections for the following reasons:

- There are increasing economic benefits from 'fast track' construction to which steel is admirably suited. Sections are manufactured and fabricated under factory conditions and can be rapidly assembled on site without expensive 'falsework'.
- Steel prices are competitive, due in part to international competition.
- Factory fabrication conditions permit the use of fine tolerances so that there should be fewer problems of 'fit'.
- Introduction of a wider range of section types together with new forming techniques has greatly increased the versatility of the material such that structural steels are now considered to be architectural materials in their own right.
- Steel is available in a number of structural forms which can be chosen to suit required applications. Examples are as follows.

Framing   Cold rolling of thin sections produces members of low mass and high rigidity. They are now being widely used for production of partitions, frames for domestic construction and 'podded' preformed assemblies such as bathrooms for large scale commercial applications. In many cases fixings can be easily made by self-tapping screws. Sections of this type should also be amenable to re-use (rather than recycling).

Small/medium spans   Standard hot rolled sections can be used as stanchions and beams with bolted connections. With bolted connections the use of high strength friction grip (HSFG) bolts permits load transfer of load by friction through steel washers rather than by shear on bolts. This permits holes to be somewhat oversized allowing increased flexibility while permitting full load transfer at each bolt. The process is further assisted by use of load indicating washers which permit tightening to the correct torque. Larger spans can be obtained by butt welded connections and use of fabricated sections in regions of higher stress.

Large spans   Three-dimensional space frame structures can can give clear spans of up to 200 m.

Specifications for steel sections

Structural steel sections have the following essential specification requirements:

Weldability   Welding is a simple fast way of achieving full structural transfer between members. Weldability requirements place compositional constraints upon the steel.

Impact performance and ductility   The resistance of steel structures to impact and earthquake damage is a major attribute. Steel structures tend to fail safely when overstressed. These properties are again achieved by appropriate composition and heat treatment processes.

Strength   The strength of steel should be the maximum which can be economically achieved within the constraints imposed as above. A number of strength grades are available for specific applications though most economic results are normally achieved by use of a small range of 'preferred' specifications.

The modulus of elasticity of steel does not vary greatly with its composition and heat treatment. The value of 210 kN/mm^2 is assumed for most purposes. Where higher strength types of steel are employed these would therefore operate at higher strains and in the case of beams, lead to higher deflections.

Hot rolled structural steel sections

Sections such as universal beams and columns, angles and channels continue to form a most important part of the market especially in commercial and industrial building, or where large clear spans or multi-storey construction are required.

The former BS 4360 steel grades have now been replaced by the BS EN 10025 equivalents. For example

S 275 J2

where

S refers to structural steel
275 refers to the minimum yield strength in N/mm^2
J2 refers to the impact performance

This specification forms the basis of most applications and is roughly equivalent to the previous grade 43D of BS 4360. The newer description of tensile performance is more appropriate since 275 refers directly to yield performance whereas the earlier 43 was one tenth of ultimate strength which has much less relevance to structural design.

Higher grades can be obtained by adding alloying elements such as manganese or niobium, or by heat treatment such as normalising. For example S 355 operates at a yield stress about 30 per cent higher than S 275. This would permit more slender design or larger spans where required.

Various impact strength options are available according to requirements and these have codes such as

JR;   JO;   J2;   K2; etc.

in order of increasing performance, which would be combined with the above designations.

All structural steels are of low carbon content, typically 0.2 per cent, to permit welding and to produce impact resistance and ductility.

Many other specifications are available, some with high tensile and impact performance, though these would normally need to be ordered in quantities above a stated minimum at increased cost.

Table 8.6 indicates minimum yield stress, tensile strength range, impact strength requirements and compositional details, for the steel sections in most common use (up to 16 mm thick).

Note the following:

■ Carbon content must generally be decreased to obtain increased impact resistance. In each case, impact strength is indicated by the energy absorbed by

Table 8.6   Properties of BS EN 10025 structural steels in common use

Grade	Min yield str N/mm^2	Tens str N/mm^2	Impact energy Charpy V-notch Joules at 0 °C	Chemical composition (max) %					
	Max thickness 16 mm			C	Si	Mn	P	S	N
S 275 JO	275	410/560	27	0.20	–	1.5	0.035	0.035	–
S 355 JO	355	490/630	27	0.20	–	1.6	0.035	0.035	–

standard notched specimens at a given temperature. Steels with higher impact strength will absorb more energy at a given temperature (or equal energy at a lower temperature).

- Hot rolling of thicker sections tends to be completed at relatively high temperatures – for example, 1100 °C. Hence, the austenitic and, ultimately, the ferritic grain structure are coarser than those of thinner sections, resulting in lower yield stresses.
- Ductility of steels decreases as their strengths increase. This factor is measured by the elongation of steel samples during tensile testing, ductile samples showing greater elongation.

Structural hollow sections

These are formed from strip by rolling it into a circular shape and then high frequency induction welding along the length to form a seam. This is followed by final shaping to give sections which are circular, square or rectangular.

Structural hollow sections have the following chief advantages:

- They are more efficient than universal beams or columns in a structural sense since material is on average further from the centroidal axis, leading to improved sectional properties.
- Use of structural hollow sections provides greater opportunities for architectural use of steel.

In *hot finished* sections the hollow section is then heated to around 900 °C to remove residual welding stresses and allow final forming of the section. Hot finished sections can be obtained in relatively large thicknesses (hence load bearing capacities) since the material is more ductile at high temperature.

In *cold formed* sections there is no such treatment prior to finishing. The maximum thickness obtainable in cold formed sections is 12.5 mm.

Hollow sections are covered by standards as follows:

Hot formed   BS EN 10210
Cold formed   BS EN 10219

In each case grades S 275 J2H and S 355 J2H are readily available (similar to structural sections as above). H stands for hollow section.

The materials composition requirements are very similar to those given above.

In constructional terms hot formed sections currently form the bulk of the market because:

- Relatively sharp corners can be produced, facilitating, for example, welding of equal square sections to each other (Figure 8.38).
- EC3, the design code for structural steel, allows higher stresses in hot formed sections in compression, where resistance to buckling is a major criterion.
- The structural performance of the more uniform material in a hot formed section is easier to predict.

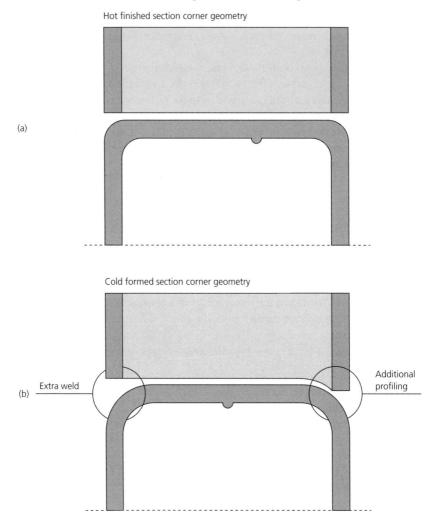

Figure 8.38   Tighter corners produced by hot finishing facilitates welded connections

■ There is a risk that, in impact, residual stresses in the corners of cold formed sections may trigger failure.
■ There is a bigger risk of distortion in cold formed sections.
■ When galvanising is used the risk of delamination at corners of cold formed sections is greater.

The use of cold formed sections is nevertheless likely to increase since there are cost and energy savings in production. BS 5950, the code of practice for fire resistant design, does not currently cover cold formed sections, though it is currently being revised. Subsequent inclusion will further encourage wider use.

Where there is any safety consideration the use of products manufactured to a quality management system such as ISO 9002 standard is essential. For less critical applications other cost competitive products are available. These might not comply, for example, with impact requirements of BS EN 10219 but can carry somewhat higher stress and might typically be used for partitions or wind bracing purposes.

WEATHER RESISTANT STEELS (COR-TEN)

These are low alloy steels containing small amounts (less than 1 per cent each) of copper, chromium and nickel. Like ordinary steels they are subject to rust formation, but whereas with ordinary steels the rust layer becomes detached, permitting further corrosion, with weathering steel the rust coating has both low porosity and high adhesion to the underlying metal. The corrosion film development is therefore exponential (see Figure 8.22). These steels are weldable, a typical code being grade S 355 JO WP (BS EN 10155). They may be used for exposed steelwork without painting, a uniform rust coating being developed. They rely upon ready access of oxygen so should not be used underground, in prolonged wet conditions or in aggressive environments.

STEEL FOR REINFORCEMENT OF CONCRETE (see also page 330)

Reinforcement may be used for the following reasons:

- To take the tensile stresses in concrete beams or slabs.
- To withstand shear stresses in beams which, for uniformly distributed loads, are greatest near the supports. These give rise to complementary stresses in such regions which require the use of additional reinforcement, either in the form of stirrups or bent-up bars at the end of simply supported beams.
- To carry a proportion of the compressive stress and to withstand tensile stresses which may arise due to eccentric loading, as in columns.
- Reinforcement may be used near the surface of mass concrete structures to control cracking by drying or carbonation shrinkage.
- In some cases, secondary reinforcement is used to prevent spalling of concrete surfaces due to fire.

The essential characteristic of reinforcement for concrete is that the steel is placed in position in the mould and concrete is then cast around it. The steel is initially passive since no stress is applied directly to it. If any movement takes place in the concrete the steel will respond and should be designed to carry associated stresses. Stresses can be transferred either by *anchorage* or by *bond*.

Anchorage occurs towards the ends of reinforcing bars and is achieved by some form of deformation – usually bending of bar ends so that the steel is mechanically locked into the concrete. Anchorage is an effective means of communicating stress from the concrete to the steel, but offers little control over the distribution of stress along the bar.

Bond refers to the transfer of stress along the length of the bar via the interface between concrete and steel. Bonding is valuable because it regulates the distribution of any tensile cracks in the concrete.

Bond between concrete and steel

The precise nature of the bond between steel and concrete is not fully understood but bond strength depends on the following factors:

The area of contact between the steel and the concrete   This may be increased by use of larger numbers of smaller diameter bars, which would have a higher surface area for a given sectional area. In the case of higher design stresses in the steel, the use of hooked ends to bars may be required, since these provide more effective stress transfer than straight bars.

Strength of the concrete   Bond strength increases with the crushing strength of the concrete, though the relationship is not linear.

Nature of the steel surface   The presence of a thin layer of adherent rust increases the bond characteristics of plain round bars, though excessive rust, resulting in a reduced cross-sectional area, would be clearly unacceptable. Deformed bars have greater bond strength but this is reduced by the presence of rust. Mechanical impact, or handling as when fixing reinforcement, should normally reduce rust to acceptable levels. Wire brushing is ineffective; it reduces the bond by polishing the rust coating. The presence of oils on bars severely reduces the bond.

High strength bars for reinforcement do not have a greater modulus of elasticity than mild steel bars, so that strains will be greater at the larger stresses which are permissible. In these cases, therefore, deformed bars are used to provide the increase in bond strength necessary.

Table 8.7 shows selected properties of the most commonly used reinforcing steels. In addition to these requirements British Standards specify chemical composition, impurities, ultimate tensile strength and rebend test requirements. BS 4449 now includes a fatigue performance test – deformed bars should be able to withstand 5×10^6 cycles of stress in axial tension at a stress range 150–200 N/mm^2,

Table 8.7   Properties of reinforcing steels

Grade	Yield strength (N/mm^2)	$\dfrac{\text{Max. strength}}{\text{Yield strength}}$	Elongation at fracture (%)	Total elongation at maximum force (%)
250	250	1.15	22	–
460A	460	1.05	12	2.5
460B	460	1.08	14	5

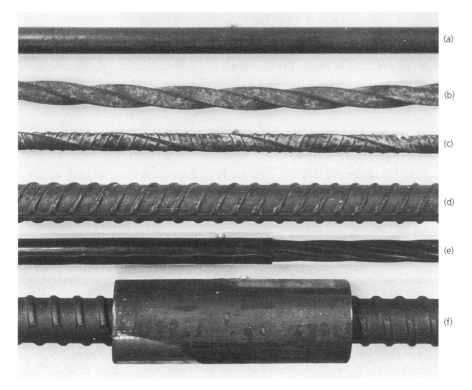

Figure 8.39 Steels for reinforcing and prestressing concrete: (a) plain round bar (BS 4449);
(b) cold worked type 1 (BS 4449); (c) cold worked type 2 (BS 4449); (d) hot rolled type 2
(BS 4449); (e) seven wire prestressing strand (BS 5896); (f) alloy steel with prestressing bar
with rolled thread and threaded connector

depending on size. This test is relevant to situations where fatigue failures are
possible, such as bridges.

Reinforcement bars manufactured in the UK can be quite easily identified
(Figure 8.39):

- **Grade 250 hot rolled mild steel bars (BS 4449)** are normally of plain round
  section. They may be preferred where rebending is necessary, for links or
  stirrups, or where unusually complex bending schedules are required. The
  maximum available size is 16 mm in view of these applications.
- **Grade 460 high yield bars (BS 4449)** are deformed and they may be of two
  types:

  Cold worked A hot rolled steel of square (type 1) or ribbed (type 2) section
  is strengthened by stretching/twisting beyond its yield point. These bars are
  recognised by the spiralling corners or ribs.
  Hot rolled These are either micro alloyed or thermomechanically treated.
  They have straight transverse ribs.

There is little to choose between the high yield types in reinforcement of concrete though each is preferred to mild steel for general applications on economic grounds. Bars with yield stresses of 500 N/mm² are available in Europe and these may be incorporated in future British Standards. When used, particular care may need to be given to avoiding excessive deflections and cracking of concrete under service stresses, which might have durability implications.

STEELS FOR PRESTRESSED CONCRETE (see also page 331)

These are fundamentally different from steels used for reinforcing concrete in that stresses are applied direct to the steel by means of hydraulic jacking systems. Once the desired level of stress has been reached the steel is anchored by:

1 tightening nuts onto threaded ends of prestressing bars, reacting against the concrete surface
2 use of wedges in a conical socket (wire or strand), reacting against the concrete surface
3 local bonding with the concrete internally along its length

An important difference between prestressing and reinforcing steels is that whereas reinforcing steels are often bent to form hooked ends or for other reasons these processes are not required in prestressing steels so that the ability to withstand plastic deformation is not required. In consequence, carbon or alloy contents can be increased, giving much higher working stresses. These are, however, linked to another equally important attribute of prestressing steels, that of extensibility. Since moduli of elasticity of prestressing steels are either similar to, or slightly lower than, those of reinforcing steels, working stresses correspond to much greater strains. Values in the region of 5000×10^{-6} are typical, with the result that prestress losses due to shrinkage, creep and other causes which might total, say, 500×10^{-6}, do not seriously reduce the operating stress in the steel (though they must be allowed for).

Prestressing steels can be broadly categorised into high tensile steel wire (either plain or in the form of strand) and alloy steel bars.

The use of carbon steel in the form of wire has the attraction that, by a combination of heat treatment and cold drawing, very high tensile strengths can be achieved. In a typical process, high carbon steel is quenched from the austenitic state into a lead bath at about 500 °C. This gives a very finely divided network of cementite particles in a ferrite matrix, which imparts sufficient ductility to permit extensive cold drawing. Depending on the load carrying capacity required, wires, which may have diameters of up to 7 mm, may be used individually or in stranded form. The latter comprise seven wires spun in helical form as hexagonal sections about a straight core. (The hexagonal sections are the same as those which form the basis of the HCP and FCC crystal structures; Figure 8.1.) A further advantage of wire and strand is that they can be coiled in long lengths for storage and transport, therefore obviating the need to form joints in tendons for large structures.

Table 8.8 Properties of steels for prestressing concrete

Material	Example of sizes available (mm)	Specified characteristic load (kN)	Equivalent stress (N/mm^2)	Minimum elongation on gauge length stated	Relaxation at 70 per cent of characteristic stress for 1000 hours	
					Normal	Low
High tensile steel wire BS 5896	4	21.0 ⎫ 22.3 ⎭	1670 ⎫ 1770 ⎪	3.5 per cent on g.l. 200 mm (max. load)	8	2.5
	7	60.4 ⎫ 64.3 ⎭	1570 ⎪ 1590 ⎭			
Seven-wire strand (standard) BS 5896	9.3	92	1770 ⎫	3.5 per cent on g.l. ⪍ 500 mm (max. load)	8	2.5
	15.2	232	1670 ⎭			
High tensile alloy steel bars BS 4486 (hot rolled)	26.5	568 ⎫	1030	6 per cent on g.l. $5.65\sqrt{S_0}$ (on fracture)	3.5	–
	40	1300 ⎭				

Note Characteristic strengths are based on breaking load with 5 per cent failures, though minimum proof stress values are also specified.

Wire and strand are available in the *as drawn* or *as spun* conditions, respectively, though these forms have the drawback of not paying-out straight after coiling and undergoing relaxation (plastic distortion) at stresses over approximately 50 per cent of characteristic strength. Products are, therefore, normally either heat treated at about 350 °C to overcome the problem of paying-out straight, or subjected to a similar process while under stress, the latter leading to low relaxation wire or strand. Two slightly different grades of wire are available in most sizes, while seven-wire strand may be in the form of *standard*, *super* (having higher tensile strength) or *drawn* (having been passed through a die and heat-treated before coiling). Selected properties of available sizes of wire and seven-wire strand are given in Table 8.8.

In some cases and particularly in pretensioned prestressed concrete (see page 331) bonding with the concrete is essential. For this purpose, indented wire, crimped wire or strand may be employed. The transmission length is the length of member required to transmit the initial prestressing force in a tendon to the concrete. According to BS 8110 it may vary between 50 and 150 diameters; seven-wire strand requiring the smallest length followed by crimped wire, indented wire requiring the greatest transmission length. Gradual decrease of prestress towards the end of pretensioned lintels, for example, is not normally a problem, since bending stresses are lower at these positions.

Alloy steel bars rely on significant quantities of alloying elements, such as manganese, rather than high carbon content or drawing, to produce high strength. Bars are available either smooth or with ribs, according to bonding requirements.

To obtain large lengths, smaller transportable lengths must be joined together. Joints and anchorages are provided by means of threads which are rolled on to avoid the loss of strength associated with a cut thread. Two grades are available – *hot rolled* or *processed* – the latter involving cold working and possibly, in addition, tempering. Selected properties of hot rolled bars are indicated in Table 8.8. Examples of prestressing strand and an alloy steel bar are shown in Figure 8.39.

APPLICATION OF HEAT FOR JOINING AND CUTTING OF STEEL

One of the major attractions of steel is that it is amenable to joining by a number of methods.

Welding

Welding of mild steel involves fusion of the parent metal to form continuous metallic bonding and therefore offers a means of producing joints which are as strong as the metal itself. Most mild steels should not pose problems since they weld easily and are not likely to be severely affected by the process. There is, however, much skill involved in the welding process, and brittleness, cracking and distortion will result if the weld is of poor quality. The situation is illustrated in Figure 8.40 which shows the principle of gas welding a single V butt weld. Heat from the oxyacetylene flame is supplied to the *weld pool* – a pool of molten metal

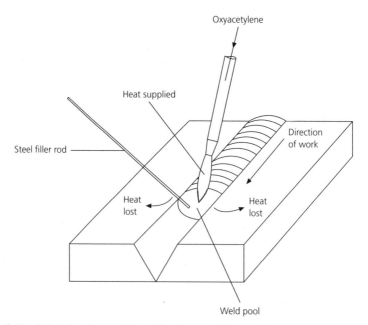

Figure 8.40 Principle of gas welding. The heat applied must be adequate to cause fusion of the filler rod and sides of the metal to be joined, while avoiding overheating

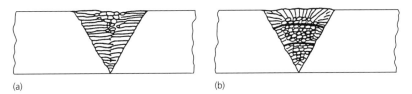

Figure 8.41 (a) Columnar crystals, resulting in weakness produced by single-run butt weld; (b) refined crystal structure resulting from building up the butt weld in three stages

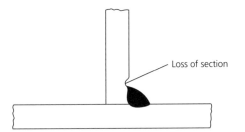

Figure 8.42 Undercutting in a butt weld caused by poor technique/excessive heat

which fuses each side of the metals to be joined. Weld pool metal is provided by repeatedly dipping the filler rod into the pool, producing the slightly rippled appearance of the finished weld. The heat supplied by the flame must exactly balance the heat lost by conduction into neighbouring metal. If too much heat is applied the neighbouring metal (heat affected zone, HAZ) becomes overheated, causing overall weakness in the weld (see 'Annealing', page 399) and weld metal itself tends to develop large orientated crystals which produce planes of weakness where they meet (Figure 8.41(a)). In fillet welds – used for example to join surfaces at right angles – overheating may lead to excess fusion (undercutting) (Figure 8.42), causing loss of section in the parent metal. If insufficient heat is applied there may be inadequate fusion especially at the base of the weld (known as poor penetration). The weld may then look satisfactory but would fail prematurely under stress. Multi-run welds have the advantage that previous runs are normalised by subsequent runs, refining grain structure and reducing brittleness (Figure 8.41(b)).

Structural steel is more commonly welded by arc welding in which the filler rod metal is melted and carried to the weld by a high voltage arc which also provides all the thermal energy needed. A weld pool again forms which must fuse with, but not undercut, the parent metal. Oxidation of the weld is prevented by a flux coating on the filler electrode which produces a gaseous shield around the arc. It also produces a slag which forms on the deposited weld metal, preventing access of oxygen and nitrogen. Alternatively the arc may be protected by an inert gas such as argon, the principle of MIG welding.

The susceptibility to damage of the heat affected zone depends on its composition and the term *carbon equivalent content* (CE) is used to summarise the effects of the various alloying ingredients:

$$CE = \%C + \frac{\%Mn}{6} + \frac{\%(Cr + Mo + V)}{5} + \frac{\%(Ni + Cu)}{15}$$

where:

C = carbon
Mn = manganese
Cr = chromium
Mo = molybdenum
V   = vanadium
Ni = nickel
Cu = copper

The numerical factors applied to metal, other than carbon, reflect their decreased tendency to produce brittle martensitic zones. Few problems should be encountered if the CE is less than 0.25. Higher values of up to, say, 0.5 per cent may be tolerated if cooling is controlled and precautions are taken to keep down the hydrogen content of the weld metal and the HAZ. Hydrogen can be introduced by moisture in some fluxes and it tends to result in cold cracking unless dispersed by heat treatment. Welding of high yield steel may present problems and preheating of the heat affected zone may be necessary to avoid metallurgical changes during the welding process. Cold worked steels will recrystallise on welding so that yield strength will be reduced. In all cases, particular care is needed when members are in tension. Prestressing steels are not normally welded – this should not be necessary and, in any case, their high carbon alloy content makes welding difficult.

Brazing

Brazing of steel is a much lower temperature process which does not involve fusion of the parent metal. The metal used is brass and the heating level applied is sufficient to melt the brass only. For the process to work, the molten brass must 'wet' (adhere to) the steel surface and fluxes are used for this purpose in order to remove the oxide layer. (It should be remembered that provided use of fluxes dissolves oxide films to reveal a metallic surface, the application of any liquid metal to steel, or any other metal, will produce a good metallic bond.) Brazing offers the possibility of producing a joint of moderate strength between steel components with minimal risk of overheating the parent metal. The joint strength is, however, much lower than that obtained by welding.

Soldering

Soldering is yet another process suitable for joining most metals including steel, though since the alloy used is normally a mixture of lead and tin, strengths are very limited. The techniques can still be useful where extended bonding areas exist, such as capillary joints in plumbing.

Heating of structural steels

Reference to Table 8.2 shows that, if heated to temperatures above about 450 °C, steels will recrystallise. Where steel has been cold worked or heat treated these effects will therefore be lost.

Heat is sometimes required to assist in bending reinforcement but if used this should be in the form of steam only.

Cutting of steel is best carried out by a high speed abrasive wheel, which avoids excessive heating of the steel, though an oxyacetylene flame will produce a cutting rather than a melting action if an excess of oxygen is used (the oxygen causes rapid oxidation of steel in the flame and, since iron oxide melts at a lower temperature than steel, the oxide is dispersed by the flame). The ends of bars or wire are not normally under stress but, in post-tensioned prestressing systems, it is particularly important that a tendon is not overheated at its anchorage.

FIRE RESISTANCE OF STRUCTURAL STEEL SECTIONS

The electronic properties of metals which give rise to their formability also result in high thermal conductivity and decrease of strength at only moderate temperatures. The high conductivity of metals results in rapid transfer of heat to metal structures in case of fire. In localised fires heat may be conducted away from the source delaying the temperature rise of the steel. However, unless in some way protected, the yield of steel in steel framed buildings may occur quite rapidly in serious fires.

On heating to a temperature of about 250 °C, the yield strength of hot rolled steels increases, but continued heating to about 550 °C weakens the steel to the point where, based on normal factors of safety, yield is likely to occur. Even before this has occurred, considerable distortion of the structure by expansion of the softening material may take place. For example, a 10 m beam heated to 400 °C will expand by approximately 50 mm. It is clear, therefore, that steel structures at risk must be protected by some means from fire.

In designing for fire there are always two considerations – protection of life, which always comes first, and then protection of the building and contents from damage. It is very rare for the steel frames themselves to undergo total failure in fires. Figure 8.43 shows the results of an experiment in which a composite floor in a framed building was subjected to a simulated typical fire and though serious distortion took place, there was no collapse. In environmental terms, however, the preservation of the structure is now regarded as increasingly important since repair of steel buildings can be very expensive where structural damage in the form of severe plastic distortion of steel has occurred.

Passive protection

Passive fire protection involves reducing the rate of temperature rise in the steel in order to achieve desired fire resistance. A large number of options is now available and the choice between them must be made by consideration of various

Figure 8.43   Performance of steel beams in a composite floor, subjected to severe fire. The beams distorted but there was no collapse. (Experiment conducted by BRE)

criteria, including building function, cost, cosmetic requirements, environmental consideration and speed of execution. In all cases the level of protection required depends upon the *section factor* of the steel. This is based upon section shape and determines the rate of heating in a fire. Possible systems include:

- Cladding, either in concrete or, better, lightweight insulating materials such as asbestos, vermiculite, lightweight concrete and lightweight plasters. Such systems occupy additional space but produce an unobtrusive 'box' appearance. Lesser protection can be achieved by block filling of columns in which the web space only is filled.
- Concrete filled hollow sections. Concrete reduces the rate of temperature rise on account of its thermal inertia and progressively takes load as the steel weakens under heat. At the same time the steel helps resist spalling of the concrete. This method might be attractive where high thermal inertia is sought, for example to control summer heat gains. However it makes recovery of steel for re-use much more difficult.
- Water filled hollow sections. Water has very high thermal capacity and in a suitably designed system can carry heat away from the fire by convection. The method can therefore be used where high risks are present, for example large scale storage of combustible materials. Systems are, however, expensive because members must permit water flow between them, all connections must be watertight, a header tank is required for water storage and corrosion inhibitors must be used.

■ Intumescent coatings. These expand to 25–30 times their volume when heated above 200 °C. Since they are spray applied, they can be applied relatively easily to more complex details. The shape of the steel frame can be retained, sometimes to good architectural effect. Coloured pigments can be included.

■ Sprayed protection. Inert lightweight coatings up to 50 mm thick are sprayed to form an *in situ* coating after completion of the structural frame. This forms a cheap and quick method of protection though the method is only really suitable for hidden spaces such as roof voids.

■ External steelwork. There have been several examples of buildings in which a part of the steel frame is exposed, thereby reducing the rate of temperature rise in fire. Guidelines exist for determining the extent (if any) to which fire protection is required.

Assessment of fire damaged steel

Most steel in existing buildings is of grade 43A (BS 4360) now referred to as S 275 in the current BS EN 10025. Where visible distortion has not occurred it can be assumed that the yield strength of the material will not have decreased significantly.

Where higher grades such as grade 50B (S 355 in BS EN 10025) are used, strength loss can be more marked even if distortion has not occurred since grain refining elements such as niobium tend to migrate under temperature gradients.

Where there is suspected loss of performance hardness tests can be carried out on affected sections and compared with those of undamaged sections to assess the possible strength shortfall. Table 8.9 shows the relationship between hardness values and ultimate tensile strengths, from which yield strengths may be inferred.

The condition of all joints and connections in affected regions should also be checked. Where there is any doubt about connecting bolts they should be progressively replaced, especially if found to be high tensile.

Table 8.9 Brinell hardness values as an indication of residual tensile strength in fire damaged steel

	Brinell hardness number	Ultimate tensile strength N/mm^2
	187	637
	179	608
	170	559
Grades S 355	163	539
	156	530
	149	500
	143	481
	137	481
Grades S 275	131	461
	126	451
	121	431

FUTURE DEVELOPMENTS IN STRUCTURAL STEELS

These may be broadly categorised into higher strength types of steel and new applications for conventional steel types.

High strength steels

With prices of raw materials rising continuously, the advantages of steels which can be used at higher stresses are self-evident and there would be accompanying environmental attractions of achieving higher performance for a given mass of material. However it will also be evident that high yield strength is normally obtained at the expense of some other important properties, such as notch ductility or weldability. In addition, when steel sections are reduced in thickness, problems of, for example, excessive deflection or greater vulnerability to corrosion may arise. Much research has been carried out into obtaining steels having higher service stresses while retaining satisfactory toughness and weldability. It is found that the most promising steels are those of low alloy content containing as little as 0.03 per cent carbon, together with low sulphur and oxygen content for good formability. High yield strength is obtained by producing a very fine dispersion of ferrite grains. This involves careful processing of steels containing the micro-alloying elements aluminium, niobium, vanadium or titanium, the processing involving controlled rolling at progressively lower temperatures. Optimum properties depend on the reduction of section per pass and on the time between passes. Applications of such techniques are likely to include tubular assemblies in constructional engineering and perhaps, ultimately, ordinary structural steel sections, where increased manufacturing costs will be offset by reduction in the mass of steel required for a section of given strength. At the present time, however, mild and high yield steels remain the most economical method of producing conventional structures, the more sophisticated high strength steels being reserved for situations such as large-span structures where very high strength/weight ratio is required.

New applications for steel

Conventional steels offer good all round performance at competitive cost and the industry has seized opportunities to broaden the market for the material by new approaches to its use in the construction industry. Examples are as follows.

Floor construction   By use of *slimfloor* asymmetric main supporting beams, which have a widened soffit flange, floors can be constructed rapidly using precast flooring components resting upon the lower flange of these main beams. This offers the possibility of a reduced overall floor thickness, additional fire protection (provided by the precast beams) to the main beams and a simple means of incorporating services within the floor section.

Bi-steel units   These are designed units comprising steel sheets separated by welded spacers for production of wall panels, beams or other units. The units can be simply jointed to form panels, the sheets acting as permanent formwork for concrete which is poured between the sheets after assembly. The steel carries tensile stress while the concrete carries shear/compressive stress.

Modular light steel framing   Galvanised, cold formed units can be used to produce simple frame units such as bathroom 'pods' or larger assemblies, resulting in fast assembly to good tolerance levels. By incorporating the thermal insulation on the exterior side of external frames condensation and damp problems should be minimised. Perhaps a very important attraction of this type of approach is the possibility of re-use with minimal loss of material at the end of the building life.

ALLOY STEELS

These contain substantial proportions of alloying elements. They may be included for the following reasons:

- To increase strength and hardness. For a given cooling rate, the BCC elements tungsten, molybdenum and vanadium enhance these properties.
- To increase hardening properties. It is very difficult to harden effectively large steel sections by quenching since, although martensite will be produced on surface layers, the bulk of the metal will tend to form pearlite. In any case, quenching is drastic and results in a distorted, brittle structure. By incorporating alloying elements in steels, TTT diagrams are significantly altered (see Figure 8.37), transformation times being increased so that a quenching effect is obtained with much slower cooling. Hence martensite can be obtained with reduced risk of thermal stresses and the material can be hardened uniformly throughout its section. Elements having this effect are manganese, molybdenum and chromium.
- To increase corrosion resistance. Aluminium, copper and chromium have this effect; see 'Weather resistant steels'. Most important, at present, however, are stainless steels which are based on the effect of chromium (see below).
- To preserve the austenitic state. Metals such as manganese and nickel, which are FCC, tend to preserve the austenitic structure of steel down to temperatures as low as room temperature. For example, a 13 per cent manganese steel is used on the teeth of mechanical shovels to resist the severe abrasion encountered. Impact or abrasion changes the surface layers of the austenite structure into martensite, which resists wear and damage.
- To produce special properties. Steels containing 36 per cent nickel, for example, have a very low thermal movement (approximately $2 \times 10^{-6}/°C$ at normal temperatures, compared to $11.6 \times 10^{-6}/°C$ for ordinary steel). They are known as *invar* steels and are used in surveyors' tapes and similar applications.

A very large number of alloy steels can be produced by the incorporation of alloying elements as above, though these are mainly used in mechanical engineering applications. In such applications surface hardening may be used, for example, flame-hardening in which, by heating the surface of a metal component and then quenching immediately, martensitic surface layers are produced on a tough, resilient core. A similar effect may be obtained by nitriding or carburising, which involve heating components in nitrogen- or carbon-rich atmospheres which combine with surface layers, giving a hardening effect.

STAINLESS STEELS

Owing to their corrosion resistance, these are the most widely used alloy steels in the construction industry. Resistance to corrosion is achieved by incorporation of chromium in steel. On immersion in water or dilute acids which contain oxygen, an impervious chromium-oxide film is formed and any faults in the coating are quickly repaired. Chromium in quantities up to 27 per cent of the steel may be used, 18 per cent being a common figure. Best results are obtained with a single-phase metal, since two phases (for example, ferrite and pearlite in ordinary steel) will tend to corrode electrolytically with one another.

The addition of nickel preserves the austenitic state and, for this reason, up to 10 per cent nickel is included in the austenitic steels. The most common types of stainless steel are 18-8 (percentages of chromium and nickel, respectively) and 18-10-3 (the last figure indicating 3 per cent molybdenum). The second type is more expensive but has greater corrosion resistance and is more suitable for external use, for example, in cladding or internally where cleaning is not practicable and, therefore, differential aeration is a possibility. Caution is necessary when welding austenitic stainless steels since, at the high temperature involved, chromium carbide forms at grain boundaries which may cause corrosion of chromium-depleted regions nearby. This problem may be avoided if steels of very low carbon content are used or if the steel is stabilised with niobium or titanium, which reduce the tendency for chromium to migrate. Austenitic stainless steels are easily distinguished from ferritic stainless steels by use of a magnet – they are non-magnetic.

Ferritic stainless steels are also available. They are cheaper than austenitic stainless steel, due to their low nickel content, but have relatively poor corrosion resistance, especially if welded, hence they are used internally. They have low carbon content and, therefore, good formability: applications include balustrades, lift fittings and pressed steel sinks.

A harder steel is obtained by quenching from about 1000 °C producing martensitic stainless steel if significant carbon is present. However, although used for cutting implements such as knives, corrosion resistance and formability are not as good as for austenitic steels. The impact strengths of ferritic and martensitic stainless steels are poor, especially at low temperatures, since they exhibit a ductile–brittle transition.

Since stainless steels depend for their protection on a chromium oxide coating, resistance to corrosion will be reduced in oxygen-remote situations or where there is differential aeration as, for example, in underground pipes. The effect will be

Table 8.10 Classification of some common stainless steels available in plate, sheet or strip (BS 1449 Part 2)

BS classification	Approx. composition	Type
305 S19	18% Cr 11% Ni	Austenitic
316 S33	17% Cr 11% Ni 3% Mo	Austenitic
405 S17	13% Cr	Ferritic
410 S45	13% Cr	Martensitic

exaggerated in the presence of chlorides or sulphates. The 18-10-3 grade is most resistant to such conditions but longest life will, in general, be obtained where oxygen has ready access to all parts of the surface, in order that the passivating oxide film can re-form if, for any reason, it is disrupted. The presence of nickel enhances corrosion resistance in less oxidising acids, such as hydrochloric and sulphuric acids.

Use of stainless steel

Stainless steel is manufactured mainly in the form of steel plate, sheet and strip (BS 1449, see Table 8.10) or light-gauge tubing (BS 4127). The material is now used for a wide variety of pressed applications internally and externally; for example.

Plumbing   Austenitic stainless steel tubing for domestic purposes is comparable in price with copper, prices fluctuate less than those of the latter, and an attractive finish is obtained especially if chrome finish fittings are used. Although work hardening makes bending more difficult, stainless steel tubes have much greater bursting pressures than copper, so that when compression fittings are used, freezing of the tube will tend to cause sliding at the joints rather than bursting of the tubes. This obviates the need for pipe replacement, though it may result in the emission of a full bore of water on complete failure of a joint. When soldering tubing, the use of chloride based fluxes is not recommended, since they cause corrosion of the tubing, especially if jointing is completed some time before the system is brought into use or flushed out. Special fluxes, such as those based on phosphoric acid, should be used.

Roofing (BS 1449; BS EN 10088)   A major attraction of stainless steel for roofing is that the weight of the material is very low on account of the low thickness which can be used – either 0.38 or 0.46 mm compared with a typical thickness for other metals of:

copper	0.45–0.6 mm
zinc	0.65–1.00 mm
lead	1.32 (code 3)–3.55 mm (code 8)

Figure 8.44   Production of stainless steel roofing. Insulation is provided between rafters, and ties are used to fix the sheeting at seams

Unfortunately thin sheet roof coverings tend to buckle slightly leading to a quilted effect if reflective. Reflection of sunlight may also be a problem. These difficulties can be overcome by use of *terne coated* stainless steel which has a coating based upon tin/lead or tin/zinc coatings. These weather to the traditional dull grey of a lead roof. To achieve this effect only thin coatings, typically 3 μm, are required, but to improve corrosion resistance the coating thickness can be increased to about 15 μm. Alloys suitable for roofing are type 305 and type 316 (see Table 8.10) the latter being more suitable for marine or industrial environments. One problem with stainless steel for such applications is that workability is much lower than that of other metals. This can make formation of more complex details such as intersections much more difficult (Figure 8.44).

Cavity wall ties   These are considered briefly as an illustration of the many small manufactured steel products used in building for which stainless steel is likely to have increasing application. The need for cavity wall ties arose immediately the benefits of cavity walls in thermal and dampproofing terms were appreciated – over one hundred years ago. Initially ties took the form of bricks laid across the cavity (headers), wrought iron or cast iron ties. However, with the need for rapid construction and reduced material costs the more convenient wire or vertical twist ties were introduced (Figure 8.45). To guard against corrosion BS 1243 initially required a galvanised coating of a specified average thickness (about 80 μm for vertical twist ties and about 40 μm for wire ties). Failures nevertheless began to occur in the 1970s,

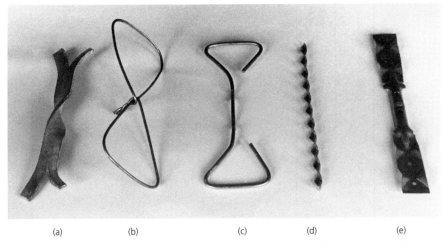

Figure 8.45   Some wall ties in current use: (a) vertical twist tie, galvanised (BS 1243);
(b) butterfly tie, stainless steel (BS 1243); (c) double triangle tie; (d) spiral tie, stainless steel;
(e) pressed steel tie, stainless steel

particularly in wire ties which contained less steel and thinner zinc coatings. Zinc
thicknesses were therefore increased to approximately 130 µm average for both
types in the 1981 edition of BS 1243 in an attempt to solve this problem. Probably
more satisfactory, especially in exposed areas, however, are more recently introduced
stainless steel ties. These are available to the original BS 1243 forms though the
extra cost of stainless steel material has led to attempts to reduce the amount of
metal in the tie with obvious rigidity and strength implications. Performance
specifications are now therefore included in the new draft British Standard on wall
ties (DD 140). They are categorised into one of four classes according to structural
performance (Table 8.11). Lighter stainless steel ties weigh about half their BS 1243
equivalents and are usually limited by their tendency to buckle in compression
though stainless steel ties in all four strength categories are available.

There are further practical factors affecting wall ties and their performance –
for example tradesmen may prefer ties that can be bent or adjusted when the two
leaves of brickwork are not built at the same time. Stiffer ties cause problems in this
respect since attempts to alter them after mortar has set may lead to damage to the
bedding and hence impairment of performance. Butterfly ties (though not as strong
as vertical twist ties) are often popular because they are flexible and because they
have no sharp edges which could cause injury to bricklayers. There are many
non-BS ties on the market. These often look virtually identical to BS ties but may, for
example, have an inadequate thickness of galvanising. Although such ties may offer a
cost advantage they could eventually cause problems especially in exposed situations
with their combination of high corrosion risk and high stress levels from wind loading.

Plastics ties are also available and these may overcome the corrosion risk but they
have limited structural performance and may be subject to creep deformation and
damage in fire.

Table 8.11 DD 140 requirements for cavity-wall ties with examples. Deflection limits are also laid down

Type	Example	Type of structure	Location	Ultimate tensile load (N) Type (i) mortar	Ultimate tensile load (N) Type (iv) mortar	Ultimate compressive load (N) Type (i) mortar	Ultimate compressive load (N) Type (iv) mortar
1 (Heavy duty)	Vertical twist ties (BS 1243)	Most masonry cavity walls. Not very flexible. Not suitable for large building movement or low strength masonry units	Anywhere in British Isles. Windspeed up to 56 m/s	5000	2500	5000	2500
2 (General purpose)	Double triangle ties (BS 1243) Stainless steel pressed plate	Not more than 3 storeys. Not greater than 15 m height	As heavy duty	–	1800	–	1300
3 (Basic)	Spiral stainless steel	As general purpose	Full exposure – max windspeed 44 m/s In towns – max windspeed 52 m/s	–	1100	–	800
4 (Light duty)	Butterfly polypropylene (plastic)	Not more than 10 m height	As basic type	–	650	–	450

As a last precautionary point, a tie can only perform well when it is correctly installed. The practice, for example, of pushing a tie into a completed joint, if it were omitted when forming it, would be very unlikely to give satisfactory performance.

8.10 CAST IRON

Carbon dissolves to the extent of 1.7 per cent in austenite and cast iron may be conveniently defined as a carbon/iron alloy containing carbon in excess of this percentage. It will be recalled that the compound cementite contains 6.67 per cent carbon, so that cast irons containing up to this percentage of carbon might be expected to exist in the form of gamma iron plus cementite above 723 °C and pearlite plus cementite below 723 °C. This may be the case but carbon in the form of graphite may also occur in the metal due to decomposition of cementite, resulting in much reduced strength if the graphite is present in flake form. There are several forms of cast iron, depending on the condition of the carbon in the metal, but they are all eminently castable (hence the name) and, as a result of their high carbon content, have a melting point as low as 1130 °C – much lower than that of steel. In fact, the solid/liquid portion of the iron–carbon equilibrium diagram is of eutectic form, 1130 °C being the eutectic temperature obtained with 4.3 per cent carbon.

Grey cast iron

This contains silicon which tends to cause the breakdown of cementite into iron and carbon. Hence, the metal is graphitic and, therefore, not as strong as other forms, having a tensile strength of about 200 N/mm^2. The presence of flakes of graphite also gives rise to stress concentrations and brittleness, fractured surfaces having a grey colour. However, in the thicknesses cast, corrosion is not normally a problem and grey cast iron has been extensively used for boiler castings, radiators, pipes, baths, gutters, staircases and architectural work. Its use for external rainwater systems is diminishing, as a result of developments in plastics, but it remains the only material for the purpose, other than lead, which with periodic maintenance has proved to have first-class durability over the course of many years. Use for boilers and baths is similarly diminishing on account of both product cost and installation cost, though there is still a market for such items due to their ruggedness, stability and long life. A further attraction of the material is that it is eminently recyclable.

8.11 NON-FERROUS METALS

The most important non-ferrous metals are aluminium, lead, copper zinc and titanium, together with associated alloys. Table 8.12 summarises the principal properties of these metals in the commercially pure state.

Table 8.12 Properties of non-ferrous metals (commercially pure). Metals are listed in order of increasing ultimate tensile strength. Iron is shown for comparison

Metal	Relative density	Melting point °C	Elastic modulus kN/mm^2	Ultimate tensile strength N/mm^2	Coeff. of thermal expansion $\times 10^{-6}$ per °C
Lead	11.3	327	16.2	18 (short term)	29.5
Zinc	7.1	419	90	37	up to 40
Aluminium	2.7	659	70.5	45	24.0
Copper	8.7	1083	130	210	16.7
Titanium	4.5	1668	109	450	8.4
[Iron	7.8	1537	210	540	11.6]

ALUMINIUM

It is seldom appreciated that aluminium, in the form of Al_2O_3, is the most common metallic element in the earth's crust, being present, for example, in most types of rock and clay. Unfortunately the extraction of aluminium is expensive and is not normally economical in the case of rocks and clays. Bauxite, $Al_2O_3.2H_2O$, is the chief mineral used for extraction of aluminium. Extraction is not possible by means of reducing agents, such as carbon, used for steel since the oxygen is too tightly bound. An electrolytic method is used, after conversion of the bauxite to aluminium hydroxide, the electrical energy requirement being about 15 units of electricity per kg of metal. However in some cases this energy is supplied as hydro-electric power and the energy required to recycle the metal is only about 5 per cent of the initial production energy so that as quantities recycled increase, its environmental rating should improve. The price of aluminium is about three times that of steel in volume terms and eight times in weight terms.

The widespread use of a reactive metal such as aluminium stems directly from the fact that a coherent, impervious oxide film forms on the surface of the metal immediately on exposure to air. The oxide coating conducts positive aluminium ions but not electrons (hence, its action as a rectifier) so that, once a certain thickness is reached, the resistance of the corrosion film will prevent further corrosion. If increased protection to corrosion is required, anodising may be carried out. This involves immersion of aluminium in chromic or sulphuric acid solutions and inducing corrosion electrolytically, using a direct current source of about 50 V, with the aluminium as the anode and a steel cathode. The oxide film forms as in normal corrosion but in a more uniform and greater thickness, due to the increased voltage. The films obtained using some acids are initially porous so that dyes can be used, producing a coloured finish and the surface then sealed. Film thickness may vary between 35 μm, suitable for aggressive environments, and 1 μm, which must be painted. The oxide coating is stable in the pH range 4–8 but corrosion may occur in acidic or caustic environments. Aluminium oxide films are very vulnerable in the presence of OH^- ions. For example, anodised aluminium is corroded by fresh or damp cement paste with production of hydrogen, hence it should not be used in contact with damp concrete: indeed, the same reaction is used for generation of hydrogen during production of aerated concretes.

The unprotected metal also corrodes in the presence of metals such as copper or iron, especially if chloride ions are present. Coatings of bitumen will be effective in such circumstances but it should, in general, be isolated from contact with other metals. Aluminium should also not be allowed to come into contact with damp brickwork, plaster, timber or soil. Paints containing copper, mercury, lead or graphite may be harmful and should be avoided; resin-based primers containing chromates and iron oxide may be used. Magnesium oxychloride, as used in flooring, should not be in contact with aluminium as it will also cause corrosion.

Pure aluminium, though not of high strength, is very suitable for weathering and foils, remaining a light colour in clean environments but becoming a dark grey in polluted situations. Aluminium in sheet roofing at the 0.8 mm thickness recommended is lighter than any other metallic roof covering. A further common use is as a lining for plasterboard, since it has low emissivity and very high vapour resistivity. Aluminium has high electrical conductivity which, combined with its light weight, makes it preferable to copper for transmission wires, in spite of the higher electrical conductivity of the latter. Most aluminium, however, is manufactured in alloy form.

Aluminium alloys

There is a very wide range of alloys which may have enhanced strength or corrosion resistance. Alloys containing magnesium are, like the pure metal, corrosion resistant and, since magnesium is of similar atomic size to aluminium (they are adjacent in the periodic table), it will dissolve to a limit of about 5 per cent in aluminium at room temperature. The single phase alloy so formed, therefore, has considerable ductility which, combined with its corrosion resistance, makes it suitable for pressed components of moderate strength, for example, corrugated roofing, or where superior corrosion resistance is essential. However, since pure magnesium alloys are single phase, they do not benefit from heat treatment. Incorporation of combinations of copper, magnesium, silicon, manganese and chromium produces alloys which can be welded and extruded and yet have 0.2 per cent proof stresses up to 435 N/mm^2 on heat treatment (cf. normal steel sections with minimum yield stress 275 N/mm^2). The extrudability of such alloys makes them ideal for relatively complex extruded sections, such as complete window frames.

The presence of silicon or copper in aluminium increases its strength by the process of precipitation or age hardening. At high temperatures, these elements dissolve to the extent of several per cent in aluminium. However, on cooling, the solubility decreases so that metal crystals tend to precipitate at grain boundaries. If aluminium is cooled quickly by quenching, there will be insufficient time for precipitation and a super-saturated (solid) solution of silicon or copper in aluminium is obtained. The impurities will then, over a period of 4–5 days, diffuse within the aluminium lattice to form clusters which oppose dislocation movement,

Table 8.13   Properties and uses of common forms of aluminium/aluminium alloy

Alloy 1987 BS designation	Essential alloying elements (per cent maximum)				Typical 0.2 per cent proof stress N/mm²	Application
	Mn	Si	Mg	Cu		
1050 A H14	0.05	0.3	–	0.05	–	Flashings, weatherings, can be hand formed
		(Al content 99.5% min.)				
3103 H14	1.5	0.6	0.1	0.1	–	Roofing, general applications; good durability
6063 AT6	0.1	0.7	0.9	0.1	160	Extrusions for window frames
6082 T6	1.0	1.3	1.2	0.1	255	Structural sections

strengthening and hardening the metal. By heating to between 100 and 200 °C for 2–20 hours, the diffusion process leads to even higher strength. These processes are known, respectively, as solution and precipitation treatments. Alloys containing copper should be protected if used externally. Protection may be in the form of sprayed pure aluminium, or, for aluminium-alloy sheets, pure aluminium cladding.

Table 8.13 shows the BS 1474 codes, proof strengths and applications of some alloy products used in building. A four-digit number code is given. The first digit refers to the alloy group: for example, 1 – virtually pure; 3 – manganese; 6 – magnesium and silicon. The other digits refer to impurities/modifications. The A, where used, is a national identification. The final code refers to temper designation, O standing for annealed, H standing for work hardened and T for thermally treated.

Aluminium is also available in cast form, denoted LM, which is used in components such as door handles and ornamental work. Castings normally have much more substantial section thickness than extrusions so that small levels of corrosion are not a problem. They combine aesthetic properties with long life and excellent recyclability.

Structural use of aluminium

The chief advantage of aluminium lies in its low density compared with that of steel: densities are in the approximate ratio 3:1, steel:aluminium. Some of this advantage is, of course, lost, owing to the reduced stiffness of the latter but, even so, an aluminium structure is not likely to be more than half the weight of its equivalent in steel. A further advantage is that in mild, unpolluted atmospheres, aluminium may be left unpainted. The embrittlement which occurs in steels on cooling is also absent in aluminium owing to its face-centred cubic structure (see page 370). At temperatures above 150 °C, loss of strength is rapid and aluminium alloys melt at about 600 °C. This clearly has implications for fire, although in a localised fire sections may rapidly conduct heat away from the fire source. Temperature movement of aluminium is almost double that of steel, though its modulus of elasticity is only one-third; therefore, although allowances for expansion are necessary, temperature stresses in a restrained member will be only about two-thirds of those in a similarly restrained steel member. The low modulus of elasticity of aluminium is a disadvantage with respect to lateral stability in compression members: this should be carefully checked during design.

Aluminium-alloy sections for structural use are normally extruded rather than rolled, since the former is the more versatile. A large number of sections is available so that, provided care is taken to select the best section for a given purpose, there is some compensation for the extra cost of aluminium compared to steel, which is only available in standard sections. Design codes are, therefore, inevitably different from those of steel and this is, perhaps, one of the reasons why the structural use of aluminium is not more widespread.

Welding of aluminium has, in the past, presented problems, partly due to the oxide film which must, of course, be removed before joining components. This may be achieved by electric arc welding in an inert gas, such as argon, which excludes oxygen. The welding process is somewhat different from that of steel on account of the much greater electrical conductivity, as a result of which, heat is conducted away from the weld pool much more quickly. Greater power inputs are therefore required. Porosity in welds also reduces strength and *joint efficiency* factors are used when designing welded connections. Alloys, if welded in a T condition, will lose performance in heat affected zones though the original properties can be restored by heat treatment. Hardened (H) metals will revert to the annealed O condition in welded zones.

Aluminium framed structures might be considered for the following applications:

- Structures comprising mainly self-weight such as large-span single storey buildings.
- Additions to increase the height of existing buildings or construction from existing foundations or slabs of limited load bearing capacity.
- Structures where access for maintenance is limited.
- Curtain walling and glazing.

At the domestic end of the market, aluminium is at a cost disadvantage to PVC in applications such as glazing and conservatories but for larger scale application aluminium offers the advantages of:

- greater rigidity
- slimmer frames
- less risk of thermal instability due to heat/solar radiation
- harder wearing surface provided by anodising

One possible problem of the material in such situations is cold bridging, leading to heat loss and condensation. The provision of a thermal break may be required.

LEAD

The excellent durability and ease of working of lead are responsible for the former widespread use of the metal in weatherings, flashings and pipes. Its use in these situations is now declining, since the material is expensive and the lower ductility of other metals is no disadvantage for components which are formed by machine. Lead occurs naturally in the form of galena (lead sulphide), which is imported from the USA, Australia or Canada. The metal is extracted by conversion of lead sulphide to lead oxide, which is then reduced by carbon to the metal.

Lead is the densest common metal, having a relative density of 11.3. It also has a low melting point, equal to 327 °C and high thermal movement $-29.5 \times 10^{-6}/°C$. The strength of the metal is low, short term ultimate strengths ranging from about 20 N/mm^2 for impure leads to 11 N/mm^2 for very pure lead (see Figure 2.2).

On exposure to the atmosphere, lead forms a protective coating of lead carbonate which is of innocuous appearance and does not stain adjacent brickwork or stonework. Although lead is by no means the most reactive of metals used in building, it may undergo electrolytic corrosion with metals such as copper if, in damp conditions, the metals touch. However, corrosion will only be serious where relatively large areas are in contact and lead coverings may be secured by copper nails, due to the small area of the latter and protection afforded by the corrosion product (galvanised or steel nails are not suitable for use with lead, since they may corrode rapidly as a result of the relatively reactive nature of zinc or steel and the small anodic area provided by the nails). Lead is also attacked by some types of organic acid, notably acetic acid, which forms the basis of vinegar, and is not highly suitable for domestic waste systems in soft-water areas. Moisture which has been in contact with peat, oak or teak and, to some extent, pitch pine, may also have a corrosive effect. Lead components are best protected by bitumen-impregnated tape in such situations.

Uses of Lead

Lead pipes   As would be expected with a metal fairly near its melting point, lead creeps, above a certain stress. Silver-copper lead is available containing 0.003–0.005 per cent of silver and copper. This has greater tensile strength, fatigue-resistance and creep resistance, so that it can be used with slightly reduced wall thicknesses for a given water pressure. A further type, which contains 0.05–1 per cent tellurium and 0.06 per cent copper, is amenable to work hardening and has greater tensile strength and fatigue resistance than ordinary lead.

Waste pipes and gas pipes which are not subject to hydrostatic pressure are normally thinner than water service pipes. The working properties of lead pipes depend on the grain size distribution, the grains being visible if a section of pipe is polished and etched. The earlier British Standards (BS 602, 1085) gave qualitative requirements for the size and uniformity of grain structure in lead pipes, to obtain a suitable balance between creep resistance (improved by coarse-grain structure) and impact strength, to resist pressure surges (improved by fine-grain structure) caused, for example, by water hammer.

Use of lead pipes is decreasing due to cost, competition from plastics and the possibility of contamination of water though there are still many in use. Pure or soft waters tend to dissolve lead, since they contain carbon dioxide which combines with the metal to form a soluble form of lead carbonate. In hard water containing calcium carbonate, a lead carbonate film forms on pipes, preventing dissolution of the metal. Local authorities normally restrict the lead content of mains water to 0.1 parts per million and soft water may require treatment to keep lead contamination to this limit. Owing to their flexibility, water filled lead pipes will withstand freezing several times though, on the same account, the infill around underground lead pipes may, over a period of time, cause leaks due to abrasion. Lumps of brick or concrete should not be used in infill materials.

Lead sheet and strip (BS 1178)   Lead sheeting for roofing purposes has, to some extent, been replaced by other materials but, even so, there are many modern examples of lead roofing, cladding and fascias which combine great durability with an aesthetically pleasing effect, particularly in prestige building. When used in such large areas, joints must be provided to allow for thermal expansion. The maximum dimension in any direction should not exceed 3 m, sheet normally being used in strips about 1 m wide. BS 1178 gives code numbers from 3 to 8, corresponding to thicknesses between 1.32 and 3.55 mm, together with a colour code. Thicker sheets should be used where dressing is required or if mechanical damage is likely, for example, due to foot traffic. The admirable working properties of lead sheet are, in part, due to the fact that, at room temperature, lead is above its recrystallisation temperature, hence it does not work harden greatly without triggering recrystallisation. These properties are further enhanced by the ease with which joints can be made by *lead burning*, provided surfaces to be joined are cleaned and a reducing flame is used. No flux is necessary for this process, which uses ordinary lead as the filler metal. Cast lead has the greatest resistance to creep since it has a larger grain size than rolled lead – the latter recrystallises during rolling.

Lead strip is still widely used for dampproof courses, since it has unsurpassed dampproofing qualities. It should be protected from contamination by cement mortar by coating both sides with a bituminous paint. Lead-cored bituminous felts overcome this problem. Further aesthetic uses are in leaded-light windows and in thin coatings to reduce the reflectance of stainless steel sheeting (see 'Stainless steel').

Lead solders (soft solders)   These normally consist of a proportion of tin and lead, possibly with a small percentage of antimony. Tin-lead alloys produce the type of phase diagram shown in Figure 8.12, the melting points being:

337 °C – pure lead
232 °C – pure tin
183 °C – eutectic composition consisting of 37 per cent lead and 63 per cent tin

When a low melting point is required, as for spigot-type joints on lead pipes, the eutectic composition (tinman's solder) may be used, though this solder is expensive, due to its high tin content. Solders for general uses are about 60:40; lead:tin, combining relative cheapness with a reasonably low melting point. Solders for wiping joints in lead pipes are about 70:30; lead:tin, in order to give a long plastic stage.

COPPER

The element occurs naturally in the form of ores containing copper, iron and sulphur. Since the metal is very corrosion resistant, a good deal is also recovered as scrap and reprocessed. Some impurities are removed by flotation, and then by oxidation in a converter. The metal is then refined either by furnace or electrolytically, according to use, the latter producing the purest metal.

Copper is the noblest metal commonly used in building – it behaves as a cathode to the others, with the exception of stainless steel, hence it will be protected by almost any metal it touches, at the expense of the other metal. It forms an attractive green patina of the basic sulphate on exposure to damp atmosphere though in cleaner atmospheres this may take up to 15 years to form. Washings from it may stain adjacent materials. Copper is resistant to most common acids and to sea water (but see below under 'Copper tube').

The tensile strength of copper depends on its purity, being about 200 N/mm^2 for the pure metal. Its main applications result from either its ductility or its very high electrical and thermal conductivity, the latter being 385 W/m°C (pure metal), the highest of any building material. Pure copper has a melting point of 1083 °C, so that creep is not a problem at normal temperatures.

Uses of copper

Copper tube   Copper is the most commonly used material for central heating and domestic water tubing, owing to its lightness and admirable working properties. It is available in annealed, half-hard or hard forms which are specified in BS EN 1057 as follows

annealed	R220
half-hard	R250
hard	R290

The figures above refer to the tensile strength of the material which rises as the tube is hardened by drawing. The type R220 is most suited to underground use where movement might occur. R250 is for normal uses in which bending may be required. R290 can only be used for straight runs (unless heated).

Copper tubing for water supplies has generally very high corrosion resistance, though it has been known to undergo corrosion in some acidic waters, such as those obtained from peaty catchment areas or containing carbonic acid (although, unlike ferrous metals or lead, pitting is only slight). Such water should be treated with sufficient lime to neutralise acidity without causing pipe scaling. Dirt or impurities in pipes may cause electrolytic corrosion, so that cleanliness during installation is essential, especially in soft water areas. Pitting corrosion has also occurred due to remnants of mineral oils used as drawing lubricants, which become converted to graphite on annealing. BS tested tube should not be affected in this way; it can be identified by regular markings on the tube. It is advisable to protect copper tubes chased into gypsum plasters with a coating of bitumen, particularly if damp conditions are prevalent. No protection should be necessary when buried in ordinary soils.

Copper tubing is also used in waste and soil systems, pipes over 50 mm diameter normally being bronze welded or silver soldered. Copper fittings may be used with copper tubing, capillary end-feed or solder ring-type fittings producing cheap, unobtrusive joints. They are also used with stainless steel tube in the absence, at present, of any similar product made in stainless steel.

Copper sheeting (BS EN 1172)   This is used for a variety of purposes in building, in dampproof courses, for weatherings and flashings, and as a roofing or cladding material. A number of forms are available, work hardened to various degrees. Annealed copper is essential for roofing, since manipulation is necessary but, where sheet is to some degree self-supporting, a half-hardened temper should be used. Copper sheeting should be fixed with copper nails, since other metals would be prone to rapid electrolytic corrosion. Welding is most easily carried out on deoxidised copper which contains a small percentage of phosphorous; the small amount of cuprous oxide which occurs in non-deoxidised (*tough pitch*) sheet tends to react with hydrogen in the flux or gas, causing unsoundness due to steam formation. The presence of arsenic in copper increases its strength at 200 °C and above, and also improves corrosion resistance. Copper may be alloyed so that it benefits from solution treatment and precipitation-hardening where higher strength sheet is required.

Copper alloys

Brasses   Zinc has an atomic radius similar to that of copper and dissolves to form a single phase FCC alloy in contents up to 36 per cent. The brasses obtained are known as alpha brasses; they are ductile and have high tensile strength, particularly, if the zinc content approaches the solubility limit or if work hardened. Since, during the process of work hardening, most distortion takes place at grain boundaries, these may corrode electrolytically with the grain body, leading to *season cracking*, unless stresses are relieved by annealing. Alpha brasses are used for pressing (such as in door furniture), stamping or drawing.

Alpha–beta brasses contain 36–46 per cent zinc, the beta phase (BCC) increasing as percentages approach the upper figure. They are less ductile than alpha brasses but are suitable for high temperature extrusion, rolling or casting. They are used in window sections, often having a bronze colour due to incorporation of a small percentage of manganese. Alpha–beta brasses are used for plumbing fittings, screws and general brassmongery.

Beta brasses contain more than 46 per cent zinc, though brasses containing more than 50 per cent zinc are not used. These brasses have melting points lower than 900 °C, compared to 1083 °C for pure copper, and are used for brazing.

Copper and copper alloy tube fittings   These are classified as capillary fittings; non-manipulative fittings, which use compression rings; and manipulative fittings, in which the tube at or near the end is formed into a flange which is compressed on to the fitting. Copper capillary fittings are described under uses of copper, though capillary fittings are also obtainable in brass. Joints are, perhaps, more conveniently made using compression fittings, though they are rather more obtrusive.

Dezincification of brass   Brass is an alloy of zinc, which is at the reactive end of the electrochemical series, and copper, which is at the noble end. It is not surprising, therefore, that under certain conditions, the zinc may corrode electrolytically with the copper. This will possibly result in blockage of hot-water fittings by the corrosion product; leakage; or even breakage. The brasses worst affected by dezincification are those with relatively high zinc content, for example, duplex (alpha–beta) brasses, containing about 40 per cent zinc, which are suitable for hot stamping of plumbing fittings. In theory, any brass containing 15 per cent or more of zinc may corrode in this way but attack in alpha brasses may be prevented by incorporation of small quantities of arsenic.

The severity of dezincification depends on the properties of the water, attack being most likely when the pH value is over 8 (that is, alkaline) and chlorides are present. The presence of temporary hardness tends to reduce attack. The reaction rate increases with temperature so that hot-water fittings are most commonly affected. Where such conditions are encountered, an alternative material should be used, such as copper, gunmetal (a bronze containing about 2 per cent zinc) or a protected alpha brass. Note, however, that since oxygen is involved in the corrosion process, brass fittings in recirculatory systems, such as central heating circuits, are largely immune from dezincification.

Bronze   Phosphor bronze is an alloy of copper and tin. Most bronzes have a tin content slightly over its solubility limit of about 6 per cent, so that they are generally harder and stronger than brasses. They have great durability, so that they are used for sculpture and ornamental work, and in place of brass where corrosion resistance is essential. Phosphor bronze contains up to 13 per cent tin and 1 per cent phosphorous, has high strength and corrosion resistance and is used for fixing of stone cladding.

ZINC

The element occurs in the form of zinc sulphide (zinc blende), extraction being carried out by concentration of the ore using a flotation process, conversion to zinc oxide and then reduction, using carbon, to the metal.

The metal has a low melting point (419 °C) and high thermal movement – up to $40 \times 10^{-6}/°C$ in sheet form, the actual value depending on the direction of rolling, since the HCP crystal structure is non-isotropic.

The tensile strength of pure zinc is low but it may be increased by hot working or alloying, though the ductility of zinc is, owing to its HCP structure, less than that of copper, lead and aluminium. Cold weather or severe working may make it necessary to warm the metal. It is, however, self-annealing since, at room temperature, it is above its recrystallisation temperature.

Many uses of zinc are based on the fact that it forms a coherent protective coating of zinc oxide or zinc carbonate on exposure to the atmosphere, though polluted or marine atmospheres increase the corrosion rate. Nevertheless, zinc is claimed to have good durability in most environments and rolled zinc sheeting has been used for roofing, weatherings and rainwater goods.

Flashing material may be alloyed with lead and has good working properties, though with relatively low tensile strength and creep resistance.

Roofing or cladding forms (BS EN 988)   These may be alloyed with copper and titanium to give higher strength and creep resistance together with lower thermal movement than the pure metal – typically $0.22 \times 10^{-6}/°C$. The metal cannot be burnt or formed like lead but similar details can be obtained by soft soldering (the correct flux, for example Bakers fluid, must be used together with low antimony solder). Zinc weathers to a colour rather similar to that of lead but has very little scrap value so that it can be used in accessible situations with little risk of theft. Adequate allowance for thermal movement must be made by means of joints in large areas of metal and better durability is obtained when the slope is sufficient for foreign material to be washed off. The metal forms an impervious corrosion product when in contact with cement but will be corroded by the acids which occur in timbers such as western red cedar or oak.

An important use of zinc results from its position in the electrochemical series (Table 8.3). A coating of the metal will protect any metal below it in the series, even if scratched, though durability of the product naturally depends on the thickness of zinc and the aggressiveness of the environment. However, zinc or zinc coatings should not be allowed to contact copper or other more noble metals, which would greatly accelerate corrosion. As well as in the processes described under 'Electrolytic corrosion', zinc is often used in zinc chromate form in anti-corrosive paints, though such paints, being porous, should be regarded as primers.

The low melting point of zinc is advantageous in the above applications but particularly so for die-casting in which, when alloyed with aluminium and magnesium, a wide variety of hardware components and fittings is produced. The metal is also used in paint pigments, such as zinc oxide and lithopone.

TITANIUM

Titanium has been used for many years in the aircraft industry as it can be used to form a range of alloys, has good corrosion resistance and combines reasonably high strength with low density (Table 8.12). The metal is quite abundant, being after hydrogen (Figure 5.1). It is only recently that it has found applications in construction, notably as a sheet cladding for roofs. The attractions appear to be:

- It can be anodised to produce very good corrosion resistance in any environment.
- Attractive colourings can be created by interference of light in the anodised surface.
- The metal has lower thermal movement than most other metals so that thermal stresses are reduced.
- Its quite high strength permits low metal thickness to be used (typically 0.6 mm).
- Raised jointing seams can be produced in the usual way or flat seams can be produced, with an underlying membrane. This permits the material to be used almost like a skin, emphasising the shape of the roof rather than the material itself.

Titanium crystallises to a hexagonal close packed structure (Table 8.1) which makes it a difficult metal to work – similar to austenitic stainless steel. This, combined with the high materials cost, tends to restrict application to high-profile buildings.

CHAPTER 9

ORGANIC MATERIALS

The term 'organic' traditionally refers to materials which are derived from living materials, for example, timber, leather and cotton or wool. More recently, the term has come to be used for a wider range of materials which, in common with the above, have properties dependent to a large degree on the presence of carbon in their structure. Plastics, paints, adhesives, mastics and bitumens fall into this larger group.

9.1 PLASTIC MATERIALS

While use of some of the more traditional organic materials in the construction industry continues at a steady rate, others, such as plastics, have enjoyed a spectacular rise in application and this trend is likely to continue. Plastics offer the advantages of low density and good resistance to the environment compared with competitive materials, for example, PVC has a density less than half that of concrete and about one seventh that of steel, while many forms require no maintenance, at least in the short term. The major problems of plastics are low stiffness (that of PVC is one sixth the value of concrete and one fortieth that of steel) and virtually all forms are combustible (if not flammable). With large investments being made into development of plastics, these and other problems are likely to be at least partially overcome if not eliminated. In energy terms, plastics may look expensive (see Table 3.1) and subject to uncertainties arising from oil price fluctuations. However, when their low density is considered, they become much more attractive in energy terms which, together with increasingly efficient usage, should assure the future for this materials group.

The backbone of each of these materials is normally the carbon atom, element number 6 in the periodic table, hence the low density of organic compounds, especially when they incorporate the lightest element, hydrogen. Carbon has two s and two p electrons in its outer shell. These electrons become hybridised (see Chapter 5) and, with a valency of 4, carbon is very largely covalent in character – it seeks to gain a stable octet of electrons by sharing with those of hydrogen, chlorine and fluorine (valency 1), oxygen (valency 2) and nitrogen (valency 3). There are a great many compounds which have a relatively simple arrangement of carbon and other atoms. The nomenclature for many organic compounds has recently been changed in school chemistry courses; the more traditional names will be used here,

437

though some of the new names are given in brackets to aid identification. The paraffin (alkane) series is the simplest hydrocarbon series:

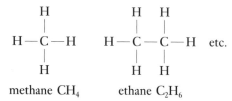

methane CH_4 ethane C_2H_6

In the paraffins, each carbon atom shares one of its electrons with another atom.

It is also possible for carbon atoms to share two or even three electrons with one another. For example,

$$H_{\diagdown}_{\diagup}H$$
$$C=C \qquad H-C\equiv C-H$$
$$H^{\diagup}^{\diagdown}H$$

ethylene (ethene)   acetylene (ethyne)

Such compounds are said to be unsaturated because double and triple bonds are often quite easily broken to form single bonds with other atoms or molecules.

Many organic substances have a relatively simple molecular structure and, therefore, a small molecular weight. They usually exist, however, in gaseous or liquid form, resulting from the short range bonding that they contain. In order to produce solids, the molecular size of these simple units must be increased. The process is known as polymerisation and, by it, many small molecules (monomers) can be linked to form large molecules – polymers. In order to produce a suitable degree of rigidity, polymers may need to have very large molecular weights, often over 100 000. There are two principal ways of polymerising materials and the resulting products tend to have characteristic properties.

Additional polymerisation

In this process, monomer molecules containing $C=C$ double bonds can be made to combine by means of initiator substances, such as benzoyl peroxide. The polymerisation of ethylene will be taken by way of example. The initiator first dissociates into unstable free radicals:

$$I-I \rightleftharpoons I-+I-$$

initiator   two free radcals
molecule

A free radical can now combine with an ethylene (ethene) molecule:

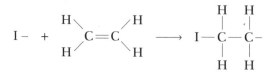

The product again contains a free radical, causing it to combine with another ethylene (ethene) molecule:

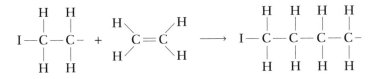

This is the propagation stage of polymerisation. It will continue until, as ethylene (ethene) molecules become used up, two growing chains combine or the free radical at the end of a chain combines with a free radical formed by an initiator molecule. By this process, ethylene is converted into polyethylene (normally called polythene), having the formula:

$$[-CH_2-CH_2-]_n$$

where n is a large number; n is sometimes referred to as *the degree of polymerisation*, and would typically be over 10 000 for most practical polymers.

At each end of such a chain, there would be an initiator fragment:

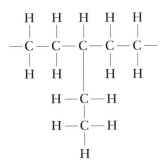

It is found, in practice, that there are often many more end groups than a simple chain structure would suggest and this is caused by branching during polymerisation:

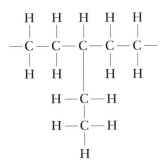

Such branches may be formed when, for example, a growing molecule collides with the side of a completed molecular chain and transfers its free radicals, the growing molecule, therefore, becoming 'dead' instead. Branches will affect the properties of the polymer, as explained later. In a further type of addition polymerisation, monomer molecules of different types may combine, for example, butadiene and styrene:

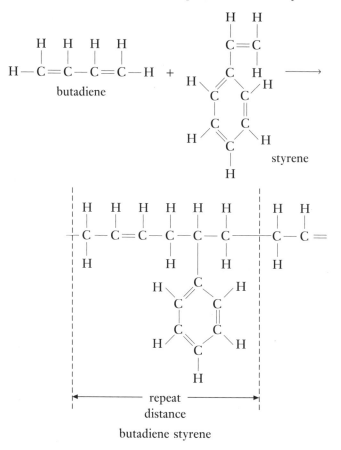

butadiene styrene

Double bonds are opened in both the butadiene and styrene molecules, though one double bond remains in each repeat unit. A chain structure is again produced but, since two different monomers are involved, the polymer is described as a *copolymer*. There may be three or more monomers in a chain, for example, acrylonitrile butadiene styrene – ABS. Where double bonds are present in the polymer, they can, in some circumstances, form links between chains leading to a more rigid *cross-linked* structure. Substances such as butadiene, in which there are two positions where additions can occur, are described as *bifunctional*.

Condensation or step polymerisation

In this type of reaction, a small molecule, often water, may be produced as a by-product. For example, polyesters are produced when a dihydroxyl alcohol reacts with a dicarboxylic acid (the di prefixes imply two alcohol (–OH) and acid (–COOH) groups in the respective compounds):

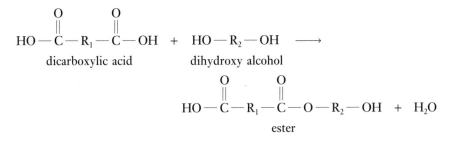

where R_1 and R_2 are hydrocarbon groups.

The reaction can be repeated, producing a long-chain molecule which is the basis of polyester fibres. Alternatively, polyfunctional acids and alcohols can be used, these having three or more reactive groups. In such cases, a three-dimensional molecular network is formed by the condensation process – one way of producing rigid polyesters. The term *step polymerisation* is sometimes preferred to *condensation polymerisation*, since polymerisation occurs by multiplication of ester molecules rather than addition of ester monomers singly. In some cases, also, there may not be a small molecule produced from the reaction.

STRUCTURE OF THERMOPLASTIC MATERIALS

When polymers retain a chain structure, there being no primary bonds between the chains, the material is described as *thermoplastic*. Thermoplastics usually exist in the solid state at room temperature, due to the effect of van der Waals forces between chains but, on heating, softening soon occurs, as thermal energy overcomes these weak bonding forces. Such polymers are therefore easily moulded by heating, although unless, by drawing, the chains are orientated in the direction of stress, thermoplastics usually have limited strength. The strength of individual thermoplastics depends on the extent of bonding between chains, which depends, in turn, on the polarity of the polymer molecule. Polyethylene is, for example, non-polar, since bonds are entirely covalent and chains are bonded only weakly but by, for example, replacing one hydrogen atom by a chlorine atom, forming polyvinyl chloride $(-CH_2-CH.Cl-)_n$ interchain forces are increased, producing a stiffer, stronger polymer with a higher softening point. This effect is obtained because chlorine is electronegative and inclined to form polar bonds, thereby leading to dipoles at regular points in the polymer chain. Thermoplastics are generally produced by addition polymerisation.

STRUCTURE OF THERMOSETTING MATERIALS

These are usually more rigid than thermoplastics on account of the extended three-dimensional bonding which they contain. Unlike thermoplastics, they are not softened by heat but, rather, they decompose if heated sufficiently. Thermosets, as explained above, are most likely to be obtained by polymerisation of polyfunctional monomers, although it is also possible to produce them by cross-linking of thermoplastics.

CRYSTALLISATION OF POLYMERS

The polymers most likely to crystallise are those having a molecular structure comprising simple regular chains. Linear polyethylene with its simpler hydrocarbon back-bone has, for example, a strong tendency to crystallise. However, even polyethylene does not exhibit 100 per cent crystallisation, since the long chains must be folded back on each other and there will always be some chain entanglement, causing amorphous regions. Crystal growth in polymers is initiated by nuclei, as in other materials and, on account of the chain structure of linear polymers, folded chains grow radially from the nucleus as polymerisation occurs. Spherical regions of semi-crystalline material (*spherulites*) may, hence, form and these grow until they meet rather like metal crystals (Figure 9.1). The size of the spherulites and, hence, the properties of the polymer, can be controlled by varying the number of impurity nuclei in the melt material. Crystallisation improves mechanical properties such as strength and stiffness since it results in a higher concentration of bonds in the material.

Branched polyethylene will show a lower degree of crystallisation, as will polymers having less regular molecular chains such as polyvinyl chloride. In many thermoplastic polymers, the extent of crystallinity is, therefore, small and properties are characteristic of amorphous materials; for example, many thermoplastics are transparent to light. Crystallinity in thermosetting polymers is generally negligible, the extensive three-dimensional framework being rather similar to that of ordinary glass.

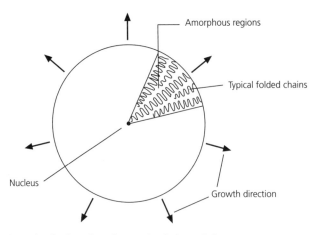

Figure 9.1   Growth of spherulites from polyethylene chains

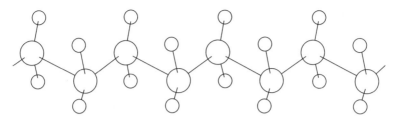

Figure 9.2 Alternation of bond directions in a polyethylene chain so as to minimise repulsion between hydrogen atoms

MECHANICAL PROPERTIES OF POLYMERS

The physical form which a polymer takes depends in a complex way on a number of factors, of which the most important are its molecular weight, its degree of crystallinity and the temperature. In the case of polymers of low molecular weight, the situation is relatively simple, polymers behaving as either amorphous (glassy) or crystalline solids; or, on heating, viscous liquids. In the solid state, molecular energy is purely vibrational while in the liquid state a translational component is possible. As molecular weight is increases (degree of polymerisation 10 000 or more) an intermediate stage is introduced in which molecular chains are entangled, as in solids, but flexibility resulting from bond rotation is possible. Since the bond directions in tetravalent carbon are well defined, adjacent hydrogen atoms in, for example, a single hydrocarbon chain are staggered in order to minimise repulsion of hydrogen atoms (Figure 9.2). Bond rotation involves the approach of hydrogen atoms and it will, therefore, not be possible, except where thermal energy is sufficient to overcome the temporary repulsion associated with it. When, on heating, bond rotation becomes possible, it provides an intermediate state – the 'rubbery' state between the brittle solid and viscous liquid state of a polymer of high molecular weight. The temperature range over which this state is exhibited increases with molecular and crystal size and it is found that there is, in addition, a diffuse transition zone between rubber and liquid, due to the fact that there is a statistical distribution of molecular sizes in the material – some molecules effectively 'melt' before others (Figure 9.3). The essential characteristics of a rubber are low elastic modulus combined with great extensibility under tensile stress, which results from straightening of the former random zig-zag chains. In the case of crystalline polymers, there is a further intermediate flexible state between rigid solid and rubber, on account of the extra rigidity produced by crystalline regions (Figure 9.4), the rubbery state being obtained when, on heating, crystallites melt. The temperature at which bond rotation becomes possible is known as the *glass transition temperature*, T_g, while the approximate temperature at which the rubber changes to liquid is described as the *flow temperature*. The term *melting point*, T_m, is generally used in relation to crystallites only. Both T_g and T_m depend on the molecular weight of the polymer. Table 9.1 shows typical values for the common thermoplastics, together with relative densities, coefficients of thermal expansion,

Table 9.1 Properties of common thermoplastics, given in order of increasing elastic modulus

Polymer	Degree of crystallisation %	T_g °C	T_m °C	Relative density	Expansion coefficient $\times 10^{-6}$ per °C	Short-term tensile strength N/mm²	Tensile elastic modulus kN/mm²
LD Polyethylene	60	−120	115	0.92	220	8	0.5
PTFE	95	−120	327	2.10	110	20	0.5
HD Polyethylene	95	−120	143	0.95	130	27	0.9
Polypropylene	60	−27; 10	165	0.90	110	30	1.3
ABS	0	71–112	–	1.05	80	40	2.5
Nylon 66	100	57	265	1.14	90	70	2.6
Polycarbonate	0	150	–	1.22	55	60	2.7
Polymethylmethacrylate	0	80–100	–	1.18	65	70	2.9
Polystyrene	0	80–100	–	1.05	70	50	3.0
Polyvinylchloride	0	80	–	1.40	70	50	3.2

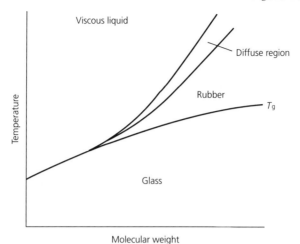

Figure 9.3 Physical states of amorphous polymers

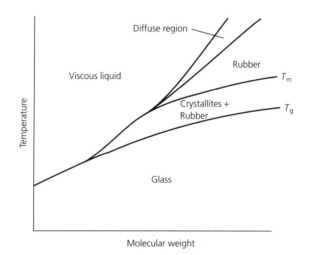

Figure 9.4 Physical states of crystalline polymers

tensile strengths and tensile elastic modulus values. Note that crystalline melting points are only given for those polymers which crystallise to a significant extent.

MODES OF USE OF POLYMERS

These fall into three areas:

1 Crystalline polymers may be employed over a wide temperature range, since they have considerable rigidity both above and below their glass transition temperature, there being only a small change of stiffness during the transition.

Above their T_g, such polymers are more likely to be leathery, while below their T_g, they are more likely to be brittle.

2 Amorphous polymers are rubbery when above their T_g and are, therefore, used as elastomers in this state.

3 Amorphous polymers are inclined to be harder, though more brittle, below their T_g. Many such polymers are utilised in this state.

While the above are generalised properties, it should be appreciated that the behaviour of individual polymers depends on their molecular structure, size and, in many cases, the rate of cooling since, as in metals, this affects the degree of crystallisation and the size of crystallites, if formed. It is also possible to reduce the T_g of thermoplastic polymers by use of solvents which interfere with attraction between molecular chains – especially in polar molecules such as polyvinyl chloride, resulting in this case in plasticised PVC. The science of molecular engineering in polymers is still advancing rapidly and it is possible that some of the less satisfactory properties of polymers, such as low rigidity and softening temperature, may be improved, resulting in new applications.

THERMOPLASTIC MATERIALS

Thermoplastics offer, in general, the advantages of flexibility and toughness over thermosets, though strength, stiffness and heat resistance are usually inferior. The ability of thermoplastics to soften on application of moderate heat is also an advantage, since:

1 The polymerisation and fabrication stages – each specialist operations – for thermoplastics can be separated. The polymers are often produced in convenient bead form and can then be moulded or extruded by heat into the desired product.

2 Moulded plastics can easily be joined by thermal fusion welding.

Polyethylene (polyethene or polythene) $[-CH_2-CH_2-]_n$

Polyethylene has the simplest chemical structure of any polymer and is also one of the cheapest and most widely used plastics. The plastic is manufactured in two basic forms involving either high or low pressure processes. The high pressure process, which was the first to be developed, gives a branched polymer with a low degree of crystallinity and which, therefore, has a low relative density (0.92), fairly low softening point (90 °C) and high flexibility. The more recent low pressure process produces a largely unbranched polymer of high crystallinity, having a higher relative density (0.96) and softening point (120–130 °C). This is produced by using ionic catalysts instead of free radical catalysts. An aid to identification of polyethylenes is that both types float on water.

The absence of polarity in the polymer results in great chemical stability; polyethylene does not dissolve in any solvent at room temperature. It is also unaffected by acids or alkalis. A further consequence of non-polarity is electrical insulation – polyethylene has extremely good insulating properties. One property resulting from the relatively weak intermolecular bond is, however, high thermal movement – the coefficient of linear expansion is approximately $220 \times 10^{-6}/°C$. The polymer is well above its T_g (−120 °C) at ordinary temperatures and is therefore tough, even in cold weather. It undergoes embrittlement, however, on exposure to ultraviolet light, due to cross-linking. This can be substantially reduced by incorporation of carbon black. Permeability to gases is relatively high, caused by diffusion of molecules from one void to another, assisted by bond rotation. Permeability of the high density variety is, however, only 20 per cent that of low density polyethylene.

The chief use of low density polyethylene is for sheeting for dampproof courses and membranes for vapour barriers and temporary shelters, although, on account of its crystallinity, transparency is not as good as that of, for example, clear PVC sheeting unless very thin films are used. Further uses include tubes, down-pipes and cisterns, though the polymer is not suited to hot-water services. Carbon black should be incorporated for external purposes. Polyethylene products have excellent resistance to impact, unless embrittlement occurs. One disadvantage of the non-polar structure is, however, that surfaces have a very low surface tension, having a rather greasy touch and are not therefore amenable to painting or to joining with adhesives. Mechanical roughening assists in providing a bond, while fusion welding is the most common way of forming joints. Epoxy resins may be used if the surface is oxidised by means of concentrated acid.

The property of refusing to bond with most other materials can be put to good use where bonding is be avoided. Polyethylene can be used with almost any adhesive as a debonding agent, allowing ready separation once adhesives have hardened.

In situations where higher strength or stiffness is required, high density polyethylene is preferred. Further applications of polyethylene are in insulation of coaxial cables and in underground pipes such as gas and water mains where the resilience and flexibility are an advantage, though substantial wall thickness must be used in the case of water mains on account of their limited tensile strength.

Polypropylene (polypropene) $[-CH_2-CH.CH_3-]_n$

This has a structure very similar to polyethylene, apart from the extra methyl group. In most commercial forms, the chain structure is stereo-regular and *isotactic*, that is, the configurations of all carbon atoms attached to the methyl groups are the same. In a simplified planar projection this may be represented:

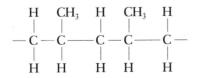

The polymer crystallises in a helical form with three methyl groups in each rotation, giving rise to a fairly high softening point (150 °C). This, together with higher stiffness than that of polyethylene and excellent chemical resistance, forms the basis of most uses of polypropylene. The polymer is used for rigid units, such as manhole mouldings and waste systems, including laboratory wastes, and may be a possibility for hot-water pipes and cylinders, though thermal movement is very high (Table 9.1) and movements must be carefully allowed for. Polypropylene is an example of a polymer which undergoes several transitions on cooling, caused by changes in different modes of movement in the molecular chain. The dominant change occurs at 0 °C and it leads to brittleness in the polymer at low service temperatures. Impact resistance can be improved by block copolymerisation with small amounts of ethylene. Polypropylene is also more susceptible to oxidation and ultraviolet rays than polyethylene – it must be stabilised for external use. A further use of the polymer is in fibre reinforcement and general purpose string since, on drawing, it produces a fibre of fairly high strength and flexibility (though the stiffness of polypropylene fibres is much lower than that of ceramic or metallic fibres). Being a pure hydrocarbon, polypropylene has a particularly low relative density (0.9) and, like polyethylene, floats on water.

Polyvinyl chloride (PVC) $[-CH_2-CH.Cl-]_n$

In this polymer, every fourth hydrogen atom is replaced by a chlorine atom, the arrangement being partly syndiotactic (adjacent Cl atoms on opposite sides of the chain):

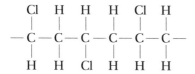

There is not, however, sufficient stereo-regularity in the distribution of the chain atoms to enable crystallisation to occur, hence the polymer is amorphous and is obtainable in clear sheet form. The unplasticised polymer has considerable strength and rigidity below its T_g of approximately 80 °C (more commonly referred to as the *softening point temperature*), as a result of its polar character. A further use of PVC is, however, based on the plasticised form: polar plasticisers, such as esters, interfere with interchain attraction and, hence reduce the softening temperature of the polymer.

PVC in unplasticised form (uPVC or, alternatively PVC-U) accounts for about 80 per cent of the total usage of plastics in the construction industry since it offers an attractive combination of good mechanical properties, low cost and is easily extruded or moulded to the desired shape. Main applications are:

■ uPVC is now widely used in extruded window and conservatory frames, usually incorporating double glazing. Larger size frames may be stiffened by steel inserts sealed into the frames, though a major advantage of uPVC frames over solid metal frames is that they contribute to the overall insulation of the window and do not attract condensation. The commonest colour of frames is white, produced by incorporation of titanium dioxide pigments. uPVC may also incorporate pigments producing a maintenance free surface with built-in colour. Brighter pigments may fade with age and it is found that darker coloured frames tend to get hotter when subjected to sunlight. Experience of long term weathering of uPVC is limited but rates of deterioration are likely to increase in sunnier climates, especially on south-facing aspects. Deterioration appears to take place by a number of mechanisms including:

1 The titanium dioxide catalyses degradation of adjacent material due to UV light. This effect is controlled, though not eliminated, by stabilisers such as lead, cadmium or barium, the latter giving better colour stability.
2 Hydrogen chloride is eliminated from the long chain molecules. This tends to produce surface acidity, yellowing and cross-linking of the polymer chains, leading to embrittlement.
3 Calcium carbonate fillers are attacked by the acid from (2) leading to loss of gloss.
4 As a result of (3) surfaces become receptive to dirt.

Gloss surfaces can be obtained but since they contain more polymer, they tend not to be used since degradation may be accelerated. Performance of any one brand depends on formulation but faster deterioration is likely with coloured frames. Another possible long term problem is that of distortion, especially of opening lights under the substantial weight of double glazing since all thermoplastics are subject to creep. BS 7413 relates to white uPVC hollow profiles for windows and includes an artificial weathering test to check for colour stability and retention of impact resistance.

In spite of uncertainty in the long term, uPVC glazing systems for small to medium size openings are so competitive in price that they remain a popular choice even when life times of perhaps around 20 years are taken into account. Figure 9.5 shows a typical section, formed by extrusion.

■ uPVC rainwater goods now dominate the domestic market. It is important to give adequate support to gutters and long pipes and to allow room for expansion – the expansion coefficient of uPVC is approximately $70 \times 10^{-6}/°C$. Expansion joints normally take the form of rubber gaskets in gutters. Some maintenance may be required, for example in long lengths, joints may open due to repeated movement and gaskets may leak if contaminated by dirt. BS 4576-1 specifies half-round gutters, circular pipes and accessories.

■ The use of uPVC for claddings, fascias, soffits and similar applications is now widespread. They are easily fixed and under the low stresses encountered in such situations, should give good performance over many years. The factors affecting performance are similar to those given under windows above.

■ Pipes for waste systems should be sleeved when in contact with other materials to avoid noise or damage on repeated movements. When rigid joints are acceptable, they are easily and cheaply made by solvent welding, there being,

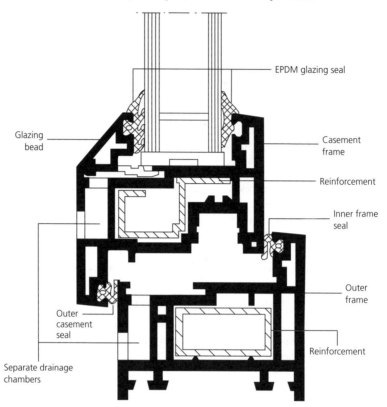

Figure 9.5   Typical section through an opening casement manufactured from extruded PVC sections, reinforced with steel inserts. (Source: Ostermann and Scheiwe, Münster)

unlike polyethylene, a number of (polar) solvents which will dissolve PVC. The two units are joined by a collar and the joint is made by solution and re-solidification of surface layers in contact. Note that the joint must be accurately made as it cannot be reopened as with metal jointing systems. uPVC waste systems are common, although where the effluent is at a high temperature, such as in the case of automatic washing machines, a polymer having a higher softening point would be recommended. The presence of chlorine in the PVC family of plastics reduces flammability, though like all polymers they are still combustible, hence pipes and soil pipes must be ducted when passing between elements of a structure in order to satisfy *Building Regulations*. A further application is underground waste pipes where uPVC offers the advantages that long lengths can easily be handled and smooth internal surfaces reduce friction, enabling smaller pipes to be used. They can also accommodate considerable ground movement without damage.

■ Clear uPVC is often used in roof lights and corrugated roofing, though light transmission and impact resistance invariably deteriorate with time.

■ Stabilised plasticised PVC is widely used in floor coverings, electrical insulation and fabrics for roofing.

It is very likely that further development will improve long term performance, leading to an even wider range of uPVC goods.

There has been serious concern that the use of chlorine in the manufacture of PVC constitutes an environmental hazard since chorine is liberated as part of the manufacturing process and there is therefore a risk of harmful chorine emission in wastes. Another area for criticism is that recycling of PVC still only takes place at low volumes, most material even then usually finding its way into lower grades of product such as garden furniture. When disposed of as landfill, it is estimated that breakdown could take in excess of 100 years. Consequently, PVC products may attract relatively poor environmental ratings.

The industry is taking steps to reduce all such emissions and demonstrate compliance with emissions recommendations. As increased quantities of PVC find their way into the waste stream there are hopes that the recycling situation will improve. Recycling plants now exist in Germany that will recover complete uPVC window assemblies, mechanically separating them into their four components – uPVC, glass, steel (from reinforcement where used) and rubber (glazing seals). Clearly the investment involved in operations of this type can only be viable when reprocessing takes place on a very large sale. In recycling terms one possibility is that recycled uPVC may be incorporated into the core of extrusions, thereby reducing the amount of virgin material required. Another is that the material could be re-used and a protective coating given to improve weathering and stability properties.

Polyvinyl fluoride (PVF) $[-CH_2-CH.F]_n$

This polymer has a similar structure to PVC, except that, since the fluorine atom is smaller than the chlorine atom, it has a much greater tendency to crystallise. PVF combines high temperature resistance and toughness at low temperatures with very good weather resistance. It is used as coating material for wood and metal, being much more durable than paint films.

Polytetrafluoroethylene (PTFE) $[-CF_2-CF_2-]_n$

This has some of the characteristics of polyethylene, being a crystalline linear polymer. However, since fluorine atoms are larger than carbon atoms, they are arranged in a spiral formation about the zig-zag carbon backbone. There is close interlocking between chains, leading to a polymer with a high melting point, 327 °C, in spite of the fact that bonding between the chains is not strong. Since the polymer is non-polar, it has no solvents at ordinary temperatures and PTFE has a very low coefficient of friction – less than that of wet ice upon wet ice. This property forms the basis of most applications – the polymer is used to line pipes which carry solid materials, for sliding bridge bearings and as a non-stick coating on cooking utensils. It is also available in tape form for sealing threaded joints in water or gas pipes. PTFE has excellent resistance to weathering and to moisture, though it is degraded by ultraviolet light. More recent uses include fabric roofing which is based upon its light diffusion, weathering resistance and self-cleaning properties.

Ethylene tetrafluoroethylene (ETFE)

This is a copolymer related to PTFE with similar chemical stability but of lower crystallinity, resulting in very good light and UV transmission properties. The material cannot be reinforced so that it tends to be used in small panels. It is used in low thicknesses, described as a *foil*, insulation being achieved by utilising it in two or more skins, an air gap being maintained by applying a steady small air pressure between the skins. Like PTFE, it has excellent dirt resisting properties.

Styrene based polymers

Styrene consists of a simple combination of ethylene and benzene in which the double-bond structure of the former is retained. This is responsible for the ease with which it polymerises. The chief polymers based on styrene are polystyrene and acrylonitrile butadiene styrene, which are both thermoplastic.

Polystyrene

This polymer has a chain structure similar to polyethylene, except that it is *atactic*, that is, there is no regular configuration of the carbon atoms containing the benzene rings in the chain. The polymer does not, therefore, crystallise but forms an amorphous thermoplastic with a softening point (T_g) of 80–100 °C. The polymer degrades at high temperatures and has a characteristic odour on burning. It is quite rigid, though brittle at ordinary temperatures due to mechanical interlock between the benzene side groups in adjacent chains. The polymer is attacked by a number of solvents and has poor weathering resistance.

Many applications are based on its cheapness and rigidity: these include containers and lining materials for refrigerators; for such purposes, the material can be toughened by incorporation of rubber. A large amount of the polymer is produced in the form of expanded polystyrene, which has one of the lowest thermal conductivities obtainable – approximately 0.033 W/m°C. The polymer in the form of small beads is heated so that great expansion occurs. The beads are then softened by steam and pressurised so that they stick together. Expanded polystyrene is widely used as an insulating material for general household purposes, in rolls for application to wall surfaces, in floor and roof insulation, as a preformed cavity infill and for ceiling tiles. The unmodified plastic burns readily, however, and may therefore constitute a serious fire hazard in sheet or foam form. Fire retarded varieties are available at increased cost. Expanded polystyrene may, in sufficient thicknesses, have vapour resistant qualities but it is not as effective as a vapour barrier of foil or polyethylene. It is also used as a void-former in structural concrete.

Acrylonitrile butadiene styrene (ABS)

ABS polymers are two phase systems comprising copolymers of styrene – acrylonitrile and the rubber, butadiene styrene. A variety of materials can be obtained and these are characterised by good impact strength, temperature resistance and solvent resistance. They are more expensive than common plastics but have, nevertheless, found application in window fasteners, rainwater goods and moulded articles, such as telephones. They have a higher softening point than uPVC and are therefore more suitable for domestic waste systems. To be solvent-welded, they require a special cement – PVC solvent cement does not have a solvent action on ABS. They are also used in gas piping.

Acrylic plastics

These are based on acrylic acid, $CH_2=CH.COOH$ (note the ethylene (ethene) group). The commonest polymer is polymethyl methacrylate:

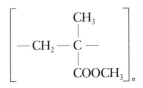

The methyl and ester groups are not interchangeable and, since the polymer is atactic, it must also be amorphous, the polar groups producing a rigid solid with a T_g of 80–100 °C. It is commonly known as *Perspex*, a clear, hard sheet material. Methyl methacrylate polymerises readily by application of heat, so that acrylic sheet is formed simply by pouring a partly polymerised syrup into a mould and then warming it. Very good surface finishes can be obtained with good weathering properties – they are not degraded by ultraviolet light. The plastic is used for the diffusion of light, as in illuminated ceilings and signs. Baths are now made out of the polymer and are much cheaper and easier to install than the traditional heavier, cast iron or steel equivalents, though of inferior durability and stability. Scratches may be polished out but the plastic is susceptible to burns and to some organic solvents. The use of water based low VOC paints formed from emulsified acrylic resins is greatly increasing.

Nylons

The chemical name for this polymer is *polyamide* and it may be obtained by polymerisation of amino acids or lactams, or by reaction of diamines with dicarboxylic acid. Amino acids have the formula $NH_2R.COOH$, where R is a hydrocarbon radical, the corresponding polymer being $[-NHRCO-]_n$. Note than an H is removed from one end of the repeat unit and an OH from the other; the acid polymerises by a condensation reaction.

Lactams have the form:

$$R \diagdown \begin{matrix} \diagup NH \\ | \\ \diagdown CO \end{matrix}$$

and polymerise when the ring is opened. Nylons obtained from amino acids or lactams are characterised by a single number representing the number of carbon atoms in the repeat unit. For example, caprolactam contains six carbon atoms and polymerises to nylon 6:

$$\left[\begin{matrix} -C-(CH_2)_5-N- \\ \parallel \qquad\qquad | \\ O \qquad\qquad H \end{matrix} \right]_n$$

When hexamethylene diamine:

$$\begin{matrix} H \diagdown \qquad\qquad\qquad \diagup H \\ N-(CH_2)_6-N \\ H \diagup \qquad\qquad\qquad \diagdown H \end{matrix}$$

is reacted with adipic acid:

$$H-O-\underset{\underset{O}{\parallel}}{C}-(CH_2)_4-\underset{\underset{O}{\parallel}}{C}-O-H$$

nylon 66 is produced by a condensation reaction in which a hydrogen atoms is taken from each end of the hexamethylene diamine to combine with a hydroxyl group from each end of the adipic acid, forming water. The chemical formula for this nylon is, therefore:

$$\left[\begin{matrix} -N-(CH_2)_6-N-C-(CH_2)_4-C- \\ | \qquad\qquad | \quad \parallel \qquad\qquad \parallel \\ H \qquad\qquad H \quad O \qquad\qquad O \end{matrix} \right]_n$$

Nylons are linear polymers and therefore thermoplastic and, since the polar –NHCO– groups are regularly spaced, the polymers crystallise, the strong interchain reaction resulting in high melting points. Nylon 6 has, for example, a melting point of 225 °C, while nylon 66 has a melting point of 265 °C. Nylons are also characterised by high strength, flexibility, toughness, abrasion resistance and solvent resistance. They tend to become brittle in sunlight, however, unless they are stabilised or carbon black is added. Products may be moulded, extruded or drawn. Moulded articles, which are normally manufactured from nylon 66, include catches, latches and rollers: extruded products include central-heating tube, drawn products and nylon rope. The ease of cutting nylon central-heating tubes makes installation very simple, though thermal movement should be allowed for and sharp objects could cause damage in service.

Polycarbonates

These are obtained by reactions of polyhydroxy compounds with carbonic acid
derivatives. The polymer has excellent transparency and is tough and rigid up to
temperatures of over 100 °C, though in common with other plastics it scratches
relatively easily and optical properties soon deteriorate. Weathering resistance
is good and sheets of the polymer are used as a glazing material of high impact
resistance. The material is not easily ignited, a valuable advantage when used as
glazing; it will tend to melt away from flame.

Modified thermoplastics for hot water services

The introduction of plastics for water service pipes and particularly hot water has
been slower than might have been expected, in view of the rapid advance of plastics
technology. This may be due, in part, to conservatism in the industry and the belief
that the plastics are no substitute for the traditional metal based systems with their
greater feeling of 'permanence'. Further disadvantages are the high thermal
movement and low softening point of many plastics. A number of modified
plastics which offer possibilities for the future are given below.

Modified unplasticised PVC (MUPVC) has improved stiffness, chemical
resistance and high temperature performance and would appear to be an attractive
proposition for applications in hot water services/wastes.

Chlorinated PVC (CPVC) is also a possibility in this area. Both the above types
cause problems when used in long pipe runs or under screeds where restrained
expansion can lead to failure.

Polybutylene is a further contender for hot water piping with the possible
advantage that it is slightly less stiff than other high temperature polymers, allowing
some of the strain of thermal movement to be absorbed within the pipe, as well as
permitting easier installation. Such types would, however, need more support clips
when suspended, to avoid sagging.

Post-chlorinated PVC is now marketed for domestic hot water systems (not
central heating systems which might, if faults occur, have to withstand temperatures
as high as 100 °C). The material can be solvent welded.

ELASTOMERS

The rubbery state has already been briefly described as a state between solid and
liquid, in which bond rotation of carbon–carbon bonds in the linear polymer
backbone is possible. Since zig-zag arrangements are statistically more likely than
straight polymer chains, elastomers exist as randomly entangled (amorphous) chains
of this type (Figure 9.6). On stretching such a polymer, chains tend to become
straighter, causing bond rotation within them and, hence, slight resistance to stretching
(Figure 9.6). Elastomers must have the ability to recover fully when stress is removed
and this is achieved either by slight cross-linking (vulcanisation) of chains to
prevent their becoming permanently re-orientated or, alternatively, ensuring that

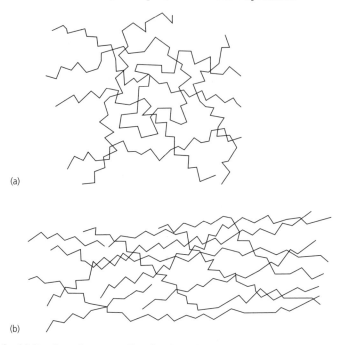

(a)

(b)

Figure 9.6   (a) Random zig-zag profile of carbon chain in unstressed elastomer; (b) chain straightening caused by stretching

the molecular size is large enough to produce sufficient chain entanglement. It may be an advantage if, on stretching fully, some degree of crystallisation occurs, since this increases resistance to further stretching at this stage. The effect is obtained on blowing up a balloon or stretching a rubber band; a definite increase in resistance is obtained at higher strains. Many elastomers exhibit this property as strains reach their maximum value (in the region of 500 per cent; Figure 9.7) and molecular chains tend to become closer and straighter. Elasticity is, of course, only exhibited over a limited temperature range. On cooling below their T_g, elastomers become brittle, while on heating, softening and ultimately melting occurs. Some elastomers, such as natural rubber, styrene butadiene rubber (SBR) and nitrile rubbers contain double bonds and are, therefore, susceptible to oxidative degradation, due to the action of heat and light in the presence of oxygen or ozone. Degradation takes the form of chain cleavage leading to tackiness and ultimately embrittlement. Such rubbers must be protected by antioxidants. It is also possible to increase the abrasion and tear resistance of elastomers. For this purpose, finely divided reinforcing agents, such as carbon black, are used – these form secondary bonds with the elastomer molecules. Elastomers which include carbon black will, of course, be dark in colour.

   The elastic moduli of elastomers lie generally in the range 1–20 N/mm^2 at 300 and 400 per cent elongation – roughly one per cent of the moduli of elasticity of corresponding polymers when below their T_g.

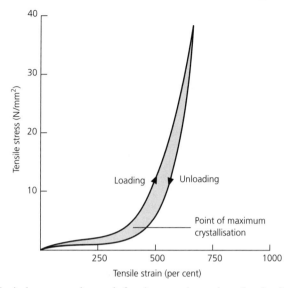

Figure 9.7 Typical stress–strain graph for elastomer in tension, showing increase of elastic modulus when crystallisation reaches a maximum. The hysteresis effect is also shown; the area enclosed represents the heat generated

Natural rubber and synthetic polyisoprene

Natural rubber is obtained from latex trees in tropical regions, including parts of Africa, South America and Malaysia, the most important species being *Hevea braziliensis*. Chemically, natural rubber consists chiefly of Cis 1.4 polyisoprene:

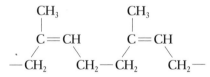

The polymer adopts an entangled helical conformation which may be stiffened by vulcanising with sulphur and other ingredients. Sulphur atoms link adjacent polymer chains by the influence of double bonds. When less than 5 per cent sulphur is used, natural rubber has great flexibility but when more than 30 per cent is used, a highly rigid structure, known as *ebonite*, results. The cross-links also help avoid crystallinity in the unstretched state, while still permitting crystallisation on stretching fully. Natural rubber is often reinforced with carbon black and must be protected from oxidation. The material is relatively cheap and has very good elasticity and dynamic properties (hysteresis losses, which could generate excessive heat in vibration conditions, are low – see Figure 9.7). It is used widely for bearings of all types, for example, bridge and machine bearings. In some cases, whole buildings have been supported on rubber bearings to stop vibration transmission, for example, from underground trains. Latex silicone compositions have been injected into masonry for dampproofing purposes. Natural rubber is also used in the tyres of large earth-moving vehicles, since less heat is generated during the severe flexing which such tyres undergo.

Styrene butadiene rubbers SBR (Buna S rubbers)

The term *Buna S* relates to the fact that polymerisation of butadiene was formerly initiated by sodium, Na, styrene being used to form a copolymer. These are cheaper than natural rubber, though tear resistance and resistance to oxidation are poor, unless treated. They have only a slight tendency to crystallise on stretching, though abrasion resistance is good. SBRs are used widely for general purposes.

Butadiene acrylonitrile rubbers (NBR Buna N or nitrile rubbers)

These are similar to SBR rubbers but have better resistance to oils and hydrocarbon solvents. They are widely used in contact adhesives, gaskets and seals.

Butyl rubber

This is a copolymer of isobutylene and isoprene and has good sunlight and oxidation resistance, due to the smaller number of double bonds present. The elastomer is also impermeable to gases and resistant to acids and oils. It is used in mastics and sealants, and landfill site liners.

Polychloroprene (Neoprene)

This has very good all-round chemical stability and, since it crystallises on stretching, it combines high extensibility with high strength. It has relatively good fire resistance. It is used in contact adhesives and sealing strips and for glazing and O rings, though it is only available in black.

Ethylene propylene diene monomer (EPDM)

This material has more recently challenged neoprene for gaskets such as in glazing (see Figure 9.5). It can be obtained in any colour, its chief disadvantage being its relatively inferior chemical and fire performance.

Silicone rubbers

Silicon, like carbon, has a valency of four, though polymers with a pure silicon backbone are unstable. Instead, the siloxane link forms the basis of polymers:

$$-\overset{|}{\underset{|}{Si}}-O-\overset{|}{\underset{|}{Si}}-$$

Silicone elastomers are normally polydimethyl siloxanes, often reinforced with silica. Cross-linking in certain forms (room temperature vulcanising, RTV, types) can be activated by atmospheric moisture. Resistance to extremes of temperature is good and they are widely used in caulking compounds. Note that polydimethyl siloxanes are non-polar and, therefore, have very low surface tension and good water repellant properties.

Polysulphides

These contain at least four sulphides per monomer molecule, two of which form part of the polymer backbone. They have very good all-round chemical stability and are widely used in mastics and sealants.

THERMOSETTING MATERIALS

Applications of these are based upon their strength, stiffness and heat resistance compared with thermoplastics. They are usually rather brittle, though the degree of cross-linking can be controlled to improve toughness at the expense of other properties, if required.

Thermosets cannot be softened by heat; they usually decompose chemically first, so that they must be moulded at the time of polymerisation. Manufacturers of thermosets would therefore be supplied with resin and hardener which must be mixed precisely in controlled conditions to obtain a reliable product. Other materials such as fillers may also be added to modify properties.

Although thermosets do not have a glass transition temperature on account of their cross-links, they do become less rigid as temperature rises due to thermal energy. This may be measured by the *heat distortion temperature* (HDT) which involves the response of the material to stress. For example, HDTs for polyesters and epoxies are both about 70 °C.

Phenolic resins

Phenol formaldehyde was one of the earliest synthetic resins to be produced and was commonly used in the manufacture of 'plastic' cases for pre-war radio sets. It was made from coal tar and was given the name Bakelite. The chemical formula of phenol is C_6H_5OH, which, on reaction with formaldehyde, gives methylol derivatives:

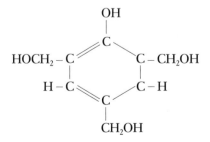

These can be linked by condensation reactions involving the methylol groups:

$$-CH_2O\boxed{H + HO}H_2C- \longrightarrow -CH_2OH_2C- + H_2O$$

Extensive bonding gives a hard, brittle, heat resistant polymer. Being thermosetting, extrusion cannot be carried out: products are manufactured from resins of low molecular weight which are normally ground to a powder compounded with a filler, hardener and other ingredients, and moulded by heat, which causes cross-linking. The manufacturing process produces a dark colour, normally brown or black. The widest use of phenolic resins has been electrical goods, such as switches and plugs, though use for such compounds has decreased and laminates now form the largest market for the resin. Such laminates are formed from phenolic resin-impregnated paper and have a sufficient degree of flexibility for use in protective coverings to working surfaces and wall boards, while retaining a hard surface. The laminates themselves are brown and resemble mica but usually have a coloured plastic coating. Phenolic resins are widely used as wood adhesives.

Amino resins

These are resins based on the amino group or amide group. The most important are urea and melamine formaldehydes:

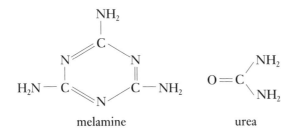

melamine   urea

Each amino group reacts with formaldehyde, by addition, to form methylol compounds:

$$-NH_2 + HCOH \rightarrow -HNCH_2OH$$

These compounds are then polymerised by condensation to form rigid polymers. Urea formaldehyde resins are similar in properties to phenol formaldehyde resins, except that they are clear, hence a wider range of colours is obtainable. They are also more moisture resistant and have greater impact strength. They are used in electrical fittings, for rigid moulded articles, such as toilet seats, and in adhesives. In expanded form, urea formaldehyde is, on account of its low cost, used widely as a cavity wall infill material, though, since substantial quantities of water are involved, surfaces must be porous to allow drying out.

Melamine formaldehyde resins have extremely good resistance to water and chemicals and can be coloured. They are also resistant to combustion, ignition and flame spread. They are widely used as surfacing coatings for decorative laminates based on phenolic resin and in adhesives.

Polyester resins

The production of linear polyesters used in fibres has already been described on page 441. Polyester moulding resins are based on mixtures of unsaturated and saturated dicarboxylic acids, the former providing sites for cross-linking, while the latter help reduce brittleness. Styrene is commonly used as the cross-linking agent – its own double bond opens during the process and it also reduces the uncured resin viscosity to assist handling. A further advantage is that water is not given off during curing. Moulding resins are polymerised until they form viscous liquids and then an inhibitor is added to prevent further polymerisation. On adding the hardener – usually an initiator such as an organic peroxide – the resin hardens at a rate dependent on temperature. The amount of hardener used is small – only a few per cent of the resin – and is not critical, though it affects the hardening rate. Polyesters often have a limited shelf life and will polymerise on their own if left for months or years. Fully cured resins are hard and tough and are commonly used with glass fibre (GRP), which bonds well, improves rigidity, and gives good impact and strength properties. Their main problem is that they shrink considerably after gelation (about 7 per cent by volume) which effectively reduces adhesion due to surface stress. Shrinkage can be reduced by use of fillers and/or reinforcement. When used in adhesives, polyester resins have been shown to have greater fire resistance than the other commonly used adhesives, such as epoxy resins, especially if a limestone filler is employed. Polyesters are used in resin cements (see Chapter 7) and, on account of their low cost, form the basis of many resin based fillers for the DIY market, such as wood fillers and 'plastic metal'.

Epoxide (epoxy) resins

The chemical components of these resins are complex, involving phenolic and other groups. The epoxide group is, however, the most important. It contains oxygen in the form:

$$R - \overset{\displaystyle O}{\overset{\displaystyle \diagup \diagdown}{CH - CH_2}}$$

The phenolic compounds are used to produce, by condensation, a linear polymer with epoxide groups at each end. Curing is then obtained by cross-linking the ends of each chain through opening of the epoxide ring. A wide range of cross-linking agents (hardeners) can be used, the main groups being amines and acid anhydrides; the mechanism may involve coupling or condensation. The polymerisation is quite different from that of polyesters, the resin and hardener having an action rather like

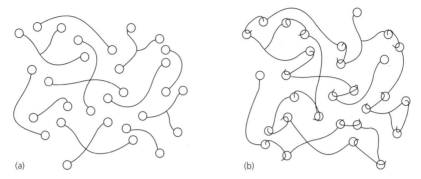

Figure 9.8   (a) Simple representation of an epoxy resin, epoxide groups being represented by 'eyes'; (b) hardener provides 'hooks' to link up with the 'eyes'. For good results the number of hooks and eyes must be equal

hooks and eyes (Figure 9.8). To obtain optimum performance, resin and hardener must be precisely mixed to provide the right number of hooks and eyes. The quantities used are approximately equal – they are best supplied in clearly labelled tins which give an accurately correct ratio when the entire contents of each are combined. In view of the smaller number of reactive sites (the ends of each chain), the hardening of epoxy resins is less sudden than that in polyesters. Most of the shrinkage in epoxy resins occurs prior to gelation, post-gelation shrinkage being only about 1 per cent and this contributes to good adhesion. They have good chemical resistance, particularly to alkalis, though they are more expensive than polyester resins. Resins are prepared in the form of partly polymerised liquids, which polymerise more fully on curing. They have been used in flooring compositions, paints, adhesives and for glass-reinforced composites.

Polyurethanes (see also 'Paints')

These are formed as a product of isocyanates and polyesters and polyhydroxy materials. It is not possible to classify polyurethanes as thermoplastic or thermosetting, since many different forms exist, some with chain-like molecules and therefore thermoplastic; others are heavily cross-linked and therefore thermosetting. Polyurethanes are often used in foamed form, for example, in cavity filling for domestic dwellings and in a more flexible form for spraying on pipes for thermal insulation purposes. In each case, they bond well to the background. Most polyurethane is foamed, using carbon dioxide, which eventually diffuses outwards from cells, being replaced by air. Polyurethanes may be used up to temperatures of about 120 °C, depending on grade, an advantage over polystyrene foams. They are also less flammable, though in common with polystyrenes more flexible types dissolve in some organic solvents and acids.

Flexible polyester and polyether foams come under the general heading of polyurethanes, being used for upholstery, sponges and cushions. Other polyurethanes are used for resilient gaskets in clay pipes and in surface coatings.

PLASTICS IN FIRE

Since virtually all plastics are organic in nature, they decompose readily in fire and, consequently, many of them constitute a hazard, possibly for the following reasons:

1 They may emit toxic gases – usually carbon monoxide.
2 They often contribute to the development of fires by flaming and/or heat emission.
3 They may emit dense smoke, thereby making escape more difficult.
4 Melting of sheet glazing materials may vent the fire, increasing the rate of spread.

The fire hazard associated with plastics depends, in a complex way, on the type of plastic and its mode of use. BS 4735 relates to fire properties of specific plastic types. It is found that many plastics perform badly in one or more of the above respects, for example:

- Polyethylene, polypropylene, polystyrene, acrylic resins, ABS, polyurethane, GRP and rubber all burn readily.
- Acrylic polymers are an additional hazard because their flash ignition temperature is lower than that of any other common polymer.
- Some plastics which do not burn readily, nevertheless produce dense smoke in fire, for example, polyvinyl chloride, which also produces hydrogen chloride gas. Polystyrene and GRP produce dense smoke, while polyurethane produces hydrogen cyanide gas on combustion.

The greatest fire hazard is, in general, caused by plastics in foamed or sheet form, since oxygen is much more readily available in this state and their temperature rises very quickly due to their low thermal inertia. Disasters have been caused, for example, by combustion of vertical acrylic sheeting, due to the extremely rapid spread of flame in this situation, and by polyurethane foams in cushions, due to rapid combustion and the toxic fumes emitted. Standard foam may reach temperatures exceeding 700 °C within two minutes of ignition with fatal levels of hydrogen cyanide emission. High resilience foams containing more than two isocyanite groups are more heavily cross-linked and, therefore, decompose less easily but the combustion modified high resilience (CMHR) foams are much preferred. They may, for example, contain a melamine resin which acts as a flame retardant. Another type contains expandable graphite particles which quickly form a thick protective layer when foam is exposed to heat.

There are many situations in buildings in which use of plastics may be regarded as safe, for example, guttering, sheet roofing, floor coverings, decorative laminates, cavity insulation, water pipes, cisterns and underground pipes.

Expanded polystyrene ceiling tiles are not recommended in areas where there is a fire risk, since they melt easily and may feed the fire, especially if the adhesive is used in dabs rather than as a continuous film. Fire retardant grades of these polymers, as well as GRP, polypropylene and polystyrene, are available but have not in the past been widely used on account of increased cost.

Many thermoplastic adhesives contain flammable solvents, constituting a considerable hazard during use and until solvent evaporation is complete. In some cases, plastics may help retard the development of fire, for example, collapse of PVC downpipes may prevent their transmitting fire vertically. Collapse of thermoplastic roof lights could reduce the build-up of heat in an enclosure, hence, delaying flashover. However, such collapse might also vent the fire and, therefore, increase local rates of burning or, in the case of external fire, the spread from a neighbouring building.

9.2 BITUMENS AND RELATED PRODUCTS

Although these are not plastic in the normal sense of the word, they are in many ways similar and may be conveniently described here. Bitumens are essentially hydrocarbons and their derivatives but they include elements such as sulphur, oxygen and nitrogen. At room temperature, the material may have a consistency varying between that of a hard, brittle solid and a thick viscous liquid. The liquid form can be regarded as a two-phase material containing solid compounds of high molecular weight, dispersed colloidally in a fluid phase of lower molecular weight, the solid fraction consisting of an amorphous network of linear hexagonal aliphatic hydrocarbon chains of fairly low length. Involved in the definition of bitumens is the requirement that they dissolve in carbon disulphide.

Bitumens occur naturally in the form of asphalt, for example, Trinidad Lake asphalt, and in certain rocks, and have been used successfully over many centuries in a wide variety of applications. They are now more commonly produced synthetically, either by distillation or by air-blowing of oil, the bitumen being left as a residue after the evaporation of volatile fractions. The properties of the final bitumen depend on the composition of the crude oil, the type of production process and the temperature to which it is heated. Bitumens may be of the *straight-run* type, which have high viscosity and will need to be heated to become workable, or the *cut-back* variety, which have lower viscosity, due to the addition of fluxing oils such as kerosene or creosote.

Bituminous materials have chemical stability characteristic of the paraffin family of materials, being resistant to acids and alkalis. There is, however, some degree of unsaturation in them so that oxygen from the atmosphere tends to cause cross-linking by oxidation and subsequent embrittlement, particularly when exposed to ultraviolet light. Resistance may be increased by using surface coatings of already oxidised bitumen in the case of roofing felts and providing protection from sunlight wherever possible with light-reflecting mineral aggregates such as limestone. Note that many oils, from which bitumens are derived, will seriously soften or erode bituminous asphalts.

Bitumens are examples of visco-elastic materials – they are able to flow plastically under gradually applied stresses but may undergo brittle fracture if subjected to sudden stress. Hence, there are two prerequisites for successful use:

- Materials must not be subjected to significant sustained stress which would cause creep.
- Sudden movements should be avoided.

The precise extent of each of these effects depends, in practice, on the viscosity of the bitumen (measured by *penetration values*), the temperature and the presence of fibres or stabilising materials such as aggregates. Blown bitumens are, in general, less temperature susceptible than distilled bitumens.

Tars are related to bitumen, these being mainly of aromatic composition and derived from the condensate obtained from the distillation of coal during the production of smokeless fuels. The crude tar or pitch is fluxed back with oils to give required properties. Applications of tars are often based on their low viscosity at high temperatures, such that they can be sprayed – surface dressings for roads are an example. Tars also resist softening by fuel oils better than bitumens, though they are generally more temperature susceptible than the latter.

The following applications are based on the flexibility and waterproofing qualities of bitumens.

Roofing felts

These contain blends of bitumens incorporated in mineral fibres, such as asbestos or glass; or organic fibres, such as wood pulp. Felts incorporating organic fibres have been used for lower layers of built-up roofing, or in flat roofs where a bitumen covering is to be applied. Types to which an oxidised bitumen coating has been applied during manufacture are in all cases more satisfactory than unsurfaced saturated grades. A granular mineral surfacing, when applied, increases protection from the sun, allowing use as the external surfacing material on sloping roofs. Reinforced grades are available where felts have to be self-supporting; for example, under tiling.

Other types of felt include sheathing felts containing long fibres which are, therefore, dimensionally stable, being used under asphalt roofing and flooring; and asbestos and glass fibre felts, which are more durable than organic fibre felts. Flat roofs and built-up felt roofs, in particular, have given widespread problems in the past, these largely being associated with moisture penetration. The critical requirements of a built-up felt roof are:

- Moisture must not 'pond' on the roof. Minimum falls of 1 in 80 are required to achieve this end, though with undulations and settlement, a minimum fall of 1 in 40 is probably more realistic.
- Successive layers must bond well together (though felts should not be rigidly fixed to the substrate). The safest method is probably hot bonding, though good workmanship is required if air pockets are to be avoided and the operation should not be carried out in wet weather. Hot applied felt can be subjected to light foot traffic immediately on cooling but will embrittle with age, especially if solvent is lost by overheating at the time of application. Cold applied (cut-back) mastics can also be used, though up-stands and edges tend to debond unless pressure is applied until excess solvent has evaporated. This can take some weeks and felts should be trafficked with great care in this time. However, if correctly applied, cold bonded felts should give equal life to hot bonded varieties.

Torching grades of felt are also obtainable in which the felt itself is softened and bonded without additional adhesive. These are thicker and more expensive than traditional felts but offer a significant advantage in saving time and labour.

The current version of BS 747 for roofing felts omits the former organic fibre or asbestos based felts, referring now to felts with a glass fibre or polyester base. Polymer modified versions offer even better performance.

Bituminous mixtures

The most important bituminous mixtures in building are *asphalts* – these consist of bitumens with inert material which increases their rigidity, stability and abrasion resistance. The mineral may already be present in naturally occurring bitumen, for example, lake asphalt, or it may be added, in which case, it is usually graded, crushed limestone. Asphalts which are essentially solid and impermeable at room temperature, but which soften on heating, are known are *mastic asphalts*. The limestone aggregate is of maximum size 2.36 mm, graded down to 150 µm. Important uses include roofing, tanking and flooring and they are hand applied, a process involving a good deal of skill. Roofing or tanking grades are normally softer than flooring grades, the former requiring flexibility while the latter require hardness and abrasion resistance. Although bitumens are acid-resistant, limestone is susceptible to acids, hence siliceous fillers and aggregates should be used in such conditions.

Asphalts in rolled form are also widely used as road surfacing materials, combining a smooth riding surface with excellent skid resistance and impermeability to moisture. For this purpose, some small degree of oxidation is advantageous, since it results in weathering so that the coarse aggregate remains exposed. Lake asphalt and pitch bitumen are valued because they oxidise in this way. Lake asphalt also has lower solubility in certain solvents than synthetic bitumens. (Since bitumens have a tendency to soften when in prolonged contact with oil, bituminous surfacings subject to heavy contamination, such as near bus stops, often deteriorate in this way; tar bound surfacings are preferred in such situations.) Asphalts for roads must be carefully blended mixtures of binder, filler and fine and coarse aggregates, in order to combine flexibility and strength in all weather conditions.

Most bitumens used in road construction are of the straight-run type and these are heated before placing. They can, therefore, be brought into use immediately on cooling.

Cut-back bitumens are, however, useful for such applications as repair work and footway surfaces and car parks, since they can be placed without heat. Provided excessive thicknesses are avoided, the fluxing oil evaporates reasonably quickly, causing stiffening. Nevertheless, heavy traffic or abrasion must be avoided for a week or more after placing, depending on weather conditions.

Macadams are a further form of bituminous mixture widely used in civil engineering; they have, in general, a coarser particle grading than asphalts. Whereas asphalts rely on a stiff mortar for rigidity, macadams rely on particle interlock.

Hence, asphalts contain a higher proportion of mortar, distances between aggregate particles being greater.

Further uses of bitumens include dampproof courses, paints, adhesives and thermoplastic flooring tiles. The latter also contain inert fillers, resins and asbestos fibres.

Bituminous emulsions

Emulsions comprise small particles of bitumen dispersed in water by means of an emulsifying agent. Emulsifying agents may be anionic or cationic. The former are negatively charged and, since they are soluble in bitumen, impart a negative charge to the surfaces of particles, causing them to repel one another, thus stabilising the emulsion. Stearates are examples of anionic emulsifiers, the aqueous phase being alkaline in this case. Cationic emulsifiers operate on a similar basis except that they impart positive charges to the surfaces of the bitumen particles, the aqueous phase being acidic. Cetyl trimethyl-ammonium bromide is an example. The speed of 'break' of emulsions can be controlled and it also depends on temperature. Cationic emulsions generally break faster than anionic emulsions, possibly because many solid materials are thought to be charged negatively, thus having a slight neutralising effect. Anionic emulsions are usually cheaper, however.

Bitumen emulsions are widely used for dampproofing purposes, such as the provision of dampproof membranes on concrete floors or walls. On breaking, by evaporation of water, the film changes from dark brown to black in colour. Anionic emulsions are normally used for this purpose. For satisfactory results:

- They must be applied to a smooth, continuous substrate.
- Ponding must be avoided – it delays drying.
- Three or more coats should be used, allowing drying in between.
- If walls are to be plastered, the final coat should be sanded to form a key.

Bitumen emulsions are also used extensively in road construction; for example, as a tack-coat during resurfacing, good adhesion being obtained to damp surfaces. Further applications include use as a curing membrane for lean concrete road bases and as an adhesive for porous materials, though they are not suited to conditions involving tensile stress.

9.3 JOINTS/SEALING COMPOUNDS

Joints are present in all types of building; they are essential in the formation of openings, they may be included to permit movement in larger structures or, especially in modern construction methods, they are the result of site assembly of preformed or precast units. They have a number of functional requirements:

- They must be resistant to weather in all its forms, including rain, snow, wind and dirt.

- They must accommodate any movement of elements joined.
- They should be durable.

In many traditional joints, these functions were not completely fulfilled but penetration of small amounts of water would not be serious – it would be absorbed by porous materials and evaporate during dry spells. In more recent construction, the use of claddings of low or zero porosity, such as metals, combined with large lengths of joints and severe exposure in taller buildings, has led to a much higher incidence of failure. Research continues to address this problem and there are now various alternative means of weatherproofing. They include sealing compounds, gaskets and two stage joint systems.

There is an enormous range of materials available for sealing purposes in construction, representing the widely differing requirements of individual situations. When assessing the suitability of a sealant for a particular situation, consideration should be given to the following:

- size of the gap to be filled
- movement of the gap to be filled
- surface nature of the materials to be joined
- degree of resilience and abrasion resistance required
- amount of maintenance envisaged
- degree of exposure to the environment or chemicals
- life time required
- methods of application available
- temperature range in service

Materials will be considered under the headings: brittle materials, putties, mastics and elastomeric sealants. Properties are summarised in Table 9.2.

Table 9.2   Typical properties of common mastic/sealant materials, with linseed oil putty and cement mortar for comparison

Type	Application method	Time to full cure	Mode of operation	Abrasion resistance	Max. extension (%)	Life (years)	Cost index
Cement mortar	Hand	1–3 days	Brittle	Good	0	1–50	0.01
Linseed oil putty	Hand	6 weeks	Plastic	Fair/good	0–1	5–20	0.5
Hot-poured bitumen	Pour	On Cooling	Plastic	Fair	10–15	20–50	1
Oil-based mastics	Gun	(skin time) 1–3 days	Plastic	Poor	5–10	5–10	1.25
Acrylic/butyl	Gun	20 days to 6 months	Plastic/ elastic	Fair/good	10–15	10–15	1.5
Polysulphide	Gun	1pt 3 weeks 2pt 1 week	Elastic/ plastic	Good	20	25	5
Silicone	Gun	1pt 7 days 2pt 1 day	Elastic	Good	25	25	7

BRITTLE MATERIALS

Mortars based on cement, gypsum, lime and other materials should only be used in situations where movement is minimal or where small cracks are no disadvantage, for example, pointing of brickwork and stonework. Common applications of these materials in situations where they are not likely to give long life are in roofing flaunchings and for fixing WC pans to the soil pipe. In each case, cracking of mortar or jointing materials is likely, due to stresses imposed by relative movement. Ductility of these mortars can be improved by addition of lime or emulsified resins such as polyvinyl acetate.

PUTTIES

These are used primarily for glazing and are designed to act as a bedding material, as well as a filler and sealant. Linseed oil putties comprise processed vegetable oils and whiting (ground chalk). A hard skin forms by oxidation within four to eight weeks, depending on temperature. In timber frames, stiffening is assisted by loss of oil into the wood, though frames should be primed to prevent excessive loss and subsequent embrittlement. Note that some of the newer types of primer may stop oil loss to the extent that putty may be very slow to stiffen, making painting difficult. In metal casement windows, there is negligible oil loss so that putties having a slightly stiffer consistency and containing accelerators such as gold size should be used. The use of metal casement putty may also be preferred in timber framed windows where, as a result of the use of preservatives or primers, the wood has less ability to absorb oil.

Putties will shrink and ultimately crack unless excessive oxidation is prevented by painting after the original hardening period. Excessive hardening also impairs the sealing and bonding properties of the putty. The shelf life of putties is limited because the linseed oil tends to separate and harden in the container, unless oxygen is excluded.

MASTICS

These may be regarded as of a basically non-hardening nature, although films may be formed. The mastics must be sufficiently viscous to resist sagging in the thickness required but sufficiently ductile to flow plastically when joint movement occurs. Mastics generally have excellent adhesion to virtually any surface including plastics but since they absorb movement by plastic flow, they are not suited to large movements that would result in eventual displacement or wasting of material. Mastics may be based on bitumen, vegetable oils, or synthetic polymers.

Bitumen or bitumen/rubber blends are widely used for joint sealing in civil engineering applications, such as roads and bridges. They are normally used in horizontal joints, being liquefied by heating and then poured into the joint. Stiffening occurs on cooling and quite large joints can be sealed in this way, giving long life, provided adhesion to the joint surface is not lost. Brittleness may cause sealant failure at low temperatures. Bituminous mastics are relatively cheap.

Oil based mastics comprise vegetable oils and resins blended to give suitable oxidation characteristics, together with fibres and fillers. They form a skin within a few days, though hardening continues over a period of years, and life is not normally longer than 10 years, due to the cracking that ensues. Oil based mastics have poor resistance to abrasion. They are, however, relatively cheap and well suited to sealing of window and door frames in domestic situations. Impregnated tapes are also available, particularly suitable for joining sheet materials, as in roofing. For greatest durability, such tapes should have a protective aluminium covering.

In addition to oil based mastics, compounds based on acrylic and butyl polymers are available, these becoming semi-elastomeric as a result of solvent evaporation. This process may result in shrinkage of larger thicknesses but these mastics are somewhat superior in terms of abrasion resistance and durability to oil based mastics. They are intermediate in cost between oil based and elastomeric types.

ELASTOMERIC SEALANTS

These have the advantage of being much more resilient than mastics, together with the ability to withstand greater joint movement (in the region of 20 per cent). Anticipated life is longer – about 25 years in the case of some polysulphide and silicone sealants. Polysulphide sealants are most effective in the two-pack form, though curing can, as with silicone sealants, be achieved by means of atmospheric moisture. One- and two-part polyurethane sealants are also available. Elastomeric sealants are considerably more expensive than the above types but, nevertheless, find wide application, especially in cladding and certain walling systems which often contain vast lengths of joints.

With all types of sealant, adequate preparation is essential – surfaces should be clean and dry. They may need sealing. Priming always increases sealant adhesion, especially in porous substrates, since it prevents water access to the sealant/ substrate interface, a common cause of failure. Primers should be applied by qualified personnel and, for the best results, the sealant is tooled after application, to ensure adhesion to joint sides.

Tests for sealants

There are many respects in which a sealant may prove to be unsatisfactory, the complexity of the problem being reflected by the variety of tests detailed in BS 3712 Parts 1–3 (*Methods of Testing Building Sealants*). The tests measure homogeneity, relative density, extrudability, penetration, slump, seepage, staining, shrinkage, alkali resistance, shelf life, paintability, application time, change in consistency, skinning properties, tack-free time and adhesion of fresh material to mature sealant. Selected tests can be applied, as required, to particular sealants and, in addition, there are BS specifications for silicone based building sealants (BS 5889) and two-part polysulphide based sealants (BS 4254). It must be emphasised that these tests are essentially laboratory tests of a standardised nature and that *in situ* performance will depend additionally on the detailing of the joint, the quality and cleanliness of surfaces to be joined, the standard of workmanship during application, the extent of joint movement and the influence of the environment. Such performance can only be accurately determined from long term experience of use.

JOINT DESIGN – ELASTOMERIC SEALANTS

In order to appreciate the principles of joint design, it is vital to understand the mechanism by which sealants absorb movement. Their flexibility has been explained in molecular terms but, in common with many soft solids and liquids, they are subject to the requirement that, to remain intact, their volume must remain constant during deformation. Examples of this in practice are shown in Figure 9.9. Note that, in compression, the material must *barrel* while in tension, it must *waste* – the material is accommodating movement by changing its shape. Therefore, to allow maximum movement, joints must be designed to permit a maximum change of shape to occur. This is generally the case when:

- The width of the sealant is about twice its thickness.
- Bonding occurs at the sides but not the front or back of the joint. (Polyethylene strip may be used to prevent bonding; Figure 9.10).

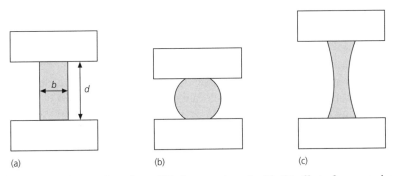

(a)   (b)   (c)

Figure 9.9 (a) Elastomeric sealant of ideal proportions $d = 2b$; (b) effect of compression; (c) effect of tension. In all cases the volume of sealant remains the same

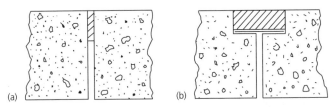

(a)   (b)

Figure 9.10 (a) Ineffective way of sealing a crack still subject to movement; (b) correct method. The crack has been opened out and a debonding backing strip provided

In order to maintain a weather seal, the ends of the sealant must bond well to the two sides of the joint which may, therefore, require a minimum thickness of 5–10 mm, depending on substrate. Hence, the minimum sealant width will be 10–20 mm for maximum accommodation of movement.

As an illustration of joint design, consider a 5 m lightweight concrete cladding panel, installed soon after manufacture into a reinforced concrete framed building. If the panel has an expected drying shrinkage of 200×10^{-6}, thermal movement of $10 \times 10^{-6}/°C$ and a possible temperature differential with the column of -20 °C (it will cool more quickly in winter than the frame), the total movement of the 5 m panel would be:

$$\text{Panel shrinkage} \qquad 5 \times -200 \times 10^{-6} = -0.001 \text{ m}$$
$$= -1 \text{ mm}$$

$$\text{Panel thermal contraction} \qquad 5 \times (-20) \times 10 \times 10^{-6} = -0.001 \text{ m}$$
$$= -1 \text{ mm}$$

The total movement is 2 mm. This will tend to open a gap between the panel and the frame.

If this movement is distributed at one joint at each end of the panel, then each joint must accommodate 1 mm contraction. If the joint thickness is chosen to be 5 mm, the ideal width would be:

$$5 \times 2 = 10 \text{ mm}$$

and this would accommodate:

$$\frac{20}{100} \times 10 = 2 \text{ mm}$$

of movement assuming a movement capability of 20 per cent. Hence, this joint would be satisfactory. Note that in each case the worst scenario must be identified in determining the size of joint needed.

A filler such as expanded polyethylene should be used as a backing material for joints – this is essential to achieve good contact with the sides of the joint during tooling.

It is quite possible to seal small gaps such as cracks, but application of sealant as in Figure 9.10(a) would be useless if continued movement is expected. It would be better to open the crack to give the shape shown in Figure 9.10(b) and seal, using the principles given above.

In a large modern framed building the length of joints employed may add up to several miles and it is clearly of great importance that they are designed and constructed to be reliably waterproof. This calls for an understanding of the operation of elastomeric sealants as described above and also care in the detailing of joints so that effective sealing can be carried out. Figure 9.11 shows a detail in which windows/cladding units are fixed to a concrete frame which is itself clad in brickwork. The joint as designed should operate, but it is important in a situation such as this to ensure that each operation can be carried out and that tolerances are

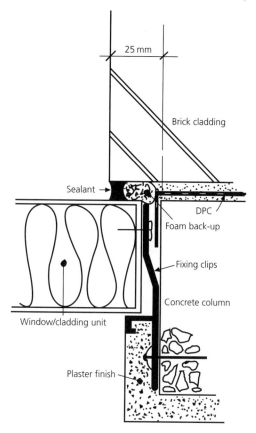

25 mm

Brick cladding

Sealant →

DPC

Foam back-up

Fixing clips

Concrete column

Window/cladding unit

Plaster finish

Figure 9.11 Complex junction between cladding unit, brick cladding and concrete frame. The detail must be executed accurately to permit the sealant to work

sufficiently good to provide for the sealant to be installed as shown. For example if the brick cladding were 25 mm short, it would be very difficult to seal the joint.

GASKETS

These are made from elastomeric compounds similar to those used in sealants, the essential difference being that they are preformed and held in place by friction between the compressed gasket and the surfaces to be sealed. Considerable flexibility can be achieved by inclusion of voids in the gasket, combined with carefully designed shapes which minimise internal stress resulting from changes in width.

Such gaskets require smooth surfaces in cladding units to produce an effective seal – metal, glass and plastic are ideal, though joints in concrete, if primed, can also be treated in this way. Gaskets can simply be pushed into place, hence, they are easily renewable. They are widely used as double glazing seals (Figure 9.5).

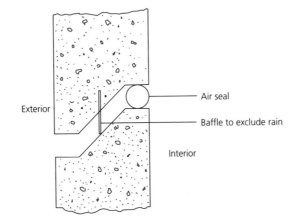

Figure 9.12 Principle of two stage horizontal joint in a concrete cladding panel. The functions of excluding rain and preventing air penetration are performed separately

Dirt may affect performance, particularly in horizontal joints and drain holes are necessary to permit any trapped water to escape. Gaskets can also be made from foamed elastomers, the pores having a similar stress relieving effect to the voids in solid types. Principles of joint design are as for elastomeric sealants though movement capability may be increased by careful design.

TWO STAGE JOINT SYSTEMS

These use carefully designed baffles (rain screens) to shed all moisture except wind-blown rain which is allowed to drain out, air penetration being prevented by a separate internal seal (Figure 9.12). Some systems are sufficiently open to avoid air pressure differentials across the cladding panels, minimising wind-blown rain admission and reducing the stresses due to wind on the panels themselves. Drainage provision must again be made. Two stage systems have the advantage that the sealant material is protected from rain and ultraviolet light. Treatments of vertical and horizontal joints of this type are different and particular care is necessary where the two joints meet.

9.4 ADHESIVES AND ADHESION

The advent of synthetic resins has completely revolutionised the field of adhesives and they are quite rapidly finding use in situations where, previously, there would have been no chance of success. Resins can be used, for example, in load bearing situations, where moisture, heat or chemical pollution is likely, and where there is little in the way of a physical key. It must be emphasised that two surfaces cannot be joined without the establishment of bonds between them. In more traditional types of glue and cement, there is little true bonding between the adhesive and the

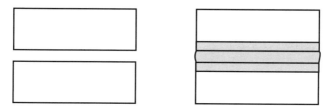

Figure 9.13   The traditional method of bonding absorbent materials. The 'adhesive' penetrates the surfaces and, on hardening, forms a mechanical key. There is little true adhesion. Examples include inorganic cements and plasters

$$\bigcirc + \bigcirc \longrightarrow \bigcirc + \text{Energy}$$

Figure 9.14   Surface energy in liquids. On touching, two globules coalesce, forming a single large globule of surface area less than the original total surface area

surface. The material depends for its action on penetration of the surface so that, on setting such that chemical bonding *within* the adhesive is established, the portions of set material beneath the material surface interlock (rather than bond) with surface layers, resulting in tensile strength (Figure 9.13). Such adhesives are really *cohesive* rather than *adhesive*. Hence, non-porous surfaces, such as glass or metals, are very difficult to stick together by this method and the advent of plastic materials has led to requirements for joining these also. Many resins now produced are, however, in the true sense of the word, adhesive – they stick to the surface by bonding with it, whether or not it is porous. The exact nature of the bond depends on the adhesive and the surface to be bonded and, though the bonding mechanism in most cases is not fully understood, it is true to say that the existence of primary bonds is not a prerequisite for successful adhesion: van der Waals bonds will result in satisfactory properties, provided a sufficient number is established. The bonding properties of materials can be predicted from a knowledge of their surface energies. Surface energy may be defined as the energy required to form unit area of new surface. It is always positive, due to surface tension effects in solids and liquids. It is well known, for example, that two drops of the same liquid on contact will merge to form a larger drop, since in so doing the total surface area of liquid will be reduced, resulting in a more stable arrangement (Figure 9.14). Solid objects behave similarly, though the effect is less obvious. Adhesion between a solid and liquid may be obtained either if the solid is *soluble* in the liquid, in which case the surface solid material migrates into the liquid forming a mixture or alloy (Figure 9.15), or if one *wets* the other, in which case adhesion will occur at the interface (Figure 9.16).

The solubility of one material in another depends on the molecular structure of each and, in particular, the polarity of molecular groups, that is, the charge eccentricities of the molecular bonds. A term which summarises polarity is *solubility factor*, which may be used for organic materials and solvents to predict the type of

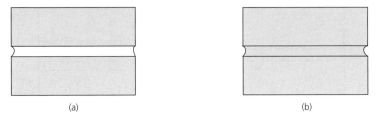

Figure 9.15   Joining of surfaces by solvent welding: (a) immediately after application of solvent; (b) some time later. The solid dissolves and diffuses into the solvent until the solvent itself solidifies. The layer of solvent should be thin

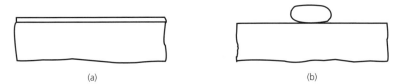

Figure 9.16   Adhesion of a liquid to a solid: (a) if the surface tension of the solid is higher than that of the liquid, the liquid wets the surface; (b) if the surface tension of the solid is lower than that of the liquid, the liquid does not wet the surface. The same effect occurs if there is a thin coating on the solid of a material of low surface tension, for example, grease

solvent required for, say, solvent-welding, or to safeguard against use of plastics in contact with certain organic liquids. To be soluble in a certain solvent, the polymer should have a similar solubility parameter. Table 9.3 shows typical values for some common polymers and solvents. Hence, chloroform may be used for solvent-welding PVC or polystyrene; phenol will dissolve nylon. (There are exceptions where, for example, due to crystallinity, polymers are insoluble in solvents of similar solubility parameter. Polyethylene is, for this reason, insoluble in any organic solvent at room temperature.)

This type of solubility should not be confused with, for example, the dissolution of metals in acids; the latter type of reaction is normally irreversible and due to chemical attack. Unlike the above example, the product is entirely different, having a new molecular structure.

Since solution of organic polymers is principally due to separation of molecular chains, heavily cross-linked polymers will not dissolve, though swelling may occur if such a material is subjected to potentially active solvents; that is, solvents which would dissolve less heavily cross-linked varieties of the polymer.

Closely related to the solubility parameter of polymers are their surface tension values and this is to be expected, since both derive from molecular properties. Polyethylene, for example, has a much lower surface tension than more polar polymers. The requirement for wetting (and hence adhesion) of a liquid to a solid is that the liquid has a surface tension less than that of the solid, since, if it is greater, the liquid would be cohesive, tending to form globules on the solid

Table 9.3   Solubility parameters of some common polymers and solvents. Polymers usually dissolve in a solvent of approximately equal solubility parameter

Polymer	Approximate solubility parameter
PTFE	6
Polyethylene	8
Polypropylene	8
Polyisoprene	8
Polystryrene	9
Polymethylmethacrylate	9
PVC	9
Amino resins	10
Expoxy resins	10
Phenolic resins	11
Nylon	13
Solvent	
Carbon tetrachloride	8.6
Benzene	9.2
Chloroform	9.3
Acetone	10.0
Phenol	14.5
Water	23.4

surface (Figure 9.16). Surface tensions of solid materials are roughly in the order of solubility parameters, with inorganic materials having much higher values. Hence, water wets inorganic materials and epoxy resins will adhere to them, as well as to some thermosetting plastics. On the other hand, many of the polymers with low surface tension are water repellent and cannot be joined even by epoxy resins. Surface treatment of polyethylene is required, for example, if epoxy resins are to adhere to it. The surface tension of the material must be increased and this can be achieved by oxidation, by an oxidising acid, for instance. It is important when joining materials with true adhesives that the surfaces be completely clean, since even a very thin coating of grease or similar material may reduce considerably the effective surface tension of the surface, impairing the bond. Appropriate solvents should be used for this purpose. Similar arguments apply to solvent-welding.

   There are, of course, cases in which water repellent properties are advantageous. If, for example, the surface tension of stonework could be reduced, it would become water repellent and, therefore, more resistant to the effect of moisture. This can be easily achieved using silicone resins, which have an extremely low surface tension. Release agents for concrete work on a similar principle. If, on the other hand, wetting is required, this can be achieved by reduction of the surface tension of the liquid, as is obtained by adding surface-active substances such as soap, to water. Some workability acids for concrete work on this principle – the water is able to wet particles more effectively, decreasing interparticle friction.

TYPES OF ADHESIVE

It is not easy to classify adhesives according to use, since many varieties are multi-purpose in their application. Hence, they are here divided into natural adhesives (glues), and thermosetting and thermoplastic synthetic adhesives, though the first four types to be described are used mainly for timber products. In this context, durability is of extreme importance and BS 1203 and 1204 give four classifications as to the suitability of timber glues (non-structural) for various situations. They are as follows:

WPB stands for weather- and boil-proof types.
BR indicates boil resistant types that fail on prolonged exposure to weather.
MR refers to moisture resistant but not boil resistant adhesives.
Int adhesives are resistant to cold water but, unlike the above, are not required to be resistant to micro-organisms.

A further general point which should be considered when selecting an adhesive is the mechanism of curing or hardening. While most of the adhesives described below are truly adhesive (that is, they bond to the receiving surface), the curing processes of some still require an absorptive surface – to allow removal of water or solvent, for example. Such adhesives would not be effective if used to join two non-absorbent surfaces.

Animal glues

These have been used for centuries in carpentry and joinery, and are obtained from the skin or bones of cattle and sheep. The major constituents of animal glues are proteins – large molecules occurring in gelatinous form, softening to viscous liquids at temperatures above about 40 °C and gelling to form a solid at lower temperatures. Animal glues are normally sold in powder form and are melted in warm water to give a solution of suitable viscosity. Cooling of the liquid after application results in rapid gelling, producing some strength and drying finally produces a tough and rigid product. Note that the removal of water from the glue during curing is essential; the wood assists in this process, due to absorption. Glued joints in wood should be as strong as the wood, though damp conditions will, of course, reduce strength and animal glues are not suitable for external use unless adequately protected from moisture (BS 1204, Int rating). The glue will harden gradually in the pot but can be softened by reheating with water. Animal glues are specified in BS 745.

The water solubility of animal glues may be a distinct advantage for some applications such as antique furniture. If a failure occurs in such items it is easy to carry out repairs by softening the glue with water. It would be very difficult to remove modern synthetic adhesives from joints in the event of a failure since they are largely non-soluble.

Casein glues

Casein glues are of rather similar structure to animal glues, though they are derived as a precipitate from skimmed milk by the action of acids. Glues are obtained in powder form which also contains an alkaline solvent, necessary to dissolve the glue on addition of water. Mixing is carried out cold and setting is partly by evaporation or absorption of water and partly by a natural gelation process. The latter reaction results in a limited pot life of glues, once mixed – usually of about 6 hours. Some degree of water resistance can be obtained by incorporation of formaldehyde but use is normally confined to dry situations (Int rating).

THERMOSETTING ADHESIVES

Urea formaldehyde

This is one of the commonest adhesives in joinery for general purposes, producing a strong rigid joint if the glue-line is thin. To make the adhesive, formaldehyde and urea in aqueous solution are allowed to react together to a certain stage and the reaction is then stopped while the resin is still liquid. These resins would then be of two-pack form, since a hardener is required to initiate the final stage of the hardening process. Alternatively, the water may be evaporated, giving a powder which has a longer shelf life than the liquid forms. On addition of water, the original properties are restored and setting commences on adding the hardener. Some varieties contain both the resin and hardener in powdered form, mixed together so that water only need be added. Urea formaldehyde adhesives have little natural tack, so that they are best suited to joining porous materials such as wood or cork. Pressure is essential to hold surfaces together during curing.

One problem which occurs with this and other resins is crazing, which tends to occur with large volumes of adhesive as might be used when gaps of, say, 1 mm have to be filled. Crazing is caused by shrinkage during the condensation reaction. It can be reduced by incorporation of fillers which are used in gap-filling glues. If restricted pot life is a problem, joints can be made by the separate application method in which the resin is applied to one surface and the hardener to the other, so that setting does not commence until the joint is made. In this way, strong bonds can be obtained as quickly as 10 minutes after the joint is assembled, though there is still time to position the components accurately, unlike contact adhesives. Urea formaldehyde adhesives are used for fabrication of flush doors, laminated timber and decorative laminates. Hardening of glue-lines will occur at room temperature but it may be accelerated by hot pressing, strip heating or, most recently, radio-frequency heating, which may enable curing times as low as a few seconds to be obtained. Although urea formaldehyde has good water resistance at normal temperatures, very hot water or prolonged wetting breaks the resin down – it has the MR rating. Hence, laminates, for example, around the kitchen sink, should be bonded by a more resistant adhesive. Further important uses of urea formaldehyde resins are for particle board and plasterboard partitioning.

Melamine formaldehyde resin adhesives

These are usually supplied in powder form and are mixed with water to give a colourless resin. They set on heating to 100 °C and give good weather resistance (BR rating), though they are more expensive than urea formaldehyde adhesives.

Phenol formaldehyde

These are available as liquids which polymerise on heating to temperatures over 100 °C and are used for assembly of plywood sheets. The resin is applied, the sheets formed and the laminates then subjected to a hot press, which causes hardening in a period of about 5 minutes. Film varieties have also been used; these bond the surfaces on warming. Cold-curing phenolic adhesives are also available. These are insoluble in water and set by addition of strong acids so that the glue is acidic, having pH values as low as 1. They have been used for assembly gluing of wood, though the setting reaction is exothermic and too much hardener tends to reduce pot life on account of the accelerating effect of heat on the curing rate.

Phenolic adhesives are hard but brittle, so that fracture in wood joints is possible, especially if thick glue-lines exist. Moisture resistance is excellent; they have the WPB rating of BS 1204. Correct mixing, application and curing are important and they must be handled with care, so that they are best suited to factory use.

Resorcinal formaldehydes

These are related to phenol formaldehyde adhesives but, unlike the latter, will cold-cure under neutral conditions by addition of formaldehyde. Adhesives consist of water-soluble liquid resin obtained by mixing resorcinol with a quantity of formaldehyde which is insufficient to cause cross-linking. The hardener is, or contains, formaldehyde, which completes the process. Curing can be carried out cold or accelerated by moderate heat.

Resorcinol adhesives are important in glued laminated timber, since they have very good durability in extremes of weather (BS EN 301 type I rating), though, since timber may not stand up to such conditions unprotected, preservatives should be used; most preservatives do not affect the bond obtained. These adhesives are also useful for joining wood products and laminates to brick, concrete and asbestos backgrounds and are tolerant of a certain amount of moisture. Grades containing fillers should be used when bonding to uneven substrates. In common with phenolic adhesives, resorcinol types are strong but brittle.

Epoxy resins

These are a most important development since, with the exception of some thermoplastics, they form a good bond with almost any material. This is considered to be due, at least in part, to the low post-gelation shrinkage of these resins, so that surface shear stresses do not arise. A further contribution to adhesion is probably due to the presence of hydroxyl groups in the polymer. There are many different curing agents and, as these become built into the final cross-linked molecule, one particular curing agent will probably be best suited to the particular physical and chemical properties of each situation. With certain hardeners, the use of the correct quantity is important, although the polyamide hardener for two-part general household purposes is chosen partly because, in this case, the effect is minimised. Greatest bond strength with this hardener is obtained by high curing temperatures – 100 °C or over.

An alternative to the two part method is to produce limited reaction between resin and hardener during manufacture and then to stop the reaction, which results in a one part adhesive that can be hardened by heating. Such resins are used as film adhesives.

Epoxy resins can also be modified by combining with other polymers, which may, for example, improve flexibility or impact strength, usually at the expense of tensile strength. Polysulphide and vinyl polymers can be used for this purpose. Epoxy resins have, in general, excellent mechanical strength and creep resistance, and are resistant to weather, acids, alkalis and most hydrocarbons. They are now widely employed for structural joints between many types of material, including glass, metal, thermosetting plastics and concrete. Although the adhesive is expensive, applications are likely to increase as the construction industry exploits more fully the advantages of glued joints.

Alkyl cyanoacrylate adhesives

These have the advantage of polymerising very rapidly when spread in thin films, the reaction being catalysed by water which, in the very small quantities required, is present on most surfaces. The adhesives are, however, expensive, so that uses are confined to bonding of small objects having surfaces which closely conform to one another. They form the basis of *superglues* and must be handled with care to avoid bonding of skin tissues.

Polyurethane adhesives

Though not widely used, these have a useful property not possessed by the other adhesives: they will join unvulcanised rubber to metal. Water resistance properties are between those of urea and phenolic adhesives. Modified forms are now being used in fibreboards and particle boards where they have the advantage that there is no formaldehyde emission as with the other common adhesive types used with these products (formaldehyde emission in buildings has been known to cause human discomfort and is considered by some to pose a health risk). The adhesive can also be easily blended to vary its moisture resistance, higher isocyanate forms having greater resistance.

THERMOPLASTIC ADHESIVES

The setting action of these may occur as a result of cooling, solvent evaporation or by emulsion coalescence (see also the comment on film formation on page 490). Hence, the term *curing*, implying chemical changes, is not always appropriate. As is characteristic of thermoplastics themselves, adhesives so based are more flexible but weaker and more prone to creep than thermosetting adhesives, so that they are not normally used for structural purposes. However, when large contact areas are possible, gap filling varieties can act as a substitute for nailed or screw fixings for lightly stressed applications. Solvent based versions set more quickly, though since they contain VOCs they cannot be considered to be environmentally friendly.

Polyvinyl acetate

This is probably the most important thermoplastic adhesive in building, chief applications being as a wood adhesive for internal use and as a bonding agent for concrete. The polymer is obtained from acetylene and acetic acid and is most commonly used in emulsion form. The monomer liquid is emulsified in water, forming very small droplets which are then polymerised by the action of heat and a catalyst. The adhesive sets when the solid particles, on evaporation or absorption of the moisture, cohere to form a tough, clear film which is no longer water soluble. The chief advantage of PVA adhesives is that they are water miscible and do not require the use of a hardener, though one of the materials to be joined must be absorbent. Thin glue-lines with absorbent materials set quickly, though gap filling properties are less satisfactory. Very high strength joints are obtained in dry timber, though joints quickly fail by softening in the presence of moisture. Other properties are characteristic of thermoplastic adhesives – they have better impact resistance than thermosetting adhesives but prolonged high stress will cause creep and ultimate failure.

A further common use of PVA emulsions is as a bonding agent between new and old concrete. The resin may be diluted and applied to the substrate or added to the new mix, in which case it enables thinner sections and screeds to be applied with reduced danger of cracking. PVA emulsions are often used to seal surfaces such as plaster before tiling and as an internal ceramic tile adhesive, in which case gap filling properties would be imparted by incorporation of fillers. Some two part types are claimed to be proof against boiling water – these may find application in exterior joinery. More recently introduced one part types become slightly cross-linked on curing and hence form a convenient one part glue with some heat and water resistance. They are, nevertheless, not suited to situations for permanent dampness, high stress or high temperature. Gap filling varieties are available in cartridges for general bonding purposes, though time must be allowed for the emulsion to break.

Polystyrene adhesives

These normally operate on the solution principle, containing some dissolved polystyrene also. Polystyrene, PVC and polymethyl methacrylate can be joined in this way, though the solvents are highly flammable.

Bituminous adhesives

These form good bonds with a number of materials and are moisture resistant and flexible. Natural or synthetic rubber and solvents may be included to give the desired combination of elasticity and strength. Adhesives are obtainable in the following forms.

Water based emulsions   These form an effective bond between porous materials in situations where long term tensile stresses are not present. They are also used for laying wood-block flooring and PVC or thermoplastic tiles (see also page 467).

Solvent types   These are used in laying linoleum and for PVC and thermoplastic tiles, though tests should be carried out to ensure that staining does not occur. Both solvent and emulsion types rely on some degree of absorbency in the substrate to allow absorption/evaporation of the water or solvent.

Hot-applied varieties   These are used for wood-block flooring and in built-up felt roofing. Application temperatures are in the region of 150–200 °C.

Rubber based adhesives

Many rubber based adhesives rely on a very high degree of tackiness, which results in good adhesion to most surfaces, including, in some cases, thermoplastics such as polyethylene. Rubbers may also possess the property of auto-adhesion, in which two films bond immediately on touching. The origin of such behaviour is complex but it depends on the ability of molecules in the two films to diffuse and interlock when the surfaces are pressed together. This depends, in turn, on the viscosity of the adhesive.

The most important rubber based adhesives in the construction industry are *contact adhesives* – the adhesive is applied to both surfaces and, after solvent evaporation, the surfaces are carefully brought together, giving immediate auto-adhesion. If one of the surfaces is porous, they can be brought together immediately, the solvent dispersing subsequently by diffusion. Neoprene is widely used in contact adhesives.

Latexes such as styrene butadiene rubber (SBR) are widely used for floor coverings while latex-modified cements are used for bonding ceramic tiles to a variety of backgrounds, including metal.

When bonding materials such as polyethylene, in which adhesion is due solely to dispersion forces, the cohesive strength of the adhesive must be kept low in order to distribute stresses as evenly as possible at the interface. Flexible polyisobutane, of carefully chosen molecular weight distribution, can be used to give a reasonable bond between polyethylene sheets in this way.

Although the bond strength in these adhesives is low, large bonding areas, combined with flexibility in the adhesive, can result in satisfactory performance in a very wide range of porous and non-porous building materials, provided only small long term stresses are applied.

Solvent based adhesives based upon SBR rubbers are now available quite heavily filled and marketed in cartridges for general application as gap filling adhesives in the industry. They have the advantage of being relatively quick to set (by solvent evaporation).

9.5 PAINTS

Paints are surface coatings generally suitable for site use, marketed in liquid form. They may be used for one or more of the following purposes.

- To protect the underlying surface by exclusion of the atmosphere, moisture, chemicals, fungi and insects.
- To provide a decorative, easily maintained surface.
- To provide light- and heat-reflecting properties.
- To give special effects, for example, inhibitive paints for protection of metals; electrically conductive paints as a source of heat; condensation resistant paints.

Painting constitutes a small fraction of the initial cost of a building and a much higher proportion of the maintenance cost. It is, on this basis, advisable to pay careful attention to the subject at construction stage. Furthermore, there are a number of situations in which restoration is both difficult and expensive once the original surface coating has failed and weathering has affected the substrate; for example, clear film-forming coatings on timber. In these situations, particular care is necessary.

CONSTITUENTS OF PAINT

Paints consist essentially of a vehicle (or binder) and pigment, the former being responsible for setting, gloss and impermeability, while the latter is responsible for opacity, colour and, to some extent, strength. Once applied, the coating must harden within a few hours. The hardening process may be due to one of the following:

1 Polymerisation by chemical reaction with a hardener or by oxygen (or, in some cases, moisture) in the air.
2 Coalescence of an emulsion.
3 Evaporation of a solvent.

Categories (1) and (2) may be referred to as convertible coatings since, on hardening, coatings cannot easily be restored to their earlier liquid state. Note, however, that there are usually no new primary bonds produced when emulsions in category (2) coalesce. Type (3) are non-convertible, since the liquid state can be restored simply by adding a suitable solvent.

The properties of the paint may be modified to a large extent by the pigment. Pigments are fine insoluble crystalline particles which give colour-hiding ability and

Table 9.4 Types and properties of pigments

Property	Type	Comment
White paints	Inorganic	Titanium dioxide is the whitest pigment. No organic white pigments are available
Black paints	Inorganic	Carbon is the blackest pigment. No organic black materials are available
Brilliance and clarity	Organic	The most attractive and cleanest looking colours are made with organic pigments
Light fastness	Inorganic	The valency bonds in inorganic compounds are usually more stable to ultraviolet light than those in organic compounds
Non-bleeding	Inorganic	Inorganic compounds are almost insoluble in organic solvents
Heat stability	Inorganic	Very few organic compounds are stable above 300 °C

body to the paint. They may be organic, inorganic, natural or synthetic. If the resulting hardened paint film is to be glossy, all the pigment must be below the surface. The amount of gloss is determined by the type and shape of pigment and the ratio of pigment to binder, usually on a volume basis. In general, very glossy paints have less pigment and, therefore, less hiding power, while matt ones are underbound. Increasing the pigment proportion increases hardness but decreases flexibility. Other properties, such as corrosion resistance and exterior durability, are affected by the quality and quantity of pigment. Some desirable properties and types of pigment are shown in Table 9.4. It will be appreciated that films of paint are, by normal everyday standards, very thin at around 25–40 μm. A pigment particle will be about 1/100 of this thickness, intentionally near that of the wavelength of visible light to obtain maximum opacity. Other materials may be incorporated, as follows, for example.

Extenders are used to improve some of the properties of the paint, although they have little or no pigmentary value. They can, for instance, be used to control the amount of gloss. A semi-gloss paint might need so much pigment to achieve the correct gloss characteristic that it would not pour and would leave heavy brush marks. The addition of a suitable filler can considerably reduce cost and at the same time improve viscosity and finish. The amount of pigment should be the minimum amount required to give colour and hiding power. The amount of extender used varies, depending on the paint, but may be as much as 45 per cent. Materials used as extenders do not affect the colour because their refractive index is very close to that of the medium. Extender particles are much larger than pigment particles.

Solvents are volatile liquids added to assist in manufacture and they also improve storage properties and lower the viscosity of wet paint. They must evaporate very rapidly when the paint is applied to a surface, so that high viscosity is obtained and, hence, freedom from runs. The solvent can affect the final result and it is vital that the correct type is used for any particular resin system. Solvents are now described as VOCs – volatile organic compounds. They are regarded as environmentally undesirable because they contribute to atmospheric pollution and may be a factor in certain complaints such as asthma.

Driers are added to oil-bound paints. These are usually oxides which will give out oxygen and thereby increase the rate of oxidation of the binder and thus the rate of drying. The rate of drying can also be increased by using *blown* linseed oil – linseed oil which has had air blown through it.

Plasticisers are added to some paints to make the hardened film more flexible.

Light stabilisers are added to make the paint colour more stable under sunlight.

Fungicides and insecticides are added to prevent attack by insects and to prevent the formation of moulds. These are needed with household vinyl emulsion paints to prevent mould growth feeding on the cellulose or other colloids in the dry film.

It will be appreciated that the formulation of paints is highly complex, manufacturers continuously seeking to improve the performance and competitiveness of their products. Most modifications are evolutionary – it is rare for completely new paints to be produced. There are no standards at present relating to undercoat or gloss paints, partly due to resistance by manufacturers, who are not in favour of compositional standards as they prefer the freedom to make changes in formulation. Performance standards, which describe the ability of paints to carry out essential functions over a long period of time, are now becoming available. Nevertheless, the relative merits of different brands of paints for the same application tend, at the moment, to be largely assessed from individual user experience.

OIL BASED PAINTS

These were traditionally based on linseed oil (obtained from flax seeds) or tung oil (obtained from soya beans). Although satisfactory properties can be obtained by refining these oils, they are now usually modified by alkyd resins. These paints are traditionally solvent borne; the resins are combined with organic solvents which provide suitable flow properties. Some understanding of the drying action of oils used in paints can be gained from an examination of the organic acids which form their basis. Vegetable oils, such as linseed oil, consist of combinations of triglycerides of fatty acids. Glycerol has the formula:

$$
\begin{aligned}
&CH_2-OH \\
&| \\
&CH\ -OH \\
&| \\
&CH_2-OH
\end{aligned}
$$

while fatty acids have a variety of formulae, each containing the $-COOH$ (acid) radical, for example:

$CH_3(CH_2)_{16}\ COOH$ stearic acid – no double bonds

$CH_3(CH_2)_7CH{=}CH(CH_2)_7COOH$ oleic acid – one double bond

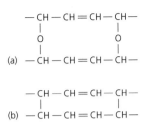

Figure 9.17   Cross-linking of molecular chains in a drying oil by means of oxygen: (a) oxygen directly involved in link; (b) oxygen not directly involved in link

On reaction, these form products of the type:

$$CH_2 - OOCR_1$$
$$CH \ - OOCR_2$$
$$CH_2 - OOCR_3$$

where R_1, R_2 and R_3 are the hydrocarbon sections of the acids. Water is also produced, the OH coming from the glycerol and H atoms from the acid radicals.

Quite large molecules may result from this process and the oil may be made to polymerise if double bonds in the hydrocarbon sections give rise to cross-linking. For example, the chain $-CH_2-CH=CH-CH_2-$ combines with a similar chain by means of oxygen. The oxygen may become directly involved (Figure 9.17(a)) or may simply initiate C–C bonds (Figure 9.17(b)). These linkages may occur between any neighbouring hydrocarbon chains, so that oxygen from the atmosphere causes gradual cross-linking of molecules until a solid film is produced. The strength and stiffness of the solid depends on the number of unsaturated groups in the original glyceride and, hence, on the acid type present. Stearic acid, for example, is saturated; therefore it would not form a drying oil. Acids containing a smaller number of unsaturated groups would result in oils which produce soft films, more suitable for mastics, while acids containing a large number of unsaturated groups might produce hard, brittle films. Driers based on compounds of lead, cobalt and manganese can be incorporated in the oils, accelerating curing. They are characterised by having more than one valency value, so that they can effectively carry oxygen into the molecular network to assist in polymerisation. Boiled linseed oil, which has hardening properties superior to those of ordinary linseed oil, contains metal compounds incorporated by heating. All oil based paints have the disadvantage that continued oxygen access after hardening tends to result in gradual embrittlement.

Once hardened, oil based paints behave as thermosetting plastics, being resistant to solution in the oils from which they are formed. Such paints form films in the can unless a protective coating of, say, white spirit is applied.

Alkyd resins

These are a form of polyester resin and play an important part in modern oil paints. They are produced by mixing a polyhydric alcohol with phthalic anhydride

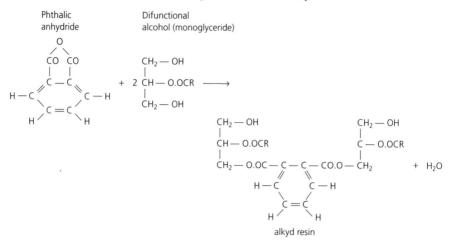

Figure 9.18 Formation of a linear alkyd resin by reaction between phthalic anhydride and a difunctional alcohol. If a trifunctional alcohol were used, a rigid cross-linked structure would be obtained

(so-called because it is derived from phthalic acid by extraction of a water molecule). Figure 9.18 shows the reaction with a difunction alcohol containing a drying-oil residue. A long chain molecule is formed with a number of unsaturated groups in the drying-oil residue (R). These dry by the action of oxygen, as in oil based paints. Gloss retention, durability and colour retention properties of alkyd-modified drying oils are much better than those of pure alkyd or pure oil and, since the polymers are more polar, they are stronger and more adherent to the substrate than pure oils. Vehicles containing a high proportion of oil are known as *long-oil* vehicles and these are tough, flexible and suitable for external use. Medium- and short-oil types are more suitable for internal use.

Polyurethane paints

These are available in a variety of forms. The hardest films are obtained from two pack versions of the resin. The resin itself is a polyester, as are the alkyd resins, though produced by a different process, involving a dibasic acid:

$$R \Big\langle {{-COOH} \atop {-COOH}}$$

a diol:

$$R \Big\langle {{-OH} \atop {-OH}}$$

and a polyol, for example:

$$R \underset{\diagdown\; OH}{\overset{\diagup\; OH}{\rule{0pt}{0pt}\!-\!OH}}$$

The resulting ester has several –OH groups which are reacted with isocynates containing NCO groups (–N=C=O). A typical reaction is:

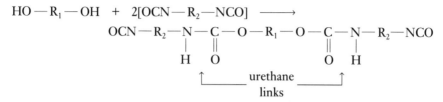

The isocynates cause polymerisation by linking the hydroxyl groups belonging to esters.

When an excess of isocynate is used, the polymers terminate in –NCO groups, which can result in cross-linking, due to moisture in the atmosphere. This process also gives rise to carbon dioxide gas and is used in the formation of expanded polyurethane.

Two pack polyurethane resins may be used, polymerisation occurring if there is an excess of –OH groups in one pack and of –NCO groups in the other. The two pack varieties produce extremely good film properties, though toluene diisocynate vapour can be dangerous if inhaled. Also, temperature tends to affect the curing rate and two pack paints, which require accurate mixing are themselves a disadvantage for general site use. Various types are available; the hardest resins may be used on furniture, while more flexible grades are required for external use or flooring. Note that curing of two pack polyurethanes is not dependent on the atmosphere.

Air drying polyurethane paints and varnishes

These are commonly used in building for a wide variety of purposes. They are really modified drying oils. For example, by means of glycerol, hydroxyl groups are introduced into an air-drying oil, producing a monoglyceride.

$$2 \begin{array}{c} CH_2-OH \\ | \\ CH-OH \\ | \\ CH_2-OH \end{array} + \begin{array}{c} CH_2-COOR \\ | \\ CH-COOR \\ | \\ CH_2-COOR \end{array} \longrightarrow 3 \begin{array}{c} CH_2-OH \\ | \\ CH-COOR \\ | \\ CH_2-OH \end{array}$$

oil   monoglyceride

The addition of polyisocynate to the monoglyceride will cause some polymerisation by action on hydroxyl groups, as previously. The polyurethane 'oil' will then dry by the effect of oxygen on unsaturated bonds, the film having similar properties to two pack varieties, though not normally as hard or tough. In varnishes in particular, some polyurethane coatings may not withstand the thermal and moisture movements which occur when applied to wood. South-facing aspects exaggerate the problem.

Moist-curing polyurethane varnishes

These are sometimes claimed to react with water in damp timber, forming an impervious film. Although the polymerisation process involves moisture, quantities required are only small – approximately 1 per cent by weight of the paint. Hence, they will not 'dry' wet timber. Paints of this variety cannot be produced, since pigments normally contain sufficient moisture to cause them to set in the can.

Saponification of oil based paints and varnishes

It has been explained that oil based paints are glycerides of organic acids, that is, they are formed from glycerol. If an alkali, such as calcium hydroxide, contacts such an oil based film, there is a tendency to revert to glycerol with the production of the corresponding salt.

$$\begin{array}{l} CH_2 - COOR_1 \\ | \\ CH \ - COOR_2 + 3MOH \\ | \\ CH_2 - COOR_3 \end{array} \longrightarrow \begin{array}{l} CH_2 - OH \\ | \\ CH \ - OH + MCOOR_2 + MCOOR_1 + MCOOR_3 \\ | \\ CH_2 - OH \end{array}$$

This leads to the breakdown of films and formation of a scum. Hence, oil-containing paints (including polyurethanes) should not be used on alkaline substrates, such as Portland cements. Alkali resistant primers, such as PVA emulsion paints, should first be applied.

EMULSION PAINTS

These are now very widely used in interior decorating. Examples are polyvinyl acetate (PVA) emulsions, which are suitable for application to new cement or plaster. The molecules are very large but are dispersed in water by colloids to give particles of approximately 1 μm in size. Hence, these paints have the advantage of being water miscible, although, on drying out, coalescence of polymer particles occurs, resulting in a coherent film with moderate resistance to water (Figure 9.19). The film is, however, not continuous, so that the substrate can, if necessary, dry out through the film. Emulsions must have a certain amount of thermal energy to coalesce, hence, there is a *minimum film formation temperature* (MFFT) for each type. Typical MFFT values are, for PVA 7 °C and for acrylic copolymers, 9 °C. They should not be used below these temperatures.

Water based paints have more recently been produced for a much wider variety of applications in view of their much lower VOC (volatile organic compound)

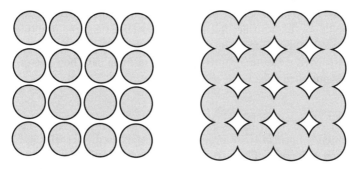

Figure 9.19 Coalescence of an emulsion to form a coherent but non-continuous film

Table 9.5 Volatile organic compound content of some common paints as declared by retailer B&Q

VOC category	VOC content %	Example of product
Minimal	0–0.29	PVA emulsion paints
Low	0.3–7.9	Water based wood primer
Medium	8–24.9	Water based interior varnish
High	25–50	Gloss finish
Very high	Above 50	Exterior wood stains

contents. In particular many paints or other protective coatings for wood are now water emulsion based. Table 9.5 shows the 'self-declared' VOC contents for a range of paints for surfaces including wood and plaster supplied by a major UK retailer.

SOLVENT BASED PAINTS

From some points of view, non-convertible coatings are an advantage, since:

- Such paints do not form films in the can.
- Subsequent coats bond into one another.

The hardened films may have good chemical resistance but are much less tolerant of certain organic solvents than convertible types. Example of solvent based paints are cellulose and bituminous paints.

Cellulose paints

The cellulose constituent is a form of nitro-cellulose dissolved in a solvent such as acetone. Plasticisers are added to give elasticity and synthetic resins to give a gloss, since pure cellulose gives little gloss. Drying usually occurs rapidly but well ventilated areas are essential and the paint is highly flammable. Cellulose paints are most suited to spray application (though retarded varieties for brushing are available). These properties, together with the fact that the paints give off a penetrating odour, tend to restrict the use of cellulose paints to factory application. In these conditions, high quality finishes can be obtained and the resulting coat has good resistance to fungal attack and to chemicals, including alkalis. Very high VOC levels are associated with these paints and many of them have now been replaced by water based equivalents.

Bituminous paints

These are intended primarily for protection of metals used externally and have poor gloss-retention properties. Thick coats give good protection, though the solvents used sometimes cause lifting, if applied over oil based paints, or bleeding, in subsequently applied oil based coats. Sunlight softens the paint, though resistance can be improved by use of aluminium in the final coat. Chlorinated rubber paints have similar properties; uses are often based on their resistance to alkalis.

PAINTING SPECIFIC MATERIALS

Ferrous metals

Steel forms the largest bulk of metals used in building and is one of the most difficult to maintain. The need for application to a good substrate cannot be overemphasised; the best time to paint steel is immediately after production, though mill scale (iron oxide film produced during hot rolling) should be removed because:

- It behaves cathodically to the bare metal and may lead to local corrosion.
- It may eventually flake off due to differential movement.

Hand and power tool cleaning may only be about 50 per cent effective and will not lead to the best protection. Grit blasting may be applied to remove any corrosion, though a very rough finish makes it more difficult to achieve a uniform paint film. Pickling – treatment with hydrochloric/phosphoric acid – is a factory process mainly used as a pre-galvanising treatment.

Paint systems normally consist of primers, intermediate coats and finishing coats. Primers are designed to grip the substrate, to provide some protection (possibly by inhibition) to the steel and to act as a suitable base for subsequent coats. Inhibitive primers include red lead (lead oxide), zinc dust, zinc chromate and zinc phosphate. Of these, red lead, though toxic, is often still preferred because it is fairly tolerant of poorly prepared surfaces and is amenable to application in thick coats. Metallic lead primers, though non-inhibitive, are fairly tolerant of poor surfaces and may be easier flowing, and quicker drying than red lead. They also have superior chemical resistance. On account of their toxicity, lead based primers are not recommended for use in domestic situations.

Special (*prefabrication*) primers about 15–20 μm thick are often applied to steel soon after production to afford weather protection prior to and during fabrication. The thickness is limited because the paints must not interfere with subsequent cutting or welding operations.

They must be quick drying to permit almost immediate handling. They have good short term weathering resistance without the need for further paint coats. They are typically based on epoxy resins and may contain inhibitors such as zinc.

Intermediate coats are designed to build up paint thickness. For best protection two or more coats are recommended since successive coats hide defects in previous coats.

Where a decorative finish is required, paints based upon alkyd, acrylic or polyurethane binders may be used. Two pack versions give thermosetting characteristics to the film resulting in increased hardness and better chemical resistance. Red oxide (micaceous iron oxide), as well as being a moderately effective primer, may be used in undercoats and finishing coats.

For any one application it is recommended to specify a system supplied by one manufacturer to avoid possible compatibility problems.

Generally, the wetter the situation and the more aggressive the climate or atmosphere, the more coats should be given. In extreme situations or where extended life without maintenance is required, protection is only likely to be achieved by impregnated wrappings, bituminous or coal tar coatings, thick, factory-formed films or prior treatment, such as galvanising.

A further alternative for protection of steel is airless sprays which permit higher solid (and therefore lower VOC) contents. They are suited mainly to use under factory conditions.

Non-ferrous metals

Zinc and aluminium are the non-ferrous metals most likely to require surface coatings and each provides a poor key for paint, unless surface treatment is first carried out. Zinc, in particular, reacts with most oil based paints, forming soluble salts which reduce adhesion. Zinc should be degreased with white spirit, followed by roughening of unweathered surfaces with emery paper or etching treatment. Primers containing phosphoric acid are available for this; they often also contain an inhibitor, such as zinc chromate. Other suitable primers contain calcium plumbate, zinc dust or zinc oxide. For aluminium, etching is again an advantage, followed by zinc chromate or red oxide primers. Lead based primers are not suitable.

Wood

Preliminary treatment includes stopping holes and treatment of knots with shellac. A primer is essential to penetrate and yet block the pore structure. Undercoats are unsatisfactory here, since they often do not penetrate the wood and may flake off later. Lead based primers have been replaced by newer types such as aluminium (BS 4756) and acrylic water borne primers (BS 5082), which now provide strong competition to the conventional solvent based primers. Acrylic primers are tolerant of higher moisture contents in the timber, though experience shows that permeability is too high for use in single coats as a protection for joinery timber exposed on site prior to installation. Where exposure of primed timber is a possibility, it is recommended that the specification should require that priming paints comply with the appropriate British Standard, since factory based primers in particular, often do not comply with the six months exposure test requirements of current standards.

Undercoats are used to obtain the desired colour, contributing to the film thickness and, therefore, protection, though the gloss coat provides the bulk of the protection.

Alkyds and polyurethanes form the vehicle in most traditional solvent borne paints and these have the advantages of being rather more tolerant of dampness together with less pronounced brushmarks. Newer types, such as acrylic emulsions, are now available for internal and external use. Acrylic paints have demonstrated a number of advantages including:

- low odour
- rapid drying
- ease of application
- ease of cleaning equipment, brushes, etc.
- good durability, particularly cracking resistance
- resistance to yellowing

The resistance to cracking mentioned above is due to the fact that setting does not involve oxygen, hence embrittlement due to continued oxygen penetration does not occur. Possible problems of acrylic paints include the fact that they may inhibit hardening of fresh putty, they may be affected by rain during drying and gloss levels are not as good as those of conventional alkyd paints. With increasing emphasis being placed upon the environmental effects of materials, there are obvious attractions of the use of low solvent paints. Table 9.5 shows relative values for the VOC levels of some paints and other treatments for wood. Figure 9.20 shows the symbol used by a major UK retailer to indicate the VOC level based upon Table 9.5. See also 'Exterior wood finishes'.

Figure 9.20   Display symbol used by major retailer B&Q to indicate VOC content of paints and other finishes (courtesy B&Q plc)

Plastics

Most plastics in common use do not require painting, and paint coats, once applied, cannot be removed by normal techniques. Paints, on the other hand, will reduce the rate of degradation of plastics such as polyethylene and uPVC. Adherence is poor unless the surface is first roughened to give a mechanical key. The impact strength of some plastics, such as uPVC, may be adversely affected if painted, by migration of solvent into the paint.

RHEOLOGY OF PAINTS AND RELATED MATERIALS

It is well known that the behaviour of solids under stress is described by the use of stress/strain diagrams and, at least at low stress values, a given stress defines a given strain for a particular material. The relationship is clearly different for liquids since, in the absence of long range order, the application of stress will cause immediate relative motion within the body of the liquid; there is no elastic stage. The equivalent of a stress/strain diagram for liquids would, therefore, be a graph relating shear stress to strain rate and this relationship will depend on the nature of short range bonding, practically observed as the viscosity of the liquid. For many liquids, usually of low molecular weight, a straight-line graph would be obtained, at least for low shear stresses (Figure 9.21). Such liquids are described as Newtonian. A typical consequence of this relationship between shear stress and shear strain is that the velocity of a liquid in a pipe varies parabolically from the maximum value (at the centre) to zero (at the walls), assuming flow is non-turbulent.

As the molecular weight of a liquid is increased, viscosity increases and interaction between molecular chains alters the relationship between shear stress and shear rate. At small stresses, some such liquids shear at a lower rate than would be expected, while others have the characteristics of a solid, supporting a small shear stress without continuous movement (Figures 9.22 and 9.23). Such behaviour is non-Newtonian. The form of Figure 9.22 is obeyed by most paints, in which large molecules tend to become tangled together. Small shear stresses are insufficient to cause alignment, while larger stresses achieve this, resulting

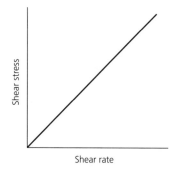

Figure 9.21   Newtonian liquid. Shear rate is proportional to shear stress

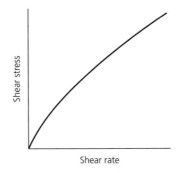

Figure 9.22   Non-Newtonian liquid. Shear rate is lower than would be expected at low shear stresses

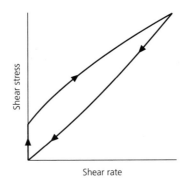

Figure 9.23   Non-Newtonian liquid. Semi-solid properties are displayed at low shear stresses. Such liquids are thixotropic, the degree of thixotropy being proportional to the area enclosed by the increasing and decreasing shear stress cycle

in decreased viscosity. The viscosity of paints must be carefully controlled by regulation of molecular chain size and use of solvents, since a paint must be sufficiently liquid to brush or spray and yet sufficiently viscous to resist sagging or curtaining on vertical surfaces after application. It is found that these properties are best achieved when there is a grading of molecular weights in a way similar to that by which aggregates for concrete are graded.

Some types of paint exhibit colloidal properties and obey the form of Figure 9.23. This is caused by van der Waals bonds and can be useful, since the gel structure of the paint breaks down under shear stresses due to brushing and re-forms afterwards, improving the stability of the fresh film. Such paints, known as thixotropic, are attractive to the amateur, since they are tolerant of poor application technique, and to the professional, since they allow thicker coats to be applied, where required. The area enclosed by the complete stress/strain curve of Figure 9.23 indicates the degree of thixotropy. They have the slight problem that the physical effort of application may be higher that that of the traditional liquid paints owing to the increased stress needed to fluidise the material.

CHAPTER 10

TIMBER

Timber is one of the earliest materials to be used in building. When correctly used it has extremely good durability. The material still plays a major part in general building, particularly in flooring and roofing. Advances in the field of adhesives and fabrication techniques have increased enormously the potential of the material and better methods of testing, combined with increased understanding of timber properties, have enabled timber to be used at quite high stress levels. At the same time, the material can be worked with simple hand tools, is resistant to many types of chemical and atmospheric pollution and can be finished to give aesthetically pleasing effects. This section does not attempt to deal exhaustively with all the properties and applications of the wide range of varieties that exists. Instead, the fundamental properties and applications will be described.

10.1 CHEMICAL AND PHYSICAL STRUCTURE OF TIMBER

Wood is a naturally occurring fibrous cellular composite which has evolved over thousands of years to produce a material of great strength and rigidity in relation to its self-weight. The fibres are composed of cellulose, a polymer of glucose $(C_6H_{10}O_5)_n$. There may be as many as 10 000 glucose molecules in one linear polymer chain. Chains crystallise to form microfibrils, bonded by hemicellulose – a much lower molecular weight form of cellulose – and lignins – polymers based on phenol – to form hollow cells which account for the light weight of timber.

Wood grows by the process of photosynthesis which is similar in all green vegetable matter. A brief description is given to illustrate how the growth of wood:

- releases oxygen into the atmosphere
- binds atmospheric carbon dioxide into solid organic material

1 Chlorophyll in the leaves absorbs energy and uses it to convert water into hydrogen ions and oxygen.
2 The oxygen is released into the atmosphere while the hydrogen ions react with phosphates and nitrates in the sap (derived from the soil) to produce complex phosphate and nitrate based compounds.

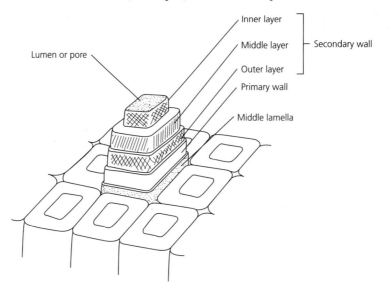

Figure 10.1   Cell wall components in wood. The wall thickness has been exaggerated for clarity

3 These compounds, in conjunction with enzymes, reduce carbon dioxide to carbohydrates.
4 Further reaction with enzymes results in the production of the constituents of wood as above.

The detailed make-up of cells varies greatly from species to species and type (softwood or hardwood), though the cell wall structure is basically as shown in Figure 10.1. Each new cell is formed by division of cells in the cambium layer (beneath the bark) and grows in stages:

1 middle lamella (mainly lignin)
2 primary wall – random microfibrils
3 secondary wall – S_1 outer layer (narrow) crossed, spiralling microfibrils
           – S_2 middle layer (thick), microfibrils in a common spiral at an angle of 10–30° to the long axis
           – S_3 inner layer (narrow), crossed spiralling microfibrils

The S_2 layer forms about 3/4 of the cell wall in thickness and the *microfibrilar angle* of microfibrils in this section plays a major role in determining the mechanical properties of the timber. The building up in layers of the cell wall, rather like winding of layers onto a car tyre, produces a structure of immense rigidity: the longitudinal E value of the cell wall itself is similar to that of concrete and the tensile strength is many times greater – about 150 N/mm^2.

Woods can be conveniently divided into hardwoods and softwoods according to detailed size and shape.

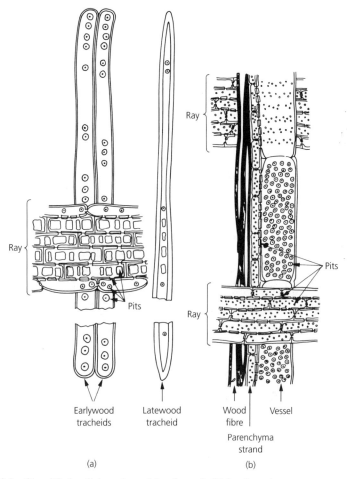

Earlywood   Latewood   Wood   Vessel
tracheids   tracheid   fibre

Parenchyma
strand

(a)   (b)

Figure 10.2   Simplified radial sections: (a) softwood; (b) hardwood

Softwoods

Softwoods may be regarded as the most primitive form of timber. They have been in existence for over 18 000 years, since the end of the ice age. Softwoods comprise a single basic cell type, tracheids, which fulfil both good conduction and support functions in the tree (Figure 10.2). They have an appearance rather like drinking straws, though the wall thickness of each tracheid is of variable thickness, being thin in springwood and thicker in summerwood. Springwood has a density typically of 350 kg/m^3 and summerwood 700 kg/m^3 when air-dry. The fibres themselves have a much higher density – of around 1500 kg/m^3 – hence, completely saturated wood may sink in water. Figure 10.3(a) shows a scanning electron micrograph of these cells. Softwoods also contain rays running in the radial direction, though the rays are rarely visible in the bulk timber. Rays contain hollow food storage cells

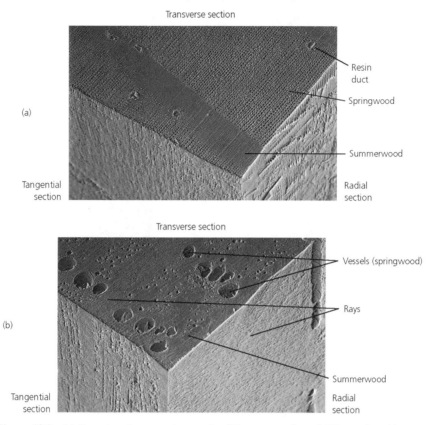

Figure 10.3   (a) Scanning electron micrograph of European redwood (*Pinus sylvestris*), a common softwood. The resin ducts form mainly in summerwood. (×10 magnification.) (b) Scanning electron micrograph of oak (*Quercus robus*), a ring of porous hardwood. There are broad rays between each group of about four vessels in springwood. Most of the radial section in this picture is occupied by a ray with a clear horizontal grain. (×10 magnification)

called parenchyma. Pits in cell walls permit conduction of food (in the form of sap) between them. Figure 10.4 shows radial, tangential and transverse sections through a five year old softwood log. Note also the following properties of softwoods:

■ They are coniferous (cone bearing) and have needle-shaped leaves that are retained in winter.

■ They grow quickly to give wood of generally low density and fairly low strength which is easily worked. Trees may be felled at ages of as little as 30 years.

■ They are not normally highly durable unless protected by preservatives (exceptions to this are Western red cedar and sequoia).

■ Since they grow quickly, softwoods are relatively cheap and are widely used in all forms of construction.

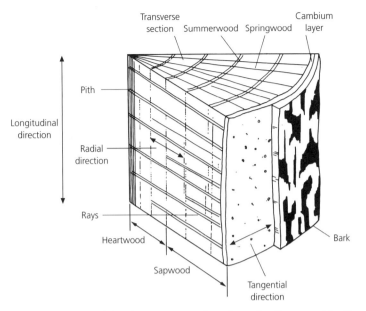

Figure 10.4 Radial, tangential and transverse sections through a five year old softwood log

The most commonly used species in the UK are European redwood and European whitewood. Easily obtainable at extra cost are Douglas fir and Western red cedar. Sitka spruce, a fast growing, low density whitewood, is now being widely grown in the UK and should be increasingly marketed over the next ten years.

Hardwoods

These have a more complex cell structure than softwoods, structural support being provided by long, thick walled cells known as fibres, while food conduction is by means of thin tubular walled cells called vessels (Figure 10.2). Parenchyma (food storage cells) are again present as longitudinal strands and in rays, the latter often being clearly visible in hardwoods; the silver grain in oak is caused by light reflection of the rays. Cells are again interconnected by pits.

Hardwoods have the following properties:

- They are obtained from broad leaved (deciduous) trees which lose their leaves in winter. Sap requirements in spring are often, therefore, considerable in order to form new leaves. In consequence, large vessels may form in spring wood. These woods, of which oak is an example (Figure 10.3(b)), are described as *ring porous*. When pores are more evenly distributed, a *diffuse porous* wood results. When sawn, large pores give a characteristic grainy texture to the wood surface. Both oak and mahogany, a tropical diffuse porous hardwood, show this effect.
- Hardwoods grow slowly to give wood of generally high density and high strength. Hardwood trees may not mature for 100 years or more.

- Many hardwoods are in general durable, and some such as oak may last for centuries without use of preservatives (though, exceptionally, balsa and poplar are also hardwoods and these are neither strong nor durable).
- Since they generally grow slowly, most hardwoods are relatively expensive and their use is restricted to applications calling for high strength, hardness or durability.

Some hardwoods such as African walnut, rosewood and mahogany are becoming increasingly scarce due to destruction of the rain forests though others such as keruing are readily available at reasonable prices. Conversely some indigenous species such as sweet chestnut are now available. This has a grain structure rather similar to oak but without the rays, is easier to work and has an attractive golden brown colour.

Heartwood and sapwood

Most of the chemical activity in a growing tree occurs just below the bark cells in the cambium layer dividing to form new wood cells and bark. Food conduction and storage are fulfilled by outer layers of the tree described as sapwood. Inner layers (heartwood), though still important with regard to structural support, no longer store food, the cells undergoing chemical changes to assist in preservation of the tree. Heartwood is generally more acidic than sapwood and, in the absence of stored food, is more resistant to fungi and insects in service. Heartwood can often be recognised, as it is usually visibly darker than sapwood. The proportion of sapwood in a tree decreases continuously with the age of the tree, young trees being almost all sapwood while in mature trees, the proportion may be less than 20 per cent. Since heartwood is preferred for many applications, it will be evident that older trees are more valuable in this respect, though commercial pressures often result in felling of trees that are mainly sapwood. Hardwoods in particular normally contain much higher proportions of heartwood since they are not felled until relatively great ages.

10.2   MOISTURE IN TIMBER

Cellulose contains many OH– groups which are hydrophilic and, therefore, attract large quantities of water, though the effect is to some degree controlled by lignin which coats the microfibrils. Moisture content in timber is defined as:

$$\text{moisture content} = \frac{\text{mass of water present in a sample}}{\text{mass of that sample when oven dry}} \times 100$$

It will be noticed that on the basis of the above definition, if the mass of water contained exceeds the mass of the dry timber, a moisture content of over 100 per cent is obtained and this is, in fact, not unusual in newly felled (green) timber. The sapwood tends to contain more moisture at the time of felling, the average moisture

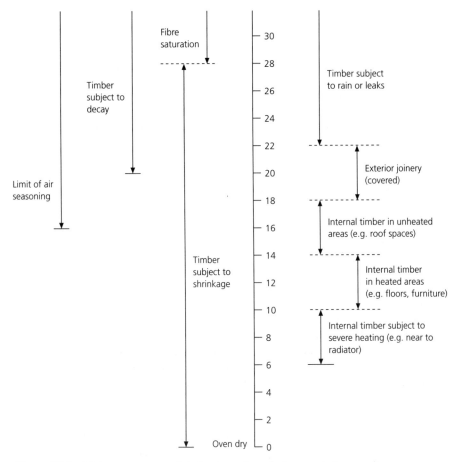

Figure 10.5 Moisture contents of timber in various environments (per cent)

content being about 130 per cent, though values of over 200 per cent are possible. In heartwood, the moisture content is more likely to be around 50 per cent. The moisture content of green timber varies considerably with species. Dense timbers, which include most hardwoods, have smaller cavities and larger cell walls, hence they can store less water.

The moisture content of timber in service is dictated by the relative humidity of the environment, a dynamic equilibrium being established in any set of conditions. Figure 10.5 shows equilibrium moisture contents in a variety of conditions in buildings. At a moisture content of about 28 per cent, the timber is said to reach *fibre saturation* – the cell walls cannot accommodate any more moisture at this stage, additional water occupying cell pores. This state will only occur in buildings in the presence of leaks. It will be clear that, since relative humidities vary both on a daily and seasonal basis, the moisture content of timber will fluctuate correspondingly, though the changes may be slow, especially in timber of large section size or if protected by paint. In centrally heated buildings the moisture content of timber appears to reach a minimum value in late spring since during the winter cold air (which cannot accommodate large quantities of water) is drawn into the building and heated, producing a low relative humidity. It has been recently found that in some buildings with high all round air temperatures such as hospitals, moisture contents of timber may drop to surprisingly low values such as 8 per cent.

A further, most important, property of timber is that the cell walls become swollen on wetting, so that wet timber invariably occupies more space than dry timber. At moisture contents below fibre saturation value the volume of timber varies linearly with its moisture content. At moisture contents over saturation value, there is no such variation, since additional water occupies cell cavities where there is no swelling effect. The amount of moisture movement depends on the direction considered, typical average shrinkages between the saturated green state and 12 per cent moisture content being 4–6 per cent tangentially, about two-thirds of this radially and negligible (about 0.1 per cent) longitudinally. These values reflect the fact that wood is most stable in the longitudinal direction, since most cells are aligned in this direction, while some stability in the radial direction is afforded by the rays. As would be expected, the moisture movement of timber varies considerably according to type. An important consequence of the variation of movement of timber with direction is that distortion occurs when green timber is dried. Since tangential shrinkage is greater, growth rings tend to be thrown into tension and, therefore, become straighter on drying. Figure 10.6 shows exaggerated sketches of the effect of drying a number of sections cut from a log. Subsequent movements between, say, 90 and 60 per cent relative humidity are

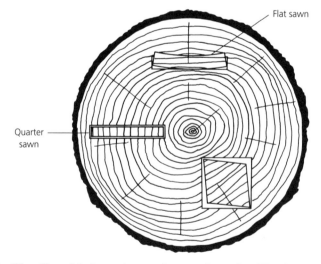

Figure 10.6   The effect of drying various sections cut from a log. The cheapest way of converting timber is to make parallel cuts through the log, known as through-and-through cutting. Most of the resultant timber will then be flat sawn rather than quarter sawn

much smaller. Classification may be based upon the sum of radial and tangential movements:

large   more than 4.5 per cent
medium   3–4.4 per cent
small   less than 3 per cent

MEASUREMENT OF MOISTURE CONTENT

Moisture contents can only be accurately measured by oven drying a representative sample at 100–105 °C, though there are other methods which can be used *in situ* and some may be able to give indications of moisture gradients in the wood. Of these, the electrical resistance meter is probably best known, forming the basis of many survey instruments. The electrical resistance of wood varies greatly with moisture content up to 28 per cent, fibre saturation, and this can be correlated to moisture content, though the correlation may vary by several per cent according to species. The resistance range varies enormously, increasing from about 1 kΩ to 100 MΩ measured over a separation of about 50 mm, as the timber dries. Complex circuitry is therefore required to record moisture contents on a simple meter or digital scale. Most instruments are limited to moisture contents in the range about 8–28 per cent. Below 8 per cent the resistance is too high to measure, while above 28 per cent, the correlation alters. Contact with the wood is by stout needles which can be pressed or hammered into the material. Insulated needles can be used to give moisture contents beneath the surface and the value measured is normally the wettest region between the needles. Temperature changes the calibration by about 1 per cent moisture content per 7 °C, though, with careful use, moisture content can be obtained to an accuracy of within 1 per cent moisture content. Permanently monitored computer operated installations based on the electrical resistance technique have been used, especially in listed buildings with inaccessible timber frames. Figure 10.7 shows such an installation being used to monitor moisture contents during creep testing of a glued laminated timber section.

SEASONING

Seasoning may be defined as the controlled reduction of the moisture content of timber to a level appropriate to end use. Correctly seasoned timber should not be subjected to further significant movement once in service. Seasoned timber is also immune to fungal attack, is stronger, has lower density and is, therefore, easier to handle or transport, is easier to work, glue, paint or preserve than wet timber. The importance of a controlled reduction of moisture will be clear when it is appreciated that moisture loss always takes place at or near the surface of the timber; and internally, drying can only take place if there is a moisture content gradient in the wood causing it to migrate to the surface. Hence, shrinkage occurs preferentially at surfaces. Surface layers are, therefore, thrown into tension as they contract, resulting in *fissures* in the form of checking (longitudinal surface rupture) or

Figure 10.7   Stainless steel pins being used to monitor moisture contents of a glued laminated timber beam during creep testing. The pins are wired back to a data logger which records results

splitting (longitudinal rupture passing through the piece). Serious splits in large timbers are sometimes called *shakes*. Evaporation tends to occur more quickly from the ends of timber during seasoning since wood cells are open to the atmosphere at this position. Checking and splitting are therefore often more pronounced near the ends.

Traditionally a large amount of timber was air seasoned by stacking in open formations under cover (Figure 10.8). Drying takes place due to air percolation around the timber. To avoid damage the timber must be protected from direct rain or sun. The process may take some years depending on the type and section size of timbers, the size of air gaps between them and climatic conditions. (The traditional rule for hardwoods used to be one year for each inch thickness.) The minimum moisture content obtainable is between 15 and 20 per cent dependent on species. Its use in the UK has now, therefore, been largely replaced by kiln seasoning in the case of softwoods, although air seasoning may still be used for initial drying of hardwoods if damage is to be avoided.

Figure 10.8 Air seasoning of timber. Pieces must be well stacked, separated by stickers and protected from sun and rain

Conventional kiln seasoning involves heating the timber in sealed chambers, initially under saturation conditions, the humidity then being progressively reduced to produce drying. Water has a much lower viscosity and therefore flows more freely through the timber at higher temperatures, so that drying can take place more rapidly without damage. Alternative techniques are dehumidifiers, in which water is condensed from the air by cooling, and vacuum kilns, in which water is encouraged to evaporate by evacuation of the kiln, condensed onto cooled plates and then drained off. Drying schedules depend on the type and size of timber sections and their initial and target moisture content. Kiln seasoning takes about one week per 25 mm thickness for softwoods and about two weeks per 25 mm thickness for hardwoods. The process has the advantage that it sterilises the wood and that it can reduce moisture contents to lower levels than air seasoning – values of about 12 per cent can be reached quite easily. Timber is *converted* prior to seasoning, that is, logs are cut into sections appropriate to end use but which can be seasoned in a reasonable time. Drying of complete logs would be uneconomic, taking several months and even then, with a risk of serious shakes. Seasoned timber is generally supplied at a moisture content of 12–15 per cent. Lower levels could be obtained at extra cost for use in air conditioned buildings and this would be recommended where high heating levels referred to above are encountered. Once timber has been seasoned it is important to avoid significant moisture content changes. Hence timber must be protected when being stored on site. Where possible (for example in remedial work), it is a good idea to condition timber in its service environment prior to use by stacking in open formations for a time. This will reduce the risk of subsequent distortion.

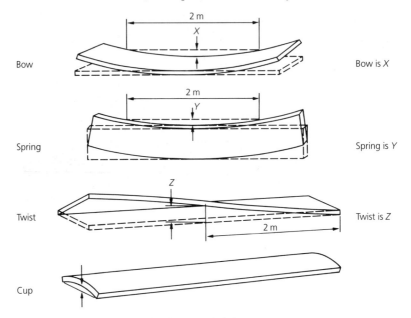

Figure 10.9   Types of distortion in timber and their measurement

10.3   FACTORS AFFECTING THE DISTORTION OF TIMBER PRIOR TO OR IN USE

Distortion is a serious problem in many applications of timber and failure to recognise the potential difficulties when specifying or selecting timber can lead to either poor quality work or rejection of large quantities of unsatisfactory material. Figure 10.9 shows the main types of distortion in timber. Distortion is affected by the following:

- Species of timber. Some timbers having similar moisture movements in the radial and tangential directions, for example, Douglas fir and utile are inherently less prone to distortion than timbers having high relative movements, for example, European redwood and beech.
- Method of conversion. Quarter sawn timber distorts less on moisture content change than flat sawn timber, though the latter has a more interesting grain pattern. Pieces of timber which contain the pith are also more prone to distortion, especially springing. Where the piece is not cut parallel to the longitudinal axis, bowing, springing and twisting are likely. Young trees with a high proportion of sapwood are much more prone to distortion than heartwood of older trees.
- Slope of grain. Twisting is often caused when the grain is not straight or when the density of the wood varies, for example, due to unequal growth on different sides of the tree. Timber having uniform and straight grain is less likely to undergo distortion.

- Stacking procedure. Each piece of timber in a stack must be adequately supported by *stickers* prior to, during and after seasoning, in order to avoid distortion due to self-weight. Twisting of timber can be largely avoided if it is held firmly in stacks during seasoning.
- Moisture change following seasoning. Rapid moisture changes by water borne preservatives or severe drying on occupation of new properties can lead to serious distortion. Such changes should be avoided.
- Final application. On certain types of door, for example, panels may split on drying if assembled and glued at high moisture content. Veneered doors should be of the same veneer each side to avoid bowing by differential shrinkage or movement.

10.4 DURABILITY OF TIMBER

Timber is extremely resistant to deterioration by natural agencies such as sun, rain and frost, presumably by adaptation through evolution over thousands of years. Even when wet it can in some situations perform as a structural material. An amazing early use is in the form of piled foundations in reclaimed land for the city of Venice. It is recorded that the Santa Maria Della Salute church by Baldassare (Figure 10.10), commenced in 1631, was founded upon 1 156 627 alder stakes of 4–5 m length sunk into the clay (Howard *The Architectural History of Venice*). The building survives to this day, though many Venetian buildings have gradually settled on their piled foundations. By far the worst problems with timber are living organisms – fungi and insects, for which the material serves as food. There are, however, other modes of deterioration which should be considered.

SUN AND RAIN

These are responsible for natural weathering processes by which the timber changes colour and, over a period of years, may show surface deterioration. The rate is usually small, averaging about 5 mm per hundred years, though higher and lower values may occur. The sun is probably the main cause of these changes. Ultraviolet light degrades the lignin causing bleaching of the natural pigment to a grey colour. Staining may occur if fungi develop, though this requires the wood to be subject to periods of dampness. As the lignin adhesive deteriorates at the surface, cellulose fibres are subsequently abraded by wind and rain, exposing more lignin. The process may be accelerated if checking, due to rapid surface moisture content changes, increases access of sun and rain. A few species are considered to benefit from this process: weathered oak is sometimes preferred to the smooth, naturally pigmented, freshly exposed material – the appearance is more 'restful'. Generally, however, protection from weather is advisable, not least to prevent moisture contents from rising to unacceptably high levels. Paint adhesion to weathered surfaces is much reduced.

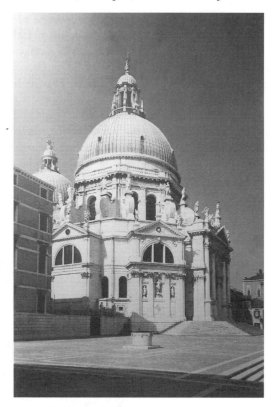

Figure 10.10   Church of Santa Maria Della Salute, Venice, supported by 1 156 627 alder stakes (Source: Howard, *The Architectural History of Venice*)

ACIDS AND ALKALIS

Wood is naturally mildly acidic, having a pH value of between 2 and 5.5. In consequence, timber is resistant to many mild acids, though strong or oxidising acids will cause damage. Alkali resistance is much lower and in such environments, especially if damp, timber must be protected. The most common instance of decay resulting in this way is by use of nails or other ferrous fixings in damp environments. The nails corrode electrolytically, hydroxyl ions collecting at cathodes (see page 381). The associated alkalinity then causes deterioration of the timber. The process may be accelerated due to the presence of highly acidic iron salts, also a product of steel corrosion. The presence of salt-type preservatives increases decay rates by promoting corrosion of metal fasteners, if damp conditions prevail.

FUNGAL ATTACK

The growth of all fungi in timber requires a moisture content of at least 20 per cent, since the enzyme action necessary cannot take place in completely dry timber.

Fungal attack is almost invariably linked to high moisture contents resulting from poor design/construction/maintenance. These may be in the form of:

- leaking roofs
- water crossing defective cavities
- clogged or leaking gutters and downpipes
- permanent dampness as in rising damp
- admission at joints; water then trapped by capillarity
- admission at end grain; water is absorbed many times more quickly by end grain than by radial or tangential faces

There are numerous species, some attacking the growing tree or the unseasoned green timber, while others are able to attack wood in service. The former are mainly the concern of the timber producer, hence attention here is given only to fungi which are active in seasoned timber. Where the cellulose only is destroyed, the wood breaks into small cubes and the rot is known as *brown rot*. Where both cellulose and lignin are destroyed, the wood becomes soft and fibrous, the rot being known as *white rot*. Rots can also be divided into dry rot and wet rot. There are many species falling into the latter category but they are separated from dry rot because remedial treatment of dry rot must be much more carefully carried out.

Dry rot (Serpula lacrymans, formerly called Merulius lacrymans)

Dry rot, a brown rot, is so named because it finally leaves the wood in a dry, friable condition. Growth begins when rust-red spores come into contact with damp timber in poorly ventilated positions, the fungus developing as branching white strands (hyphae), which form cotton wool like patches (mycelium) and finally, soft, fleshy spore-producing, fruiting bodies (sporophore). Once established, the hyphae can grow into adjacent timber, brickwork and plaster in search of moisture or food. They may become modified into vein-like structures (rhizomorphs) 2 or 3 mm in diameter, which are able to transport moisture from damper areas to continue the attack in drier timbers. Dry rot thrives at normal indoor temperatures but is killed at temperatures of over 40 °C, hence it does not grow in positions subject to solar radiation. It becomes dormant at temperatures approaching freezing. Drying out of timber renders the fungus dormant and it may die after one year in this state, though the process may take longer if the ambient temperature is reduced.

Wet rots

As the name suggests, wet rots require higher moisture contents in order to thrive, optimum values being in the region of 50 per cent. There are many types of wet rot, the most common in the UK being:

Coniophora puteana (formerly Coniophora cerebella)   Coniophora puteana (cellar fungus) is commonly found in very damp situations in buildings, especially basements or cellars. The fungus is of the brown variety and leads to cube

Figure 10.11 Wet rot in window joinery caused by nail fixing of broad glazing beads externally, which trapped moisture. In dry weather it is found that drying is lower in painted areas – the paint delays drying

formation, especially in large timber sections, though a skin of sound timber may be present. Surface undulations in timber may be the first sign of damage. Affected wood turns dark brown or black but fruiting bodies are rare.

Phellinus contiguus (alternative name Porio contigua)   Phellinus contiguus is responsible for a considerable increase in window joinery decay, which has occurred over the last 20 years or so (Figure 10.11). It is a white rot, hence wood breaks into soft strands. The cause of the increase in this type of rot is probably the use of sapwood in joinery, which, together with poor design and inadequate glues, leads to water penetration at joints. Figure 10.12 shows the type of joint typical in modern window joinery. Note the complex section with high surface area and low wood thickness. In this particular piece, there is a finger joint present at its tip (these types of section are often produced in continuous lengths by finger jointing to reduce wastage). All things considered, this type of joinery has little chance of success unless adequately preserved and protected against moisture ingress. Since most joinery of this type is painted, the first signs of attack are often undulations in or splitting of the paint film.

Damage from these fungi is more serious in the southern half of Great Britain, due to the added effect of certain wood boring weevils which are found in association with wet rots in these regions, increasing the rate of deterioration.

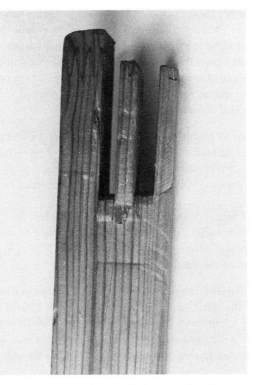

Figure 10.12 Typical joint in modern joinery. This particular joint, part of a new opening casement, pulled out easily, indicating inadequate adhesive performance

Treatment for fungal attack

Dry rot, though less common than wet rots, is far more serious because:

- It thrives in damp masonry, soil or plaster and can remain dormant for some time when growing conditions are poor.
- It tends to occupy cavities or inaccessible places, so that it can remain undetected until serious damage has occurred.

The fungus is more common in wetter parts of Great Britain, such as the North and West, though in any region it is most likely to occur in old, derelict or damaged buildings which are more prone to dampness problems. Many older buildings have undergone alterations and this can result in debris being lodged in cavities or under floor spaces. London has a high incidence of dry rot outbreaks, probably for this reason.

The first task in tackling dry rot is often to establish to what extent it is present since it often grows completely concealed from view. Some indication of risk may obtained by a musty smell and a recent method of detection involves the rot hound, in which dogs are trained to respond to the characteristic odour of the fungus.

Fruiting bodies are not always present. They tend to form when the fungus is threatened with poor growing conditions, the mycelium producing spores in order to safeguard the future of the fungus. Traditionally, quite drastic remedial methods were employed since removal of affected timber may not be sufficient – indeed the presence of plaster and corroded metal appears to provide nutrients which are required for the enzyme action involved in growth to proceed.

Traditional remedial measures are:

- Cut out and burn all visibly affected timber, together with an extra 600 mm of apparently sound neighbouring timber which may contain hyphae.
- Remove any affected plaster, again working beyond the visibly affected area.
- Preserve sound exposed timber and treat surrounding brick or concrete with a fungicide.
- Replace affected timber with preserved, seasoned timber.
- Use zinc oxychloride in paint or plaster coatings applied to affected wall surfaces. This has a fungicidal action.
- Rectify the cause of the dampness.

The essential difference in remedial treatment of dry rot and wet rots is that the latter do not normally infect neighbouring dry timber, brickwork or plaster, hence it is necessary only to cut away and replace affected timber. It is nevertheless still important to remove the cause of dampness and preservation of new timber is generally recommended in case any temporary dampness should recur.

INSECT ATTACK

Insects, such as the beetles that commonly infest timber in construction, have a characteristic life cycle: egg, caterpillar (larva), chrysalis and finally adult beetle. Beetles then emerge from the timber and fly to new timber to lay eggs and repeat the cycle, which may take several years. Virtually all the damage is done by the larvae, which burrow through the wood leaving fine powder (*frass*) behind. This material is in fact their faeces and when the powder is light in colour this is a sure sign of an active infestation – the larvae do not have highly efficient digestive systems so the wood passed through them without any great change in appearance. The larvae spend their lifetime boring within the wood – they cannot move around on the wood surface so that quite advanced deterioration can result with very few flight holes. Table 10.1 shows some common types of insect.

Note that sapwood is much more susceptible to insect attack than heartwood.

The powder post beetle is mainly a problem in timber yards, while the others can thrive in indoor, seasoned timber. The seriousness of attack also varies according to species; for example, the common furniture beetle is extremely widespread but normally takes many years to produce a noticeable number of flight holes or advanced deterioration. Instances of advanced deterioration by *Anobium punctatum* are rare.

Table 10.1 Some types of insect attack on seasoned timber

Beetle	Life cycle	Visual signs	Timber attacked
Common furniture (*Anobium punctatum*)	3–5 years. Emergence, May–Aug.	Beetles 5 mm long, granular dust, $\frac{1}{2}$ mm diam. flight holes	Sapwood of untreated hardwoods or softwoods
House longhorn (*Hylotrupes bajulus*)	3–11 years	A few large holes up to 10 mm, surface swelling, beetles 10–20 mm in length	Sapwoods of softwoods (Surrey)
Death watch (*Xestobium rufovillosum*)	4–10 years. Emergence, March–June	Large bun-shaped pellets, ticking during mating period, May–June	Mainly hardwoods, especially if old or decayed
Powder post beetle (*Lyctus* species)	1–2 years. Emergence, June–Aug.	10 mm exit holes, fine dust	Sapwood of wide-pored hardwoods, partly or recently seasoned

Much more serious is the house longhorn beetle. It is relatively rare in the UK but in parts of Southern England, and particularly in West Surrey, attacks have been severe because local climatic conditions appear to suit the beetles at the flight stage of their life cycle. Beetles lay their eggs in favoured pieces of timber and larvae can then completely consume sapwood portions while still leaving little evidence of their presence, since their life cycle is up to 11 years and there may be very few flight holes in destroyed timber. Quite a lot of damage can be done by one beetle which may lay up to 100 eggs. The larvae have cannibalistic tendencies but even if only say 10 survive, much damage can result since they will consume many times their own volume of wood during the larvae stage. It is estimated that one grub can consume about the volume of a matchbox during its cycle. The larvae are relatively large – up to 20 mm in length having a corrugated profile Figure 10.13. They produce quite large oval flight holes. Roof timbers are mainly affected and UK *Building Regulations* now require effective protection of roofing timbers to be used in certain local authority areas in these regions.

The death watch beetle is chiefly a problem in partially decayed oak members of older churches and other historic buildings where its incidence is at present increasing. Treatment in such cases can be difficult, due to access problems and the very large sections often employed.

DURABILITY CLASSIFICATIONS

Durability classifications now exist for timber, based on the results of tests on 50 mm square section pegs of the heartwood of many different species driven into the ground. They are rated as follows:

Figure 10.13   Rafter badly damaged by house longhorn beetle. The larvae destroyed the sapwood which fell away (lower part of picture) leaving heartwood intact (upper section). The size of the two larvae shown below can be judged from the 10p piece

very durable	more than 25 years
durable	15–25 years
moderately durable	10–15 years
non-durable	5–10 years
perishable	less than 5 years

It is interesting that several hardwoods, for example, European ash, beech and chestnut, are perishable when judged in this way, though the test is very severe. Most softwoods are classified as non-durable. Heartwood of timber, such as Afromosia, African mahogany, European oak, sapele, teak, utile (hardwoods) and Western red cedar and sequoia (softwoods), are classed as durable and need no preservative.

PRESERVATIVES FOR TIMBER

Preservatives should be toxic to fungi and insects, chemically stable, able to penetrate timber and non-aggressive to surrounding materials, particularly metals. Note that preservatives are also often unpleasant to handle and some are harmful on contact with skin or inhalation. They should be handled with extreme care. Increasing attention should also be given to the longer term – for example the prospects for re-use or recycling of timber containing preservatives. They may be classified according to type.

Tar oil preservatives

Creosotes are the best known examples. They are derivatives of wood or coal, though coal distillates are superior. These are only suitable for external use, since they give rise to a noticeable colour and smell and render timber unsuitable for painting. They are, however, cheap, can be applied to timber with a fairly high moisture content and reduce the rate of weathering. They tend to creep into porous adjacent materials such as cement renderings and plaster.

Water borne preservatives

These consist of salts based on metals such as sodium, magnesium, zinc and copper and also arsenic and boron. Copper chrome arsenate (CCA) is a common example; it colours the lighter softwood a pale green. They are very tolerant of moisture in the timber, are odourless, non-flammable, non-creeping and do not stain timber, which may be painted on drying out. In some varieties, insoluble products may be deposited in the wood, resulting in excellent retention properties.

Although water based preservatives are fairly cheap, they have the disadvantage that they swell the timber and will cause corrosion of metals in contact until drying-out has occurred. For this reason, they are best used as a preliminary protection followed by kiln drying, rather than as an *in situ* treatment. Machining of CCA treated timber is hazardous because the resultant dust contains arsenic compounds.

Organic solvent preservatives

These comprise organic based fungicides/insecticides such as lindane, metallic naphthanates and pentachlorophenol in volatile organic solvents. Penetration is excellent provided timber is fairly dry, hence they are often preferred where only brush or spray application is feasible. They are non-creeping, non-staining, non-corrosive to metals, non-swelling and quick drying. Painting is usually possible, once dried. They are widely used for remedial treatment, though many organic solvents are highly flammable and present a fire risk until dried, especially when used in confined areas such as roof spaces. They are more expensive than the other types. The manufacture, transportation, storage and use of organic preservatives all pose quite serious health and environmental hazards. For example they contain high VOC (volatile organic compound) levels and stringent steps must be taken to avoid contamination of humans, animals and plants at each stage.

Application methods

The technique of application is just as important as the choice of a suitable preservative, since any preservative, poorly applied, will fail to penetrate well into timber sections. The difficulty increases with the size of timber sections and some timbers, such as Douglas fir, have naturally low permeability to preservatives.

Where machining or drilling operations are to be carried out, they should be completed prior to application of preservatives in order that maximum penetration is obtained at all exposed surfaces. Some hardwoods, such as teak, have an oily surface which should be degreased with cellulose thinner prior to treatment or application of surface finishes. The following techniques can be used:

- Brushing or spraying. These give only limited penetration of timber, although they may be the only feasible methods for *in situ* remedial work. However, if several coats are applied and the treatment is repeated periodically, a useful degree of protection is achieved.
- Dipping. This involves immersion in the preservative for a period, which, if long enough, can give considerable penetration of preservative. In the *open tank* method of dipping, the preservative is heated and then cooled while the timber is submerged; this gives even better results. It is, nevertheless, not suited to hardwoods or low permeability softwoods.
- Double vacuum process. A vacuum is applied to timber in a sealed chamber and the preservative then introduced. The vacuum is released, forcing the preservative into the wood. After a soaking period, the vacuum is reapplied to eject excess preservative. This process is now widely used in conjunction with organic preservatives for external joinery such as window frames.
- Pressure impregnation. This is often used for injection of water based and tar oil based preservatives. In the empty cell process, the air surrounding the timber is compressed, the preservative introduced and then a still higher pressure applied, forcing preservative into the wood. On releasing the pressure, excess preservative is ejected. If the timber is not pressurised prior to injection of the preservative (in some cases, a vacuum is first applied), the retention of preservative is increased (full cell method). This method is more expensive in terms of the quantity used and, since saturated, will take longer to dry afterwards. Good protection is, nevertheless, achieved.
- Diffusion process. In the boron diffusion method, freshly converted green timber with a moisture content exceeding 50 per cent is dipped into a concentrated borate solution. The timber is covered to prevent evaporation and left for some months for the solution to diffuse through the timber. The timber is then seasoned in the normal way. This method provides effective penetration of difficult timbers, though applications are limited by the fact that, in damp conditions, the preservative leaches out by the same diffusion process which is used for its injection. Some preservative capsules designed to be inserted into joinery at high risk areas such as joints are based on boron.

PRESERVATION REQUIREMENTS AND STANDARDS

Current *Building Regulations* focus upon chief risk areas in health and safety terms as regards preservation requirements in buildings. These include situations where there is embedded timber or where there is risk of attack from the house longhorn beetle. *Building Regulations* do not require most structural timbers to be preserved.

There may be additional situations where there is commercial benefit in terms of customer satisfaction or reduced risk of claims which are considered to justify preservation requirements and the National House Building Council applies much more widespread requirements to domestic buildings constructed by its member companies. These include preservation of tiling battens, flat roof timbers and external joinery.

Traditional specifications have focused upon a *prescriptive* approach to preservation, whereby a preservative type and method of treatment are identified for a specific situation with its assumed risk; for example:

BS 5268 Part 5: *Structural Timber*
BS 5589: *Non-structural Timber*

In BS 5268 the risk and treatability of decay in specific situations are identified, leading to a *risk assessment* for each situation. Recommended treatments are then given on this basis.

When preservatives are to be used, they should be applied in compliance with recommended standards, for example, BS 4072 for copper chrome arsenate. The NHBC regulations work on a similar basis.

The European approach to wood preservation is based upon *performance* criteria in which toxicity levels for given preservatives are identified and then in a given situation, the required penetration and retention levels of the preservative are stipulated. This has involved much research on the performance of the wide range of preservatives on the market in terms of toxicity levels, together with methods of measurement of penetration and retention levels. European standards now exist and directives call for these to be applied in construction. In the long term this approach should lead to appropriate use of preservatives where they are really needed. At the present time there is evidence to suggest that the analysis methods which are essential for assessment of performance of preservatives against stated criteria are not yet sufficiently accurate to permit confident application of the approach.

ALTERNATIVE MEANS OF AVOIDANCE OF FUNGAL AND INSECT ATTACK

Problems of fungal attack can be avoided in the first place by careful attention to roofing and flooring details and ventilation. It is, in general, poor practice to allow timber to contact brick or concrete without some impervious separating membrane, particularly in solid walls or foundation concrete. Careful maintenance of roofing and guttering should avoid problems in roofing timbers. In timber framed buildings, it is wise to treat all timber with preservative before construction and the use of an effective vapour barrier near to the internal side of the construction is essential. This prevents diffusion of water vapour from inside the building, where its pressure is often higher, into the structure where, on cooling, it could result in condensation. A moisture barrier should be provided externally to prevent rain penetration and yet allow small quantities of moisture in the structure to evaporate. In timber framed flat roofs, it is unwise to enclose timber between an internal vapour barrier and the impervious bitumen coating unless adequate ventilation is ensured.

There is now much greater awareness of the environmental hazard posed by wood preservatives and attention is being given to other means of ensuring durability. Clearly, there are conflicts of interest since building owners will be anxious to minimise the risk of attack and building societies will also require safeguards against problems occurring. While *Building Regulations* only require wood preservation in a few components of buildings such as embedded timber and structural timbers at risk from the house longhorn beetle, other organisations may prefer to take the safe approach of requiring preservatives where risks are higher, such as in flat roofs.

Possible ways of reducing the hazards associated with preservatives are:

■ In new building consider the use of durable species of timber. (Note that sapwood of all timbers is considered to be at risk from insect attack.) There may be a cost penalty of using less common timbers.

■ Take adequate steps to ensure moisture contents of timber remain within acceptable levels.

■ Where preservatives are required consider less hazardous varieties. For example, boron based compounds are considered to carry lower risks (though they have limited applicability).

■ It is possible to deal with fungal attack in older buildings by control of the environment. This might also avoid undue intrusion into the fabric of the structure. Methods of electrically monitoring moisture in timber in inaccessible positions (Figure 10.7) might be considered so that when problems do occur, they can be detected before serious damage is done.

EXTERIOR WOOD FINISHES

In view of the very large quantities of non-durable timber used externally, it is not surprising that there are many products on the market, with vigorous competition between manufacturers, sometimes leading to claims that need to be regarded with some caution. Some of the main types of finish are outlined below with chief characteristics, though performance of individual types may vary greatly, according to individual manufacturer's formulations. Agrément certificates are rarely issued for coatings of this type, since manufacturers' formulations tend to evolve with time and cannot be easily checked.

Oil or resin based paints

These are based on air-drying oils or synthetic resins such as acrylics (see the section on 'Paints') and their chief properties are gloss finish and relatively low permeability (although all paints possess some degree of permeability). They have been used successfully for many years, though some may be slightly intolerant of trapped moisture in the wood which may cause bubbling or flaking, especially on south facing aspects. Oil (alkyd) based paints fail due to embrittlement as explained on page 487. Since permeability is not relied upon, stripping old paint is not necessary on redecoration, provided areas of damaged paint are made good.

Special wood paints

These paints may claim to be microporous or to accommodate wood movements or both. Tests on microporosity have shown that many such paints do 'breathe', though often not to a markedly greater extent than conventional paints. There is some evidence that, since cracking and blistering are less likely in such paints, they should have greater life. However, it should be appreciated that such paints would not be able to overcome high moisture contents caused by water admission at defective joints or breaks in the paint film. In some cases, there may be a possibility of higher resultant moisture contents in the wood, due to extra penetration through microporous paints during wet spells. Such porous paints produce a sheen rather than the high gloss finish which may be preferred by some clients. Hence, although advantages may occur in some situations, the need for adequate preparation for painting, together with good maintenance, still applies. The microporosity of paints is also lost if more than about three coats are applied and this should be considered when redecorating such coatings.

Varnishes

These are essentially drying oils or resins without any pigment which enhance the natural colouring and grain of the timber. Formulations vary greatly, with some being suited to external use, though the main problem with all varnishes is that exterior surface maintenance must be meticulous. Cracking or pealing of varnish very quickly leads to bleaching or staining of underlying surfaces due to exposure to water and/or ultraviolet light. Once affected, it is difficult to restore wood to its former state. The problem is especially severe on horizontal surfaces and south facing aspects and it is not, therefore, recommended that varnishes be used in such situations. When varnishes have microporous or other properties such that they are guaranteed not to flake or peel, they can be regarded as stains and may be treated as such. Solvent borne varnishes have low solid content and therefore tend to have very high VOC levels (Table 9.5) which should be considered as an environmental disadvantage.

Exterior wood stains

The essential features of stains are:

- They penetrate the wood so that texture and grain remain, though some (high-build) types form a surface film in addition.
- They have a low solids content.

- The coatings breathe quite freely, though water repellents are added to reduce admission of liquid water. Fungicides are also usually added to control mould growth.

Moisture contents of stained timber may rise from time to time, so that non-corrosive metal fixings should be used. Since many forms preserve the texture and sometimes the colour of underlying wood, imperfections may be seen and putty is not recommended for glazing – gaskets or glazing beads are preferred. Stains cannot cover cracks or gaps which might be covered by a paint.

An important advantage of wood stains is that coatings generally erode rather than flake or crack, so that maintenance is confined to periodic washing and application of extra coats, though colours become progressively darker with age.

Low build, transparent forms help retain the natural character of the wood and may leave very little surface film so that protection to wood is correspondingly less and wood may begin to weather comparatively quickly. Many types are now available with coloured pigments which are becoming aesthetically more popular. These increase protection to the wood, though there is the risk that, if a thick surface film is applied, failure characteristics of traditional varnishes, cracking and peeling, may occur. Higher build varieties are recommended on difficult woods, such as some hardwoods, with careful preparation, such as removal of oils, if present. Adhesion is improved on rough sawn wood finishes, especially of difficult timbers, though these have limited aesthetic acceptability at present.

Not surprisingly, with the variety of finishes obtainable and relative ease of maintenance, the use of wood stains is rapidly increasing.

Wood stains for exterior use tend to have very high VOC levels (Table 9.5) which must be considered as an environmental disadvantage.

10.5   FIRE RESISTANCE OF TIMBER

The thermal conductivity of timber is low, varying from approximately 0.12 W/m°C for softwood to 0.16 W/m°C for some hardwoods. As a result, heat losses through the material are small (though this has, in the past, caused pattern-staining in otherwise uninsulated roof structures) and timber resists the formation of condensation better than metals. The same property also contributes to the relatively good performance of timber in serious fire. Burning of wood occurs due to ignition of flammable gases, which are given off when the temperature of the material exceeds about 300 °C (pyrolysis). The rate of burning is, however, remarkably slow for the following reasons (Figure 10.14):

- Charcoal, like wood, is an excellent insulator, though charcoal itself burns at temperatures over 500 °C.
- Gases produced by pyrolysis help cool the charcoal as they diffuse to the surface before burning.
- Moisture must be evaporated; some of this diffuses inwards, maintaining the moisture content at the core.

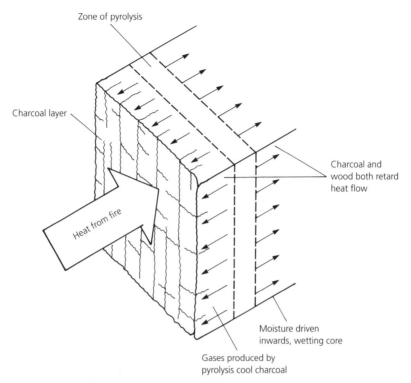

Figure 10.14 Mechanism by which heat flow into timber exposed to fire is impeded

The charring rate of a flat timber surface is predictable and for fire resistance purposes is taken as:

0.64 mm per minute (density less than 650 kg/m³)
0.50 mm per minute (density more than 650 kg/m³)

The lower rate applies to common hardwoods such as oak, utile and keruing. The higher rate applies to most softwoods, although a higher figure of 0.83 mm per minute is indicated in BS 5268 Part 4 for Western red cedar. Rates may be increased when fire encroaches from more than one side.

It is, therefore, possible to design timber structures for fire resistance by provision of sacrificial timber of thickness appropriate to the resistance required. In fact, since short term strengths can be used for fire purposes, it would, in any case, take approximately 20 minutes in a serious fire before the short term design stress was reached. Attention in structures should also be given to fixings, for instance, split ring connectors still transfer load if loosened by fire damage, whereas bolted joints may fail completely. In general all metal fittings should be recessed beneath the char zone or protected from fire.

It is found that laminated beams, which are usually bonded with the WBP adhesives phenol or resorcinol formaldehydes, have fire resistance similar to that of solid timber of the same dimensions. Plywood box beams are less fire resistant than laminated or solid timber, since the thickness of wood is usually smaller. Although timber floors might also be considered unsatisfactory, the normal first floor construction in houses has usually quite good fire resistance, largely due to its enclosed nature and the underlying plaster ceiling. In this respect, however, plywood is better than tongue-and-groove boards, which are better than flush boards, since a large number of joints or gaps increases ventilation. In timber framed buildings, satisfactory fire resistance can be achieved by use of plasterboard on walls and ceilings, though firestops are also necessary in wall structures containing large vertical cavities.

A number of fire retardant preparations are available. Some types impregnate the timber, increasing the proportion of charcoal and decreasing combustible gases in a fire. This treatment is carried out on seasoned timber cut to its final size. Typical chemicals include aqueous solutions of monammonium phosphate, borax and ammonium chloride, and preservatives are often included, though strength may be reduced by as much as 50 per cent due to cell wall damage, if timber is not correctly dried afterwards. Alternatively, intumescent paints, varnishes or pastes may be used. These form protective films under the action of fire, preventing access of oxygen and they do not adversely affect the strength of timber on treatment. Fire retardants are valuable in reducing surface spread of flame in cladding materials and partitions. Class 1 performance under BS 476, Part 7 can be obtained (Class 3, untreated). Charring rates are, however, not greatly reduced and the best protection against structural failure is the provision of extra (sacrificial) timber to protect underlying material.

10.6   MECHANICAL PROPERTIES OF TIMBER

Since timber is an anisotropic material, its strength properties are heavily dependent on the orientation of stress in relation to the grain direction. To describe fully the mechanical properties of timber, a number of different tests must, therefore, be applied. Traditionally, standard tests have been carried out on perfect (*clear*) specimens of timber in order to give a maximum strength for each type of species. In bulk timber, this strength must be modified by factors depending on the stress situation, section size and the presence of imperfections, when calculating allowable stresses. Clear specimens are normally small, typically, of 20 mm square section, in order that adequate numbers of clear specimens be obtainable from smaller trees. BS 373 describes methods of testing small, clear specimens of timber.

In contrast to ceramics, the tensile strength of timber is much higher than its compressive strength when measured parallel to the grain, since compression causes buckling of the fibres. The short term tensile strength of most softwoods is, for example, in the region of 100 N/mm^2 (tested dry) while compressive strength is about 40 N/mm^2. Failure in tension is mainly due to shear failure between fibres or cells. The tensile strength of individual tracheids may be several times the figure given above.

Bending stresses are very commonly applied to timber in service and, as would be expected, flexural strength, as measured by the modulus of rupture, is between tensile and compressive strength, values for dry softwoods being typically in the region of 70 N/mm^2 in the short term. In simple bending, the upper compression layers buckle, causing the neutral axis to move downwards during the test so that, ultimately, the lower part of the specimen fails in tension. The modulus of rupture is normally quoted, on the assumption that simple bending theory applies: if allowance were made for movement of the neutral axis, the magnitude of the modulus of rupture would be similar to the strength measured in pure tension.

The shear strength of timber parallel to the grain is low – about 11 N/mm^2 for dry softwoods – though this is not normally a problem, since shear stresses in beams of commonly used sections and spans are much smaller than bending stresses.

Hardness of timber is another important property, since it affects indentation and wear of flooring. BS 373 describes a test which measures the load required to press a steel ball of 11.3 mm diameter into timber to a depth equal to its radius. Typical dry softwoods give loads of approximately 3 kN, while hardwoods, such as oak, may give twice this load.

In addition to the above properties, others such as cleavage strength, compressive strength perpendicular to the grain, impact bending strength and moduli of elasticity may be measured as required. Some hardwoods will give results up to twice those for softwoods, depending on the wood type and the test carried out. The hardwood greenheart performs particularly well in all the tests.

FACTORS AFFECTING STRENGTH

There are a great many factors affecting the strength of bulk timber and the nature of the material is such that widely differing results can be obtained from differing specimens of the same species. Hence, the factors to be given are only one aspect of variations: in order to make the best use of sound timber and, at the same time, identify samples of low strength, current practice depends on carrying out simple tests on each piece of timber used.

Density

Despite the fact that the solid density of the fibrous component of wood is approximately 1500 kg/m^3, the density of dry bulk timber is rarely over 1000 kg/m^3, softwoods having densities in the region of 500 kg/m^3, while hardwoods have densities averaging 700 kg/m^3. These densities give an indication of the air void content which, as with ceramic materials, can be correlated with strength. Timber strength is, however, much less sensitive to air voids than that of concrete, an approximately linear relationship existing between compressive strength and the relative density of air-dry timber. As a rough guide, the compressive strength of clear timber parallel to the grain can be taken to be

100 times its relative density; for example, a softwood having a relative density of 0.5 would have a short term compressive strength of about 50 N/mm^2.

Rate of growth

When softwoods grow rapidly, the proportion of springwood tends to increase and, since this is of relatively low density, strength is adversely affected. Rate of growth is, therefore, a factor which is considered in visual stress grading. The situation is much less simple in hardwoods, summerwood also growing faster, in general, when conditions are suited to more rapid growth. Ring porous hardwoods, such as mahogany, can be denser and stronger as a result of increased growth rates (though these are still much lower than growth rates in softwood).

Moisture content

The strength of clear timber rises approximately linearly as moisture content decreases from fibre saturation and may increase threefold when the oven-dry state is reached. At more common moisture contents in the region of 15 per cent, the strength would be approximately 40 per cent higher than in the saturated state, depending on the type of wood. The mechanism of the strength increase is rather similar to that of shrinkage in concrete, the contraction resulting in decreased inter-fibre spacings and, therefore, stronger bonding between fibres. However, strength increases in timber containing defects are smaller, and large sections with large defects may become weaker on drying due to shrinkage damage.

Slope of grain

Significant deviation of the fibre direction from the longitudinal axis of timber may result from inaccurate sawing but is more commonly due to irregular growth of the tree. The tensile strength of timber drops rapidly as the slope of grain increases, increasing the risk of shear failure. Tensile strength perpendicular to the grain direction is very small, typically 3 N/mm^2. As would be expected, the effect on compressive strength is much less marked and that in bending is, therefore, intermediate, a 10° slope of grain producing a bending strength fall of about 20 per cent. Strength also varies with microfibrillar angle. As this increases, strength decreases, following the slope of grain principle. Poor growing conditions may increase the microfibrillar angle and, hence, affect strength.

Knots

Knots correspond to branches of the tree and result in severe distortion of growth rings in the trunk. There are two basic types: live knots, which correspond to branches which are living at the time of felling, and dead knots corresponding to dead branches and often found lower down on the trunk. Both types reduce strength but dead knots cause a larger strength reduction because, in these, the cambium layer is discontinuous and the knots tend to contract, often falling out of thin sections, such as boards. The effect of knots depends on their position in the section, being generally more serious in tensile areas than compression areas, though, in many cases, the effect of knots is due rather to their effect on the local slope of grain than to the presence of what is effectively a hole in the timber. The effect in bending is much more severe when knots are near to one edge of the timber. The difficulty of visually assessing the weakening effect of knots is one of the disadvantages of visual stress-grading of timber. They are defined by the term *knot area ratio* (KAR), the proportion of the cross-section they occupy at any point (this must be estimated by measuring interceptions on faces or edges of each piece).

Section size

Experimental evidence indicates that smaller timber sections have, on average, higher strength. (This may follow the arguments given for the same effect in concrete cube tests, linked to the possibility of larger defects in larger pieces.) An approximate guide to the effect is given by the ratio:

$$\frac{\text{strength at section depth } h}{\text{strength at section depth 200 mm}} = \left(\frac{200}{h}\right)^{0.4}$$

For example, a 300 mm depth section would, on this basis, have strength

$$\left(\frac{200}{300}\right)^{0.4} = 0.85 \text{ times the depth of a 200 mm depth}$$

Creep

All timber is subject to time dependent deformation and, since the material appears to fail when a certain strain is reached, failure may eventually ensue. This is, however, rare in practice since permissible stresses in timber are very much lower than short term stresses, mainly on account of strength variability in the material. It is found that, in the steady state, timber subject to a stress of less than about half its short term strength should have indefinite life at constant relative humidity. A major factor in creep levels of timber is its moisture environment. The principles of *mechanosorption* are now widely accepted in the context of timber in flexure:

1 Any change (increase or decrease) producing a moisture content experienced for the first time following loading of the timber leads to an increase of deflection.
2 Thereafter *increases* of moisture content *reduce* creep, while *decreases* of relative humidity *increase* creep.

This might seem surprising since it is known that drying timber increases its strength. The above behaviour is, however, supported by much experimental evidence.

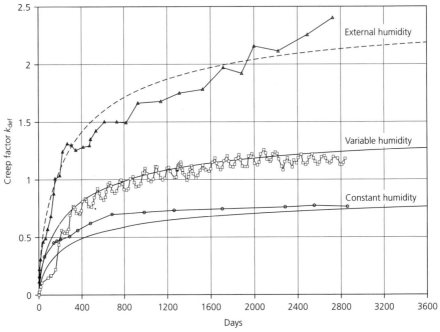

Figure 10.15   Creep of timber in various environments. The creep in the external environment is over three times that in a constant humidity environment. Mechanosorptive effects take place on each cycle in the variable humidity environment and seasonally in the external environment, giving greater deflection in drier conditions

The creep factor k_{def} (in flexure) is defined in Eurocode 5 for the design of structural timber as:

$$\frac{\text{creep deflection}}{\text{initial elastic deflection}}$$

Figure 10.15 shows experimental k_{def} values of glued laminated softwood timber sections subjected to bending loads corresponding to allowable flexural stress, over a period of years. The graphs represent the extremes of constant relative humidity, severe external exposure, and an artificial environment in which the relative humidity was cycled. The creep levels are seen to vary greatly from a very low level at constant humidity to a creep factor approaching 2.5 for the external environment.

Eurocode 5 gives k_{def} allowances to be made for creep deflections according to load duration and service class.

Load durations are classified follows:

short term:	less than one week (0–7 days)
medium term:	one week to six months (7–180 days)
long term:	six months to 10 years (180–3600 days)
permanent:	more than 10 years (> 3600 days)

Table 10.2 Creep factors for deflection, k_{def}, recommended by Eurocode 5

	Load Duration Category	Service Class		
		1	2	3
1	Permanent	0.6	0.8	2.0
2	Long term	0.5	0.5	1.5
3	Medium term	0.25	0.25	0.75
4	Short term	0.0	0.0	0.3

Service classes are defined as follows:

■ Class 1 is characterized by a moisture content in the materials corresponding to a temperature of 20 °C and relative humidity of the surrounding air only exceeding 65 per cent for a few weeks per year. (Average moisture content not exceeding 12 per cent.)
■ Class 2 is similarly defined except that a maximum relative humidity of 85 per cent applies. (Average moisture content not exceeding 20 per cent.)
■ Class 3 refers to conditions leading to higher moisture contents than service class 2.

The corresponding allowances are given in Table 10.2. These give the allowance that must be made for deflection *in addition* to the initial elastic deflection. Hence a creep allowance of 3 means that a deflection of 3 times the initial elastic deflection should be allowed for, the total deflection being *four* times the original elastic deflection. It will be noted that much larger allowances are required for service class 3 conditions (which would correspond to a covered external environment).

It is known that in practice larger deflections will occur in repeated deep moisture cycles especially in timber of small section, or without a surface coating since moisture content changes will be larger in such cases. The blanket figures of Eurocode 5 do not include such effects. In view of the fact that creep does not lead to failure, a timber section can be pre-cambered to compensate for the estimated long term deflection, though this increases costs. Given that moisture content changes have a marked effect on the rate of creep, it is not surprising that timber structures which become very wet during construction exhibit higher creep levels. In order to minimise this effect timber frames should be protected from rain until they have the protection of a roof.

In practical terms the important outcome of creep is that deflection levels can be significantly increased. The risk of failure is very low but such deflections can be problematic as follows:

■ Visible deflections in timber tend to cause concern.
■ Deflections can cause cracking in supporting structures such as walls.
■ Some finishes such as *in situ* plasters can be damaged.

STRESS GRADING OF STRUCTURAL TIMBER

The attractive properties of timber for structural purposes are high specific strength (strength comparative to weight), ease of construction and finishing, aesthetically pleasing appearance, fire resistance and durability. These are offset to some extent by the variability of timber properties which require careful grading, combined with a statistical approach to obtain the best safe performance. Indeed, the *Building Regulations* now require stress-graded timber for all structural applications, so that most timber is graded before use. There are two methods of grading: visual and mechanical.

Visual stress grading of softwoods (BS 4978)

The object of visual grading is to make a rapid visual assessment of the principal factors affecting the strength of each piece, that is, knot characteristics, fissures, slope of grain, wane rate of growth and distortion. Other less common defects, such as worm holes and fungal decay, are also covered by the standard. Each piece is categorised as:

- special structural (SS)
- general structural (GS) or
- reject

In the case of knots (measured by knot area ratio), fissures, slope of grain, rate of growth and distortion, there are separate requirements for the (SS) and (GS) grades.

The total knot area ratio (TKAR) is defined as the proportion of any one section which is occupied by knots. Margin knots are those occupying more than one half of the top or bottom quarter section of the timber (MKAR $> \frac{1}{2}$); (Figure 10.16).

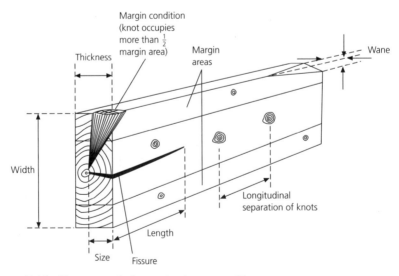

Figure 10.16   Key terms relating to visual stress grading

Table 10.3 Visual stress grading requirements for softwoods. Data taken from BS 4978

Property	Type	SS	GS
Knots	Margin condition (MKAR $> \frac{1}{2}$)	TKAR $\frac{1}{5}$ or less	TKAR $\frac{1}{5}-\frac{1}{3}$
	No margin condition (MKAR $< \frac{1}{2}$)	TKAR $\frac{1}{3}$ or less	TKAR $\frac{1}{2}-\frac{1}{3}$
Fissures	Not through thickness	Not longer than half the length of the piece	Unlimited
	Through the piece	Not longer than twice the width of the piece	Not longer than 600 mm on any running metre
Slope of grain	–	Must not exceed 1 in 10	Must not exceed 1 in 6
Wane	–	Wane shall not reduce the full edge and face dimensions to less than two-thirds of the dimensions of the piece. Length of wane is unlimited	
Rate of growth	–	Average width of annual rings not greater than 6 mm	Average width of annual rings not greater than 10 mm
Distortion	Bow	Not greater than 10 mm over a length of 2 m	Not greater than 20 mm over a length of 2 m
	Spring	Not greater than 8 mm over a length of 2 m	Not greater than 12 mm over a length of 2 m
	Twist	Not greater than 1 mm per 25 mm width over a length of 2 m	Not greater than 2 mm per 25 mm width over length of 2 m
	Cup	Unlimited	Unlimited

Where margin knots are present, acceptable TKAR values for any grade are reduced since they have a more marked weakening effect on the timber (Table 10.3). The only totally accurate way of measurement of KARs is to cut the piece at every knot and as this is clearly impracticable, graders must estimate the KAR values by consideration of intercepts of knots on the wood surface. This is considerably assisted by inspection of the end grain since this helps establish the position of the centre of the tree from which the wood is cut. In any one piece the grader is likely to focus upon the worst knot only. Then it is sufficient to identify if a margin condition is present and then to classify the worst TKAR value into the ranges:

TKAR < 1/5
TKAR > 1/5, < 1/3
TKAR > 1/3, < 1/2
TKAR > 1/2

This will permit a grade on the basis of KAR to be assigned. The grade of the piece will be that corresponding to the worst defect found in each piece.

Considerable care is often necessary in relation to distortion which accounts for a substantial proportion of rejects in grading. Even timber which passes this assessment may later be subject to distortion if there is a significant change of moisture content.

Grading is a skilled process and graders should be properly trained to carry out this important operation, even though it only takes a few seconds on each piece.

Visual stress grading of hardwoods (BS 5756)

The stress grading of hardwoods is in many respects similar to that of softwoods though there are a number of important differences, for example:

- There are no rate of growth requirements since spacing of growth rings does not have the marked effect on mechanical properties that occurs in softwoods.
- The effect of knots is assessed by their *limiting dimension* which is based upon the width of each knot in relation to the dimension of the surface on which it occurs.

Grading designations are assigned as follows.

Structural tropical hardwood HS (single grade)   This grade is only assigned to sections of minimum cross-sectional area 2000 mm^2 with thickness at least 20 mm.

Structural temperate hardwood   These are divided into general structural or heavy structural temperate hardwoods according to section size:

THA or THB (THA better grade):   cross-sectional area > 20 000 mm^2 and thickness not < 100 mm
TH1 or TH2 (TH1 better grade):   other sizes

The timber with THA or B grades is described as *heavy structural temperate hardwood*.

Machine stress grading (BS EN 519)

This is based on the correlation which has been found between bending strength (modulus of rupture) and modulus of elasticity, as determined by deflection tests on samples subject to bending (Figure 10.17). The 1 per cent lower strength value is obtained directly from the correlation of Figure 10.17, the only additional factor required to obtain grade stresses being the factor of safety. The figure applies to European redwood (*Pinus sylvestris*) and European whitewood (*Picea abies*), which together form a very large proportion of the UK market at present.

Modern stress grading depends heavily on computer operated machines. These rapidly measure the relationship of load to deflection on successive portions of each timber piece as it passes through. Timber is normally loaded on face rather than on edge for practical reasons and the machine automatically compensates for any out-of-straightness. Visual inspection is also carried out for defects, such as fissures, distortion, wane, resin pockets, etc. The grading result corresponds to the worst

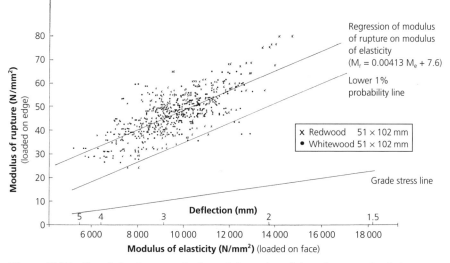

Figure 10.17 Correlation between elastic modulus and modulus of rupture for timber
(Reproduced by permission of BRE)

deflection measured by the machine along each length of timber and the grading is
given in the form of the corresponding strength class of BS 5268 direct, simplifying
design. The fact that it has been machine graded is indicated by the BS EN 519
marking on the piece. An advantage of machine grading is that better yields are
obtained than by visual grading of the same batch and there is less reject material.
The machine can also, if desired, be set to a stress matched to that of the timber
being graded to make the best use of it, provided the design stress is adjusted
accordingly.

Stress grading machines must comply with an approving authority if grading in
accordance with BS EN 519 and machines are subject to unannounced periodic
inspection in order to ensure that standards are maintained. Machine grading is
now widely used for timber to be employed in structural elements, such as roof
trusses and general laminated timber.

10.7 DESIGN OF STRUCTURAL TIMBER SECTIONS FROM VISUAL STRESS GRADE RESULTS

Timber design is now covered by Eurocode 5, though most UK design is probably
still carried out using BS 5268. This standard is based upon strength properties of
practical timber sizes rather than small clear specimens, although the latter still play
a part where experimental information on larger sizes is inadequate. To illustrate,
the procedure for calculating safe bending stresses for SS and GS grades of timber
under long term load is:

1 From test results on SS grade dry timber (where available), calculate the characteristic bending stress based on 5 per cent failures. A Weibull distribution is used for this, since it gives more accurate predictions of behaviour at the lower end of the strength range than the normal distribution (see page 136).
2 Obtain *grade stress* values by multiplying the characteristic stress from (1) by factors allowing for duration of load, size and factor of safety. For bending, these factors are:

■ Section depth factor:

$$0.849 = \left(\frac{200}{300}\right)^{0.4}$$

(Grade stresses are based on 300 mm sections, whereas much experimental information is on 200 mm depth sections.)
■ Duration of load factor: 0.563. This gives the long term loading stress comparative to short term test results.
■ General safety factor: 0.724.

This gives an overall reduction of 0.346 (divide the characteristic stress by 2.89).
3 For GS grade timber, a relativity factor, obtained by experiment, must be applied to the SS value. For bending, it is 0.71.
4 Other modes of behaviour, for example, moduli of elasticity and compressive stresses, are obtained in the same way, if experimental information is available on bulk samples. For some species, however, tests on clear samples still form the basis of grade stresses, conversion factors being based on known correlations between the behaviour of small clear timber sections and SS grade sections.

To take an illustration:
For European redwood or whitewood, the SS grade characteristic bending stress with 5 per cent failures of a 200 mm × 50 mm section is 21.6 N/mm².
Therefore, the SS grade stress is

$$\frac{22.0}{2.89} = 7.5 \text{ N/mm}^2$$

GS grade stress = 7.5 × 0.71 = 5.3 N/mm²   (long term loading, 300 mm section)

These are seen to be well below the mean short term bending strength of clear timber, typically 70 N/mm², the various causes of the large reductions having been outlined above.

STRENGTH CLASSES IN BS 5268

Until recently the design procedure for structural timber could be quite complex, since the designer might need to be aware of timber species available, together with

Table 10.4 Strength class and properties of structural timber. Data taken from BS 5268 (all values in N/mm²)

BS5268 strength class	Bending parallel to grain	Tension parallel to grain	Compression parallel to grain	Compression perpendicular to grain		Shear parallel to grain	Modulus of elasticity	
				(a)	(b)		Mean	Min.
C14	4.1	2.5	5.2	2.1	1.6	0.6	6 800	4 600
C16	5.3	3.2	6.8	2.2	1.7	0.6	8 800	5 800
C18	5.8	3.5	7.1	2.2	1.7	0.6	9 100	6 000
C22	6.8	4.1	7.5	2.3	1.7	0.7	9 700	6 500
C24	7.5	4.5	7.9	2.4	1.9	0.7	10 800	7 200
TR26	10.0	6.0	8.2	2.5	2.0	1.1	11 000	7 400
C27	10.0	6.0	8.2	2.5	2.0	1.1	12 300	8 200
C30	11.0	6.6	8.6	2.7	2.2	1.2	12 300	8 200
C35	12.0	7.2	8.7	2.9	2.4	1.4	13 400	9 000
C40	13.0	7.8	8.7	3.0	2.6	1.4	14 500	10 000
D30	9.0	5.4	8.1	2.8	2.2	1.4	9 500	6 000
D35	11.0	6.6	8.6	3.4	2.6	1.7	10 000	6 500
D40	12.5	7.5	12.6	3.9	3.0	2.0	10 800	7 500
D50	16.0	9.6	15.2	4.5	3.5	2.2	15 000	12 600
D60	18.0	10.8	18.0	5.2	4.0	2.4	18 500	15 600
D70	23.0	13.8	23.0	6.0	4.6	2.6	21 000	18 000

Notes: 1. The higher compression perpendicular to grain figures (a) can only be used where specifications prohibit wane in bearing areas.
2. 'C' prefixes are for softwoods, 'TR' for trussed rafters and 'D' for hardwoods.
3. These figures apply to service classes 1 and 2.
4. Density requirements are also included in the standard.

stress grades for each species. Working stresses vary from species to species for a given stress grade. If, therefore, one species is no longer available, there may be some difficulty in finding a replacement with similar structural properties. The introduction of strength classes in BS 5268 helps alleviate such problems by allowing many different species/grades to be grouped together, so that the designer need no longer specify an individual species (unless there are other reasons for such a requirement). Table 10.4 shows the main properties of the strength classes defined. Note that these are long term working stresses, derived from characteristic short term strengths, for service classes 1 and 2. The strength codes derive their number from the characteristic bending strengths of the timber represented, for example C14 indicates a characteristic bending strength of 14 N/mm². The strength class designation is now normally marked on timber as well as its stress grade, in order to simplify its selection for specific purposes. By grouping strengths into classes, some species will inevitably not be used at their full grade stress and the option remains to use selected species at their appropriate stress, though for most purposes, the simplification provided by strength classes should be attractive to designers.

Table 10.5   Strength classifications of common timbers

Standard name	Strength class (Earlier strength class codes in brackets)			
	– (SC1)	C14(SC2)	C16(SC3)	C24(SC4)
Imported				
Parana pine			GS	SS
Redwood, whitewood			GS	SS
Western red cedar	GS	SS		
Douglas fir-larch (Canada/USA)			GS	GS
Hem-fir (Canada)			GS	SS
British grown				
Douglas fir		GS	SS	
Scots pine			GS	SS
European spruce	GS	SS		
Sitka spruce	GS	SS		

The D prefixes refer to hardwoods; for example oak of visual grade TH1 complies with D30 while larger oak sections with THA visual grade comply with D40.

Table 10.5 shows classification of some common species, together with the corresponding strength classes in the earlier edition of BS 5268. (The timber design tables in *Building Regulations* are still based on the earlier BS 5268 strength classes SC3 and SC4.) The *Building Regulations Approved Documents* tables form an extremely simple method of design for straightforward structures such as domestic floors or roofs.

10.8   TIMBER AND WOOD PRODUCTS

Recent years have seen an enormous increase in the variety of products, mainly on account of advances in mechanised techniques and the introduction of synthetic resin binders. Attractions include:

- lower wastage of timber
- competitive cost
- reduced variability
- large range of sizes available, particularly sheet materials

The mechanical properties of products depend upon two main factors:

1 The density of the product. In some cases densities much higher than that of the original wood can be obtained by application of compression during manufacture.
2 Fibre directions in the final product. In some products fibres are equally oriented (normally within the plane of a sheet) while in others they may be arranged to give better mechanical properties in one direction.

The reaction of products to moist environments will depend chiefly upon the type of adhesive used. A number of products are formulated for a variety of applications on this basis.

Increased use of timber products has led to developments of new types of fixings which should be used for best results, for example, parallel shank chipboard screws which are preferred to traditional tapering wood screws for fibre or particle boards.

The glues in many of these products tend to require more frequent sharpening of cutters and saws. Tungsten carbide or diamond tipped cutters with modified cutting angles are preferred to minimise this effect.

Fibre building boards

These are produced, for example, by pulverising of wood down to individual fibres and then compressing them to form large sheets. Small percentages of adhesive – usually phenolic – may be used to act as a binder, though fibres have inherent adhesive properties and are, to some degree, self-binding. Several different types of product are obtained, according to the degree of compression of the fibres. BS EN 622 classifies fibre building boards as hardboard, medium density fibre board, medium board and softboard, according to density. Figure 10.18 indicates typical density values together with bending strengths. It is noticeable that strengths increase greatly with density, higher density products being obtained by high machine pressures during manufacture.

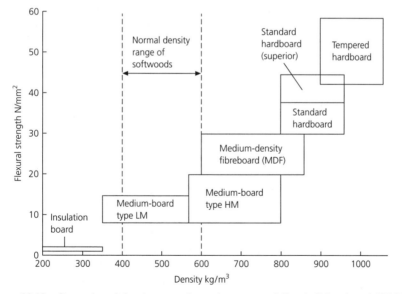

Figure 10.18 Strength and density range for various types of fibre building board. Within each range thinner sheets give higher strengths

Hardboards have a density of over 900 kg/m^3, being fully compressed and normally having one smooth side, the other side having a mesh texture. Thicknesses are generally in the range 3.2–8.4 mm and flexural strengths are quite high, as much as 60 N/mm^2. Standard hardboards are susceptible to moisture and should be conditioned by application of water to the mesh side before use. This ensures that, in conditions of service, the sheets are under tensile rather than compressive stress. Prolonged dampness will cause irreversible swelling and strength loss in standard hardboards. These boards, nevertheless, find wide application in dry situations, being used in internal flush doors, built-in units, partitions and in flooring beneath sheets of tile coverings. Tempered hardboard, having a density normally over 960 kg/m^3, may be used in damp situations, finding application in roofing, soffit boards and external claddings.

Medium boards have densities in the range 350–800 kg/m^3, being subdivided into high density and low density subgroups. These may have a variety of finishes and are used in partitions and wall linings in thicknesses of 6–12 mm. They can also be used as an intermediate sound-absorbent layer on concrete or timber floors.

Medium density fibreboard (MDF), unlike the other fibre building boards, is made by a dry process and has experienced a dramatic increase in use over the last few years. It is of similar density to medium board, but significantly higher performance is obtained by bonding with urea formaldehyde/melamine formaldehyde resins. It can be treated in much the same way as solid wood but without its inherent defects and problems of instability. Thicknesses from 3 to 60 mm are available and MDF can be cut, routed, moulded, embossed or finished to a high standard. The standard board is not waterproof and must be protected by sealing of the surface, if used in potentially damp situations. Applications of MDF include mouldings, such as architraves, window boards and skirtings and furniture. Other items such as staircases can be produced, provided design accommodates the differences in properties between MDF and solid timber. A newer moisture resistant type has, however, recently become available, identifiable by a green core, and an isocyanate (polyurethane) bonded product with even better performance may soon be marketed, these adhesives having the advantage of being free of formaldehydes.

Insulation boards have lower density, below 350 kg/m^3, with a thermal conductivity in the region of 0.05 W/m°C. They, therefore, contribute considerably to thermal insulation when used as ceilings, sarking and sheathing. The latter would normally be bitumen impregnated to improve moisture resistance. Various finishes are available when used for wall and ceiling linings, and they can be treated similarly to plasterboard, either used dry or given a skim coat of plaster. Alternatively, they may be papered direct, provided a hardboard primer is first applied. A further use of insulating boards is to accommodate irregularities in subfloors, though a further covering such as hardboard would, of course, be necessary to provide mechanical protection. All boards, especially the lower density varieties, are susceptible to fire, though surface spread of flame can be reduced by the use of fire retardants.

Particle boards (chipboards)

These consist of small particles or splinters of wood, bonded together with synthetic resins, the resin normally occupying about 8 per cent by weight of the dry board. Fibres are usually randomly orientated, parallel to the plane of the board, and the wood chips are graded, with coarser chips at the centre of the board and finer chips at the surface to give greater surface hardness than is available from ordinary softwoods. The strength and moisture resistance of chipboards depends greatly on the type of adhesive used. Those based on urea formaldehyde lose about 60 per cent of their strength and increase in thickness by about 10 per cent on wetting. Recovery does not occur on drying, hence such chipboards must be kept dry during use. Melamine formaldehyde/urea formaldehyde mixtures or phenol formaldehyde adhesives give much lower strength reduction and expansion on wetting and show recovery on drying. In BS EN 312 particle boards are classified as follows:

general purpose	P2
load bearing	P4
load bearing (damp conditions)	P5
heavy duty	P6

Chipboard is now very widely used in flooring. Grades P4 or P5 are recommended, the 20 mm thickness having a flexural strength of at least 15 N/mm^2 compared to 11 N/mm^2 for grade P2 chipboard. Where there is the possibility of dampness, Type P5 should be used, which combines the properties of both types. This type is now generally used in new building on account of the difficulty of ensuring adequate protection of boards during construction. In addition to the problems of strength loss and swelling of non-moisture resistant types, damp chipboard is vulnerable to fungal attack. Adequate support must be given to chipboard, since creep appears to be greater than in solid timber especially in conditions of varying humidity.

Advantages of chipboard for flooring include:

- Very rapid construction. Both labour and materials costs are much less than those of, for example, solid tongue and grooved (T&G) boards.
- Dimensional stability. Chipboards are stable dimensionally and do not warp, provided they are protected from damp.
- A smooth splinter free surface is obtained.

Possible disadvantages include:

- If non-moisture resistant boards are used, there is the risk of irreversible damage if, as sometimes occurs during construction, they become wet due to rain or leaks.
- Boards are difficult to nail and can be quite brittle near edges. Screws are preferred for fixing with predrilled holes.
- Boards usually have deep T&G edges for fire resistance. Access to the underfloor space can, therefore, be difficult; tongues cannot be easily cut, as in solid timber equivalents.

■ If an excess of formaldehyde is used during manufacture there is a risk of formaldehyde gas being emitted in use, with possible health consequences. A new form of chipboard bonded by isocyanate adhesives should overcome this difficulty (though the problem was mainly associated with certain imported boards).

Further uses include sarking and roof decking, although the above comments concerning moisture still apply. The material may also be used for shuttering for concrete, provided the surface is sealed before use. Once the long term structural properties of chipboards are evaluated and grade stresses are obtained, they are likely to be used for structural purposes in much the same way as plywoods and will provide increasing competition for the latter.

Plywood

Plywood consists of glued wood panels comprising outer and inner sheets, the grains in adjacent sheets being at right angles. An odd number of plies is normally used, since this enables symmetrical distribution of plies about the centre sheet. Hence, if the moisture content changes, equal and opposite bending effects occur, due to differential movement of symmetrically opposed sheets. As a result, plywood can be used in situations with variable moisture content with minimum distortion. Sheets behave more isotropically in the plane of the sheet than solid timber and, although flexural strength is not as high as in a solid timber sheet loaded parallel to the grain, it is much greater than solid timber loaded across the grain. Sheets can also be much larger than would be available in solid timber. The flexural strength and stability of plywood increase with the number of plies. Also, since the glue-lines restrain surface cracking due to the stresses imposed by moisture changes, there is a limit to the thickness of surface plies that can be used. In order to achieve greatest flexural strength, the face grain should be parallel to the span, especially if a small number of plies is used.

The working properties of plywood are similar to those of ordinary timber, though it can be nailed or screwed without risk of splitting, since there are no cleavage planes. Impact strengths are much better than those of ordinary timber, for the same reason. Performance standards for plywood tend to be rather complicated, since they are often those of the country from which they are imported, but colour codes are now quite commonly used, as they are for other timber products.

When used decoratively, the surface veneers may be of hardwood. Most plywoods have a characteristic figure in surface veneers resulting from the method of producing them, which involves rotary cutting from the log.

The moisture resistance of plywoods depends greatly on the type of adhesive used. Where significant exposure is anticipated, WBP type adhesives should be employed.

Plywoods are also very widely used for cladding, for structural purposes and as a form facing material for concrete, where their resilient nature permits the maximum number of re-uses.

Waferboard and oriented strandboard (OSB)

These are similar in essence to chipboard but, by using larger flakes of wood, less adhesive is required and improved strength and stiffness are obtained. Waferboard was formerly produced from flakes of about 75 mm square and 0.5 mm thick. This has now been replaced by OSB which is produced from narrower strands built up into a plywood-like structure with crossed strands in alternate layers (normally three plies). The boards therefore have a major and minor axis which should be apparent by inspection. For best performance the major axis should be aligned with the stress – for example, across floor joists. In strength performance terms the boards lie between chipboards and plywoods, though they are much cheaper than plywoods. For flooring, the surface obtained is not as smooth as that of a good quality chipbard. Various grades are available as follows (BS EN 300):

general purpose	OSB/1
load bearing	OSB/2
load bearing (damp conditions)	OSB/3
heavy duty (damp conditions)	OSB/4

OSBs are highly competitive with other sheet materials and their use looks set to increase.

Cement based materials

The traditional *wood wool* slabs tended to have a poor reputation due to friability and lack of durability, though there are now cement bonded particle boards available which offer the advantages of:

■ very smooth hard surfaces to which decorative coatings can be applied
■ greater stiffness than chipboard
■ fungi resistance
■ dimensional stability
■ fire resistance

They can be regarded as a form of fibre reinforced cement having a cement content of 65–70 per cent by weight and a density of about 1200 kg/m^3. Again, standardisation and experience of use should greatly increase applications of such materials at the expense of solid wood equivalents. Possible applications include flooring, partitions and ducting.

GLUED LAMINATED TIMBER (GLULAM)

This form of construction was developed in the late 1960s with the advent of durable and waterproof high strength adhesives. The high strength/weight ratio of timber and the ease with which virtually any size or shape can be produced are the chief reasons behind the success of glulam. The material offers many advantages over solid timber, for example:

Figure 10.19   Combination of curved members (20 mm laminates) and straight members
(45 mm laminates) to produce large span single storey buildings in glulam

- Much larger section sizes are possible and there is virtually no limit to the length of members obtainable.
- Laminates can be quickly seasoned to the correct moisture content while larger solid timber sections must be seasoned very slowly to avoid damage.
- Knots in any one laminate are of relatively minor importance – the term *margin knot* is not relevant in glulam.
- There is less distortion in glulam, after fabrication, compared with solid timber.
- Maximum benefit can be obtained from the better grades of material which would be placed near the edges of beams, while poorer grades can be placed at the centre.
- Curved members, such as portal frames or complex forms, can be easily obtained at extra cost (Figure 10.19).
- Section size can be varied to accommodate stress variations without significant wastage of timber.

Although a number of the above advantages apply irrespective of section size, it is unusual to specify glulam for sections of under about 150 × 65 mm (unless curved) since all laminate surfaces must be planed and the loss of material in this process, combined with the cost of fabrication, would render the material uneconomic. In curved sections, the laminate thickness should not be greater than radius/125.

The current standard laminate thickness is 45 mm, obtained by planing standard 50 mm sawn material. Thicknesses as low as 20 mm may be used for curved members.

Timber for laminating can be graded by BS EN 518 (visual) or BS EN 519 (machine) though wane is not permitted and attention should be paid to excessive distortion, resin pockets and wormholes.

Laminates are normally end joined by finger joints which are made mechanically. A good finger joint should have finely tapered fingers, should be free of knots and assembled at the correct end pressure – in such conditions joints in individual laminates should normally be as strong as the wood, which will tend to fail at knots or other defects.

A typical fabrication procedure would be:

1 Treat the timber with preservative.
2 Ensure moisture content is in the range 11–15 per cent with range not larger than 4 per cent.
3 Select suitable laminates, finger jointing as necessary and planing to produce a smooth surface on both sides.
4 Select a suitable adhesive conforming to BS EN 301 Type 1 for all service classes or BS EN 301 Type 2 for service classes 1 or 2.
5 Coat laminate surfaces uniformly with adhesive and assemble, bearing in mind the limited pot life of the adhesive.
6 Build up laminates to the required profile, clamping to produce uniformly thin, but full, glue-lines.
7 Cure: radio frequency heating can be used to cure some types of adhesive very quickly.
8 After curing, plane to smooth surfaces as required, cut to final size and drill all necessary holes. (Tolerances may have to be slightly larger than for steel, in view of the fact that there may be some movement or distortion after fabrication.)
9 If not preserved at stage (1) apply an organic preservative, taking special care with end grain.

The whole process should be factory based and success depends on good quality materials and workmanship throughout.

The use of glulam in the UK is increasing, though its use per capita is much lower than in other European countries. Most applications are for buildings, such as churches, sports halls and other single storey large span structures, since the material does not have the strength or rigidity to carry large superimposed or dead loads, especially involving masonry. Design codes are, at present, conservative, with large creep allowances and low assumed E values, such that the critical aspects of design in many structures are deflection criteria. One possibility is to precamber beams so that they begin with a negative deflection. A further aspect being investigated is that the surface of timber could be made hydrophobic (water resistant) so that the effect of relative humidity changes upon moisture content is reduced. This would then reduce creep deflection (see 'Creep in timber') and increase the structural competitiveness of the material. Most construction is in softwood but hardwood can also be used, for example, in severe environments or for structures with a long design life. The roof ribs of the Thames tidal barrier are, for example, in iroko, which is classified as very durable.

TIMBER FRAME CONSTRUCTION

Timber frame construction is by no means new and there are many such buildings, built several centuries ago, which are still in good condition. In fact when brick construction was widely introduced, a brick tax was for a time imposed in some areas. For this reason *mathematical tiles*, which are tiles fixed to battens, conveying the appearance of brickwork (Figure 6.19), were for a time popular. The frames in earlier structures were often very substantial, possibly being of hardwoods such as oak.

In current timber frame construction, relatively small softwood timber sections are employed, strength and rigidity being obtained by fixing sheet materials to a simple framework. This is known as *stressed-skin construction* and is widely used in the fabrication of beams (*box beams*), in roofing and for the frames of complete structures, such as houses and even multi-storey blocks of flats. Brick claddings may be used and these contribute to stiffness of the overall structure. Although such structures contain lower embodied energy than conventional masonry construction, further reduction can be achieved by tile or sheet materials for claddings.

It is important to allow for the likelihood of higher movement, especially during and immediately after construction, compared to other materials. If construction is carried out in wet weather, considerable shrinkage will result and this should be allowed to occur before finishes are applied. Finishes should also be capable of a certain degree of movement if unsightly cracking is to be avoided. Completely dry finishes are, in this respect, better than more traditional finishes. Sound insulation between timber framed dwellings and flats might be considered a problem but, by careful design, detailing and construction involving floating floors, structurally independent composite party walls and insulating materials, the grade 1 sound insulation standard can be achieved.

Timber framed housing received much adverse publicity in the mid-1980s, much of this due to poor workmanship, rather than to inherent technological weaknesses. Since the publication of BS 5268 Part 6 much progress has been made and the inherent advantages of timber frame are now being much more widely exploited. This standard applies to dwellings of not more than four storeys. Figure 10.20 shows a six storey block of flats produced by the Building Research Establishment in order to investigate the feasibility of larger scale timber frame construction. The rigidity of such structures depends in part on their cellular nature, each cross wall contributing to stiffness.

TRUSSED RAFTERS

The use of trussed rafters in domestic roof construction is now almost universal, since they are an economic means of producing stable and dimensionally accurate roof structures with the advantage of factory rather than site production. Timber is machine stress graded, common thicknesses being 35 mm up to 11 m spans and 47 mm up to 15 m spans. Areas to which attention should be given are:

Figure 10.20 Six storey timber frame building produced by the Building Research Establishment. The building has a cellular structure; the brick cladding and internal walls contribute significantly to the building stiffness

1 Adequate bracing is essential to prevent buckling of compression members and to permit the whole structure, including gable walls, to resist wind effects.
2 Superimposed loads, such as those due to water tanks, should be applied close to node points and spread by bearing joists.
3 Trussed rafters should not be exposed to weather for long periods – corrosion of connector plates may occur. Organic rather than water borne preservatives are used for the same reason. In damp environments, stainless steel connector plates may be preferred, though they increase cost considerably.
4 Condensation in roof spaces could cause connector plate corrosion; it can be avoided by use of warm roofs or adequate ventilation of cold roofs.

CHAPTER 11

FIBROUS COMPOSITES

11.1 INTRODUCTION

This chapter examines a number of recently developed materials, often based on traditional materials, but incorporating some form of fibre reinforcement.

The materials discussed up to this point may be divided into three main groups, the properties of which can be summarised conveniently in tabular form (Table 11.1).

Ceramics   On a volume basis, these are the cheapest materials. They have high compressive strength and are very rigid. They can be formed, using cements, into large structures of complex shape. They are, however, brittle and exhibit low tensile strength on account of microscopic cracks which are usually present in them. Hence, they cannot withstand, unassisted, large tensile, flexural or impact loads. Chemically, they are relatively stable. They have low thermal movement but, since many ceramics are porous, they are subject to moisture movement and moisture associated deterioration.

Metals   The higher strength metals have the highest modulus of elasticity and tensile strength of any commonly used building material, though they also have high density. They are, at present, a most important structural material, being the only group which will withstand, satisfactorily, high tensile stresses. They are also formable, though heavy sections or castings require high manufacturing

Table 11.1 Comparative properties of the main materials groups. They are given a star rating (maximum 5) for each property. Ratings represent an average for materials in each group

Materials group	Cost	Density	Compressive strength	Tensile/ impact strength	Stiffness	Thermal movement	Thermal insulation	Fire performance
Ceramics	★★★★	★★	★★★	★	★★	★★★★	★★	★★★
Metals	★	★	★★★★	★★★★	★★★★	★★★	★	★★
Synthetic Organic	★★	★★★★	★★★	★★★	★	★	★★★★	★

temperatures, while lighter sections, such as sheet or strip, can often be worked at ordinary temperatures. These properties result in widespread uses for pressed or craftsman-formed components. Metals are generally the most prone of the three groups to atmospheric attack. They are costly.

Organic materials Plastics, the chief synthetic organic materials, are characterised by low modulus of elasticity, variable ductility and moderate tensile properties. They soften at relatively low temperatures but are resistant to many chemicals and to attack by moisture. They are extremely versatile and, within the general limits given above, can be modified to suit specific requirements. On a cost basis, they are becoming increasingly competitive with other materials.

The shortcomings, as well as the chief attributes of the three major groups are evident. Regarding chemical or atmospheric instability, it has been possible by means of a study of the nature of degradation or attack, to select materials according to the situation required, or to modify or protect them if the environment is potentially harmful. As regards mechanical properties, there are perhaps three major problem areas which restrict the use of the ceramic and organic groups:

1 The low tensile strength of most ceramics.
2 The brittleness of most ceramics.
3 The low elastic modulus of those plastics which are sufficiently ductile to be viable for structural uses.

Although research is continuously taking place into ways of improving metals, it may be said that the above problems have given rise to the most urgent investigations, with the result that a number of new materials have been developed which are gradually finding increasing application in construction.

Ceramics and plastics can, in some senses, be regarded as opposite groups of materials, since one is rigid and brittle, while the other is often ductile and tough. Two chief techniques for overcoming the disadvantages of ceramics have been undertaken – fibre reinforcement and the incorporation of resinous materials, the latter having been described in Chapter 7. The former technique is by no means new. Lime plasters were traditionally reinforced with horse-hair, while the use of reinforcement in concrete is fundamental to its use in almost all situations. Fibre reinforcement may also be used, in certain cases, to overcome the problem of the low elastic modulus of plastics.

The principles of fibre reinforcement may be established by a consideration of the effect of incorporation of fibres upon the three properties listed above, that is, tensile strength, impact strength and elastic modulus. In each case, the type of material to which the technique is to be applied must be borne in mind.

The following notation is used. The symbols f and E stand for stress and elastic modulus, respectively. There are two suffixes for each symbol. In the first:

    f   denotes fibre
    m   denotes matrix
    c   denotes composite

In the second:

t   denotes tension
c   denotes compression
s   denotes shear

The symbol V denotes volume fraction (one suffix only).
Hence, for example,

E_{ct} denotes elastic modulus of composite in tension
V_f denotes volume fraction of fibres.

Additionally, the symbol $\wedge$ is used over a stress to indicate maximum (failure) stress or strength, for example,

$\hat{f}_{mt}$ denotes matrix failure stress in tension.

11.2   REINFORCEMENT IN BRITTLE MATERIALS SUCH AS CONCRETE

There are two quite distinct mechanisms by which incorporation of fibres could improve the tensile strength of such materials.

The first involves the fibres themselves carrying a substantial proportion of the stress, such that the matrix remains within its tensile stress capacity while the composite carries a relatively high stress.

The second produces a similar effect to that in conventional reinforced concrete, though on a smaller scale; that is, the fibres do not prevent cracking but they enable relatively high tensile stresses to be carried by control of cracking in the matrix, crack interfaces being held together by fibres. These two techniques are considered in turn.

TENSILE STRENGTH (PRECRACKED)

This requires that fibre and matrix be rigidly bonded, there being no relative movement between them. The stress carried by the composite can only exceed that carried by the matrix if the fibres have *greater stiffness* than the matrix, so that, for a given strain, the fibres are more highly stressed than the latter. Assuming fully aligned fibres, if, for example, the strain in the composite is e, then

$$\text{stress in matrix } f_{mt} = e \cdot E_{mt} \tag{11.1}$$

$$\text{stress in fibres } f_{ft} = e \cdot E_{ft} \tag{11.2}$$

Hence, stresses in fibres and composite are proportional to their respective E values.
Dividing (11.2) by (11.1) gives

$$f_{ft} = f_{mt} \cdot \frac{E_{ft}}{E_{mt}}$$

The load in a given section must be the sum of the loads in the fibre and matrix components:

$$\text{load} = A_f f_{ft} + A_m f_{mt} \tag{11.3}$$

where A_m and A_f are the cross-sectional areas of the matrix and fibre respectively.

The average stress in the composite, f_{ct} (cross-sectional area A_c), is

$$\frac{\text{load}}{A_c} = \frac{A_f f_{ft}}{A_c} + \frac{A_m f_{mt}}{A_c} \tag{11.4}$$

Note that with uniformly distributed parallel fibres:

$$\frac{A_f}{A_c} = V_f \qquad \text{volume fraction of fibres}$$

and

$$\frac{A_m}{A_c} = (1 - V_f) \qquad \text{volume fraction of matrix}$$

Hence, equation (11.4) becomes

$$f_{ct} = V_f f_{ft} + (1 - V_f) f_{mt} \tag{11.5}$$

In brittle materials, such as concretes, the matrix is likely to crack first and $\hat{f}_{ft}$ is normally much greater than $\hat{f}_{mt}$. Hence, at first cracking, equation (11.5) becomes

$$f_{ct} = V_f f_{ft} + (1 - V_f) \hat{f}_{mt} \tag{11.6}$$

(first crack: fully bonded aligned fibres).

Since

$$f_{ft} = \hat{f}_{mt} \frac{E_f}{E_m} \qquad (\text{and } not \ \hat{f}_{ft})$$

then

$$f_{ct} = V_f \hat{f}_{mt} \frac{E_f}{E_m} + (1 - V_f) \hat{f}_{mt}$$

$$= \hat{f}_{mt} \left[1 + V_f \left(\frac{E_f}{E_m} - 1 \right) \right] \tag{11.7}$$

(first crack: fully bonded aligned fibres).

To maximise first crack strength, V_f and E_f must be made as large as possible – try, for example, steel fibres in concrete:

$\hat{f}_{mt} = 3 \text{ N/mm}^2$ (tensile strength of concrete)
$E_{mt} = 30 \text{ kN/mm}^2$
$E_{ft} = 210 \text{ kN/mm}^2$ (steel fibres)
$V_f = 0.02$ (maximum possible with conventional mixing without 'balling-up')

The tensile strength of the composite is:

$$f_{ct} = 3\left[1 + 0.02\left(\frac{210}{30} - 1\right)\right] = 3.36 \text{ N/mm}^2$$

Such an increase would not normally justify the expense of adding fibres and the steel fibres in this situation are also under-utilised: they carry a maximum stress of only 21 N/mm^2 even if they bond perfectly, which is very much lower than their yield stress (the order of 1000 N/mm^2). The f_{ct} value given above is also a maximum, since smaller increases would be obtained if fibres were not fully aligned with the stress or if there were bond failure between fibre and matrix. It will be appreciated that the applications of this method of precracked strength improvement are very limited.

Fibre orientation, F_0

As regards fibre orientation, fibres will work most efficiently when aligned parallel to the stress direction. There are often difficulties in achieving this, however, and there are also instances where random fibre orientation is necessary to provide strength in more than one direction. Experiments have shown that, where fibres are oriented in a plane, their efficiency in any one direction in the plane is approximately one-third ($F_0 = 0.33$) that of fibres which are fully aligned in that direction. Where fibres are randomly oriented in three dimensions, this factor falls to about one-sixth ($F_0 = 0.17$), the material then behaving isotropically. These factors should also be applied to the precracked strength equation (11.6) which becomes:

$$f_{ct} = V_f f_{ft} F_0 + (1 - V_f)\hat{f}_{mt} \tag{11.8}$$

If the steel fibres referred to above in the concrete composite were arranged randomly in three dimensions, equation (11.7) would become

$$f_{ct} = 3\left[1 + 0.02\left(\frac{210 \times 0.17}{30} - 1\right)\right] = 3.01 \text{ N/mm}^2$$

Clearly the situation is even worse than before. It is probably against this background that it is difficult to see how steel fibre reinforced concrete could substitute for traditional reinforced concrete.

TENSILE STRENGTH (POST-CRACKED)

This may be illustrated by considering a simplified arrangement of a single long fibre embedded in a cylinder of matrix (Figure 11.1). Load/strain diagrams are used rather than stress/strain diagrams, since unlike stresses:

load in composite = load in matrix + load in fibres

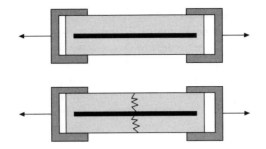

Figure 11.1 Schematic representation of fibrous composite

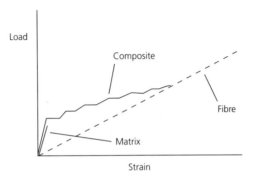

Figure 11.2 Load–strain relationship for fibre, matrix and composite

Matrix and fibre separately might have load/strain curves as shown in Figure 11.2.

The gradient of a load/strain diagram will be dependent on the area of cross-section of the specimen. Supposing, for example, matrix and fibre are to be of similar stiffness, the ratio of the gradients would be in the ratio of their cross-sectional areas. The former graph is steeper than the latter, since fibres normally form a small volume fraction and do not usually contribute the major part of stiffness. Figure 11.2 indicates much larger strains in the fibre before failure than occur in the matrix. On combining the fibre and the matrix, the load/strain curve would change as follows. The initial gradient of the curve would be equal to the sum of the individual gradients , since, at a given strain, each part would carry loads as before and the total load would be the sum of these. It will be supposed, initially, that E values are equal, so that there will be no shear force on the fibre/matrix interface. At the same strain as previously, the ceramic will crack. Assuming the fibre is bonded to the matrix then, when the first crack forms, there would be a slight, sudden extension as the fibre at the point of cracking takes all the load. The first crack corresponds to a short horizontal line representing the instantaneous increase of strain which occurs as the fibre stretches to carry the total load at the point of cracking. Assuming that the first crack occurred at the weakest point in the matrix, the load can now be increased until the next weakest part cracks, giving a further step. The process then repeats until there is a regular pattern of cracks, each crack reducing the tensile stress in the matrix at that position to a value below its tensile strength. If the matrix became useless due to extended cracking, the gradient of the load/strain diagram would approach a value corresponding to the stiffness of the fibre component, assuming the fibre was effectively anchored at its ends. In practice, however, the matrix should support some proportion of the load, due to shear/friction with the fibres, so that the dotted line corresponding to fibre stiffness would represent the minimum gradient possible. The detailed load/strain relationship will depend on a number of parameters relating to the fibre and matrix.

Volume fraction of fibre

A high volume fraction gives greater post-cracking load and smaller crack width, since the fibre would extend less in order to carry the additional load caused by cracking. The gradient of the last part of the load/strain diagram would also rise.

Modulus of elasticity of fibre

A high E value would have an effect similar to using a high fibre volume fraction (although, in practice, it would increase the probability of fibre pull-out).

Bond between fibre and matrix

A high bond strength would again reduce crack widths, thereby causing more frequent cracking; hence, the steps in Figure 11.2 would become smaller but more numerous.

Fibre strength

Increasing the strength of continuous fibres will increase the length of the final part of load/strain graph, thereby increasing ductility prior to failure (assuming bond failure does not occur). The fibre strength required will, in practice, depend on the post-cracking characteristics required as well as on the volume fraction and bonding properties with the particular matrix.

Supposing we have a small number of well bonded fibres, the ultimate strength of the composite would reflect the ability of the fibres themselves to carry load. This is represented by the first part of equation (11.8). Hence,

$$\hat{f}_{ct} = V_f \hat{f}_{ft} F_0 \tag{11.9}$$

Pull-out factor, F_p

Consideration must now be given to the effect of fibres having finite length (hence, limited bonding ability). Equation (11.9) must hence be modified to

$$\hat{f}_{ct} = V_f \hat{f}_{ft} F_p F_0 \tag{11.10}$$

where F_p is a pull-out factor as explained below.

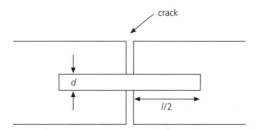

Figure 11.3 Idealised fibre/crack disposition – the chance of fibre pull-out is minimised because the fibre is equally embedded on each side of the crack

The possibility of pull-out clearly increases as fibres become shorter. Consider a fibre of length l and diameter d, which bridges a crack at its centre (Figure 11.3). It is subject to a surface shear stress f_{fs}, which, for simplicity, is supposed to be uniform, along half its length. The fibre is best utilised if its length is sufficient to enable the shear force to develop a stress equal to the tensile strength of the fibre $\hat{f}_{ft}$.

Hence, if the fibre has diameter d:

$$\hat{f}_{fs}\frac{\pi dl}{2} = \frac{\pi d^2 \hat{f}_{ft}}{4}$$

or

$$l = \frac{d\hat{f}_{ft}}{2\hat{f}_{fs}}$$

or

$$\frac{l}{d} = \frac{\hat{f}_{ft}}{2\hat{f}_{fs}} \tag{11.11}$$

The ratio l/d is known as the aspect ratio and it is simply related to the ratio of fibre tensile strength to fibre bond strength. A great deal of fibrous composite technology hinges on this simple equation. If a fibre has a high tensile strength, for example, steel, then either a high bond strength would be required to avoid pull-out well before $\hat{f}_{ft}$ is reached or fibres of high aspect ratio would be required. Plain steel fibres have, in fact, a failure (yield) stress in the region of 1000 N/mm² whereas the effective bonding strength to concrete is only about 3 N/mm². Efficient use of such fibres in concrete, therefore, requires aspect ratios of at least 160. Fibres of length 40 mm, for example, would require a diameter of approximately 0.25 mm maximum, which may tend to cause buckling of fibres during mixing. If a certain minimum fibre diameter is considered practicable, then a minimum fibre length, known as the critical fibre length, is required to ensure that the full tensile capacity of the fibre can be utilised. Carbon fibres have strengths of over 2000 N/mm² so that very high bond strength or very thin or long fibres are necessary to utilise this strength.

From the point of view of the composite, short fibres produce an additional problem, since across any one crack there will be a substantial proportion of fibres embedded to a length less than $l/2$ on one side and more than $l/2$ on the other side. The former ends would, therefore, tend to pull out, even if the aspect ratio is just sufficient as defined above. Hence, for fibres at the critical length:

$F_p = 0.5$ (approximately)

The reduction is greater when shorter fibres are used. For this reason, either aspect ratios higher than that given by the above equation would be necessary, or a higher volume fraction of fibres would be needed to compensate for pull-out of a proportion of them.

Critical fibre volume

An approximate expression has been given for ultimate composite strength:

$$\hat{f}_{ct} = V_f \hat{f}_{ft} F_p F_0$$

There is a possibility that, with small quantities of randomly orientated fibres, this may be less than the precracked strength as given by equation (11.8):

$$f_{ct} = V_f \hat{f}_{ft} F_0 + (1 - V_f) \hat{f}_{mt}$$

The critical volume fraction for the previous example (page 549), taking a two-dimensional array of fibres ($F_0 = 0.33$) at their critical length ($F_p = 0.5$), is given by:

$$V_f \hat{f}_{ft} F_p F_0 = V_f \hat{f}_{ft} F_0 + (1 - V_f) \hat{f}_{mt}$$

where

$$\hat{f}_{ft} = \frac{\hat{f}_{mt} E_f}{E_m}$$

Hence

$$V_f \cdot 1000 \cdot 0.5 \cdot 0.33 = V_f \cdot 3 \cdot \frac{210}{30} \cdot 0.33 + (1 - V_f) \cdot 3$$

This gives

$165 V_f = 6.9 V_f + 3 - 3 V_f$

and then

$161.1 V_f = 3$

so that

$V_f = 0.019$

Hence, there should be at least 1.9 per cent fibres by volume in order to avoid sudden failure at the point of first cracking. This critical volume could be

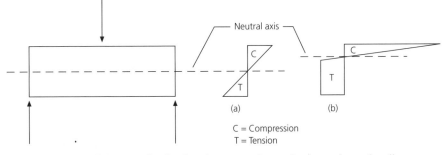

Figure 11.4 Possible stress distributions in rectangular section beam due to bending:
(a) assumed distribution in calculating modulus of rupture; (b) simplified representation
of stress distribution in fibrous composite at failure. For a given tensile failure stress,
distribution (b) gives a much increased failure load

reduced by longer or better bonded fibres which would increase F_p, or by
aligning fibres with the stress direction which would increase F_0. In an idealised
composite, failure should be gradual, involving partial fibre failure, indicating
that fibres are well utilised, and partial pull-out at high stresses, which avoids
sudden failure.

FLEXURAL BEHAVIOUR

It is often found that flexural behaviour is substantially better than the above
equations for tension would suggest. This is often because the neutral axis moves
towards the stronger zone during loading so that the area of weaker material below
the neutral axis is increased and its average distance from the neutral axis also rises
(Figure 11.4). This invalidates conventional bending theory and gives rise to
apparently better performance.

MODULUS OF ELASTICITY

The modulus of elasticity must also be considered in the pre- and post-cracked
conditions and, in each case, formulae similar to those employed for tensile strength
can be used.

The value corresponding to the *precracked* state is, for example, given by:

$$E_{ct} = V_f E_{ft} + (1 - V_f)E_{mt} \tag{11.12}$$

for long fibres oriented parallel to the stress (compare with equation (11.6)). In
order to produce a stiff composite, it will be necessary to incorporate a high volume
fraction of stiff fibres, aligned, if possible, with the stress. This is usually difficult
with ceramic type materials, which have, in any case, fairly high E values without
the need for reinforcement.

The modulus of elasticity corresponding to *multiple cracking* is given by:

$$E_{ct} = V_f E_{ft} \tag{11.13}$$

for the same situation, though this will be reduced progressively to zero if pull–out occurs. Orientation factors F_0 equal to 0.33 (2D) or 0.17 (3D) should be applied to the fibre component of equations (11.12) and (11.13) as appropriate.

IMPACT STRENGTH

The energy absorbed by a unit during a destructive test is proportional to the area under its stress/strain diagram and, although energy absorption characteristics may alter when sudden impact occurs, it is found in general that stress/strain diagrams give a good indication of resistance to impact generally. The area under diagrams of the form of Figure 11.5 will be largely dependent on the degree of post-cracking ductility. This can be increased by use of long fibres having relatively low bond strength and high failure strain. Polypropylene in cement satisfies these requirements and produces excellent impact resistance, while asbestos fibres in a cement matrix bond well and have low failure strains, resulting in poor impact resistance. Typical stress/strain graphs for each are shown in Figure 11.5. It is unfortunate that requirements for high impact strength often result in low tensile strength, so that it may be difficult to obtain both at the same time; a composite comprising, for example, a mixture of polypropylene and asbestos fibres might help in this respect.

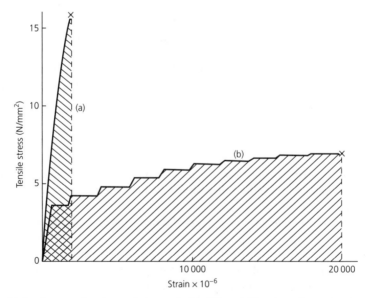

Figure 11.5   Areas under stress–strain graph for different fibre types in a cement matrix: (a) rigid, strongly bonded fibres such as asbestos; (b) softer, weaker-bonded fibres such as polypropylene

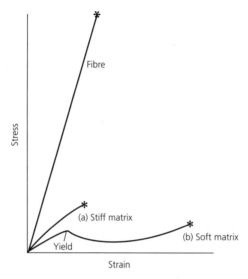

Figure 11.6 Stress–strain characteristics of glass fibre relative to those of: (a) stiff resin matrix; (b) soft resin matrix

11.3 REINFORCEMENT OF PLASTICS

The chief reason for reinforcing plastics is, as explained above, to increase their stiffness, although useful increases in strength, impact resistance, fatigue resistance and creep resistance should also be obtainable. Most plastics have E values in the range 1–5 kN/mm^2, so that use of glass ($E = 70$ kN/mm^2) or stiffer fibres should make a large difference to the precracked modulus of elasticity. The basic forms of stress/strain graph for stiff fibres such as glass, together with plastics, are shown in Figure 11.6. Note that the behaviour of plastics can be changed greatly: some softer plastics, such as thermoplastics, have low stiffness and may flow plastically before failure, while thermosets, such as polyesters, tend to exhibit very little plastic flow, though they are both subject to creep.

Reinforced plastics can, in practice, only be used up to the strain at which either the matrix or fibre breaks or yields. The softer plastics, which yield well before the fibres are fully stressed, would not produce a high stiffness composite, although failure of such a composite would be gradual and this might be of value in some applications, for example, where brittle failure may not be acceptable. The ideal situation might be that in which both fibre and matrix fail at about the same strain, since full use is then made of each component. The resulting composite would then be of relatively high stiffness but rather brittle, especially in pure tension, though flexural applications, which are much more common, would result in more acceptable behaviour on overstressing. On account of the effects of creep, working stresses in reinforced plastics should be much lower than (as little as 15 per cent of) yield or failure stress so that post-crack behaviour is of less importance than elastic behaviour. Hence, use of the relatively stiff brittle matrix of Figure 11.6 should not normally cause problems.

The precracked modulus of elasticity of a composite for strains up to the yield/failure of the matrix or fibre (whichever occurs first) will be given by equation (11.12). Applying the equation to a two-dimensional array of glass fibres ($E = 70$ kN/mm^2) at a volume fraction of 0.2 in a plastic having $E = 3$ kN/mm^2:

$$E_{ct} = 0.2 \times 70 \times 0.33 + 0.8 \times 3 \qquad \text{(fibre orientation factor 0.33 included)}$$

$$= 7.0 \text{ kN/mm}^2$$

The stiffness of the plastic has been more than doubled by incorporation of the fibres.

The first crack tensile strength of reinforced plastics will also be generally enhanced, although it is just possible that, in a soft matrix with a few fibres, the fibres could soon become overstressed, breaking and acting as voids in the composite. More commonly, first crack strength can be given by equation (11.5), using the appropriate matrix stress with the fibre stress weighted according to the relative stiffness of fibre and matrix:

$$f_{ft} = \hat{f}_{mt} \frac{E_{ft}}{E_{mt}}$$

Orientation factors should, of course, be used if appropriate. If the matrix yields, its yield strength can be used in the same way to calculate the precracked tensile strength, weighting the matrix stress according to stiffness. Other situations, such as fibre failure before matrix failure and ultimate (post-crack) strength, can be represented by careful use of the equations given above.

Impact performance of reinforced plastics depends on stress/strain characteristics of the composite in much the same way as demonstrated in Figure 11.5 for ceramics. Hence, softer plastics will perform better, although impact performance of sheet composites is generally much better than that of unreinforced plastics of comparable stiffness, since energy can be absorbed at the point of impact by fracture of fibres and/or matrix, together with local debonding. Such performance is another of the major attributes of fibre reinforced plastics.

11.4 CAUTIONARY NOTES

■ The above arguments and equations are designed to serve as an introduction to the behaviour of fibre composites and to identify critical parameters. Detailed properties vary considerably according to fibre and matrix properties and there are often marked variations in behaviour of composites of a given type. The equations should, in particular, be taken only as a guide to the order of magnitude of likely properties.

■ Fibre volume fractions are used throughout and these can sometimes be misinterpreted when batching. For example, 2 per cent by volume of steel fibres in concrete would correspond to

$$2 \times \frac{\text{density of steel}}{\text{density of concrete}} = 2 \times \frac{7.8}{2.3} = 5.6 \text{ per cent by mass (weight)}$$

■ To obtain results approximating to theoretical performance, full compaction is necessary. If incorporation of fibres leads to even a small number of air voids, performance may be impaired, rather than improved, particularly if measured in strength terms.

11.5 TYPES AND PROPERTIES OF FIBRES

The three basic groups of materials each make contributions to the range of fibres which exists for reinforcement of materials. In the ceramic group, they are glass and asbestos fibres; in the metallic group, steel fibres; and in the organic group, carbon, polypropylene, Kevlar and polyester fibres. Their properties are summarised in Table 11.2.

Ceramic fibres

The inherent weakness in tension of bulk ceramics has already been linked to the presence of flaws which, according to Griffith's theory, result in greatly amplified internal stresses. It is known that the size of these flaws reduces greatly as the

Table 11.2 Typical properties of some common fibres in order of decreasing elastic modulus

Fibre type	Relative density	Modulus of elasticity (kN/mm²)	Ultimate tensile strength (N/mm²)	Specific modulus of elasticity (kN/mm²)	Specific tensile strength (N/mm²)	Strain at failure (per cent)
Carbon high modulus	1.9	420	2100	221	1105	0.5
Carbon low modulus	1.9	240	2400	126	1260	1
Steel (low-carbon)	7.8	200	1100	26	141	Necks
Asbestos (chrysotile)	2.6	160	200–2000	62	385(Av.)	2
Kevlar	1.45	130	3000	90	2069	2–3
'E' glass	2.55	70	3000	27	1176	5
Polyester	1.38	14	1100	10.1	797	15
Polypropylene (fibrillated)	0.9	8	400	8.9	444	7

physical size of ceramic units decreases – this is presumably related to reductions in stress differentials caused, for example, by temperature gradients during manufacture. Fibres may therefore be in some cases 100 times stronger than their bulk equivalents and this, combined with much smaller bending radii which are possible with small diameters, gives fibres a degree of flexibility which is quite unattainable with bulk equivalents.

Note that, although attention in relation to health risk has centred chiefly upon asbestos fibres, it now appears that there are risks associated with all forms of man-made mineral fibres. Due precautions should therefore be taken where appropriate.

Glass fibres

Glass fibres are manufactured by drawing filaments from the base of platinum crucibles (bushings) containing molten glass. Each bushing contains several hundred holes and the filaments so formed are collected to form strands and then wound onto a drum. Individual filament diameters depend on glass properties, hole size and drawing speed, though they are usually about 10 μm. A 'size' such as polyvinyl acetate is used to bind filaments together and protect them from damage during fabrication at a later stage (these fibres should not be confused with glass fibres for thermal insulation purposes, which are much coarser and produced by a different process). The strands, which are normally of approximately elliptical section, typically 0.6×0.08 mm, may be formed into continuous lengths called roving, woven into cloth, or chopped to form matting. Cloth consists of continuous fibres and, therefore, results in much greater composite strength than chopped-strand matting (fibre length, approximately 40 mm).

It is unfortunate that ordinary borosilicate or E glass is attacked by alkalis contained in Portland cements. A special alkali resistant fibre has now been developed for use with such cements; it contains a fairly large proportion of zirconium oxide ZrO_2 and is marketed under the trade name *Cem-Fil*.

Asbestos fibres

Since they occur naturally, their use is, perhaps, more traditional than that of other fibre types. Strength values vary greatly, increasing in the order amosite, chrysotile, crocidolite, though the latter, *blue asbestos*, is no longer in current use, due to the health hazard involved. The strength of crocidolite lies in the range 200–4000 N/mm^2, though some small reduction may occur with age. Chrysotile, which is commonly used commercially, has strength in the range 200–2000 N/mm^2 with a mean strength of about 1000 N/mm^2. Shorter fibres tend to give higher strength. Elastic moduli are also variable, chrysotile having a mean value in the region of 160 kN/mm^2. Fibre diameters are very small, about 1 μm or less, and this may be an important factor in the excellent bonding between asbestos fibres and cement. It is likely that EU regulations will spell an end to the use of all asbestos for building applications.

Metal fibres

The most common metallic fibres are steel. They are usually relatively coarse, for example, 300 μm in diameter and, since it is unusual for failure of steel fibre reinforced materials to be caused by failure of the fibres themselves (failure is usually by pull-out of fibres), there is little point in seeking higher tensile strength than is obtained by drawing ordinary low carbon steel. Tensile strengths are approximately 1100 N/mm², though metals are different from ceramic fibres in that failure involves necking and, consequently, larger strains. The E value of steel fibres is similar to that of larger steel components – about 200 kN/mm². Fibres may be plated to increase corrosion resistance and best results are obtained with fibres of non-uniform section, such as deformed fibres or irregular fibres resulting from special melt extract processes.

Organic fibres

Polypropylene fibres   This material has great flexibility and toughness, combined with low density, and imparts substantial improvements to the impact strength of materials it reinforces. It is normally manufactured as a film formed by extrusion. This film is then slit and drawn so that spherulites in the partially crystalline polymer are converted into orientated fibrils. The film may, finally, be twisted into twine. The E value is the lowest of any common fibre, approximately 8 kN/mm², so that it would be of no use in increasing precracked strength or stiffening materials. Tensile strength is approximately 400 N/mm². As would be expected of a non-polar polymer, the fibre has little affinity for water but the fibrillated form produces a good mechanical bond with cement based materials. Polypropylene fibres are resistant to a wide variety of chemicals.

Kevlar (polyamide) fibres   These are a form of nylon and are very high performance fibres, having a modulus of elasticity of 130 kN/mm², significantly higher than that of glass, together with low creep. They may be used in the production of high performance composites, though applications are limited by cost.

Polyester fibres   These have a similar appearance to polypropylene fibres but are denser, stiffer, stronger and have a higher melting point. They may be used in applications similar to those where polypropylene is used but where the increased performance justifies the higher cost of these fibres.

Carbon fibres   These are relatively new and highly promising materials based on the strength of the carbon–carbon bond in graphite and the lightness of the carbon atom. Carbon fibres are produced by heat treatment of plastic fibres, such as acrylic fibres, so that the carbon atoms link together to form small graphitic crystallites, which are orientated by stretching while hot. The fibres are about 10 μm in diameter but consist of tiny fibrils stranded together in quantities up to 100 000. There are two chief varieties: high strength fibres with ultimate tensile strength of approximately 2400 N/mm² and E value of 240 kN/mm²; and high modulus fibres with ultimate tensile strength of 2100 N/mm^2 and E value of 420 kN/mm^2. With strengths of this order, it is clearly important that a good bond with the matrix is achieved if the fibres are to be used for reinforcement, otherwise they would pull out under stress. Unfortunately, the more perfect the graphitic structure, the less likely the fibre is to bond to other materials, so that treatment, for example, by etching, is required to obtain maximum benefit from this material. A further problem is that the fibres are brittle, which together with their fineness means that they could not be, for example, be mixed in a concrete mixer. They would be broken up by the abrasion. The fibres are also very expensive so that use is confined to applications largely outside the construction industry where very high performance together with low weight are required.

11.6   FIBRE REINFORCED CEMENT PRODUCTS

Materials based on Portland cements form a natural choice for application of fibrous materials since they are cheap, but they leave much to be desired in respect of ductility, impact resistance and tensile strength, the latter being 3–5 N/mm^2. Many of the above named fibres have been used in attempts to improve these properties, though the modulus of elasticity of the composite is, in most cases, similar to that of the unreinforced material, since the fibres normally form small volume fractions of the total. Table 11.3 summarises properties of typical composites based on Portland cement and, for comparison, glass reinforced gypsum.

GLASS REINFORCED CEMENTS (GRC)

Glass fibres have been included in certain types of cement product, notably precast units, with a view to increasing flexural strength and impact resistance. Manufacture is often on the lines of asbestos cement goods – by spraying chopped glass fibres 10–50 mm in length on to a perforated base and, at the same time, spraying on a cement slurry. When a sufficient thickness is built up, excess water is removed by vacuum suction and the flexible composite sheet can be shaped and then cured. Fibres may, alternatively, be introduced by mixing a small percentage of the shorter fibres directly with cement slurry before moulding but, since the resultant orientation is completely random, the tensile strength would only be about half that of products containing fibres orientated in a plane. Short term direct tensile strength is in the region of 17 N/mm^2 for 5 per cent by volume of alkali resistant glass and the strain at failure in tension is in the order of $10\,000 \times 10^{-6}$, indicative of the good bonding between fibre and matrix. Tensile strength increases with fibre length and, initially, with fibre volume fraction although density reduces as volume fractions rise. At volume fractions over 5 per cent, the density effect overrides the fibre volume effect, strength falling off as V_f increases above this value. The modulus of rupture is, for reasons stated earlier, much higher than pure tensile strength; values up to 50 N/mm^2 are attainable. GRC has very much better

Table 11.3 Properties of typical fibre reinforced cements compared to glass reinforced gypsum and semi-compressed asbestos cement

Material	Modulus of rupture (N/mm²)	Tensile strength (N/mm²)	Strain at failure-tension (per cent)	Impact strength	Effect of weathering
Glass-reinforced cement (5 per cent by volume of alkali-resistant sprayed 40 mm fibres) (air storage, 28 days)	45	17	1	High	Strength and impact resistance reduced in damp conditions
Steel fibre-reinforced concrete (2 per cent by volume of 40 mm fibres)	8	4	1	High	Surface fibres may corrode
Polypropylene fibres in concrete (0.2 per cent by volume of 40 mm fibres)	5	4	High	High	Good weather resistance
Glass-reinforced gypsum (Class B) (4 per cent by volume sprayed 40 mm fibres)	30	15	0.7	Very high	Must be protected from damp
Semi-compressed asbestos-cement sheet (15 per cent by weight of fibres)	20	14	0.2	Low/ medium	Gradual embrittlement on exposure

impact strength than either unreinforced cement or asbestos cement, 5 per cent by weight of glass fibre (about 3.5 per cent by volume) giving an impact strength about five times that of the latter. Resistance to fire is much better than that of the unreinforced material, particularly if a proportion of PFA is included. Thermal shocks are also absorbed more satisfactorily. The main problem with GRC is that ordinary (E glass) glass fibres are attacked by the hydrating cement. Figure 11.7 shows that, after a short time in air storage, flexural strength (modulus of rupture) decreases steadily until a low value, about half its previous maximum, is reached. Impact strength is similarly affected. More recent types of glass (Cem-Fil), which include zirconium oxide ZrO_2, are more alkali resistant and therefore not as severely affected as E glass. Typical curves are also shown in Figure 11.7, the newer Cem-Fil 2 performing even better. Performance can be improved further by using high alumina cement instead of Portland cements since HAC paste has a lower pH value. Pozzolanas also improve performance.

It is found that even with alkali resistant glass, impact strength reduces with age, especially when the composite is subject to wetting. This is due to an increase of bond strength between fibre and matrix resulting in a higher fracture rate of fibres under impact conditions.

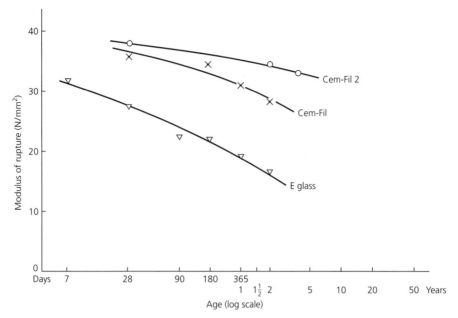

Figure 11.7   Relation between modulus of rupture and age of glass fibre reinforced Portland cement composites subject to normal UK weathering (Information derived form BRE Digests 216 and 331)

Glass fibres can also be mixed *in situ* with concrete, though the volume is limited to about 2 per cent maximum, since workability is considerably reduced, especially if long fibres are used. Flexural strength can, nevertheless, be increased up to 4 times with careful mix design. Applications for GRC are primarily non-structural, in view of the uncertainty in long term performance even with alkali resistant fibres. Uses are as follows:

■ In precast pipes, thinner lighter units can be produced for given strength/impact resistant properties.
■ Smooth faced building boards in standard (2400 × 1200 mm) sheets of thickness 4, 6 or 8 mm are now marketed. These have much greater impact resistance than asbestos cement sheets and can be used externally as well as internally. They are suitable for cladding, partitioning, roof decking and permanent formwork.
■ GRC panels are being quite widely used in cladding, partly on account of their light weight (thicknesses as low as 10 mm can be used since there is no need to incorporate steel reinforcement). Decorative surfaces can be applied.
■ Very rapid wall construction has been achieved by assembling dry concrete blocks and then spraying the surfaces with a fibre–mortar mixture. A strong, waterproof structure is formed.
■ Small precast units, such as manhole covers, ducting, fence posts and garden furniture, can be made.

- A cement and sand premix containing alkali resistant glass fibres is available for rendering purposes. High impermeability and impact resistance should be obtained with correct application.

There are still areas for development of GRC. These include:

- Development of improved fibres leading to better long term weathering properties and, hence, eventual use in load bearing situations.
- Incorporation of polymers for improving behaviour of the cement matrix. Acrylic and other polymers lead to reduced water/cement ratios on account of their lubricating effect in the wet mix. Hence, the hardened material is stronger and more impermeable and the polymer film may also protect the glass from alkali attack by the cement. A further alternative is to use sand–lime mixtures instead of cement. These are not so aggressive to glass because they are less alkaline, though products must be cured by autoclaving.

STEEL FIBRE REINFORCED CONCRETE

Steel has the advantage over other fibres of a high modulus of elasticity, though to exploit this, good bonding properties are required. It might seem reasonable to assume that a good bond would be obtained between steel and concrete in much the same way as in conventional reinforced concrete. However, the bond in the latter is known to be at least partly due to surface irregularities produced by hot rolling and the presence of thin, adherent rust films. Surface irregularities are virtually absent in cold drawn steel fibres, which are less than 1 mm in diameter. This, together with the fact that it is not possible to reproduce the corrosion effect in a fine fibre, results in a relatively poor bond between fibre and matrix in steel fibre reinforced concrete (about 5 N/mm^2 for smooth, single wires). As a result long fibres would be theoretically required in order to produce sufficient anchorage to utilise fully the strength of the steel. There are practical problems here since long fibres tend to ball up, resulting in increased porosity and reduced strength in the final composite. There are advantages in having a composite system in which failure occurs by partial pull-out of fibres, since there is often more warning than when it takes place by fracture of fibres (as, for example, in asbestos cement). However, it clearly represents inefficient use of a composite if fibres pull out when well below their yield stress. To this end, many fibres used to reinforce concrete may be deformed (crimped or indented) to increase resistance to pull-out (Figure 11.8.). Alternatively, surface oxidation by heating or controlled acid attack may be used to increase the bond, and irregular melt extract fibres are available.

Design and mixing of fibre reinforced concrete require careful attention. Balling up of fibres is likely to occur if more than 2 per cent by volume of fibres is used or if fibres are not added gradually to the mix (for example, through a coarse mesh sieve). Long, thin fibres (for example, 50 mm in length and 150 μm in diameter) aggravate the situation. In one system, fibres are bound with a size which breaks down on mixing, permitting larger volume fractions without risk of balling. The

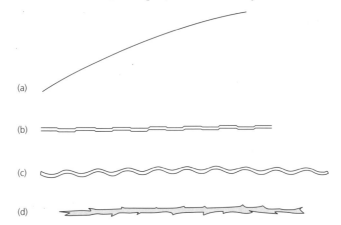

Figure 11.8 Typical steel fibres: (a) plain round; (b) indented; (c) crimped; (d) melt extract

fibres themselves make the concrete unpleasant to handle, due to their stiff, prickly nature. A high proportion of sand is preferred for incorporation of fibres to prevent balling, though rich mixes are not necessary for this purpose. High water cement ratios, crimped or deformed fibres all appear to increase the risk of balling.

Compaction is best achieved by vibration, though with higher fibre concentrations, pokers are not suitable since, on withdrawal, a cavity remains which is difficult to close. Table vibration tends to orientate the fibres parallel to the plane of the slab, which is not always advantageous, though the precise effect depends on the shape of the component and its effect should be investigated as part of the design operation. Wetter mixes may also lead to migration of fibres. Quality control is an important aspect of materials of this type and sawn core tests, for example, cylinder splitting, are preferable to smaller cast samples which may not be representative of the *in situ* material. Manufacture and placing of steel fibre reinforced concrete require, in general, specially trained personnel.

Hardened properties

It has already been emphasised that precracked tensile or compressive strength is unlikely to be improved by addition of steel fibres, although it is possible that, if fibres control microcracking, there may be some advantage on this basis. Flexural performance is likely to be more significantly improved. Load/deflection relationships of the form of Figure 11.9 are claimed by manufacturers, though the extent to which the composite remains intact after the initial point of cracking depends very much upon the mix and fibre characteristics. The chief factors are:

■ Fibre aspect ratio. Increasing values (normally achieved by longer fibres of 50 mm or more) increase flexural strength but decrease the percentage fibre content at which balling-up in the mixer occurs, so that, in practice, aspect ratios greater than about 100 are not normally used.

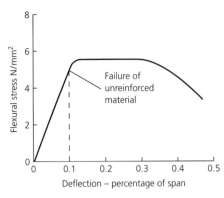

Figure 11.9 Behaviour of steel fibre reinforced concrete in flexure

- Fibre volume fraction. The problem here as indicated above is that volume fractions over 2 per cent tend to make compaction very difficult.
- Concrete strength. Higher strength concretes bond better with the fibre permitting use of shorter fibres and this in turn allows greater volume fractions to be employed.

Figure 11.10 shows the approximate relationship between flexural strength and percentage by volume of fibres. Assuming full compaction can be achieved the flexural strength of a mix which has an unreinforced strength of 5 N/mm² (high quality concrete) will be doubled by incorporation of 2 per cent by volume of certain fibres. A significant plastic stage should be obtained with concrete of this quality so that failure will be much more gradual than in unreinforced concrete, with ample visual warning. Weaker mixes would not show such an improvement.

Deformed or treated fibres would be unlikely to improve precracked strength but quite substantial improvements in ultimate strength can be achieved (hence, the bands of Figure 11.10). The effect is, however, complex; for example, indenting of wires may reduce their strength, while crimping may produce bursting stresses in the concrete, especially if fibres are close together, as would be the case when high volume fractions are employed. It is important to appreciate that the action of steel fibres in increasing flexural strength of concrete is different from that by which conventional reinforcement works. The latter is not designed primarily to prevent cracking; it carries the entire tensile load when cracking has occurred. Fibres, on the other hand, do not carry all the tensile load in concrete – they tend to bridge the microcracks which, under tension, grow to form observable cracks. However, as in the case of conventional reinforced concrete, they do carry some of the load after cracking has taken place, so that a cracked fibre reinforced concrete may be stressed further before failure occurs. In order to utilise the fibres more efficiently to improve flexural strength, experiments have been carried out in which fibres have been concentrated at positions of highest tensile stress. This could be achieved by placing of fibres separately during pouring of concrete or by post-treatment, such as is used for gunite (a cement mortar sprayed on to a solid background). The latter

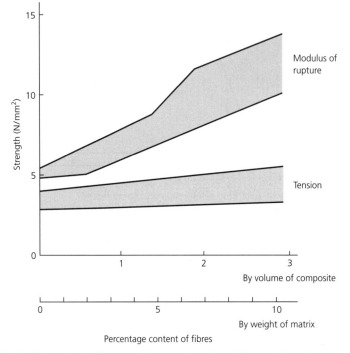

Figure 11.10   Flexural and direct tensile strengths of steel fibre reinforced mortar and concrete

could be very effective, provided safety precautions are taken against injury by the steel fibres.

One major improvement of steel fibre reinforced concrete is in impact resistance, as indicated by the area under the graph of Figure 11.9. As with flexural failure, impact failure is progressive, with ample warning of distress before total collapse occurs. Increases of 20–30 times that of plain concrete are obtainable. Impact properties in thin slabs are particularly improved. Figure 11.11 indicates that high tensile crimped fibres give much higher toughness than indented varieties, presumably due to increased friction during pull-out. Spalling and abrasion resistance are also improved.

When used in exposed situations, the steel fibres which are adjacent to the surface inevitably corrode, owing to exposure or carbonation of the concrete surface. Such exposure does not affect the mechanical properties of the uncracked composite, though it results in rust staining, which might be unacceptable if appearance is important.

When concrete becomes cracked, carbonation reaches much greater depths and, although short term strength may not be affected, long term strength would be expected to decrease as the effective fibre diameter is reduced. Corrosion of fibres can, of course, be prevented by use of stainless steel or coated fibres; brass may be used for the latter.

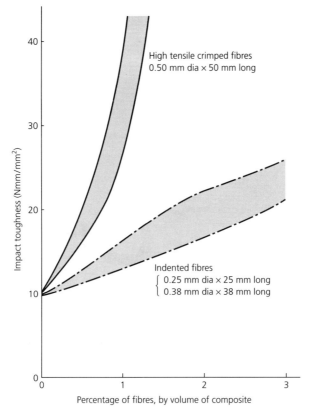

Figure 11.11 Impact toughness of steel fibre reinforced mortar and concrete

Uses

The high cost of steel has been a drawback in the use of steel fibres in concrete – a concrete containing, for example, 2 per cent by volume of fibres is likely to cost up to six times as much as an ordinary mix. Such a figure could, on the other hand, be misleading, since the material cost is usually a fairly small fraction of the cost of the structure, which might be only slightly increased. Also, the enhanced properties of the concrete may allow the use of thinner sections for a given specification, enabling some saving to be made on the quantity of concrete. Since a wide variety of fibre types is available, it is important to evaluate these for the particular use, as the optimum type and volume fraction of fibres will depend on the precise performance requirements of individual applications.

Most applications are based on situations which require a greater degree of impact and flexural strength than can be provided by ordinary concrete. Examples are in factory floor slabs, aircraft aprons or runways and surface screeds. The durability of the concrete and, particularly, resistance to extended cracking are considerably improved. Further possible applications include precast units, such

Figure 11.12   Use of guniting technique and steel fibres to stabilise uneven surfaces. Fabric reinforcement could not follow contours such as these

as manholes, pipes or panels and claddings where surface coatings can be used to prevent rusting of the fibres. The material has been used as a thin overlay to damaged or worn concrete surfaces. In such applications, careful surface texturing measures are necessary to avoid exposing fibres.

In one application, a perfectly spherical domestic dwelling was produced by inflating a large balloon, spraying a rigid plastic on the exterior to form a mould and then spraying a steel fibre/cement mix internally to form a rigid structure.

The material can also be used in guniting – a typical mix might be 3:1 sharp sand:cement with 2 per cent by volume of fibres, preferably stainless steel, giving a flexural strength of about 7 N/mm^2. At a thickness of about 50 mm, it can be used for applications such as tunnel/sewer linings, stabilisation of irregular rock faces (Figure 11.12), shell roofs, repairs to concrete structures and fire prevention.

A recent innovation is incorporation of relatively high volumes of steel fibres in high strength concrete. A typical mix might be (by mass):

3 parts binder (including super-plasticiser and microsilica)
2 parts quartz or other high strength sand
1 part steel fibres
0.4 parts water

Note that there is no coarse aggregate and this permits incorporation of a fibre volume of 6 per cent, much higher than values obtained above. Mechanical mixing produces workability that is short lived and barely sufficient for compaction. However, the hardened material has a compressive strength of around 140 N/mm^2

joint

Figure 11.13 Jointing of precast staircase units using steel fibre reinforced concrete (joint arrowed)

and a flexural strength of around 20 N/mm². It has been used to produce structural joins between precast units such as floor beams (laterally) to give strength. A further application is jointing of precast staircase sections. Laps between adjacent projecting reinforcing bars of only 5–10 diameters produces structural continuity, compared with around 40 diameters which would normally be required for full stress transfer. The action has been described as a *cold weld*. An example of precast staircase segments joined in this way is shown in Figure 11.13. The material is very expensive, but in such situations would only be used in small quantities and could well permit economies of design.

POLYPROPLYENE FIBRES IN CONCRETE

It might seem surprising that polypropylene is used in concrete at all, since its modulus of elasticity is only about 8 kN/mm², compared with at least three times this figure for most concretes, so that uncracked tensile or flexural properties could not be directly improved. The fibres are, however, cheap and may have marked effects on the plastic properties and on post-cracking behaviour and impact resistance of the hardened concrete. The extent to which modification takes place depends upon the type and quantity of fibre used.

When small quantities – around 0.2 per cent of short monofilaments – are used there may be significant improvement in the performance of the hardened concrete. Some of these are derived from modifications to the behaviour of the fresh material. For example, since bleeding is reduced, risk of plastic cracking is also reduced and water remains uniformly distributed to ensure that hydration occurs in the early

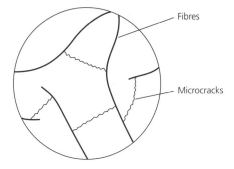

Figure 11.14   Action of polypropylene fibres in arresting microcracks in concrete

Figure 11.15   Tensile fracture surface of concrete containing polypropylene monofilament fibres

hours after placing. It is during this period that microcracks tend to form due to thermal/shrinkage effect and the fibres appear to act as crack arrestors (Figure 11.14), preventing cracks from joining together to produce more serious defects. As a result, control of shrinkage cracking is achieved without the use of steel reinforcement. This, together with improved hydration, is also claimed to give slightly improved strength and fatigue resistance, in spite of the weakening effect of soft fibres. Figure 11.15 shows a photograph of a fracture surface which gives an indication of the disposition of monofilament fibres in a typical mix. Polypropylene

Figure 11.16 Use of laser controlled hydraulic machine to produce high speed floor construction. Incorporation of polypropylene fibres reduces the need for reinforcement. Dowel bars can be laid before placing concrete, joints then being sawn after hardening

fibres are now widely used in floor slab construction, controlling shrinkage cracking and reducing the amount of steel reinforcement required. Joints would normally still be provided if a completely crack free surface is intended. In more lightly loaded applications, it may be possible to dispense with traditional contraction joints which use dowel bars, cutting instead warping joints (by cutting to 1/3 of slab depth as soon as possible after hardening). Such systems permit maximum utilisation of modern laser operated laying machines which operate without shuttering, laying up to 1000 m^2 of concrete floor per day (Figure 11.16).

A system has been devised commercially to improve the properties of concrete in the plastic stage. The addition of about 0.2 per cent by volume of short monofilament fibres leads to a heavily air entrained mix, containing up to 45 per cent air. This gives the fresh concrete thixotropic properties having, perhaps, zero slump, and yet flowing and compacting readily under vibration. At the same time, it is resistant to bleeding and segregation, which would normally occur in concrete mixes with such high quantities of air. The mix appears fatty so that finishing is easily carried out and textured finishes with fine detail can be achieved, one method being to use patterned rollers. Mixes of varying strength and density can be produced, using dense or lightweight aggregates. The hardened material is said to have improved frost resistance (due to the entrained air), lower permeability to water and increased resistance to surface crazing and impact, compared with normal concretes. This type of concrete has been used in precast products, such as cladding panels, though wider use in prefinished flooring and walling units, or structural concrete, may arise.

Higher strengths are obtained with flat open continuous networks of polypropylene film, moduli of rupture of over 30 N/mm^2 being possible, with fibre volume fractions in the region of 6 per cent. Cracking of such composites takes the form of fine cracks, often at spacings of less than 10 mm, which are invisible, except at high loads – the failed composite ultimately acquires a leathery consistency. This behaviour is much less marked at small volume fractions or when monofilaments are used – the latter tend to pull out, due to lateral Poisson's ratio contraction. Permeability should be reduced due to controlled microcracking, and surface fibres, being chemically stable, are not affected by atmospheric exposure. Any exposed fibres soon wear away in use. The material is suitable for cladding and infill panels. Polypropylene offers the advantage over steel that the amount of cover is not critical, so that low thickness sections could be employed.

Further uses of polypropylene fibres include:

- For controlling cracking in rendering materials, especially if impact loading is anticipated. The rectangular grid form can also be used for this application – it would be prefixed prior to the rendering operation.
- For corrugated roofing. The brittleness associated with mineral fibre cement sheets would be overcome, though the latter may be more resistant to high temperatures caused by fire. The post-cracked strength of polypropylene fibre reinforced cement sheets would fall dramatically at temperatures above 165 °C, the melting point of polypropylene.
- For crack control of concrete in marine applications where use of steel again has the attendant risk of corrosion.
- Improvement of fire properties of structural concrete. This might seem to contradict the effect in sheets (above), but it is found that a small percentage of fibres can reduce stresses associated with vaporisation of trapped water, especially in higher strength concrete. Although strength is not improved, the risk of exposure of steel by spalling may be reduced, leading to an improvement in fire resistance.

11.7   OTHER TYPES OF FIBROUS COMPOSITES

GLASS REINFORCED GYPSUM (GRG)

The most suitable plaster for reinforcement is class B, retarded hemihydrate. This material may have a high compressive strength, up to 50 N/mm^2 or more at low water content, with a modulus of elasticity of about 20 N/mm^2, though tensile strength is low, approximately 6 N/mm^2. Fibres can therefore be profitably used. They will not contribute significantly to stiffness but will considerably improve tensile, flexural, impact and fire resistance properties. The cheaper *E* glass fibres can also be used in GRG, since the matrix is of a non-alkaline nature. Mixing methods are as for GRC, maximum flexural strength occurring with about 7 per cent by volume for sprayed fibres. Impact strength is improved remarkably – by over twenty times – while flexural strength increases to approximately 30 N/mm^2 for 4 per cent by volume of fibres, though the flexural stress at the elastic limit is only 10 N/mm^2 approximately. Compressive strength decreases as fibre content increases, probably because fibre interference causes reductions in density. The bond strength between fibre and matrix in GRG is known to be lower than that in GRC, and this is the likely reason for the improved impact strength of the former, especially at high fibre contents. Impacts are absorbed by causing partial failure between fibre and matrix. Higher bond strength produced by better compaction is known to result in lower impact strength.

One most important property of gypsum is its high content of water of crystallisation. As a result, GRG has a high specific heat and very good fire resistance.

The material has almost no shrinkage, though gypsum is slightly soluble in water, so that GRG could not be used externally unless adequately protected. One such system was marketed as a rendering material but was withdrawn on account of the difficulty of providing a fully waterproof surface coating. Special plasterboards are available which can be used in just the same way as conventional plasterboard but with greater fire resistance due to glass reinforcement of the gypsum core. A building board with a smooth face ready for decoration is also available, containing a tissue of glass fibre just beneath the surfaces with a reinforced core. It can be screwed or nailed without pre-drilling and has uses in many fire resistance applications.

Other applications include:

- For precast components, such as ducts, GRG could be used in similar situations to asbestos cement (internally only) but without the health hazard associated with the latter on drilling or cutting.
- GRG could be used in sandwich construction, for example, in timber doors to improve fire resistance, or with foamed plastics, to give fire resistant partitions with good heat insulation properties.

GLASS FIBRE REINFORCED POLYESTER RESIN (GRP)

Great advances have been made in reinforcement of plastics and, at present, GRP forms the largest bulk of these materials. The mechanical properties of polyester resins are dependent on the polymerisation process, on the presence of plasticisers and fillers and on the temperature. The tensile strength of polyester resins lies in the range 40–100 N/mm^2 and the tensile modulus of elasticity varies between 2 and 5 N/mm^2, too low for efficient structural use. The E value of glass fibre is about 70 N/mm^2 in tension, so that considerable improvements in stiffness can be obtained by incorporation of glass fibres, the chief reason for their use. A property of glass which must be considered in the context of reinforcing resins is its high affinity for water. Water is adsorbed in a thickness of twenty or more molecules to the surface of glass on account of its polar bonding; indeed, experiments have shown that water is partly responsible for its low tensile strength. Hence, if a moderate tensile stress is applied to glass, fracture may occur over a period of time unless the glass is completely dry, presumably by the action of stress corrosion due to water in flaws. The exact nature of the glass/resin bond is not fully understood, though water appears to affect the bond even when keying agents are used, since these rarely cover the whole surface and must, in any case, penetrate adsorbed water layers. Fibres are covered with a protective size after manufacture, which is generally removed before keying agents are applied. An exception is PVA size, in which keying agents can be incorporated.

It is found that relatively high aspect ratios, for example, 2000, are required to utilise most efficiently the tensile properties of glass fibres. For example, a 10 μm diameter fibre should be at least 20 mm in length to obtain sufficient stress transfer. Fibres in chopped-strand mat generally have lengths in the range 20–50 mm.

Manufacture of GRP

Hand lay-up   The laminate is produced by building up layers of resin and fibre in an open mould, often itself made of GRP. A gel coat may then be added, which will improve appearance and weathering performance. Hand lay-up produces only one smooth face. It may be used for small numbers of mouldings or for very large products.

Pressure moulding   The composite is built up on one-half of the mould and then a second matching mould is applied, which presses the composite into shape. A slightly higher proportion of fibre can be used, giving greater strength than in lay-up processes. Volume fractions of fibre are limited to about 20 per cent. Two smooth surfaces are obtained.

Continuous process   These are similar to extrusion processes and can be used for such products as corrugated sheeting. Lengths are limited only by transport considerations and volume fractions of up to 40 per cent of glass can be incorporated.

Winding   This process can produce very high strengths with volume fractions of up to 60 per cent but is limited to cylindrical tanks and pipes.

Pultrusion   In this process continuous fibres are formed into rod or strip with volume fractions as high as 70 per cent. This results in a composite with mechanical properties to rival those of steel but at much lower weight and no risk of corrosion.

Properties of GRP

The most important factors governing mechanical performance are the volume fraction and arrangement of the glass fibres which in turn depends on the

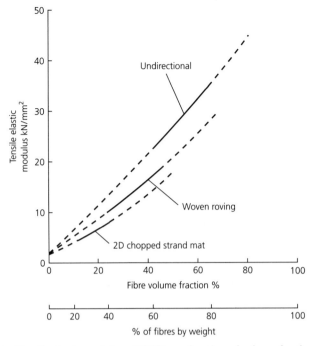

Figure 11.17   Tensile elastic modulus of GRP as a function of volume fraction

production method. Figure 11.17 shows how stiffness varies with volume fraction. The three curves reflect the efficiency of aligned fibres, compared with two-dimensional roving and random chopped fibres. The approximate working ranges are indicated in each case.

Highest performance is obtained by winding of continuous roving or pultrusion, such as in pipe or rod manufacture. E values in the range 20–35 kN/mm^2 are obtainable with tensile strengths of over 500 N/mm^2, relatively low thermal movement (5–12 × 10^{-6}/°C) and high heat resistance (up to 260 °C).

The more common lay-up processes using chopped fibres produce E values of 5–8 kN/mm^2 with tensile strength 60–120 N/mm^2, thermal movement 25–35 × 10^{-6}/°C and heat resistance up to 175 °C. On stressing of composites containing discontinuous fibres, plastic deformation, together with reduction in E value, can occur at quite small loads, due to adhesion failure at the end of fibres. The gel coat usually cracks at about 80 per cent of ultimate stress and crazing of the resin matrix further reduces stiffness before failure. Figure 11.18 shows a typical stress/strain curve which is representative of curves for reinforced ductile materials. On loading in flexure, the neutral axis moves towards the compression zone, giving apparently higher flexural strength than is obtained in direct tension. It might be expected that, since the stiffness of a resin is increased significantly by fibre reinforcement, the impact strength would be reduced. In fact, impact strength is increased and this is due to the increase in tensile strength that can be obtained

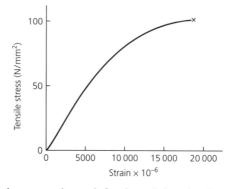

Figure 11.18   Typical stress–strain graph for glass reinforced polyester using chopped strand mat, loaded in tension. The exact relationship depends on the rate of loading

even after crazing in the resin matrix has commenced. Although creep properties of resins are improved by fibre reinforcement, creep is still a significant phenomenon and E values taken over long periods of time are invariably less than those obtained by short term measurements. Failure stresses given above may also be reduced substantially in long term loading, especially in moist conditions, though design is usually limited by deflection rather than stress considerations.

A most important property of GRP is its low density compared with similar materials (approximately 1500 kg/m³). Light transmission of the composite is about 85 per cent when new, slightly lower than that of glass, but GRP is stronger and tougher than the latter.

Weathering resistance depends on the type of resin used, the proportion of resin and the nearness of fibres to the surface. Resins shrink on curing (this is, in fact, thought to be at least partly responsible for the glass/resin bond), tending to leave fibres standing on the surface so that water may penetrate by capillarity, destroying the resin/fibre bond. Some resins themselves are able to absorb water, resulting in the same effect. Alternatively, weather may erode the surface of the resin, exposing fibres. The ingress of moisture may be prevented by use of gel coats on the surface of laminates, though these should be lightly reinforced with glass monofilaments to prevent cracking. Acrylic resins are often used as gel coatings. The gel coat thickness is important – too thin a coat reduces protection to fibres while too thick a coat encourages gel coat crazing. They are usually in the region 0.3–0.4 mm. If, after a time, fibres become exposed, it is best to rub the surface down and apply a sealer resin. The procedure should then be repeated periodically. Lives of 30 years or more may be obtained in this way.

Fire resistance of GRP

GRP gives rise to considerable smoke in fire and will be destroyed if exposed to extreme heat, since polyester resins are combustible. Decomposition is, however, retarded by the glass fibres which tend to stabilise the resin, hence GRP can be used in many situations in which fire regulations exist. As information on behaviour in specific situations becomes more widely available, increased use is very likely. One of the problems at present is that regulating authorities require expensive fire performance tests if materials such as GRP are used in an innovative manner. Some methods of complying with fire regulations are as follows.

Surface spread of flame/fire propagation index   GRP can be made to satisfy the class 0 category of *Building Regulations* by use of flame retardants, though these lower strength and weathering resistance.

Fire resistance   Panels can achieve a fire resistance of 1 hour by use of double skins with a non-combustible core, or by backing up with a non-combustible material, such as blockwork (in which case, fire stops should be built in behind panels).

Roofs   Roofing panels must perform satisfactorily when exposed to both internal and external fire. Automatic venting can be provided by fusible links, allowing lights to open when they become hot or by smoke detector-controlled operating gear. Sprinkler systems have been used to prevent fire penetration of roofing systems.

Applications

GRP is the most important plastic composite in the field of structures. Although the stiffness of the material is not high, the production of the complex shapes which are essential for rigidity is easily and cheaply carried out.

Prefabricated buildings, claddings, roofing   Small, lightweight buildings, such as filling station canopies, shelters and even dwellings, have been constructed from GRP mouldings and, on account of the low density of the material, allow substantial savings in foundation costs (Figure 11.19). The use of sandwich constructions containing cellular plastics increases stiffness where necessary. Many cladding systems have been produced, mainly for commercial buildings. The design of all structures requires careful attention to stress transfer at fixings to avoid local failure of GRP, though thickness can easily be increased locally to meet higher stresses. Thermal movements at joints can be considerable, though GRP has some ability to absorb this on account of its fairly low E value. In many systems, movements have been accommodated by allowing panels to flex or allowing structures, such as domes, to expand as a whole about a fixed position.

The translucency of GRP makes it an attractive material for roofing for buildings which have a daylight requirement. The simplest form is corrugated sheeting, which can be used in place of single sheets in an ordinary corrugated roof. Specially designed dome-lights are now often used and, by fabrication of sections, large spans, such as over swimming pools, warehouses and arcades, can be obtained without the need for a supporting frame, though light-transmission qualities deteriorate over a number of years. Joints must be carefully treated to avoid water

Figure 11.19   Two storey monocoque construction using GRP. The material forms both the frame and the cladding

penetration. They are best positioned on ridges rather than valleys, so that less reliance is placed upon sealants for waterproofing.

Some very attractive results have been obtained with GRP and cladding and roofing applications are likely to increase as the smooth, seam-free lines and corners associated with its use become more widely accepted. Their lightness permits quite large architectural features such as cupolas to be preformed and then positioned by crane or helicopter for fixing by simple bolting systems.

Concrete moulds

Concrete is being used increasingly as a facing material for structures and the high quality of patterned finishes that can be obtained with GRP moulds can be a major contribution (Figure 7.50). For small numbers of units, moulds may be hand-made while, for a larger number of slabs or columns, it may be worthwhile to use presses. Moulds may be used many times over, though choice of release agent is important and moulds may require stiffeners, especially if large components are involved. A further possibility is the use of GRP moulded units as permanent shuttering for concrete. Units containing locating ducts for reinforcement are assembled and concrete is then poured in and compacted. This provides perfect curing conditions and eliminates formwork removal. This method is most likely to be used for large numbers of *in situ* units, for example, columns, where the mould would also make some contribution to strength.

Prestressing tendons

A further quite novel application of GRP is for use as tendons for prestressed concrete. The ideal tendon is one having a low *E* value, operating at high strains, since contractions due to shrinkage or creep of the concrete would then be less significant. Strengths approaching those of prestressing steel have been achieved, though care would be necessary to allow for any creep effects. A major problem in this context is that special anchorage systems are required. Since glass fibres are brittle, conventional gripping systems such as wedges tend to cause crushing of fibres at the anchorage. It will also be appreciated that overstressing of GRP prestressed concrete could cause sudden failure, unlike steel equivalents in which the steel tends to deform before failure. The design of GRP prestressed structures is therefore fundamentally different from that of conventional steel prestressed systems. Nevertheless GRP offers the advantages of low weight and corrosion resistance over steel in such applications which are likely to increase as the technology advances.

Other applications

GRP is used in a large number of products in building. Plumbing applications include tanks and cisterns and even hot water cylinders. In all cases, stresses due to inadequate support or poorly aligned pipework should be avoided. Other uses include window frames, cladding panels, garage doors and ventilators.

USES OF GLASS FIBRE IN OTHER FORMS OF PLASTICS

Epoxy resins have also been used in reinforced form; they adhere well to glass fibres and produce a composite of superior strength and chemical resistance to GRP. They also shrink less on curing so that initial stresses in composites are reduced. They are, however, more expensive than polyester resins. Typical applications include moulds for concrete products, such as posts, when the moulds, on account of their toughness and strength, can be used time after time.

11.8 FABRIC STRUCTURES

Fabrics, in this context, are thin reinforced sheet materials, designed to operate under uniform two dimensional stress; they are unable to support any bending stress. The use of fabric structures as building enclosures is increasing quite rapidly as a result of technical innovation in the following areas:

- Development of high strength, low creep, durable fabrics.
- Developments in fabric jointing techniques, particularly fusion welding.
- Increased understanding of the complex stresses in tensioned fabrics, assisted by computer modelling.

Fabric structures may be considered for certain building types for the following reasons:

- Construction costs are often quite low; in the case of air supported structures, the cost may be as little as half that of conventional equivalents.
- Foundation loads are very low, facilitating construction on difficult ground.
- Cable stayed structures often have dramatic architectural effect.
- Uninterrupted clear spans can be obtained.
- Very uniform daylight levels can be achieved.
- Thermal insulation is easily obtained by use of double membranes incorporating an air cavity or other insulation.

A number of fabrics are available including polyester reinforced PVC with a life of 15–20 years and glass reinforced PTFE with a life of about 50 years. They are available as translucent fabrics to which pigments can be added to colour or control light transmission. The PTFE fabric has the advantage of being largely self-cleaning, satisfactory performance having been already obtained on some structures for over 10 years, though their much higher cost restricts their use to 10 per cent of the total market at present. Fire performance of the two varieties should be satisfactory, although they behave quite differently in fire. PVC fabrics, though combustible, do not support flame and fire tends to burn holes through the material, thereby venting smoke and gases. PTFE fabrics satisfy *Building Regulation* class 0; they are not punctured by fire except in extreme heat, though lack of venting may lead to toxic fume build-up from the PTFE.

AIR SUPPORTED STRUCTURES

These have a limited range of applications, for example, sports/storage enclosures and have a utilitarian appearance externally but provide a very low cost solution to the provision of such buildings. A small positive pressure is sufficient to keep the building 'inflated'. Windows are not normally provided, since daylight filters through the fabric. Door openings must include an airlock – revolving doors, for example, will suffice. Pressure is maintained by a compressor which may also provide air temperature control and ventilation. For effective insulation in winter and solar control in summer, double skin structures are preferred. Increased use of height at the building perimeter can be made by providing traditional walls with an air supported fabric roof. In the absence of rigid walls, services and fixings such as lights are generally floor mounted. Fears have been expressed as to the safety of such buildings in fire, since there is a risk that, if the fabric is punctured, the structure could collapse. There is, however, evidence to suggest that, even with large holes in the fabric, collapse is quite slow, partly due to the extra buoyancy generated by hot air/gases in a fire. There should, therefore, be adequate time for escape and it is on this basis that fire regulation approval is given to air domes.

Figure 11.20 Use of cable stayed glass reinforced PTFE fabric in retail outlet. The canopy masks the relatively simple shop units beneath

FRAMED OR TENSION STRUCTURES

These permit much greater flexibility in terms of both form and use of fabric structures. For example, glazing is much more easily incorporated and services or other utilities can be fixed to the structural frame. Framed structures may be based on tubular steel or aluminium, while in tension structures, fabrics are typically attached to stainless steel cables supported from masts, which are jacked up to provide correct tension. The form of the latter structures may be particularly dramatic where multiple masts are used (Figure 11.20) and can be used to produce large, uninterrupted areas of floor space internally, together with good internal height clearance, if required.

BIBLIOGRAPHY

Materials Science, 4th Edition, J. C. Anderson, Chapman & Hall, 1990.
Materials for the Engineering Technician, R. A. Higgins, Edward Arnold, 1987.
Introduction to Materials Science for Engineers, J. F. Shackelford, Macmillan, 1996.
Engineering Materials, Z. D. Jastrzebski, Wiley and Sons, 1977.
Engineering Materials 2. An Introduction to Microstructures, Processing and Design,
    M. F. Ashby and R. H. Jones, Pergamon, 1986.
Mitchell's Materials Technology, Y. Dean, Longman, 1996.
Engineering Materials Technology, W. Bolton, Butterworth Heinemann, 1998.
Materials in Chemical Perspective, K. L. Watson, Stanley Thornes, 1975.
The New Science of Strong Materials, J. E. Gordon, Pelican, 1991.
Construction Materials, Their Nature and Behaviour, Ed. J. M. Illston, E. &
    F. N. Spon, 1994.
Civil Engineering Materials, Ed. N. Jackson and R. K. Dkir, Macmillan, 1996.
Materials for Architects and Builders, An Introduction, A. R. Lyons, Arnold, 1997.
Corrosion of Building Materials, Dietbert Knofel, Van Nostrand Reinhold, 1978.
Materials for Civil Engineers and Construction Engineers, M. S. Mamlouk &
    J. P. Zaniewski, Addison-Wesley, 1999.

CHAPTER 1

BRE Digest 397 – *Standardisation in Support of European Legislation*, 1994 (revised
    1995).
BRE Digest 408 – *A Guide to Attestation of Conformity under the Construction
    Products Directive*, 1995.

CHAPTER 2

Vapour Resistivity Details, CIBSE Guide A, 1999.
Building Technology and Management (*Journal of Chartered Institute of Building*)
    April–June 1983, Articles by G. D. Taylor on frost.
Fire Safety in Buildings, H. L. Malhotra, BRE, 1987.

Design of Fire-resisting Structures: H. L. Malhotra. Surrey University Press, 1982.
Buildings and Fire, T. J. Shields and G. W. H. Silcock, Longman, 1987.
Guidelines for the Construction of Fire-resisting Structural Elements, W. A. Morris,
    R. E. H. Read and G. M. E. Cooke, BRE Report, 1988.
BRE Digest 225 – *Fire Terminology*, 1986.
BRE Digest 228 – *Estimation of Thermal and Moisture Movements and Stresses*, 1979.
BRE Digest 233 – *Fire Hazard from Insulating Materials*, 1980.
BRE Digest 285 – *Fires in Furniture*, 1984.
BRE Digest 300 – *Toxic Effects of Fires*, 1985.
BRE Digest 320 – *Fire Doors*, 1988.
BRE Digest 343/4 – *Simple Measurement and Monitoring of Movement in Low Rise
    Buildings*, 1989.
BRE Digest 361 – *Why do Buildings Crack?* 1991.
BRE Information Paper 4/86 – Ignition and growth of fire in a room,
    P. L. Hinckley and A. W. Williams.

CHAPTER 3

*BRE Methodology for Environmental Profiles of Construction Materials, Components
    and Buildings*, BRE, 1999.
The Real Green Building Book 2000, Association for Environment Conscious
    Building.
Renewable Energy; Power for a Sustainable Future, Ed. Godfrey Boyle, Open
    University/Oxford University Press, 1996.
Ecology of Building Materials, B. Berge, translated by F. Henley, Architectural
    Press, 2000.
Materials for Construction and Building in the UK. Report of a Working Party of
    the Materials Forum and the Institution of Civil Engineers, Institute of Metals,
    1987.
Hazardous Building Materials – A Guide to the Selection of Alternatives, S. G.
    Curwell and C. G. March, E. & F. N. Spon, 1986.
The Green Guide to Specification. An Environmental Profiling System for Building
    Materials and Components. BRE Report 351, 1998.
Planning for a Sustainable Environment. A Report by the Town and Country
    Planning Association, Ed. A. Blowers, Earthscan, 1993.
The Carbon Cycle, New Scientist, 2 Nov. 1991.
Life Cycle Energy Analysis of Buildings: A Case Study, R. Fay *et al.*, Building
    Research and Information, 2000.
BRE Digest 358 – *CFCs in Building*, 1992.
BRE Information Paper 23/89 – CFCs in the building industry, D. J. G. Butler.
BRE Information Paper 2/90 – Greenhouse gas emissions and buildings in the
    United Kingdom, G. Henderson and L. G. Shorrock.
BRE Information Paper 11/93 – Ecolabelling of building materials and building
    products, C. J. Atkinson and R. N. Butlin.

CHAPTER 4

See British Standards referred to in the text.
Statistics for Construction Students, J. A. Bland, Construction Press, 1985.
Statistical Methods in Management, Tom Cass, Cassell, 1980.
Statistics for Concrete, Part 1, C. A. R. Harris, Concrete Society (Digest No. 5).
Monitoring Concrete by the Cusum System, B. V. Brown, Concrete Society (Digest No. 6).
Using repeatability values in an aggregate testing laboratory, R. Sym, *Concrete*, **21**, No. 7, 1987.

CHAPTER 6

Soils and the Environment, S. Ellis and A. Mellor, Routledge, 1997.
Rammed Earth Structures; A Code of Practice, J. Keable, Intermediate Technology Publications, 1996.
Earth Construction, H. Houben and H. Guillaud, Intermediate Technology Publications, 1994.
Design Notes, Brick Development Association, Windsor.
The Pattern of English Building, Alec Clifton Taylor, Faber & Faber, 1987.
Stone in Building – Its Use and Potential Today, J. Ashurst and F. G. Dimes, Architectural Press, 1977.
Conservation of Building and Decorative Stone, Ed. J. Ashurst, Butterworth Heinemann, 1990.
The Building Sandstones of the British Isles, E. Leary, Department of the Environment BRE, 1983.
The Building Limestones of the British Isles, E. Leary, Department of the Environment BRE, 1983.
Tuck Pointing in Practice, J. Carey, Information Sheet 8, Society for the Protection of Ancient Buildings, London.
Mortars for Blockwork, A. W. Stupart and J. S. Skandamoorthy, BRE Information Paper, 1998.
Lime and Limestone; Chemistry and Technology, Production and Uses, J. A. H. Oates, Wiley VCH, 1998.
Factors affecting the brick/mortar interface bond strength, B. P. Sinha, *International Journal of Masonry Construction*, **3**, No. 1, 1983.
Practical Building Conservation: English Heritage Technical Handbook, **3**, Mortars, Plasters and Renders, J. Ashurst and N. Ashurst, 1988.
Lime mortars and plasters in building, K. H. Renton and H. N. Lee, *Building Technology and Management*, April/May 1989.
Flat Glass Technology, R. Pearson, Butterworths, 1969.
Technical Data, Pilkington Glass Ltd., St. Helens, Lancashire.
Technical data, Solaglass Ltd., Sittingbourne, Kent.
BRE Digest 157 – *Calcium Silicate (Sand Lime, Flint Lime) Brickwork*, 1992.

BRE Digest 177 – *Decay and Conservation of Stone Masonry*, 1984.
BRE Digest 269 – *Selection of Natural Building Stone*, 1983.
BRE Digest 362 – *Building Mortar*, 1991.
BRE Digest 410 – *Cementitious Renders for External Walls*, 1995.
BRE Digest 441 – *Clay Bricks and Clay brick Masonry*, Parts 1 & 2, R. C. de Vekey, 1999.
BRE Information Paper 10/85 – The conformance of masonry mortars and their constituent sands with British Standards, K. E. Fletcher.
BRE Information Paper 6/97 – External cladding using thin stone, T. J. S. Yates and B. Chakrabati.
BRE Information Paper 18/98 – Stone cladding panels, T. J. S. Yates and B. Chakrabati.
BRE Information Paper 17/98 – Lightweight veneer stone cladding panels, T. J. S. Yates and B. Chakrabati.
BRE Information Paper 9/99 – Cleaning exterior masonry; pretreatment/ assessment of a stone building, S. Pryke.

CHAPTER 7

Portland Cement Composition, Production and Properties, G. C. Bye, Pergamon, 1983.
Properties of Concrete, A. M. Neville, 4th Ed., Longman, 1995.
Portland Cement Paste and Concrete, I. Soroka, Macmillan, 1979.
Concrete Technology, A. M. Neville and J. J. Brooks, Longman, 1987.
Design of Normal Concrete Mixes, Department of the Environment, HMSO, 1988, 1997.
Recent research developments in abrasion resistance, R. J. Kettle and M. Sadegzadeh, *Concrete* **20**, No. 11, Nov. 1986.
Chemical Admixtures for Concrete, M. R. Rixom and N. P. Mailvaganam, E. & F. N. Spon, 1999.
Admixtures for Concrete, Cement and Concrete Association, 1991.
Properties of microsilica in concrete, P. Whale. *Concrete* **23**, No. 8, Sept/Oct 1989.
The relation between porosity, microstructure and strength, and the approach to advanced cement-based materials, K. Kendall, A. J. Howard and J. D. Birchall, *Phil. Trans. R. Soc. London* A310, 139–153, 1983.
Time dependent properties of high strength cements, C. M. Cannon and G. W. Groves, *Journal of Materials Science* **21**, pp. 4009–4014, 1986.
Diffusion through concrete, D. M. Roy and D. D. Higgins, *Concrete* **21**, No. 1, 1987.
The Thaumasite Form of Sulphate Attack: Risks, Diagnosis, Remedial Works and Guidance on New Construction, HMSO, 1999.
Specialist concrete pavement surfaces, *Quarry Management*, Dec. 1999.
Guide to Exposed Concrete Finishes, M. Cage, Architectural Press, Cement and Concrete Association, 1970.
Precast Concrete Claddings, Ed. H. J. P. Taylor, Edward Arnold, 1992.
Concrete Communication Conference, British Cement Association, Southampton, 1998.

Concrete Communication Conference, British Cement Association, Cardiff, 1999.

BRE Digest 330 – Parts 1–4 *Alkali Silica Reaction in Concrete*, 1999.

BRE Digest 363 – *Sulfate and acid resistance of concrete in the ground*, 1996.

BRE Digest 392 – *Assessment of Existing High Alumina Cement Concrete Construction in the UK*, 1994.

BRE Digest 405 – *Carbonation of concrete and its effects on durability*, 1995.

BRE Digest 433 – *Recycled Aggregates*, 1998.

BRE Digest 434 – *Corrosion of Reinforcement in Concrete: Electrochemical Monitoring*, 1998.

BRE Information Paper 7/89 – The effectiveness of surface coatings in reducing carbonation of reinforced concrete, H. Davies and G. W. Rothwell.

BRE Information Paper 3/86 – Changes in Portland cement properties and their effects on concrete, P. J. Nixon.

BRE Information Paper 21/86 – Determination of the chloride and cement contents of hardened concrete, M. H. Roberts.

BRE Information Paper 11–12/87 – Pulverised fuel ash – its use in concrete, J. D. Matthews.

BRE Information Paper 8/88 – Update on assessment of high alumina cement concrete, R. J. Collins.

BRE Information Paper 11/88 – Assessing carbonation depth in ageing high alumina cement concrete, A. M. Dunster.

BRE Information Paper 6/92 – Durability of blastfurnace slag cement concretes, G. J. Osborne.

BRE Information Paper 16/93 – Effects of alkali silica reaction on concrete foundations, P. L. Walton.

BRE Information Paper 5/94 – Use of recycled aggregates in concrete, R. J. Collins.

BRE Information Paper – Assessing carbonation depth in ageing high alumina cement concrete.

BRE Information Paper 14/98 – Blocks with recycled aggregate beam and block floors, R. Collins, D. J. Harris and W. Sparkes.

BRE Information Paper 8/00 – Durability of precast HAC concrete in buildings, A. Dunster.

BRE Information Paper 15/00 – Water reducing admixtures in concrete, J. P. Ridal, S. L. Garvin, A. M. Dunster.

CHAPTER 8

Engineering Metallurgy: Applied Physical Metallurgy, R. A. Higgins, 6th Ed., Edward Arnold, 1993.

Metals in the Service of Man, W. Alexander and A. Street, 8th Ed., Pelican, 1994.

The Structure, Properties and Heat Treatment of Metals, D. J. Davies and L. A. Oelmann, Pitman, 1983.

An Introduction to Metallic Corrosion, U. R. Evans, Edward Arnold, 1981.

The Fundamentals of Corrosion, 3rd Ed., J. C. Scully, Pergamon, 1990.

Metallic Coatings for Corrosion Control, V. E. Carter, Newnes Butterworth, 1977.

Hot Dip Galvanising, Zinc Development Association, London.

Testing the water on cathodic protection, J. Broomfield, *Construction Repairs and Maintenance*, March 1986.

The effect of rusting on the bond performance of reinforcement, F. G. Murphy, CIRIA Report 71, 1977.

Design of Structural Steelwork, P. Knowles, Surrey University Press, 1987.

Introduction to the Welding of Structural Steelwork, J. L. Pratt, Constrado, 1979.

Principles of welding technology, 3rd Ed., L. M. Gourd, Edward Arnold, 1995.

Technical Information, Corus Group, London.

High strength steel reinforcement, H. G. Trotter, *The Metallurgist and Materials Technologist*, Feb. 1977.

Special reinforcing steel, A. Marsden, Concrete Society current practice sheet 103. *Concrete*, **19**, No. 9, Sept. 1985.

Introduction to Stainless Steels, J. Beddows and J. Gordon-Parr, ASM International, The Materials Information Society, 1999.

BRE Digest 305 – *Zinc coated steel*, 1986.

BRE Digest 317 – *Fire resistant steel structures*, 1986.

BRE Digest 349 – *Stainless steel as a building material*, 1990.

BRE Information Paper 14/88 – Corrosion protected and corrosion resistant reinforcement in concrete, K. W. J. Treadaway, R. N. Cox and H. Davies.

BRE Information Paper 16/88 – Ties for cavity walls: new developments, R. C. de Vekey.

BRE Information Paper 17/88 – Ties for masonry cladding, R. C. de Vekey.

BRE Information Paper 12/90 – Corrosion of steel wall ties history of occurrence, background and treatment, R. C. de Vekey.

BRE Information Paper 11/00 – Ties for masonry walls, a decade for development, R. de Vekey.

CHAPTER 9

Thermoplastics – Properties and Design, R. M. Ogorkiewicx, Wiley and Sons, 1974.

Polymers for Engineering Applications, R. B. Seymour, ASM International, The Materials Information Society, 1987.

Polymer Materials – An Introduction for Technologists and Scientists, C. Hall, Macmillan, 1981.

Construction Sealants and Adhesives, 2nd Ed., J. R. Panck and J. P. Cook, Wiley and Sons, 1984.

Adhesives and Sealants, Vol 3, Engineered Materials Handbook, ASM International, The Materials Information Society, 1990.

Selection and Use of Thermoplastics, Materials Engineering Design Guides 19, OUP, 1977.

BRE Digest 372 – *Flat Roof Design: Waterproof Memberanes*, 1992.

BRE Digest 404 – *PVC-U Windows*, 1995.

BTE Digest 440 – *Weathering of White External PVC-U*, 1999.

BRE Information Paper 9/87 – Joint primers and sealants: performance between porous claddings, J. C. Beech and D. W. Aubrey.

BRE Information Paper 8–10/86 – Weatherproof joints in large panel systems, M. J. Edwards.

BRE Information Paper 12/97 – Plastics recycling in the construction industry, S. M. Halliwell.

CHAPTER 10

Timber – Its Structure, Properties, Conversion and Use, 7th Ed., H. E. Desch, Macmillan, 1996.

Wood, Nature's Cellular Polymeric Fibre Composite, J. M. Dinwoodie, Institute of Metals, 1989.

The Architectural History of Venice, Deborah Howard, Batsford, 1980.

Design of structural timber, W. M. C. McKenzie, Macmillan, 2000.

BRE Digest 296 – *Timbers: Their Natural Durability and Resistance to Preservative Treatments*, 1985.

BRE Digest 299 – *Dry Rot: Its Recognition and Control*, 1985.

BRE Digest 304 – *Preventing Decay in External Joinery*, 1985.

BRE Digest 307 – *Identifying Damage by Wood Boring Insects*, 1986.

BRE Digest 323 – *Selecting Wood-based Panel Products*, 1992.

BRE Digest 327 – *Insecticidal Treatments Against Wood Boring Insects*, 1987.

BRE Digest 340 – *Choosing Wood Adhesives*, 1989.

BRE Digest 345 – *Wets Rots – Their Prevention and Control*, 1989.

BRE Digest 354 – *Painting Exterior Wood*, 1990.

BRE Digest 371 – *Remedial Wood Preservatives – Use Them Safely*, 1992.

BRE Digest 373 – *Wood Chipboard*, 1992.

BRE Digest 387 – *Natural Finishes for Exterior Wood*, 1993.

BRE Digest 393 – *Specifying Preservative Treatments: The New European Approach*, 1994.

BRE Digest 394 – *Plywood*, 1994.

BRE Digest 400 – *Oriented Strand Board*, 1994.

BRE Digest 415 – *Specifying Structural Timber*, 1996.

BRE Digest 417 – *Hardwoods for Construction Joinery – Current and Future Sources of Supply*, 1996.

BRE Digest 422 – *Painting Exterior Wood*, 1997.

BRE Digest 423 – *The Structural Use of Wood-based Panels*, 1997.

BRE Digest 429 – *Timbers – Their Natural Durability and Resistance to Preservative Treatment*, 1998.

BRE Digest 435 – *Medium Density Fibreboard*, 1998.

BRE Digest 445 – *Advances in Timber Grading*, 2000.

BRE Information Paper 9/84 – Water-borne paints for exterior wood, E. R. Miller and J. Boxall.

BRE Information Paper 16/87 – Maintaining paintwork on exterior timber, J. Boxall and G. A. Smith.

BRE Information Paper 17/87 – Factory-applied priming paints for exterior joinery, T. B. Dearling and E. R. Miller.

BRE Information Paper 20/87 – External joinery: end grain sealers and moisture control, E. R. Miller, J. Boxall and J. K. Carey.

BRE Information Paper 7/88 – The design and manufacture of ply-web beams, V. Enjily.

BRE Information Paper 5/91 – Exterior wood stains, J. Boxall.

BRE Information Paper 14/91 – In situ treatment of exterior joinery using boron based implants, J. C. Carey.

BRE Information Paper 19/92 – Wood-based panel products: moisture effects and assessing the risk of decay, G. Lea.

BRE Information Paper 9/93 – Perspectives on European standards for wood-based panels, J. M. Dinwoodie.

BRE Information Paper 8/94 – House longhorn beetle: geographical distribution and pest status in the United Kingdom, G. Lea.

BRE Information Paper 2/96 – An assessment of exterior medium density fibreboard (MDF), K. W. Maun.

BRE Information Paper 5/96 – Progress in European standardisation for exterior wood Coatings, E. R. Miller and J. A. Graystone.

BRE Information Paper 4/97 – Preservative treated timber for external joinery: applying the new European Standards, R. J. Orsler.

BRE Information Paper 8/99 – The performance and use of coatings with low solvent content, J. Boxall and W. Thorpe.

BRE Information Paper 6/99 – Preservative treated timber: ensuring conformity with European standards, E. D. Suttie and R. J. Orsler.

CHAPTER 11

Composite Materials: Engineering and Science, F. L. Matthews and R. D. Rawlings, Chapman & Hall, 1994.

BRE Digest 331 – *GRC*, 1988.

BRE Digest 442 – *Architectural Use of Polymer Composites*, 1999.

BRE Information Paper 7/99 – Advanced polymer composites in construction, S. M. Halliwell, 1999.

INDEX